Biochemistry and physiology of the cell
An introductory text

Biochemistry and physiology of the cell

An introductory text

Second edition

N. A. Edwards

Principal Lecturer, Department of Biology, Oxford Polytechnic

K. A. Hassall

Senior Lecturer, Department of Physiology and Biochemistry, University of Reading

McGRAW-HILL Book Company (UK) Limited

London • New York • St Louis • San Francisco • Auckland
Bogotá • Guatemala • Hamburg • Johannesburg • Madrid • Mexico
Montreal • New Delhi • Panama • Paris • São Paulo
Singapore • Sydney • San Juan • Toronto

Published by

McGRAW-HILL Book Company (UK) Limited

MAIDENHEAD · BERKSHIRE · ENGLAND

British Library Cataloguing in Publication Data

Edwards, Norman Arthur
Biochemistry and physiology of the cell.–2nd ed.
1. Biochemistry
I. Title II. Hassall, Kenneth Arnold
574.8'76 QH631 79-041026
ISBN 0-07-084097-0

34567 M&G 83210

Printed and bound in Great Britain

Contents

Preface, second edition

The first edition, entitled *Cellular Biochemistry and Physiology*, was published in 1971 as an introductory text for college, university, and polytechnic students of biochemistry, and as a basic text for students of other biological disciplines such as agriculture, food science, botany, zoology, microbiology, pharmacy, and the medical sciences. The second edition, like the first, is intended both as an introduction to students specializing in biochemistry, and as a basic text for other biological students who are increasingly finding that some knowledge of biochemistry and cellular physiology is essential to an understanding of their main subjects. The book proved valuable to students studying HNC in Applied Biology and Biochemistry and the new edition incorporates the needs of the new TEC courses.

The rapid development of a subject often means that separate pieces of a puzzle eventually fit together to form a picture which is clearer and more readily memorized than the design on the individual pieces. This has certainly happened in several areas of biochemical progress during the last decade. Our second edition is superficially 'more advanced' than the first, but nevertheless, we believe, easier rather than harder to read and understand. We have earnestly attempted to maintain the style and objectives of the first edition, yet have up-dated many topics, including the control of metabolism and the roles of polynucleotides. In writing the book we have taken careful note of many helpful comments and suggestions from students and teachers both in the United Kingdom and abroad.

As we pointed out in the preface to the first edition, it would be presumptuous of us to assume that we have everywhere succeeded where others have failed, but there is one factor overwhelmingly in our favour: we have lectured in introductory biochemistry to students reading a variety of disciplines and have additionally some knowledge of Advanced Level courses in chemistry and biology. We know what the average student is likely to know—and not to know—when he picks up with interest his first textbook on cellular biochemistry. We know from experience his disillusionment when he finds, as he sometimes does, that the first hundred pages are more akin to classical chemistry than to the subject he had enthusiastically anticipated. We know his frustration on discovering that, only too often, the essential reactions of biochemistry are later mentioned in rapid succession, without any serious attempt to explain why they occur, and often, in the cause of scientific accuracy, encompassed with so many qualifications, that the main theme is all but obscured.

We have attempted to overcome some of these difficulties in a variety of ways. Care is taken to explain reactions which perplex biology students; this is often done by the use of analogy or by comparison with similar reactions involving simpler compounds. Special emphasis is placed on the rationale underlying sequential reactions and energetic processes, even at the expense of a little teleological reasoning from time to time ('the next reaction involved the dephosphorylation of the hydroxyl group since the free hydroxyl group is to be removed in the reaction which follows'). Recognizing that our book will be used by students with somewhat different backgrounds, we have provided an Appendix listing books for further reading.

Almost anybody can write a textbook on his research subject, but the writer of an elementary book walks a precarious pathway. He must not give the impression that everything is universally cut-and-dried; nor must he obscure the wood with the trees. We have attempted to present our material in an interesting manner, based on sound educational principles, but with the minimal sacrifice of scientific accuracy. The reader is warned, however, that even the most sacred of scientific tenets amount at most to the temporary dogma of our era and are subject to modification in the light of further knowledge. In no subject is this more true than in the realm of so young a fledgling as biochemistry.

We have tried above all to convey to our readers something of the fascination inherent in the subject, something of our own enthusiasm for it. We have departed occasionally from accepted methods of presentation in an attempt to put first things first—even if those things are, very temporarily, somewhat unfamiliar. The fate of acetyl coenzyme A, for example, is of far greater importance in providing the cell with energy than the almost incidental processes of glycolysis and β-oxidation which convert carbohydrates and fats respectively to this important source of energy. Such an approach at once presents the compartmentation of the cell as a concept central to our approach and provides the biologist as well as the chemist with that pleasant psychological satisfaction of being on home ground which arises when a book, on a subject new to the reader, is opened for the first time.

The reader may well prefer to read the book twice, on two different intellectual wavelengths. For example, the beginner would probably find, during the first reading, that Chapters 2 to 5 dealing with the basic properties of lipids, carbohydrates, nucleotides, and proteins were rather a strain on his memory. He would be advised initially to concentrate on getting a general overall impression rather than attempt to commit all the formulae to memory. When the reader subsequently encounters particular compounds in later chapters he should revise the earlier work. It will then be easier to appreciate the significance of the structure of the compounds by considering them in their biological context.

Experience in teaching introductory biochemistry and cellullar physiology has made clear to us the importance of explaining the functions of the cell and the reasons for them, and of doing this in a way sympathetic to the student's learning process. In particular, it is useful to know where two sentences of explanation are better than one. The response from readers has been very gratifying and the book has had a successful run, both in English (in the standard and Asian editions) and in translation.

For both types of student—first year biochemistry specialists and those for which the subject is to be a Service Course or a subsidiary subject—we have borne constantly in mind that we are writing an *introductory but not necessarily an elementary text.* We do not intend that our book should compete with advanced texts (except in the sense that we do not think students starting from scratch should be recommended to purchase an advanced text as their primary source of information). We have endeavoured to hold the student's interest and to use our personal knowledge of students' difficulties when conveying the rationale underlying the nature of living processes and the cellular structure in which they occur.

N. A. E.
K. A. H.

1

The cellular basis of life

1.1 Why organisms are subdivided into cells

The theory of evolution proposes that living systems evolve and develop only in the direction which leads to their success in a particular environment. Extinction is the ultimate fate of all organisms which fail to do this. Since almost all animals and plants are subdivided into small independent functional units, or cells, it would appear that, as a biochemical entity, the cell is not only successful and efficient *in its own right* but, additionally, that the *condition of cellular subdivision* of large organisms confers advantages to the organisms of overwhelming evolutionary importance.

Living organisms cannot exist as 'closed systems' independent of their environment. They must, of necessity, obtain chemical substances from their surroundings and they must eliminate waste materials. The passage into and out of cells of such important substances as oxygen and carbon dioxide occurs by the process of *diffusion.* This commences at the outer surface of unicellular or multicellular organisms but must also continue within the organism unless alternative methods of conduction exist. Diffusion of solute molecules in solution is a random process but the overall effect is always for solute molecules in a concentrated solution to diffuse towards a region of lower concentration. The *greater the distance* over which diffusion proceeds the *slower* is the net movement of the molecules (Fick's Law).

Simple diffusion therefore provides an efficient transport mechanism only when the distances involved are quite small. A cell with a diameter of 10 to 100 μm (1 micrometer = 0.001 mm) has a volume of between 10^{-6} and 10^{-3} mm^3. Diffusion through bulk phases of this magnitude presents little or no problem, and it is by this means that most substances move within bacteria, protozoa, and green algae. But higher plants and animals can have volumes many millions of times greater than this, and were they not divided into subunits it is difficult to conceive how diffusion from the outside inwards could have been rapid enough to be compatible with evolutionary success.

A related consideration of great importance is the possibility of cellular specialization, or division of labour, within a multicellular organism. If large organisms were not divided into cells, evolutionary specialization of parts of the organism for particular functions would, at best, have been much less profound than it has been in both the plant and animal kingdoms. Specialization of cells appears to have been an essential prerequisite for the evolutionary experimentation which has led to the existence of so many and diverse forms of multicellular organisms.

1.2 Cell size and composition—some vital statistics

Almost the whole mass of cells is contributed by no more than four of the ninety or so natural elements. These are carbon, oxygen, hydrogen, and nitrogen. All of these are, in fact, elements from the first period of the Periodic Table. In addition, nature has found a use for quite a number of other elements with atomic numbers up to about 30, often in a quantitatively minor yet mechanistically most fundamental way. Examples are phosphorus, calcium, magnesium, sulphur, iron, zinc, and copper. It is striking how nature has apparently rejected most of the remaining elements; very few of those with atomic numbers above 30 are of wide or universal significance—bromine, iodine, and molybdenum are exceptions.

Water accounts for 60 to 85 per cent of the mass of most kinds of cell. It acts as a solvent for small organic molecules. In addition, its polar qualities help to determine the orientation of polar and non-polar groups in protein molecules and thereby it influences enzyme topography (surface shape) in a literally vital manner.

Even very small bacterial cells contain several thousand chemically distinct types of molecule. Of these, perhaps 65 per cent are different types of protein molecule, 20 per cent are different sorts of nucleic acid molecule, and the remainder are mostly lipids and carbohydrates. A relatively trivial number of smaller molecules are involved as the machinery or the fuel of biochemical pathways described later in this book. It has been estimated that a typical animal cell of average size (i.e., about 10^{-5} mm^3) weighs about 10^{-9} g and contains some 10^{10} molecules of protein.

Cells vary greatly in size. The largest single cells are bird's eggs, certain nerve cells, or, in the plant kingdom, cells of Acetabularia. All of these may be 8 cm or more long. By contrast, an average metazoan cell has a diameter of about 30 μm and a small microorganism may be less than 1 μm diameter.

In plant cells, the cellulose cell wall accounts for 25 to 50 per cent of the cell's dry weight. Allowing for this, in both plants and animals cell protoplasmic protein accounts for 33 to 40 per cent of the residual dry weight. Of this protoplasmic protein, about half is present in various subcellular particles or *organelles* and the other half is dissolved in the soluble cell fraction or *cytosol.*

1.3 Some biologically important molecules and their part in metabolism

Many biochemical processes are occurring simultaneously within an active cell, and each of these often involves a series of interconnected chemical reactions. From small and relatively simple building bricks, or *precursors*, the cell constructs numerous more complex substances. The latter belong in the main to four chemical families, namely the *proteins*, *nucleic acids*, *polysaccharides*, and *lipids*. These macromolecular substances are characteristic of living organisms and essential to their existence. Individual cells of animals and plants synthesize these molecules, even when, as in animals, the organism as a whole takes in already-constructed macromolecules in the food. The sum total of all the *synthetic* processes of the cell is referred to as cellular *anabolism*.

Construction processes in biology, as in human technology, are energy consuming and include operations which demand special organization. Left to itself, London Bridge will tend to fall down. To construct objects such as bridges, numerous smaller parts have to be assembled by human endeavour. In one small part of the Universe, and for a short period of its history, local order has increased and energy-consuming operations have been possible. It is the same for a living cell; for the brief time of its living existence, energy, ultimately derived from *sunlight*, is used to provide energy for anabolic processes and for creating a greater local orderliness. Increasing the level of orderliness is known to chemists as *decreasing entropy*.

Side by side with synthesis, cellular components are constantly being *degraded* to simpler compounds. Sometimes this is to provide the building bricks for synthetic processes needed to overcome wear and tear, for a living cell has to be a self-repairing machine. In addition, however, cells live, grow, and reproduce, and to *provide the energy* for these anabolic processes other cellular components have to be broken down. Such energy-providing processes are called *catabolic* reactions. By and large, the breakdown of carbohydrates and fats provides the energy for the cellular synthesis of proteins, nucleic acids, and complex lipids. However, there is a lot more to general cell *metabolism* than this. Metabolism is the sum total of all the anabolic and catabolic processes going on in the cell. Food reserves, for eventual use as energy sources, have to be built up first, and stocking up a fuel shed is itself an energy-consuming process. Energy is not only used for anabolic processes but is needed for transferring substances through cell membranes, for muscular contraction, and for many other electrophysiological phenomena (Chapters 19 and 20).

The multitude of reactions occurring in the minute dimensions of the cell would more than tax the resources of the most elaborate chemical laboratory. The chemist can use one vessel for a reaction occurring in water and another for a reaction requiring a non-polar solvent such as benzene. Solid catalysts would probably be used in a third type of apparatus. The cell achieves similar 'compartmentation' by having an elaborate substructure; some parts of the cell are adapted for aqueous reactions, others accommodate surface reactions, and yet other regions provide a non-polar environment for essentially hydrophobic reactions.

The activities of the cell must be controlled so that they take place at a rate appropriate to the circumstances of the moment, for the cell must be able to adapt to changing environmental conditions. This applies to the rates of reactions leading to the anabolism or catabolism of cellular constituents, and to reactions leading to energy production. It also applies to the movement of metabolites between various parts of the cell and to the movement of substances between the inside of the cell and its environment. A special feature of cells is that their organelles are bounded by *lipoprotein membranes.* This type of membrane also forms the outer surface of most cells (the plasmalemma). These membranes constitute 'walls' infinitely more versatile than the walls of glass vessels used by the chemist. The reason for this is that many enzymes and carrier proteins are an *integral part* of membrane structure. Thus the membranes are themselves intimately involved in the cell's metabolism.

1.4 Structure of animal and plant cells

Most cells are clearly visible under a good light microscope, the resolving power of which is rather less than 0.4 μm and the classical microscopists were well aware that cytoplasm was by no means an amorphous jelly. They knew that, in addition to the nucleus (and, in plants, the chloroplasts), a variety of particulate structures of smaller size was present within the cell.

The advent of the electron microscope stimulated interest in the finer detail of cell structure and in the relation of such structure to function. It provides effective magnification of between 100 000 and 200 000 compared with only 1000 for the light microscope and the resolution (1 nm or 10^{-6} mm) reaches almost to molecular dimensions (the orbital diameter of a hydrogen atom is about 0.1 nm). For technical reasons, even the best electron microscopes do not achieve this level of resolution. Electron microscopy relies on the principle that some materials are more or less opaque to a beam of electrons so that differences in electron density of tissue particles are detected when penetrating electrons hit a fluorescent screen or photographic plate. To prepare material for electron microscopy it has to be dried and stained with electron-dense compounds such as osmium tetroxide or potassium permanganate and cut into very thin sections. This treatment is very 'unphysiological' so artefacts could well be introduced.

Some of our present ideas are possibly based on misconceptions resulting from the creation of artefacts, but, almost beyond question, many of the cellular substructures are either constructed from a particular type of membrane composed mainly of lipid and protein, or they are attached to such a membrane. Apart from electron microscopy, x-ray analyses of cell membranes, fractionation of the various parts of the cell, and studies of the electrical properties and permeability of cell surfaces all tend to indicate that *membranes made of lipid and protein* are the basic 'fabric of life', the material from which most of the subcellular particles or 'organelles' in all types of cells are constructed.

Figure 1.1 illustrates, in very diagrammatic form, some of the features of a non-specialized cell; animal cells typically lack a prominent cell wall, have no chloroplasts or plastids, and usually have no well-defined vacuole. Within most animal cells the *nucleus* is the most prominent organelle, although in many plant cells the chloroplasts are also large. Mitochondria are just visible under the light microscope and these, like the chloroplasts, are largely composed of lipoprotein membranes. The *plasmalemma* or cell membrane,

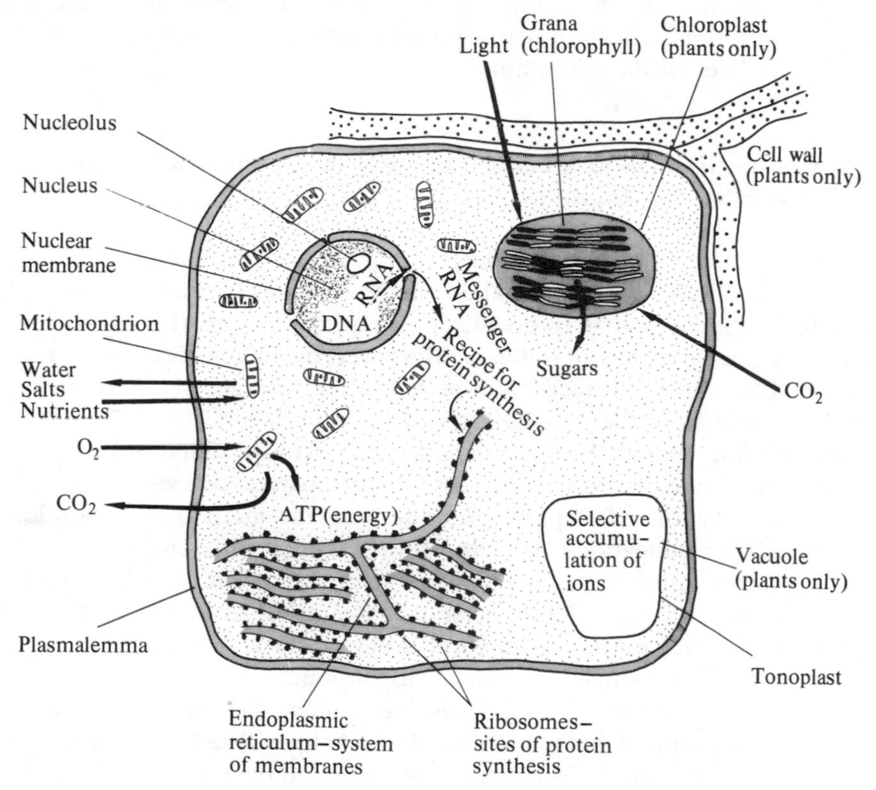

Fig. 1.1 Highly schematic structure of a generalized cell

the *tonoplast*, which surrounds the vacuole, and the nuclear membrane are all examples of lipoprotein membranes.

The background fluid in which the organelles are suspended is known as the *cytosol*; it is composed largely of water in which ions of potassium, chloride, phosphate, and many others are to be found, together with traces of organic compounds, such as glucose and amino acids and colloidal solutions of certain proteins and enzymes. Electron micrographs reveal that a complicated set of branching lipoprotein membranes apparently divides the cytosol into numerous separate compartments. In cross section, these membranes resemble a network of intercommunicating tubules, and for this reason have been termed the *endoplasmic reticulum*. These membranes probably undergo quite rapid formation and dissolution and so are less permanent than the other membranes mentioned above, although it is probably true to say that no membrane is 'permanent' in the sense that it persists unaltered for the whole life of the cell.

Several enzymes, or enzyme systems, are closely associated with the membranes of the endoplasmic reticulum. In addition, numerous small spherical bodies, the *ribosomes*, are present in the cytosol and these tend, especially during periods of active protein synthesis, to become attached to the membranes of the reticulum.

1.5 Cell fractionation as a method of studying the functions of cell organelles

Various activities occur in different parts of the cell (Table 1.1) and this facilitates the investigation of these processes, for, by employing suitable techniques, it is frequently possible to separate the organelles so that they can be *studied in isolation*. Moreover, by adding cofactors normally provided by other cell fractions, a picture often emerges of the remarkable interdependence of the various parts of the cell.

By grinding up and then centrifuging some suitable tissue or organ (e.g., animal liver, plant leaves) it is possible to separate the various kinds of subcellular particles. The principle underlying the method of separation—differential centrifugation—depends on the fact that different sorts of cellular organelles differ from one another in size and density. Consequently, they will sediment at different rates when centrifuged at different speeds. Larger or heavier particles will form a deposit in a centrifuge tube while lighter or smaller particles remain in suspension.

The grinding process is called *homogenization*; special homogenizers of various designs are manufactured for the technique. The best homogenizing buffer for any specified tissue, the optimal pH, and the precise tonicity of the medium are found empirically. Isotonic sucrose (0.25 *M*) at pH 7.0 has

Table 1.1 Location of some important cellular activities

Location	Activity	Chapter
Nucleus	Genetic control of the proteins and enzymes of the cell	16
	Also synthesizes DNA, RNA, and some essential cofactors	
Mitochondria	The inner membrane produces most of the ATP to drive anabolic processes	8
	The inner matrix enzyme systems	
	(a) Oxidize acetyl coenzyme A with the formation of NADH (TCA cycle)	9
	(b) Convert fatty acids to acetyl coenzyme A	10
	(c) Catalyse some of the reactions in the synthesis of urea	15
Chloroplasts	The seat of both light and dark reactions of photosynthesis	
	(a) Membranes—NADPH and ATP production	14
	(b) Intermembrane fluid—CO_2 reduction to form sugars	14
Ribosomes	Protein synthesis	16
Endoplasmic reticulum	(a) 'Rough' reticulum associated with ribosomes is concerned in the formation and despatching of proteins	1
	(b) 'Smooth' reticulum, ribosome free, is concerned with steps in the synthesis of lipids including long-chain fatty acids and steroids, and with the processing of extraneous lipid-soluble molecules	10, 14
	(c) Golgi bodies, lysosomes, and zymogen granules, associated with the intracellular storage of enzymes	1
Plasmalemma	Responsible for the uptake and elimination of water, certain ions, and organic molecules	17, 18
	Maintains a negative electrical charge inside nerve and muscle cells, a charge which enables conduction of impulses to occur	19, 20
Cytosol	(a) Seat of glycolysis by which carbohydrates are broken down to pyruvate	11
	(b) Hydrolysis of fats to glycerol and fatty acids	10
	(c) Locus of the enzymes of the pentose phosphate pathway, whose chief function is to produce NADPH for anabolic processes	14
	(d) Site of the reactions of gluconeogenesis	12

been used for mitochondria by many workers, while 0.15 *M* potassium chloride has been found satisfactory by the present authors for the preparation of liver microsomes. The microsomal fraction contains the remains of the rough and smooth endoplasmic reticulum.

A suitable volume of cooled homogenizing buffer is added to the tissue which is then homogenized for about 3 min, the homogenizer tube being all the while cooled in ice. After homogenizing, any incompletely homogenized

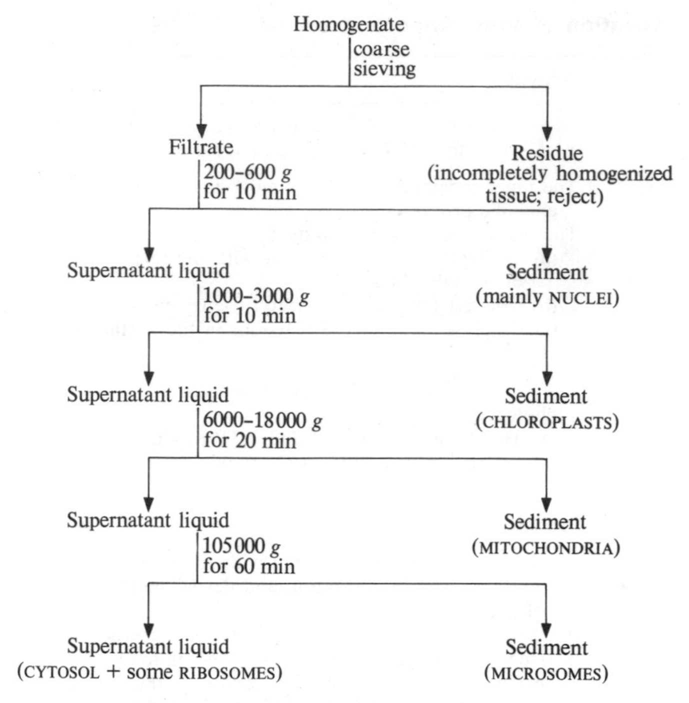

Fig. 1.2 Scheme illustrating the use of differential centrifugation for the separation of subcellular particles

material (especially plant cell walls) is removed by filtration through a cheese cloth. To separate the different types of organelles the crude homogenate is then successively spun at increasing *g* values using plastic tubes fitted with metal caps; the latter are necessary at high rotor speeds because high-speed centrifuges operate in a partial vacuum.

Figure 1.2 shows the general procedure for *differential centrifugation.* Nuclei are the heaviest particles and separate upon centrifuging for about 10 min at 400*g*. The centrifuge is stopped and the supernatant fluid is carefully decanted to a clean centrifuge tube. The process is then repeated at a greater *g* value or for a longer time. It must be remembered that cells from different tissues, or from similar tissues in different organisms, can vary widely, so the scheme is a generalization. Only experimentation, often assisted by electron microscopic examination of the sediments, can provide the optimal conditions for any individual case. Preparative work of this kind is usually performed at low temperatures (for example, 4°C) since this slows down biochemical reactions and helps to preserve the activity of the material.

A refinement, sometimes superior to differential centrifugation, is known as *density gradient* centrifugation. The principle is to create a concentration gradient of an aqueous solution, usually of sucrose or glycerol, such that the concentrated solution (for example, 50 per cent sucrose) is present at the bottom of a polypropylene centrifuge tube, the concentration gradually decreasing to perhaps 15 per cent at the top of the tube. A solution containing a mixture of organelles (or proteins) is placed gently on the surface of the liquid and the tube is centrifuged at a suitable speed, using a swing-out head. Particles move down the tube to different depths determined by their mass, weight, and shape. In consequence, a mixture of different organelles (or proteins) separates into layers in the tube. By piercing the bottom of the polypropylene tube with a hypodermic needle and affixing a narrow tube, the contents can be run out into a number of collecting vessels, so enabling the components of the original mixture to be separated.

1.6 The plasmalemma, cell walls, cilia, and flagella

The boundary of the active cytoplasm in cells is known as the plasmalemma. In plants, and occasionally in other organisms, this is surrounded in turn by a much more conspicuous cell wall. The existence of a specific surface membrane had been suspected long before the electron microscope was invented because of the characteristics of the cell boundary. For example, it was inferred that the plasmalemma was likely to contain lipid because many non-polar compounds, e.g., chloroform and higher alcohols, are able to enter erythrocytes (red blood cells) with ease. Such compounds probably gain access to cells by virtue of their solubility in the double layer of lipid molecules in the lipoprotein membrane (page 361). The cell surface has, furthermore, a high electrical resistance, a fact which is compatible with the existence of a non-polar region in the membrane. On the other hand, the surface tension between a cell and its surroundings is low, an observation consistent with the presence of a superficial layer of a polar protein but inconsistent with lipid in direct contact with the water.

Modern theories of the structure of lipoprotein membranes are considered in the next chapter (Section 2.6). For the present it is sufficient to picture the membrane as a *central layer of lipid* into which certain proteins penetrate. In addition, various other proteins are situated on its two surfaces.

It is probable that the plasmalemma is typically a continuous structure around the cell with very few, if any, pores completely penetrating it. Needless to say, the plasmalemma is a structure of great importance to the cell since substances entering or leaving the cell must pass through it. The membrane is *semi-permeable* for while water can move into, or out of, the cell, solutes usually have difficulty in passing through it unless highly selective transport mechanisms exist to facilitate their movement. The importance of selective permeability and some of the problems it raises are considered in Chapter 17, but an example will illustrate the type of phenomenon which is frequently encountered. The vertebrate intestinal mucosal cells have a great ability to absorb D-glucose, yet the closely related D-mannose is only slowly absorbed. In this case, a specific protein probably acts as a *carrier* which facilitates the uptake of glucose. Whether proteins of this type are actually the structural protein of the membrane or some other more mobile protein is still uncertain.

The plasmalemma is a delicate membrane which, when it is distorted, tends to return to its normal shape by elastic recoil. If it is slightly torn, the edges tend automatically to flow together, probably reflecting the tendency for phospholipids to aggregate in ordered layers. However, the mechanical strength of cells is limited, and they are frequently strengthened by external materials; plant cells, in particular, are characterized by the presence of a cell wall. Cells of green plants secrete fibrous polysaccharides such as cellulose while the cell wall of fungi is frequently composed of chitin. In plant cell walls substances such as lignin and pectin are often also present. Usually the cell wall is freely permeable to water and solutes but impregnation with lignins or waxes may render the walls impermeable, causing the death of the cell. Sometimes specific areas are left unlignified and these *pits* serve to allow free passage of materials. In the xylem (wood) of most higher plants the cells are devoid of cell contents and only the wall remains, thus providing a mechanism for the effective conduction of water and ions from the roots to the leaves.

In some types of cell present in organisms from almost every plant and animal phylum, the plasmalemma and certain underlying structures are modified to form projections, known as flagella and cilia, which, by waving or undulating, are often able to induce the locomotion of cells. This, at least, is their primitive function; their presence in certain protozoa and algae is well known, but it must not be overlooked that spermatozoa of higher animals also possess flagella. Furthermore, even when tissue structure precludes active movement of cells, ciliated cells are frequently to be found, and then they serve some more advanced function. They are necessary for the feeding process of many filter-feeders, including shellfish of many kinds. In mammals, they line the bronchioles and waft mucus out of the lungs. This is a cleansing mechanism, for the mucus carries with it particles of dust which have managed to enter; it is of interest that one of the adverse effects of nicotine is that it paralyses these cilia.

Electron microscopy has shown that no fundamental difference exists between a flagellum and a cilium. Each is surrounded by a membrane which is continuous with the plasmalemma of the main part of the cell. Twenty

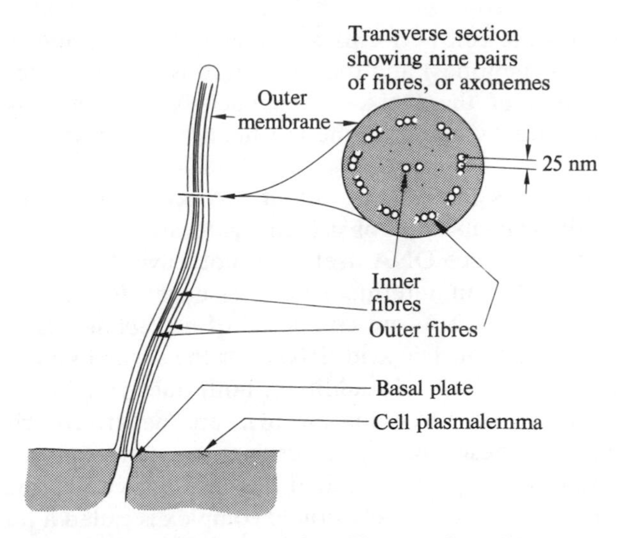

Fig. 1.3 Structure of a cilium or flagellum

fibres, or *axonemes*, always arranged in a similar fashion no matter what the origin of the flagellum, run along the length of the structure. In cross section, it is found that a central pair of axonemes is surrounded by nine outer pairs. Each fibre is about 25 nm in diameter, the electron-dense area being on the outside (Fig. 1.3).

1.7 The nucleus

All active plant and animal cells contain a nucleus which is surrounded by a membrane whose dimensions and properties strongly suggest that it has a lipoprotein structure (page 33). However, some electron micrographs of the nuclear membrane indicate that it is penetrated by pores of some 5 to 10 nm diameter. It is believed that the pores allow large molecules such as RNA (ribonucleic acid) to move from the nucleus into the cytosol.

Within the nuclear membrane is a jelly-like fluid (a gel) composed largely of water. The non-aqueous contents comprise some 70 per cent protein, 10 per cent DNA (deoxyribonucleic acid) as well as lipid and RNA. The chromosomes are not visible in non-dividing cells but, nevertheless, the essential part of their chemical structure undoubtedly persists during the resting phase. From a genetic point of view, the most important nuclear material is the DNA present in the chromosomes, for it is a carrier of information (Chapter 16), the molecular structure of DNA providing coded instructions for the proteins produced by the cell. This is one reason why the nucleus can be regarded as the control centre of the cell. Almost all the chemical reactions the cell performs are catalysed by enzymes (all of which are proteins or polypeptides) and their structure is indirectly decided by the precise composition of the DNA in the nucleus. The nucleus, therefore, plays a vital, if indirect, role in all the biochemical processes which go on in the cell.

The DNA passes its information about prospective protein structure to the ribosomes, the principal site of protein synthesis (some proteins are also made in the nucleus). Since DNA itself does not leave the chromosomes, the message is carried by an intermediary, messenger RNA. Both types of *nucleic acid*, RNA and DNA, are composed of *nucleotide bases* united with the sugar, *ribose* (in *ribo*nucleic acid, RNA), or the sugar-like *deoxyribose* (in *deoxyribo*nucleic acid, DNA). In addition, both nucleic acids contain phosphate groups which, at physiological pH, are negatively charged. The negative charges on these phosphate residues are attracted to the positive charges on certain basic proteins called histones, which are present in the nucleus; the resulting rather loosely bound complex is called a *nucleoprotein.*

The nucleus contains a number of enzymes absent from the rest of the cell, or which are in short supply there. In particular, those enzymes concerned with the synthesis of nucleotide bases and of a substance termed NAD (nicotinamide adenine dinucleotide) are located in the nucleus. Numerous oxidation and reduction reactions throughout the cell can only occur if the enzyme responsible for the reaction is itself accompanied by NAD, and this substance is therefore known as a *coenzyme. Oxidation* reactions—the removal of electrons from compounds—play a particularly vital role in cellular chemistry, the NAD being important because the electrons removed from the substance being oxidized are frequently passed to NAD. The significance of this process will be explained in later chapters.

The nucleus again illustrates the interdependence among the cell organelles. The nucleus supplies the rest of the cell with NAD, with certain other coenzymes, and with various RNAs. In return, the nucleus receives nu ous metabolites from the cytoplasm, including sugars and amino acids; it also receives, in the form of ATP, the energy which is absolutely essential if it is to perform its principal activities, almost all of which consume energy.

1.8 Mitochondria

The shape, size, and number of these bodies vary greatly in different cells. They may be spherical or elongated in form, their shape varying with their metabolic activity. There may be only a few per cell, or, as in an active liver cell, there may be up to a thousand. Typically, they are just visible under the light microscope, but their elaborate membranous substructure was only revealed by electron micrographs. Each mitochondrion is constructed from two lipoprotein membranes, the inner one of which is convoluted to form

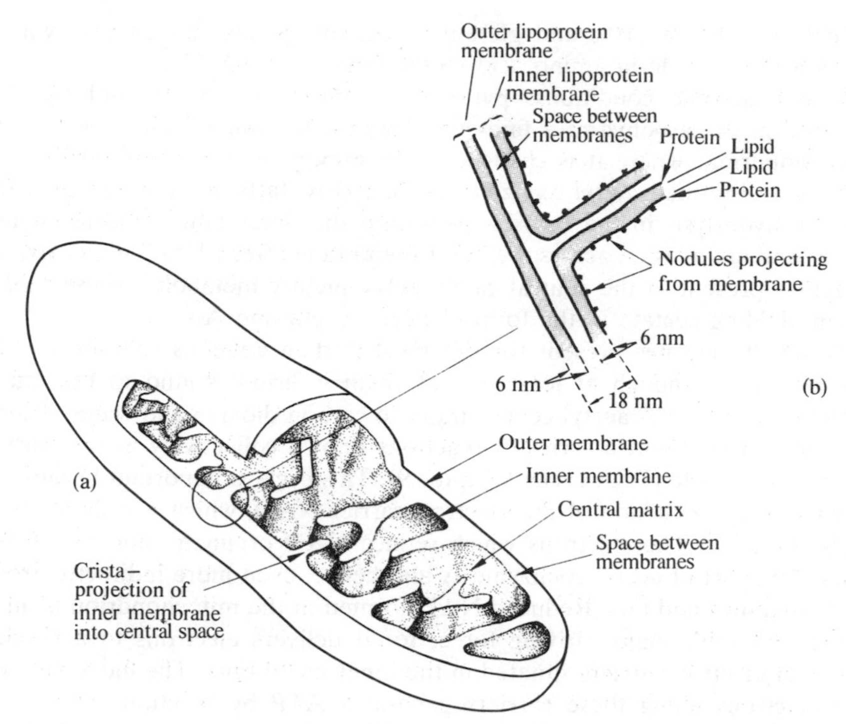

Fig. 1.4 Schematic representation of a mitochondrion (a) with a section cut away to reveal the interior and (b) enlarged to show detail of inner and outer membranes

folds or cristae which project into a central aqueous cavity, or matrix. The outer membrane is not convoluted and forms a skin-like limiting layer to the mitochondrion. Between the outer and inner membranes is a second mitochondrial cavity, which, like the central cavity, appears transparent in electron micrographs and is probably aqueous (Fig. 1.4).

The outer membrane of the mitochondrion is structurally and physiologically very different from the inner membrane. Unlike the inner membrane, which is only permeable to relatively small molecules, and then often only selectively, the outer membrane is permeable to substances with molecular weights below about 10 000. The outer membrane lacks most of the enzymes associated with the inner membrane. The intermembrane space contains a few enzymes, an interesting example being *adenylate kinase*, an enzyme often participating in reactions involved in metabolic control (Chapter 13). The central matrix of the mitochondrion is filled with a gel containing 40 to 50 per cent protein. It also contains DNA and ribosomes, so the mitochondrion can probably make a few of its own proteins without nuclear intervention. Enzymes present in the matrix include many of those involved in the *TCA cycle* (Chapter 9), all of those involved in fatty acid oxidation (Chapter 10), and glutamate dehydrogenase, an enzyme which plays a unique role in amino acid metabolism (Chapter 15).

Under aerobic conditions, pyruvate produced in the cytosol by the catabolism of carbohydrates finds its way into the inner membrane of the mitochondrion, where it is changed to an important compound containing acetate and termed *acetyl coenzyme A*. Similarly, fatty acids, liberated from fats by hydrolysis in the cytosol, penetrate the outer mitochondrial membrane and are assisted across the inner membrane after union with a carrier; enzymes present in the central cavity subsequently metabolize these acids, again yielding acetate in the form of acetyl coenzyme A.

A set of enzymes present for the most part in aqueous solution in the central cavity (though at least one of them is firmly bound to the inner membrane) oxidizes acetyl coenzyme A to carbon dioxide and water. These enzymes bring about a series of reactions known collectively as the tricarboxylic acid cycle (TCA cycle, Chapter 9). This cycle is important because it results in the reduction of the electron carrier NAD, which was mentioned in Section 1.7. The electrons which reduce NAD originate indirectly from the acetyl part of acetyl coenzyme A and hence, even more indirectly, from carbohydrates and fats. Reduced NAD formed in the mitochondrion is, in a sense, the cell's major driving force, for it delivers electrons to a special series of electron carriers situated in the inner membrane. The movement of the electrons along these carriers generates ATP by reactions known as *oxidative phosphorylation* (Chapter 8). Since most of the cell's ATP is produced in this way, the mitochondrion has been appropriately named the 'power house' of the cell.

The inner mitochondrial membrane is an excellent example of a highly organized structure, the design of which specially fits it for a particular function. It is held together by a system of phospholipids and structural protein which has been found to contain amino acids rich in non-polar side groups. These are weakly bound to the non-polar parts of the lipids. Four phospholipoprotein complexes, known as electron transport particles I, II, III, and IV (Chapter 8), are closely associated with one another in the inner mitochondrial membrane. Electrons initially derived from the acetyl part of acetyl coenzyme A pass systematically from one particle to another, so that the electrons 'fall' from a high energy level to a lower one; as they do so, *energy is tapped off and used to make ATP*. When the electrons finally reach particle IV, they combine with molecular oxygen and hydrogen ions to produce water.

1.9 Chloroplasts

The chloroplasts are the seat of the most fundamental of all biochemical reactions for it is here that the machinery exists for trapping light energy and transforming it into chemical energy. In part, this chemical energy is in the form of ATP, but reducing power is also created which eventually reduces carbon dioxide to the level of sugar. Sugars and substances derived from them, act in the first instance as fuels for driving the life processes of plants, but it must not be overlooked that in the end all animals depend upon plants for their existence.

In the green cells of plants, chloroplasts are often the largest and most characteristic organelles. Sometimes there may be one or a few, but often there may be a hundred or more in one cell; like mitochondria, their position is not rigidly fixed, and they often migrate around the cell. Their shape and size differ in different species, especially in algae, but in higher plants they are typically between 0.5 and 2.0 μm thick and up to 10 μm diameter. On a dry weight basis, some 50 per cent of the chloroplast is protein, 35 per cent is lipid of various kinds, and 7 per cent comprises pigments. The pigments (except in photosynthetic bacteria) almost invariably include chlorophyll **a**, but the nature and the amounts of accessory pigments differ considerably in organisms from different parts of the plant kingdom.

Chloroplasts, like most other subcellular organelles, are composed of lipoprotein membranes. In higher plants (algae tend to have them ordered in lamellae only), these are typically stacked together rather like a pile of coins, to provide chloroplastic subunits termed *grana.* When light strikes the chloroplast, a series of reactions is initiated which leads to the production of ATP and of reducing power. The latter is, in fact, not reduced NAD, but reduced NAD phosphate (NADPH). ATP and NADPH in turn provide the energy which drives what must undoubtedly be regarded as the most important anabolic process in nature—the conversion of carbon dioxide to sugars. Furthermore, from the sugars and their derivatives, arising from photosynthesis, the carbon skeletons of all the other major components of the plant are ultimately synthesized.

The enzymes responsible for the trapping of light energy and converting it to the energy of ATP and NADPH are largely or entirely associated with the *membranes* of the grana. On the other hand, the enzymes which bring about the reduction of carbon dioxide are present in the aqueous *matrix* of the chloroplast. The so-called 'light' and 'dark' reactions of photosynthesis are described in Chapter 14.

1.10 The endoplasmic reticulum and associated inclusions

The membranes of the reticulum, like other cell membranes, are probably typical lipoprotein double layers (Section 2.6). In some cells, the nuclear membrane is connected to adjacent endoplasmic reticular membranes.

Sometimes parts of the reticulum appear to be in the form of tubules, many of which are somewhat distended inside at what is presumed to be the internal lipid layer.

In certain areas of the cell, spherical particles of about 20 nm diameter are attached to the membrane. These particles, the ribosomes, contain some phospholipid but are mainly protein and RNA. As mentioned above, messenger RNA passes from the nucleus via the cytosol to the ribosomes where the message is deciphered. With the cooperation of a number of other constituents of the cytoplasm, including the energy of ATP, amino acids are 'zipped' together in an order or sequence dictated by the structure of the messenger RNA and hence, ultimately, by the structure of the DNA in the nucleus.

Some regions of the endoplasmic reticulum lack ribosomes and in consequence appear to have a smooth surface (Fig. 1.5). It is here that the synthesis of cell lipids probably takes place—steroids, phospholipids, and fatty acids are built up from simpler precursors. Their synthesis probably takes place in a largely non-polar environment.

Besides its two anabolic functions (the synthesis of proteins and lipids) the endoplasmic reticulum is associated with the storage of enzymes in an inactive form. In some cells, and especially well developed in cells whose

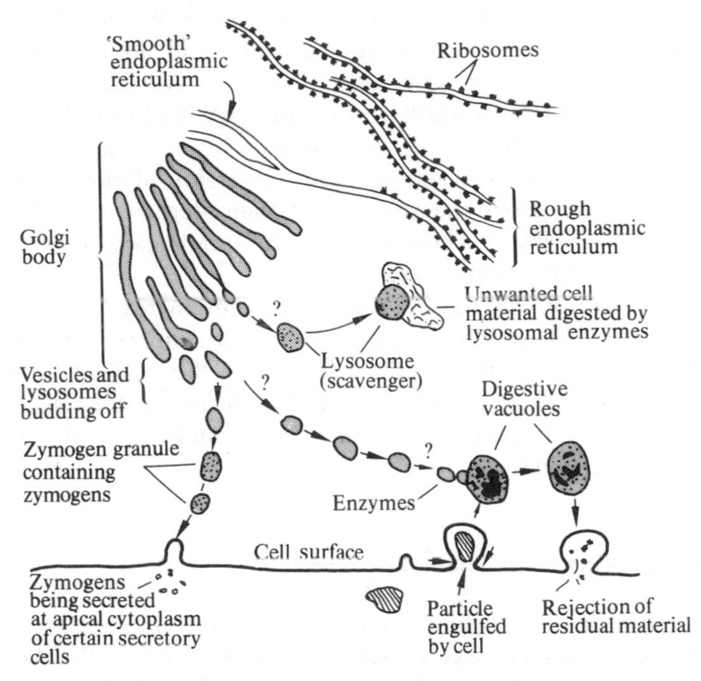

Fig. 1.5 A possible relationship among the rough and smooth endoplasmic reticulum and some associated bodies in various cells

main function is secretion, there are structures called *Golgi bodies* (Fig. 1.5). These look in electron microscope pictures like pieces of smooth endoplasmic reticulum stacked together. Here inactive enzyme precursors known as *zymogens* accumulate in vacuoles whose walls possibly arise from the reticulum. In the pancreas, for example, the zymogens trypsinogen and chymotrypsinogen are stored in vesicles. The zymogens are modified after leaving the cell, becoming the active enzymes trypsin and chymotrypsin.

A second site of enzyme storage, particularly abundant in phagocytic cells (e.g., blood leucocytes), are vesicles known as *lysosomes.* They were discovered by C. De Duve in a 'light mitochondrial' centrifugation fraction and soon shown to contain a number of enzymes with a low pH optimum, including acid phosphatase, RNA-ase and DNA-ase. Later, more than thirty enzymes were characterized in lysosomes, including the proteolytic cathepsin D and various glycosidases. Between them, these enzymes are able to degrade almost all cellular macromolecules, so clearly any damage to the membrane forming the wall of the lysosomal vesicle would cause havoc within the cell. Indeed, their release when the cell dies in a prime cause of *autolysis.* During life, however, lysosomes can take in and destroy potentially toxic foreign macromolecules, an action which is the microcosmic equivalent of amoeboid ingestion or phagocytic action.

In common with most other cell proteins, lysosomal enzymes are formed by the ribosomes and then transported by the endoplasmic tubules to the region of the Golgi body, where they are packaged in lysosomal vesicles. Hydrocortisone is able to stabilize lysosomal membranes, whereas some bacterial toxins and ultraviolet light weaken the membrane and make it more leaky. Needless to say, there is considerable speculation about the role of lysosomes in physiology and in pathology, including the phenomenon of ageing. As well as scavenging within the cell, lysosomes play an important part in some forms of extracellular digestion. For example, lysosomal enzymes probably participate in the resorption of tadpole tails which occurs during the metamorphosis of the frog.

The pathological effects of silica and asbestos appear to be due in part to the ability of particles of these minerals to damage lysosomal membranes. There is also considerable evidence for abnormal involvement of lysosomal enzymes in arthritic conditions, for rheumatoid synovial fluid contains larger-than-normal quantities of these enzymes. It has been suggested that release of lysosomal enzymes from cells near articular cartilage may be a major factor in its erosion, although lysosome membrane leakage is not likely to be the initial cause of the rheumatic condition.

1.11 Cell vacuoles, the tonoplast, and the cytosol

Many plant cells contain vacuoles bounded by a membrane, the *tonoplast.* This membrane probably has a similar structure to the plasmalemma and may, indeed, arise from it by a process of invagination. Within the vacuoles, plant cells store water, ions, and food materials such as soluble sugars. They may possibly also act as receptacles for harmful cellular waste products. The tonoplast, like the plasmalemma, is probably selectively permeable, but it seems to possess a more developed mechanism for the uptake and retention of certain ions.

The *cytosol* is the aqueous medium which lies between the intracellular organelles. Thus, assuming no disruption of organelles occurs, it is the liquid left behind when all the cell organelles have been removed. As an approximation, it is the microsomal supernatant fluid (Section 1.5), although the latter contains some ribosomes as well. It is frequently quite rich in soluble (i.e., not membrane-bound) protein; in plant leaf cells and in animal liver cells approximately 25 to 50 per cent of all the cellular proteins are recovered in this fraction after centrifugation. In many cells the cytosol also contains suspended particles of carbohydrate storage materials such as glycogen in animal cells and starch in plants.

Table 1.1 lists some of the important biological pathways where all or most of the enzymes are present in the cytosol. Of particular note are the pathways of glycolysis (Chapter 11) and gluconeogenesis (Chapter 12) which respectively split sugars to give pyruvate and build pyruvate back into sugars. The hydrolysis of fats (Chapter 10) and the sugar interconversions of the pentose phosphate pathway (Chapter 14) are also located in the cytosol.

1.12 Prokaryotic cells

The foregoing general description of the structure of cells has been primarily concerned with cells possessing a clearly defined nucleus bounded by a nuclear membrane. Cells of animals and plants are called *eukaryotic* ('good nucleus'), but the more primitive organisms (bacteria, blue-green algae, and mycoplasmas) are generally much smaller and their chromosomes are *not* clearly separated from the cytosol by any nuclear membrane (Fig. 1.6). These organisms are called *prokaryotes* ('before a nucleus') to distinguish them from organisms with more elaborate cell organelles.

Of the numerous and diverse prokaryotic species some, including various bacteria and the blue-green algae, are able to perform *photosynthesis.* Others are *heterotrophic,* getting ready-made organic nutrients from the soil or water, or from other organisms which they parasitize.

The prokaryotes are much smaller than most eukaryotic cells. One of the most highly investigated bacteria, *Escherichia coli,* is, for example, about the size of a mammalian mitochondrion ($0.8 \times 2\ \mu m$). Clearly, therefore, their cellular ultrastructure is on a different scale from that seen in eukaryotic cells.

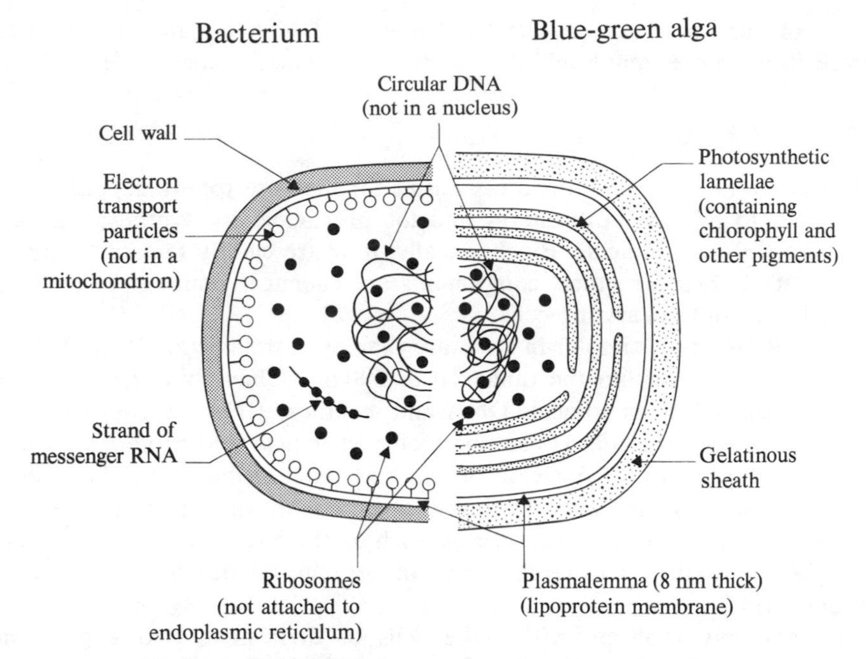

Fig. 1.6 Generalized structure of prokaryotic cells

Prokaryotic cells possess a plasmalemma composed of lipoprotein and this is usually surrounded by a cell wall which is gelatinous in blue-green algae but is generally more rigid in bacteria (page 86). In many prokaryotes the cell contains one or more chromosomes composed ot a circular double strand of DNA. In *E. coli* this would be about 1.2 mm long if it were pulled out straight, but normally it is highly convoluted to form a chromosomal mass. The ribosomes of prokaryotes are rather smaller than those of animals and plants and appear to be freely distributed in the cytosol for there is no endoplasmic reticulum.

Prokaryotes lack mitochondria. Aerobic species, however, have electron transport particles analogous to those of the eukaryotic mitochondrion, but they are situated in the cytosol. In some species (e.g., *E. coli*) they are associated with the inside of the plasmalemma (Fig. 1.6).

The blue-green algae are named because of the pigmentation in their photosynthetic membranes. These membranes are simpler than the chloroplasts of plants but the photosynthetic pigments are nevertheless situated in membranous lamellae.

One of the most remarkable features of the biochemistry and physiology of eukaryotic cells is the general similarity of many of their basic functions. This applies equally to the prokaryotes. Indeed, it is a remarkable fact that much of our knowledge of the biochemistry of higher plants and animals arose from studies made initially on simpler organisms such as *E. coli.*

1.13 Viruses

Viruses are not clearly defined organisms for they arc totally dependent on host cells to provide both energy and the precursors for synthesizing their proteins and nucleic acids. Biochemically, they are clearly *related* to organisms in that their structural components are chemically similar to those in prokaryotic and eukaryotic cells.

Viruses are much simpler in structure than even the bacterial cell. In size, most of them lie within the range 10 to 200 nm. They comprise a core of nucleic acid, which in some is DNA and in others RNA, surrounded by a shell of protein. The relative amounts of nucleic acid and protein vary, but typically only 20 to 40 per cent of the mass is contributed by the nucleic acid. The protein shell may, in simple cases, consist of only one molecular species but, in more complex viruses, such as the T-2 bacteriophage, there may be up to fifteen different types of protein surrounding the core of nucleic acid.

Viruses only replicate within the cells of their host, taking over the machinery of the host cell to supply energy and all the cofactors necessary for their replication. They can be considered to substitute a messenger coding for virus protein for that normally supplied to the ribosomes and, in addition, to utilize nucleotides and other substances present within the cell for the synthesis of more molecules of virus nucleic acid. At the moment of infection, much of the protein coat is often discarded by the virus, the nucleic acid alone being infective. Thus the bases of the nucleic acid of the virus must provide all the information needed for replication—i.e., the protein shell does not consist of enzymes essential for the process of replication.

Geometrically, viruses are of three types. Some of them show cubic symmetry, others helical symmetry, whilst those in a third group exhibit complex symmetry. Examples of each are illustrated in Fig. 1.7. Those with cubic symmetry are often based on the icosahedron, the protein subunits or *capsomers* being arranged in a geometrical pattern on each of its faces. Those with a helical structure have a central core of nucleic acid, around which capsomeres of protein are again regularly arranged. The structure of one virus of this type—the tobacco mosaic virus—has been extensively studied and has proved to be a helix of RNA surrounded by a helix of elongated capsomeres.

The T-2 bacteriophage, which exhibits complex symmetry, infects cells of *E. coli.* Despite its greater complexity, the proteins associated with its DNA probably play no part in the process of infection, for less than 3 per cent of

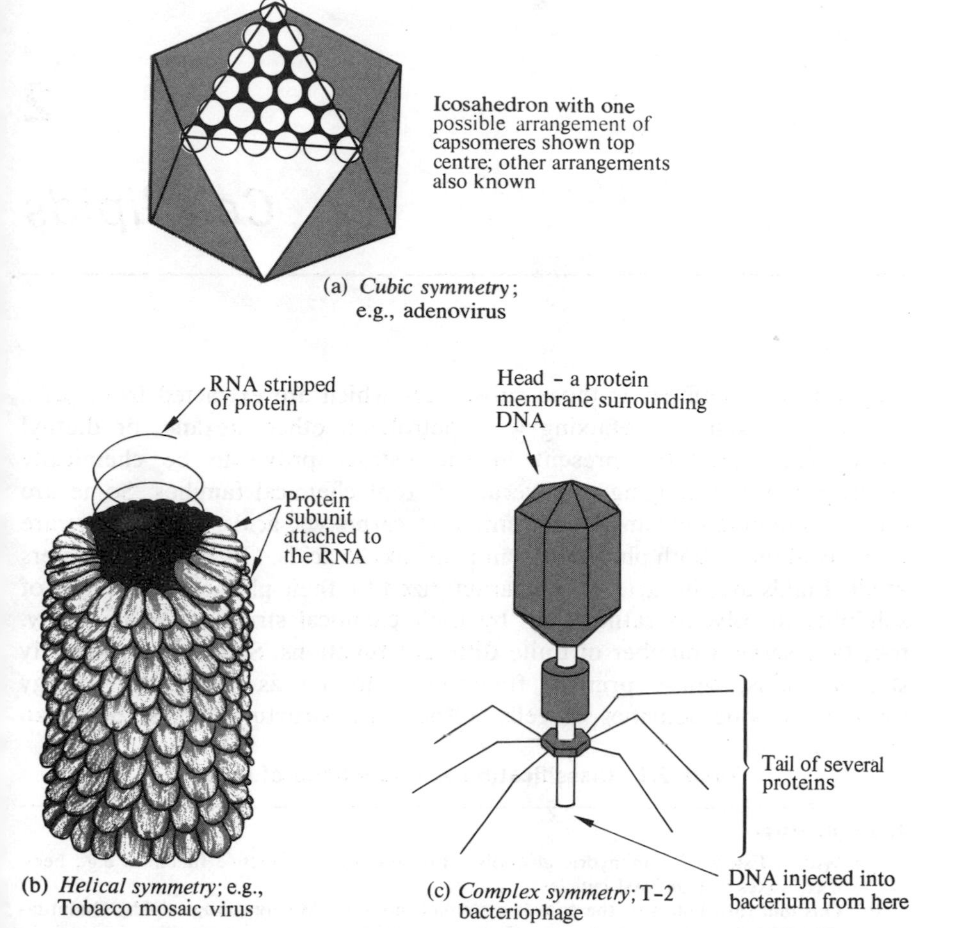

(a) *Cubic symmetry*; e.g., adenovirus

(b) *Helical symmetry*; e.g., Tobacco mosaic virus

(c) *Complex symmetry*; T-2 bacteriophage

Fig. 1.7 Structure of viruses

the protein enters the cell. In all probability the protein serves a protective role and also stabilizes the DNA in its biologically active form. In addition, however, its structure strongly suggests that the proteins in the narrow region below the head and above the base plate act as a trigger mechanism which assists the DNA to gain access to the cell.

Virus nucleic acid may be either DNA or RNA. *Animal viruses* built upon DNA reproduce in the nuclei of their host cells and those containing RNA replicate in the cytoplasm. Most *plant viruses* have a core of RNA. This is single stranded and may either occupy the centre of an icosahedron (or similar structure) or may form a regular helix which is coated with capsomeres of protein (Fig. 1.7b).

2

Cell lipids

Lipids may be defined as those substances which are extracted from cells, tissues, or organs by refluxing with petroleum ether, hexane, or diethyl ether. The substances present in the extract prove to be chemically heterogeneous, belonging to several different chemical families. Some are esters containing only an alcohol united to carboxylic acids, while others are esters containing both phosphoric and carboxylic acids. Others are not esters at all. Lipids are, in fact, more characterized by their physical properties of solubility in solvents rather than by their chemical structure. Biologically, too, they serve a number of quite different functions. Some are principally storage lipids, whose primary function is to act as sources of energy for the anabolic activities of cells. Others are structural lipids, being an

Table 2.1 Classification and functions of lipids

1. Simple esters

(a) *Waxes.* Esters of monohydric alcohols, often serving a protective function; e.g., bees' wax, waxes in plant leaf cuticles

(b) *Oils and fats.* Esters of the trihydric alcohol, glycerol. Mostly storage lipids. Quantitatively the dominant components of petroleum ether extracts of most tissues

2. Phosphate esters

(*a*) *Phosphoglycerides.* Esters containing glycerol, phosphate, carboxylic acids, and, typically, a nitrogenous base. Components of lipoprotein membranes; e.g., lecithin

(b) *Sphingolipids.* Amides containing *inter alia* a long-chain amino alcohol. Components of lipoprotein membranes

- (i) Sphingomyelins also contain base and phosphate
- (ii) Cerebrosides contain monosaccharide
- (iii) Gangliosides contain several saccharide residues

3. Non-saponifiable lipids

(b) *Steroids.* Based typically on the cyclopentanophenanthrene nucleus. Structural component of membranes; functional uses as sex hormones, glucocorticoids, mineralocorticoids, bile acids

(b) *Prostaglandins.* Cyclic derivatives of linoleic acid. Functional lipids, playing a part in regulation of metabolism as local hormones

(c) *Carotenoids.* Unsaturated long-chain hydrocarbons and their hydroxy derivatives. Light-trapping and protective functions in plant leaves

integral part of lipoprotein membranes. Yet others play functional roles, acting in animals as hormones and in plants as photosynthetic pigments. Table 2.1 lists some of the main groups of lipids, together with their principal uses.

In this chapter, the structure of representative compounds will be considered, together with the range of fatty acids present in some of them. The chapter ends with a brief account of two major techniques employed to separate glycerides or fatty acids.

2.1 Triglycerides and fatty acids

Of the several groups of substances normally classified as lipids, the best known are the edible fats and oils, the major component of which is triglyceride. They normally account for most of the petroleum ether extractable lipids of typical animal or plant products. They are important as sources of energy (about 20 per cent of human body weight is due to lipid) but are not major components of lipoprotein membranes (Table 2.3). In animals they are mostly stored in a specialized tissue called *adipose tissue.*

A well-known continuous extraction process was devised by Soxhlet, the principle of which is illustrated in Fig. 2.1. The dried tissue is placed in a parchment thimble and petroleum ether or hexane continuously distils on to

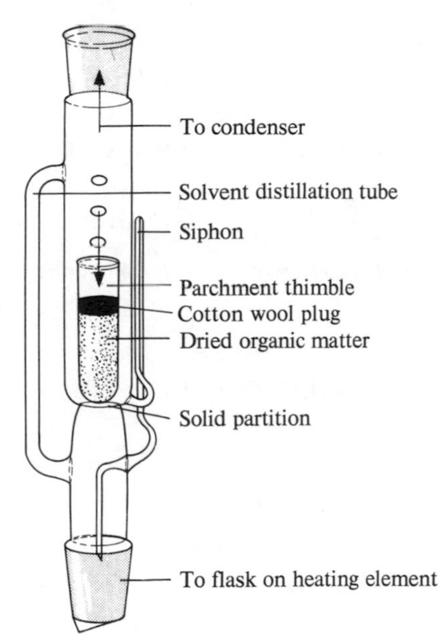

Fig. 2.1 Diagram of Soxhlet fat-extraction apparatus

the thimble contents and then discontinuously siphons back into the flask, where the non-volatile lipid accumulates.

The edible *fats and oils* must not be confused with the hydrocarbon oils. They are *esters*, formed when carboxylic acids unite with the trihydric alcohol, glycerol (Fig. 2.2a). Each of the three hydroxyl groups of the glycerol can form an ester by reaction with the carboxyl group of an acid, and, although mono- and diglycerides occur, most of the fats and oils present as major food reserves are, in fact, triglycerides. (The more exact term, triacyl glycerols, has been recommended instead of 'triglycerides', but tradition here dies hard.) The three hydroxyl groups of glycerol are frequently esterified by union with different fatty acids to give mixed triglycerides such as that shown in Fig. 2.2d.

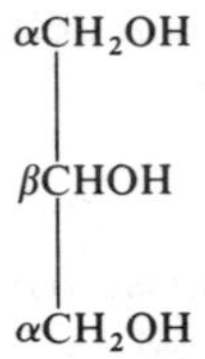

$$\begin{array}{l} \alpha CH_2OH \\ | \\ \beta CHOH \\ | \\ \alpha CH_2OH \end{array}$$

(a) *Glycerol*

$$\begin{array}{l} CH_2{-}O{-}\overset{O}{\overset{\|}{C}}{-}(CH_2)_{16}{-}CH_3 \\ | \\ CH{-}O{-}\overset{O}{\overset{\|}{C}}{-}(CH_2)_{16}{-}CH_3 \\ | \\ CH_2{-}O{-}\overset{O}{\overset{\|}{C}}{-}(CH_2)_{16}{-}CH_3 \end{array}$$

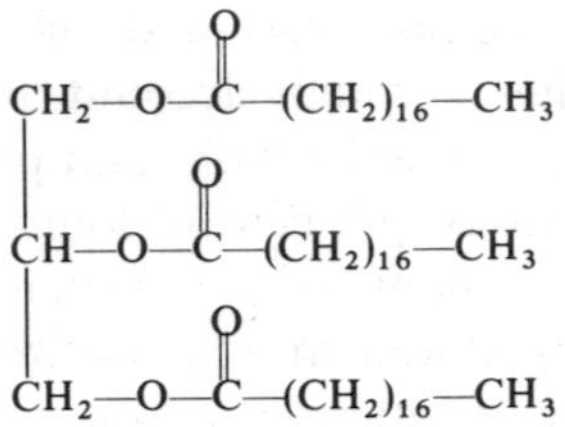

(b) *A triglyceride.* In this example, *tristearin*, each hydroxyl group of glycerol is esterified by stearic acid.

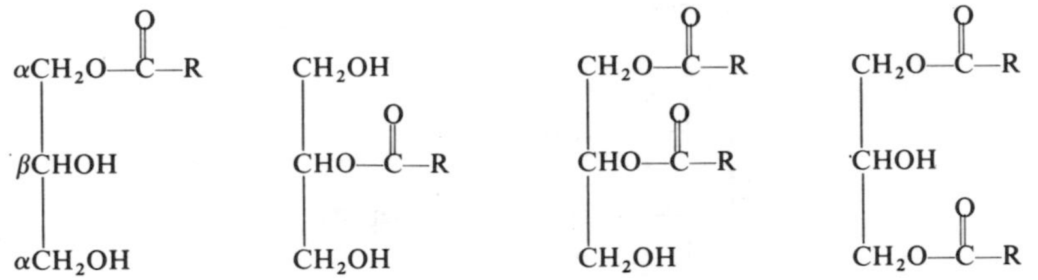

$$\begin{array}{l} \alpha CH_2O{-}\overset{O}{\overset{\|}{C}}{-}R \\ | \\ \beta CHOH \\ | \\ \alpha CH_2OH \end{array} \qquad \begin{array}{l} CH_2OH \\ | \\ CHO{-}\overset{O}{\overset{\|}{C}}{-}R \\ | \\ CH_2OH \end{array} \qquad \begin{array}{l} CH_2O{-}\overset{O}{\overset{\|}{C}}{-}R \\ | \\ CHO{-}\overset{O}{\overset{\|}{C}}{-}R \\ | \\ CH_2OH \end{array} \qquad \begin{array}{l} CH_2O{-}\overset{O}{\overset{\|}{C}}{-}R \\ | \\ CHOH \\ | \\ CH_2O{-}\overset{O}{\overset{\|}{C}}{-}R \end{array}$$

α-Monoglyceride *β-Monoglyceride* *α,β-Diglyceride* *α,α-Diglyceride*

(c) *Mono- and diglycerides*

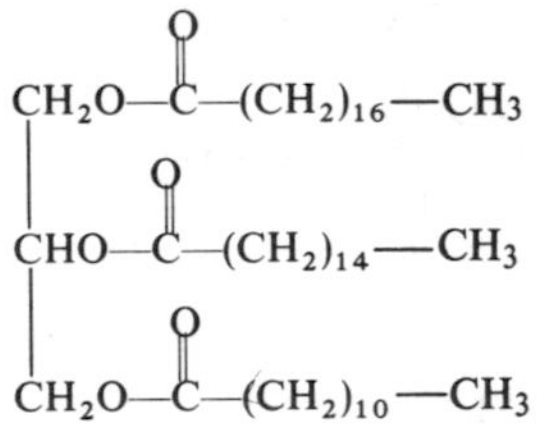

$$\begin{array}{l} CH_2O{-}\overset{O}{\overset{\|}{C}}{-}(CH_2)_{16}{-}CH_3 \\ | \\ CHO{-}\overset{O}{\overset{\|}{C}}{-}(CH_2)_{14}{-}CH_3 \\ | \\ CH_2O{-}\overset{O}{\overset{\|}{C}}{-}(CH_2)_{10}{-}CH_3 \end{array}$$

(d) *A 'mixed' triglyceride.* In this example, glycerol is esterified with stearic, palmitic, and dodecanoic acids

Fig. 2.2 Structure of glycerol and of glycerides

Two aspects of these glycerides are worthy of note. While short-chain fatty acids are occasionally present in one or more of the three possible positions (e.g., butyric acid is present in butter fat), the acids present are typically long-chain acids; it is their long, zig-zag paraffin-like chains composed of covalently-linked carbon atoms which gives the edible fats and oils their predominantly hydrophobic, lipid-soluble characteristics. Furthermore, nearly all the fatty acids occurring in cells, whether present there in the form of esters or as uncombined acids, contain an even *total* number of carbon atoms. It will be shown in Chapter 10 that this is an inevitable consequence of the way their biosynthesis is achieved by the combination of acetic acid units.

Rather more than twenty 'fatty acids' occur in various fats and oils, some common examples being given in Table 2.2. The term 'fatty acid' is used here in a biological rather than a strictly chemical sense, for by no means all the acids present in edible fats and oils are alkanoic acids of the series C_nH_{2n+1} COOH. Some, for example, contain one or more double bonds, and several more complex acids are also known. Among the commoner acids present in such materials as olive oil and palm oil are oleic acid, palmitic acid, and linoleic acid. Stearic acid, although common in glycerides

Table 2.2 Some important fatty acids present in cell lipids, and their melting points

1. *Saturated fatty acids*

Formula	Name (m.p.)
$H_3C—(CH_2)_2—COOH$	Butyric acid (−8°C)
$H_3C—(CH_2)_4—COOH$	Caproic acid (−3°C)
$H_3C—(CH_2)_6—COOH$	Octanoic acid (16°C)
$H_3C—(CH_2)_8—COOH$	Decanoic acid (31°C)
$H_3C—(CH_2)_{10}—COOH$	Dodecanoic acid (44°C)
$H_3C—(CH_2)_{12}—COOH$	Tetradecanoic acid (54°C)
$H_3C—(CH_2)_{14}—COOH$	Palmitic (or hexadecanoic) acid (63°C)
$H_3C—(CH_2)_{16}—COOH$	Stearic (or octadecanoic) acid (70°C)
$H_3C—(CH_2)_{18}—COOH$	Arachidic acid (76°C)

2. *Unsaturated fatty acids*

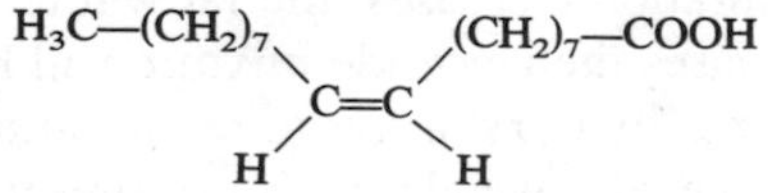

$H_3C—(CH_2)_7—CH{=}CH—(CH_2)_7—COOH$ (both H on the same side of C=C) Oleic acid (13°C)

$H_3C—(CH_2)_4—CH{=}CH—CH_2—CH{=}CH—(CH_2)_7—COOH$ (H H on the same side of each C=C) 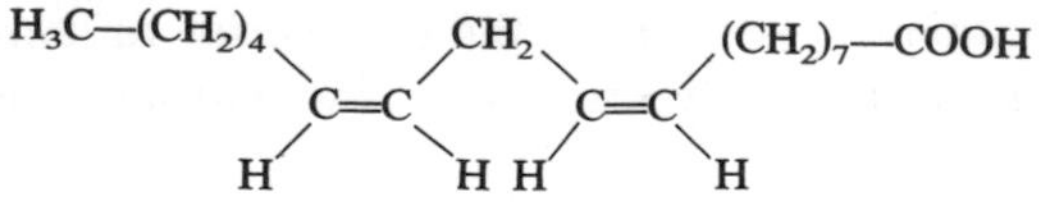 Linoleic acid (−5°C)

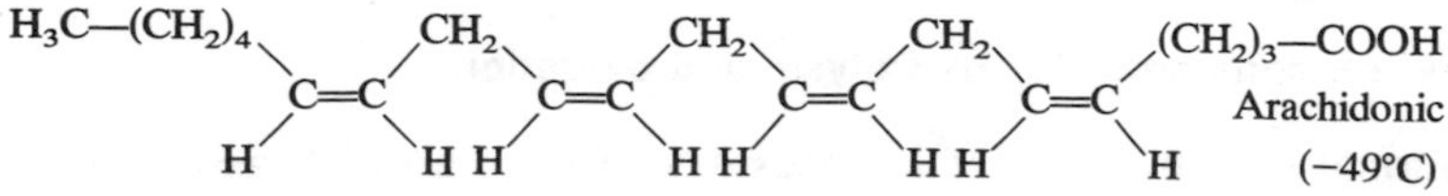

$H_3C—(CH_2)_4—CH{=}CH—CH_2—CH{=}CH—CH_2—CH{=}CH—CH_2—CH{=}CH—(CH_2)_3—COOH$ (H H on the same side of each C=C) Arachidonic acid (−49°C)

present in fats of farm animals, is not particularly abundant in plant and animal fats generally.

Mammals are able to synthesize saturated fatty acids from two-carbon units and they can also synthesize acids containing one double bond (Chapter 10). However, when two or more double bonds are present (e.g., linoleic acid) the acids must be provided in the animal diet and are therefore called *essential* fatty acids. As in the case of essential amino acids (Section 15.4) plants and most microorganisms are, of course, self-sufficient.

2.2 Fats as mixtures of mixed triglycerides

Upon hydrolysis, most fats and oils yield several fatty acids as well as glycerol—the milk of the spiny anteater is an exception in that it comprises almost pure triolein. Chemical investigations have shown that when several fatty acids are present in the total fat sample, they are usually fairly evenly distributed among the various glycerides. Molecules of a feather, so to speak, do *not* flock together. For instance, if a fat hydrolysed to 33 per cent of each of the fatty acids A, B, and C, the 'rule of even distribution' suggests that the fat would be predominantly composed of glyceride of the type

$$G\begin{cases}A\\B\\C\end{cases} \quad \text{rather than 33 per cent of} \quad G\begin{cases}A\\A\\A\end{cases},\; G\begin{cases}B\\B\\B\end{cases},\; \text{and}\; G\begin{cases}C\\C\\C\end{cases}.$$

Fats and oils, in other words, tend to be fairly complex mixtures of mixed triglycerides.

The physical properties of the mixture of glycerides which constitutes a natural fat or oil reflect the composite properties of all the fatty acids present within those glycerides. Fatty acids with lower molecular weights melt at lower temperatures than those of higher molecular weight (Table 2.2). Similarly, for acids with nearly identical molecular weights, those with double bonds have a lower melting point than those which have not. Thus, as average molecular weight falls or unsaturation increases, the fat will tend to be more and more buttery. In extreme cases the glyceride mixture will be liquid, or *oily*, at room temperature. Butter is 'buttery' at room temperature because butyric, caproic, and other low molecular weight acids are present. The oil from the laurel bush contains 31 per cent of the triglyceride of lauric (dodecanoic) acid. Olive oil contains many glycerides containing unsaturated oleic acid (Table 2.2).

2.3 Analytical constants; fat hydrolysis and oxidation

For nutritional investigations it is necessary to have a quick assessment of the overall composition of a sample of fat. Although the gas chromatograph can now often be used for this purpose (Section 2.8) the analytical constants which have historically proved valuable are still employed. Briefly, the amount of potassium hydroxide used up in hydrolysing a given quantity of fat gives the *saponification value* (a measure of the average molecular weight of all the acids present, since a single molecule of triglyceride will require three potassium hydroxide molecules for hydrolysis, irrespective of the chain length of the carboxylic acids). Similarly, the *iodine number* gives a general indication of the level of unsaturation, for two iodine atoms add on to each C=C bond.

Glycerides are *hydrolysed* or *saponified* commercially by boiling them with a solution of sodium or potassium hydroxide, the products being glycerol and a soap (the salts of the higher fatty acids). The enzyme lipase (of which there may be several variants) catalyses a similar hydrolysis *in vivo* where extremes of temperature and pH are, of course, not possible.

$$\text{Triglyceride} \rightleftharpoons \alpha,\beta\text{-diglyceride} \rightleftharpoons \beta\text{-monoglyceride} \rightleftharpoons \text{glycerol}$$

Most lipases tend to attack the ester linkages situated at the two α positions more readily than that in the central β position (Fig. 2.2). The two fatty acids attached at α positions are removed one after the other to leave a β-monoglyceride. The latter is then converted to glycerol and fatty acid, in some cases by the action of another form of lipase. The reactions catalysed by lipases are reversible but in cells where fat hydrolysis occurs and in the intestinal lumen the reaction is directed towards hydrolysis by the removal of one or more of the reaction products. Some lipases in the adipose tissue and liver are activated by hormones (Section 13.9). Hydrolysis of lipids is an essential preliminary step leading to the eventual liberation of the large amount of energy locked up in glyceride molecules, especially within the molecules of the fatty acids (Section 10.2). Fats and vegetable oils are therefore valuable stores of energy and for this reason are deposited abundantly in the *adipose* tissues of animals and within the seeds of certain plants such as castor oil. Each adipose tissue cell (adipocyte) of well-fed birds and mammals has a large fat droplet filling most of the cell, with the nucleus and other cell structures flattened around the periphery.

The double bonds of fatty acids such as oleic and linoleic acid can be saturated by hydrogenation in the presence of nickel as a catalyst. Glycerides containing them are thereby converted from oily mixtures to harder fats. The latter have advantages for storage, transportation, and for spreading on bread. This hydrogenation process is the basis of the margarine industry.

2.4 Phospholipids (phosphoglycerides) and sphingolipids

Phospholipids are chemically related to the fats and oils, but their physical properties are in many ways very different and they fulfil a quite different function within cells. If one of the α-hydroxyl groups of glycerol is esterified

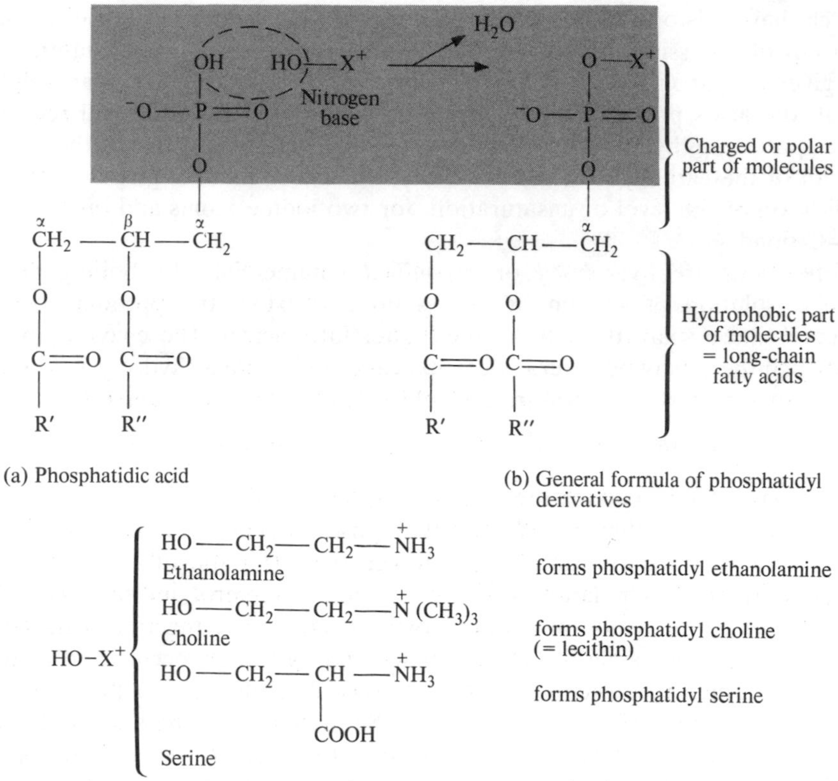

H_2O

Nitrogen base

Charged or polar part of molecules

Hydrophobic part of molecules = long-chain fatty acids

(a) Phosphatidic acid

(b) General formula of phosphatidyl derivatives

$HO-X^+$

$HO-CH_2-CH_2-\overset{+}{N}H_3$
Ethanolamine — forms phosphatidyl ethanolamine

$HO-CH_2-CH_2-\overset{+}{N}(CH_3)_3$
Choline — forms phosphatidyl choline (= lecithin)

$HO-CH_2-CH(COOH)-\overset{+}{N}H_3$
Serine — forms phosphatidyl serine

(c) Three common nitrogen bases forming phosphatidyl compounds

Fig. 2.3 Structure of phosphatidic acid and some phosphatidyl derivatives

with phosphoric acid instead of with a carboxylic acid, the phosphodiglyceride so formed is called *phosphatidic acid.* Although not formed in precisely this way within the cell, phosphatidic acid can be regarded as the precursor of most of the important phospholipids (Fig. 2.3a). It is of interest that of the two possible stereoisomers the L form greatly predominates in living organisms. It is also noteworthy that the carboxylic acid residue attached at the β position is frequently derived from an unsaturated acid.

The three main types of phospholipids in cells have a nitrogen-containing base linked to the phosphate group of L-α-phosphatidic acid. As Fig. 2.3c shows, all three bases commonly present in phospholipids are not only nitrogen compounds but also contain a hydroxyl group, and, in each case, it is the hydroxyl group which unites with the phosphoric acid residue. *Lecithin* is the best known of the many phospholipids in the cell; it has the base *choline* united to the phosphoric acid moiety of phosphatidic acid and it is therefore also known as phosphatidyl choline. The other two common bases, *ethanolamine* and the amino acid L-*serine*, give rise respectively to phosphatidyl ethanolamine and phosphatidyl serine.

An important characteristic of phospholipids in which they differ from ordinary triglycerides is that the phosphate moiety and (when present) the nitrogen bases are *ionized* and hence carry electrical charges. Consequently, whereas one part of the molecule is non-polar and has the solubility characteristics of paraffin-like compounds, the other part of the molecule is attracted to polar solvents such as water. In other words, one section of the molecule is hydrophobic and another is hydrophilic. It is this attribute of

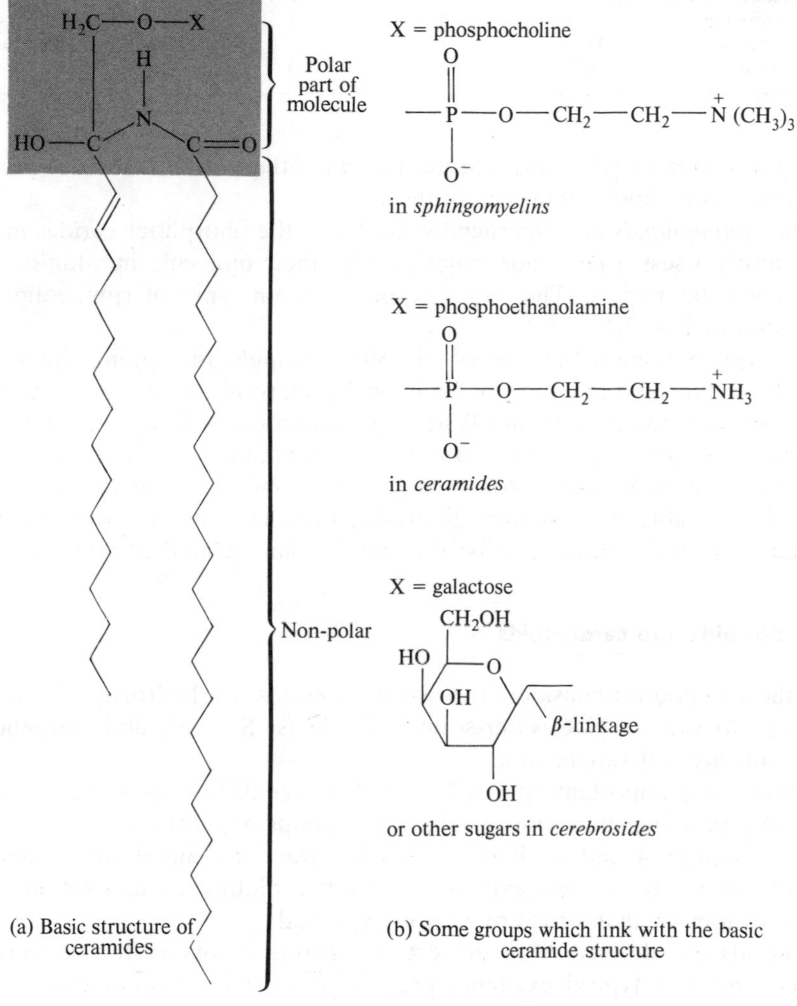

Fig. 2.4 The structure of some sphingolipids

Table 2.3 The composition of various cellular lipoprotein membranes (percentage by weight)

	Red blood cell plasmalemma	Liver plasmalemma	Liver microsomes	Kidney mitochondria	Myelin
Protein	60	60	70	76	22
Total lipid	40	40	30	24	78
Phospholipids	24	26	25	22	33
Phosphatidyl choline	6.7	8	12	8.8	7.5
Phosphatidyl ethanolamine	3.4	—	5	8.4	11.7
Phosphatidyl serine	2.4	—	2.4	8.4	7.1
Sphingomyelin	3.6	4	1.5	—	6.4
Triglycerides	Trace	Trace	3.5	Trace	Trace
Cholesterol	9.2	13	4	1.2	17

phospholipid molecules which, more than any other, renders them important components of lipoprotein membranes.

The *sphingolipids* are superficially similar to the phosphoglycerides in that they also possess a large non-polar part to their molecule in addition to a strongly polar moiety. The structure of the main types of sphingolipids is indicated in Fig. 2.4.

A major biological function of the phospholipids and sphingolipids is to help form the lipid structure of biological lipoprotein membranes. Most of these are composed of about 60 per cent protein and 40 per cent lipid. Of the latter, about 25 per cent is due to phospholipids, the remaining 15 per cent being mostly cholesterol, sphingolipids, and small quantities of triglycerides (Table 2.3). Before discussing their contribution to membrane structure we will briefly describe the steroids and related compounds.

2.5 Steroids and carotenoids

All the compounds considered in earlier sections are hydrolysed by potassium hydroxide to give water-soluble products. Steroids and carotenoids, however, are not saponifiable.

Steroids are important lipids with a wide range of biological functions. As with phospholipids and with certain other groups of substances encountered later (Chapters 4 and 5), it is as though nature, having stumbled upon a useful molecular species, explores in later evolutionary innovations how modifications of that structure can be exploited.

Steroids are derived from the cyclopentanohydrophenanthrene nucleus. *Cholesterol* is a typical example, present in most cell membranes and of importance as an intermediate in the biological synthesis of steroids generally from smaller molecular units (Sections 14.9 and 14.10). It consists essentially of four rings of carbon atoms linked together (Fig. 2.5). Of these, three are six-membered rings and one is a five-membered ring. Cholesterol (Fig. 2.5a) contains one double bond, though, like other steroids, it is essentially a saturated compound and the molecule is, in consequence, not aromatic in character. Most steroids have methyl groups attached at positions 10 and 13, and either a keto or a hydroxyl group at position 3. The cholesterol present in cell membranes is often esterified at the hydroxyl group in position 3. Cholesterol itself has a long side chain at position 17, but many other steroids have shorter substituents and in some a side chain is absent.

Several of the hormones present in higher animals are steroids. Some regulate metabolism by controlling glucose availability, others regulate the movement of inorganic ions through membranes, while yet others have a variety of functions relating to the oestrus cycle or to secondary sexual characteristics. *Cortisol* (Fig. 2.5b), secreted by the mammalian adrenal cortex, is an example of a *gluco*corticoid and aldosterone of a *mineralo*corticoid. *Oestrogen* is a female sex hormone made by the ovary while *testosterone*, made in the testis, is one of several hormones determining male secondary sexual characteristics.

Bile, a secretion of the liver, contains acids which possess the sterol nucleus. An example is cholic acid (Fig. 2.5c). These steroid acids have a carboxyl group at the end of the side chain at position 17. This group is usually conjugated with the amino acid, glycine, or with the sulphonic acid, taurine, to form glycocholic acid and taurocholic acid respectively (Fig. 2.5d). The bile acids play an important role in vertebrate digestion, for they assist in the emulsification of lipids in the small intestine.

Finally, several important poisons are based on the steroid structure. Digitoxigenin is present in the foxglove and has a powerful heart-stimulating action. Another cardiac glycoside, *ouabain*, has the sugar, rhamnose, attached to a modified sterol nucleus. It is of interest in that it is a powerful inhibitor of the 'sodium pump', a device which normally ensures that the cell content of potassium is higher and that of sodium is lower than in the circumambient fluid (Section 17.6).

The *carotenoids* resemble the steroids in two respects: they are not saponifiable and they are built up biologically *via* a pathway which, in its early steps, is similar to that by which steroids are synthesized.

The carotenoids are plant pigments which play a part in photosynthesis (Section 14.3). They are closely related to vitamin A (Fig. 14.12) which *inter alia* is a precursor of retinene, the prosthetic group of rhodopsin, a protein located in the retina and which is responsible for the photochemical reaction of vision (page 421). Also related to the carotenoids are numerous plant products, including some terpenoid oils (e.g., camphor, menthol), rubber,

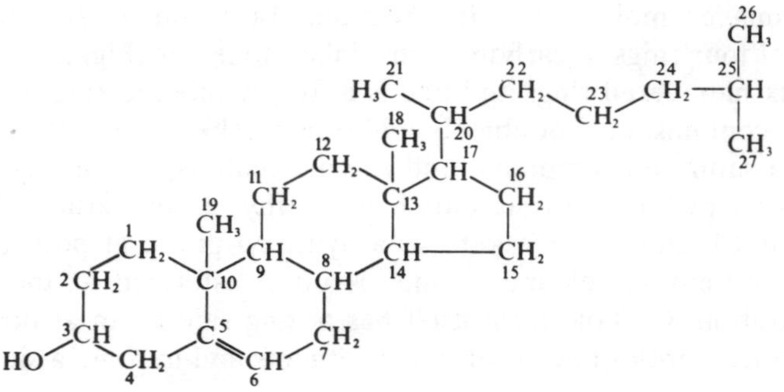

(a) Cholesterol

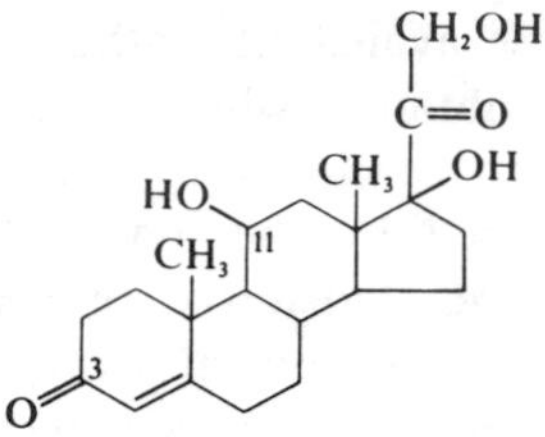

(b) Cortisol

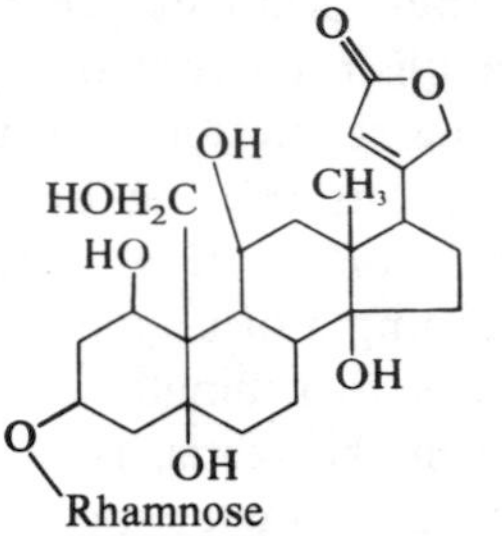

(e) Ouabain, a powerful heart poison; inhibits the sodium pump

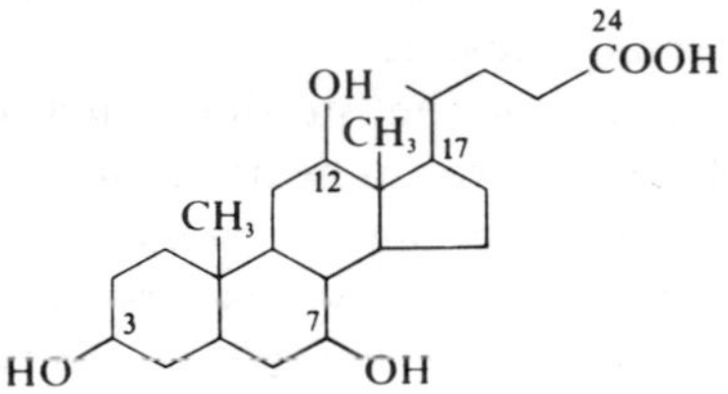

(c) Cholic acid

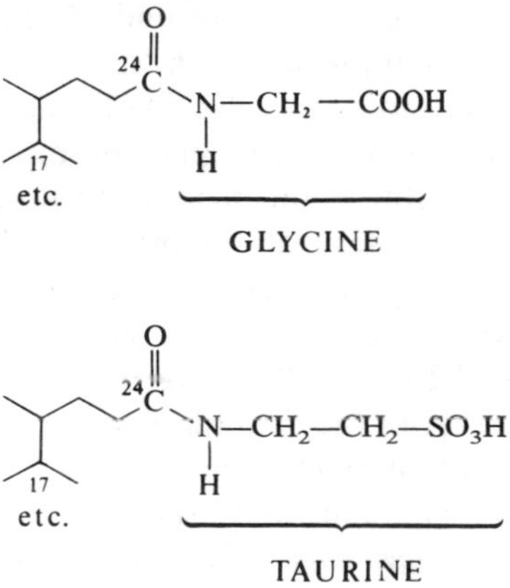

(d) Glycocholic and taurocholic acids

Fig. 2.5 Structures of some biologically important steroids. [The carboxyl group of cholic acid is usually conjugated to glycine or to taurine as shown in (d)]

and gutta percha, as well as the quinonoïd hydrogen carrier of the electron transport system (Section 8.5).

2.6 The structure of lipoprotein membranes

The predominant lipids in cell membranes are the phospholipids, sphingolipids, and cholesterol (Table 2.3). A very important property of these lipids is that their long non-polar hydrocarbon chains are attracted to one another and are sequestered away from water. Under appropriate conditions, when phospholipids are shaken up in water, they form *micelles* or submicroscopic spherical globules with the non-polar parts attracted together in the middle of the structure and the charged hydrophilic regions on the outside forming ionic and hydrogen bonds with the surrounding water molecules.

Another and more directly relevant configuration adopted by phospholipids in water is that of a membrane-type structure with the inner layer composed of the non-polar hydrocarbon chains strongly attracted together by apolar interactions, and the two outer surfaces composed of the charged hydrophilic ends of the molecules. In effect, the structure adopted is one of minimum free energy, for it maximizes the hydrophobic interactions inside and the hydrophilic ones on the surface in contact with the water. The length of the fatty acid chains determines the structure's thickness, which is about double the length of a single phospholipid molecule.

A double layer of phospholipids of this type was first proposed as a model for cell membranes as early as 1925 (Fig. 2.6a). It describes many of the known features of cell membranes including their approximate thickness, their very high electrical capacity, and some of their permeability characteristics. For example, lipid-soluble substances easily pass through them, but polar compounds tend to be excluded (Chapter 17). Since this early model some fifty others have been proposed, most of which are variants of the main types shown in Fig. 2.6.

In the 1930s and 1940s it became apparent that the membranes incorporated proteins (Table 2.3). A number of proposed models indicated that the protein was layered on both sides of a lipid bilayer (Fig. 2.6b). In the 1960s, support for this type of structure was obtained by electron microscopy using dehydrated thin sections stained with heavy metal compounds such as osmic acid. J. D. Robertson suggested that nearly all cell membranes were composed of this 'unit membrane' structure.

Several difficulties were encountered, however, in explaining the properties of cell membranes in terms of this structure. In the plasmalemma, for example, it was established that there are specific protein carriers which transport substances through the membrane. Moreover, very small polar

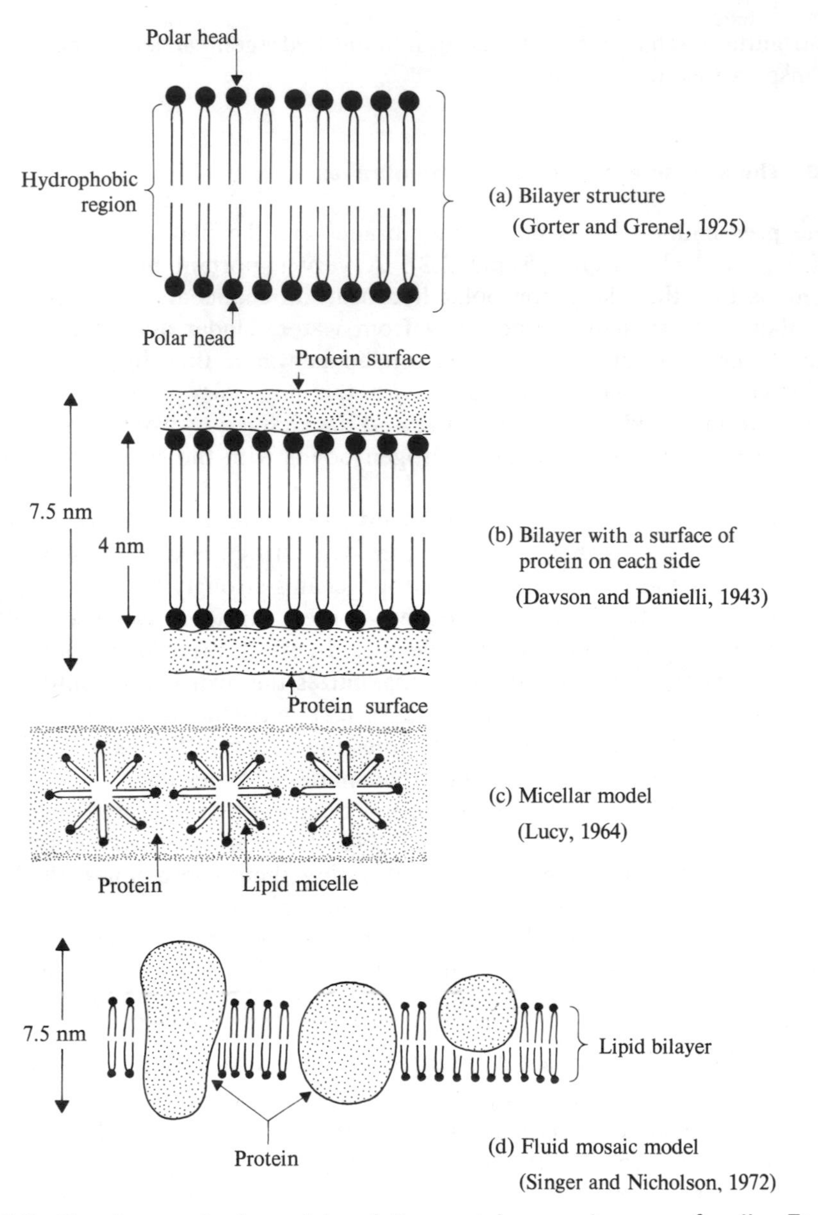

Fig. 2.6 Development of models of lipoprotein membranes of cells. Each diagram represents a cross section through the membrane

molecules such as water and urea readily penetrate the structure. This suggests some means whereby molecules of about 0.4 nm diameter could pass through (Section 17.2). Furthermore, there appeared to be two types of protein involved. Some, now known as *integral proteins*, are strongly bound to the membrane while others, the *peripheral proteins*, are easily washed away experimentally using salt solutions or chelating agents. In addition, x-ray studies suggested much of the protein was globular, a structure not easily fitted into the model shown in Fig. 2.6b.

As a consequence, several micellar models were proposed in the 1960s (Fig. 2.6c). The hypothesis was that the non-polar 'tails' of the lipids formed a close association within the micelles with their polar ends on the outside and surrounded by protein. The protein would contain the carriers and have sufficient polar channels to let water pass through while excluding the larger polar molecules. A difficulty presented by this type of model is that it would be expected to be unstable, since it lacks the strength of the attractions among the sheet of parallel non-polar fatty acid chains of Fig. 2.6b. It was also difficult to explain how highly non-polar compounds could penetrate the membrane, for they would first have to pass through the polar protein layer.

In the 1970s several *protein mosaic models* were being proposed. A modified form of the concept of the lipid bilayer has again become fashionable! The integral proteins are thought to be located within the lipid. In diagrams describing the models the globular proteins often look like icebergs floating in a sea of phospholipid molecules (Fig. 2.6d). Some proteins penetrate right through the lipid while others are located on one of the sides of the membrane and are only partially embedded. It is believed that the surfaces of the proteins in contact with the fatty acid side chains are composed of non-polar amino acid residues and therefore form hydrophobic attachments to them (Section 3.3a). Conversely, the parts of the protein projecting out of the lipid layers are believed to be charged and attracted to the water. This model would explain the difficulty of removing the integral proteins from the membranes. Some of the carrier proteins are thought to be those which fully penetrate the membrane. Water could pass through channels in some of the globular proteins.

Direct support for the protein mosaic model comes from the technique known as *freeze-etching* electron microscopy. The membranes are frozen rapidly and then fractured without dehydration or chemical staining. Next, the specimens are dried by sublimation and replicas of the surfaces are then made by spraying them with platinum and carbon; the platinum emphasizes the contours. The replicas are observed with high-resolution electron microscopes. The technique probably avoids introducing some artefacts which gave support for some of the earlier models. When the membrane is split by the fracturing process some parts reveal the protein globules sticking through the lipid layers.

Recent developments of the protein mosaic model have placed emphasis on the possibility of various areas of the membrane being more mobile or liquid-like in that they are able to flow laterally. Other regions would be more solid in nature. In effect, the membrane is thought to be *highly dynamic* rather than static, its properties changing with various demands placed upon it. The mobile structure may vary its permeability and electrical properties in accordance with the cell's functions. Thus it is probable that permeability changes with respect to sodium and potassium ions are responsible for the movement of the impulse along nerves (Section 19.3). The model suggests how one side of a membrane may be different from the other, the position of the integral proteins determining the asymmetry. So far, less attention has been given to the peripheral proteins. Perhaps these are loosely attached to the integral ones on the outsides of the membranes.

Each of the models was proposed to explain currently established observations. Each was a stimulus to further research. Doubtless the fluid mosaic model will in turn be replaced by more sophisticated and precise descriptions of the real biological membrane.

2.7 Use of thin-layer chromatography to separate glycerides

Thin-layer chromatography (TLC) is a versatile technique, illustrated in the present context by the separation of mixtures of glycerides. It can, however, be used equally successfully to separate amino acids in a hydrolysate of a protein or to separate a mixture of sugars.

Most fats and oils are, as we have seen, mixtures of several glycerides. The individual glycerides can be separated by applying a solution of the fat to a thin-layer plate coated with a silica gel preparation impregnated with a little silver nitrate (the latter is to slow down the movement of glycerides containing unsaturated fatty acids).

The 20×20 cm plate is placed in a tank containing a mobile phase comprising (for example) 66 per cent hexane and 34 per cent diethyl ether (Fig. 2.7). After the solvent front has moved upwards about 13 cm from the starting line, the plate is dried with a hair dryer and sprayed with 10 per cent sulphuric acid. On placing in an oven at 120 to 150°C the previously invisible spots of glyceride turn brown. Using this method, blood serum usually shows four or five spots.

A variant of the method allows individual glycerides to be separated and 'harvested'. If a plate is spotted with a glyceride mixture on both the left- and right-hand sides but only the left-hand side is sprayed with sulphuric acid, the corresponding (invisible) areas on the right-hand side can be scraped off and put into separate tubes. Addition of a suitable solvent will then elute the glyceride from the silica gel. After centrifugation, the

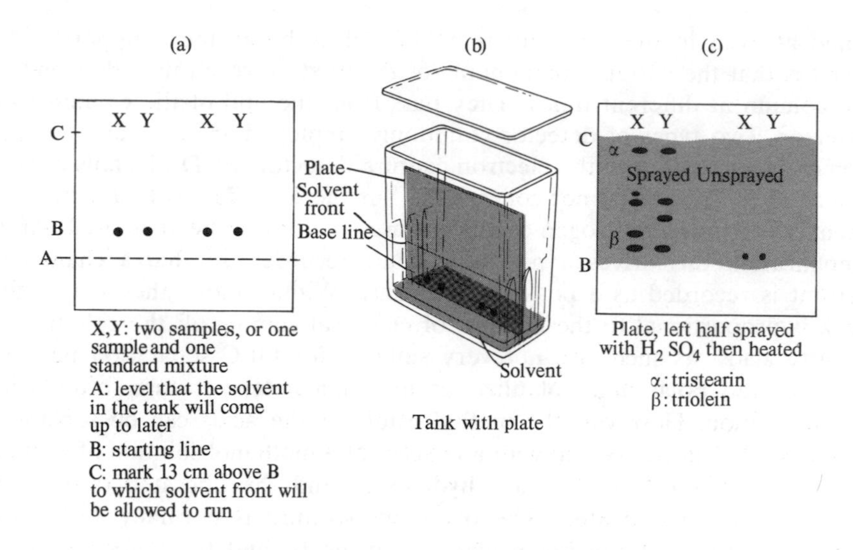

Fig. 2.7 Thin-layer chromatography

solution can then be decanted. Many of the TLC spots can be provisionally identified by running controls of known, common glycerides (e.g., triolein, tristearin, etc.), but it must be remembered that, in common with other kinds of chromatography, TLC only confirms that two samples are incontrovertably *different.* Two quite different glycerides could, by mischance, have the same R_f value in a given solvent system. The R_f value is the distance from the starting line run by the sample spot divided by the distance from the starting line of the solvent front.

2.8 Use of gas-liquid chromatography to separate a mixture of fatty acids

Gas-liquid chromatography (GLC) is a sensitive technique which can be used to separate many closely similar organic compounds. It has been used, for example, to separate metabolites of pesticides produced by animal liver preparations, to separate or identify barbiturates in forensic science, and, as in the present instance, to separate fatty acid (derivatives) obtained from fat hydrolysates.

The apparatus, in simplest form, comprises a column, packed with a suitable stationary phase, which is enclosed in an oven at a convenient temperature. A solution of the mixture to be separated is injected at one end of the column and is carried through the column by a carrier gas such as nitrogen. As it moves along the column, the various components are partitioned between the carrier nitrogen and the stationary phase (which is liquid at oven temperature but is held in place by an inert support). The result is that the various components of the mixture reach the other end of the column at different times. They pass from the end of the column to a detector. Two types of detector commonly employed are the flame ionization detector (FID) and the electron capture detector (ECD). Details of how detectors function need not concern us, but the net effect is that a trace of chemical other than nitrogen causes a minute current to be generated. After amplification this drives a pen of a chart recorder, so that a change in current is recorded as a peak on the chart. Within limits, the area of the peak is proportional to the amount of chemical coming off the column.

Fatty acids, as such, are not very suitable for GLC separation because they are too polar and volatilize at too high a temperature, often with decomposition. However, the methyl esters of the acids can be separated with ease. If fats are treated with an excess of a methanol solution of sodium methoxide ($NaOCH_3$), they are hydrolysed and the liberated fatty acids simultaneously methylated. The oven temperature is normally 180°C for separating esters of the longer-chain fatty acids, and the detection is by flame ionization. A refinement necessary when esters of smaller-chain fatty acids are present is a temperature programmer. This automatically raises the temperature between pre-set limits in a pre-set time following injection of sample. The value of GLC as a method of biochemical analysis cannot be overestimated.

3

Proteins, peptides, and amino acids

3.1 Structural versatility and primary structure of proteins

Proteins are compounds of great importance and interest because of their vital and varied roles in living organisms and their almost infinite diversity of composition. As components of cellular and subcellular membranes they are essential units of biological structure (Chapter 1). In addition the cellular catalysts, or enzymes, which enable chemical reactions to occur at a rate compatible with the existence of life, are all proteins or polypeptides. Proteins differ in their shape, size, and physical and chemical properties, and these various differences are exploited in a range of biological situations. Some proteins are roughly linear molecules which, when orientated appropriately, provide macroscopic structures of great strength, as in sinews, skin, hair, and horn. Others are so constructed that they provide the mechanism whereby muscles are able to contract or relax while others are involved in the processes by which cells recognize one another and participate in immunological reactions.

When proteins are hydrolysed chemically or enzymatically they yield a number of different *amino acids*. *Simple proteins* are exclusively constructed from amino acids but hydrolysis products of other proteins include units other than amino acids. Such proteins are termed *conjugate proteins*. In the several kinds of conjugate proteins the commonest and most important additional components are *porphyrins* (present in haem proteins), *flavins* (in flavoproteins), *nucleic acids* (in nucleoproteins), carbohydrates (in glycoproteins), and lipids (in lipoproteins).

Hydrolysis shows that some twenty different amino acids can be present in a protein and any one of them may occur many times. However, for a particular individual protein the number of amino acids present, and the order in which they are arranged, is a definite characteristic of the protein and is known as its *primary structure*. This contrasts with some other large biological molecules, among them those of starch and glycogen, where molecules differ slightly in their detailed structure and size (Section 4.6).

When amino acids are linked together in chains, as they are in peptides and proteins, they are, of course, present in a chemically combined form, each amino acid unit being linked to its neighbour by a peptide linkage (page 54). Each of these combined amino acid units is termed an amino acid *residue.*

Proteins often have very high molecular weights—the muscle protein, actomyosin, for instance, has a molecular weight of about 14 million. Protein-like substances with molecular weights of less than about 5000 are usually called *peptides.* In consequence of their large molecular weight, proteins form colloidal rather than true solutions. Thus the colloidal micelles disperse light shone into the solution (the Tyndall effect) and they carry electrical charges. When subjected to very great centrifugal force in an ultracentrifuge, large colloidal particles sediment faster than smaller ones—this use of the ultracentrifuge is clearly an extension of the technique mentioned on page 8. Differential centrifugation can, in fact, be employed as a technique for the determination of molecular weights of proteins.

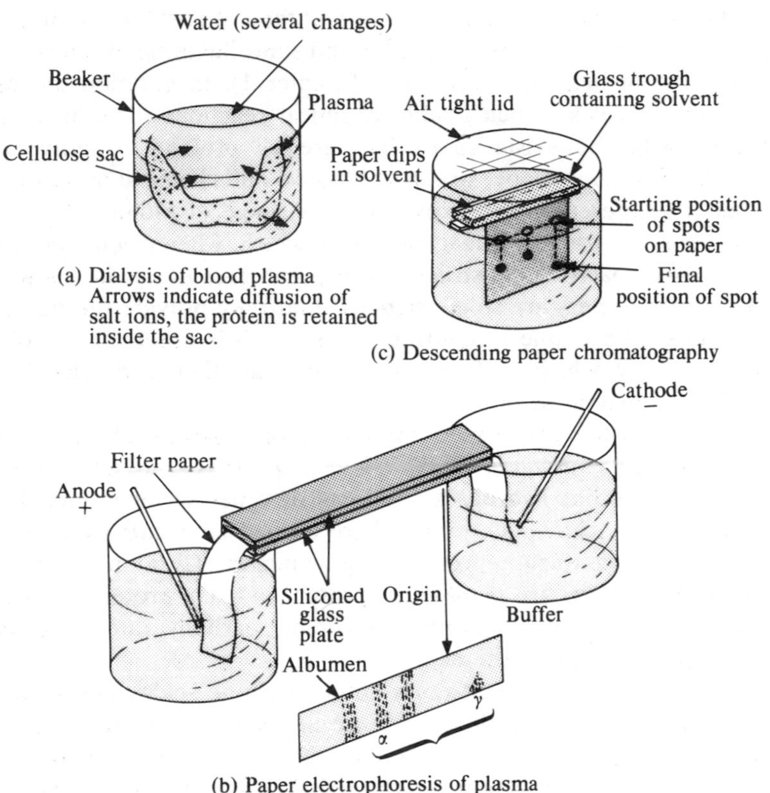

Fig. 3.1 Three useful techniques used in the study of proteins and amino acids

Another property of colloids can be used to separate proteins from smaller molecules or ions. The technique, termed *dialysis*, can perhaps be thought of as a sieving procedure, and an example will best illustrate its use. Blood plasma is an aqueous fluid containing a number of dissolved proteins as well as many inorganic ions, and especially ions of sodium, potassium, chloride, and bicarbonate. If plasma is placed in a cellulose membrane sac which in turn is suspended in a beaker of water (see Fig. 3.1a), the salt ions diffuse through the membrane leaving behind the protein molecules which are too large to penetrate the small pores in the cellulose membrane.

Mixtures of proteins such as are present in plasma can be separated by *electrophoresis*. A simple example of this valuable technique employs a strip of filter paper of about 4 by 30 cm. A thin line of some 20 μl of plasma is carefully pipetted 10 cm from one end of the paper which is soaked in buffer (pH 8.5). The paper is supported by two glass plates and each end dips into the buffer. Electrodes connected to a d.c. source produce a potential of 4 V/cm of paper length. This potential causes the charged plasma proteins to move along the paper, each protein being attracted to the oppositely charged electrode. After some 20 h the paper is removed from the apparatus and the position taken up by the various proteins is located by staining the paper with a dye such as amido black. As the Fig. 3.1b indicates, albumen moves the furthest, and the plasma globulins are separated into fractions known as α, β, and γ globulins.

3.2 Structure of α-amino acids and the effect of pH

The fundamental units from which proteins and peptides are built are carboxylic acids with an amino group attached to the carbon atom immediately adjacent to the carboxyl group. They are therefore known as α-amino acids. Alanine, for example, is derived theoretically from propionic acid:

$$\underset{\text{Propionic acid}}{H_3C—\overset{\alpha}{C}H_2—COOH} \qquad \underset{\text{Alanine}}{H_3C—CH(NH_2)—COOH}$$

Each α-amino acid has the general formula R—CH(NH_2)—COOH where R can be one of about twenty possible side groups. These are discussed in Section 3.3. For alanine, R— is the methyl group.

Since typical α-amino acids have four different groups attached to the α-carbon atom, each exists in two stereoisomeric forms. The one exception is glycine, for which R = H. These are called dextro (D) and laevo (L) forms according to their configuration—the meaning of the symbols D and L is considered in relation to carbohydrates (Section 4.1). The two forms of alanine are written as follows; all that need be noted at this point is that, by convention, L-alanine has the amino group on the left when the carboxyl group is written at the top:

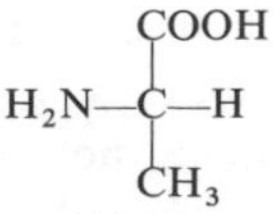

$$\begin{array}{ccc} \begin{array}{c} COOH \\ | \\ H_2N{-}C{-}H \\ | \\ CH_3 \end{array} & \begin{array}{c} COOH \\ | \\ H{-}C{-}NH_2 \\ | \\ CH_3 \end{array} & \begin{array}{c} COOH \\ | \\ H{-}C{-}H \\ | \\ NH_2 \end{array} \\ \text{L-Alanine} & \text{D-Alanine} & \text{Glycine (not optically active)} \end{array}$$

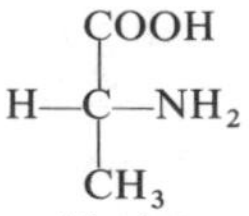

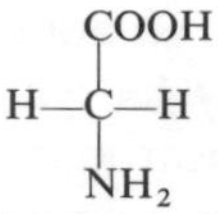

It can be shown easily with models that L and D forms are not superimposable (and therefore are different) while at the same time they are related as an object to its mirror image. In all plants, animals, and microorganisms the α-amino acids found in proteins have the L configuration. D-amino acids do occur, however, in bacterial cell walls (Section 4.7).

Many compounds encountered in relation to cell metabolism contain more than one sort of chemically reactive group. The α-amino acids, for example, have an amino and a carboxyl group. It is useful to remember that, in general, the various reactive groups present in such molecules can, in the first instance, be assumed to act independently of one another. Thus carboxyl groups in amino acids for the most part react similarly to that of acetic acid, ionizing, forming esters and amides, and so on. Similarly, amino groups in amino acids can accept a proton ($—NH_2 \xrightarrow{H^+} —NH_3^+$) or undergo acetylation. As will be seen later, many amino acids contain other active centres; hydroxyl groups, like the hydroxyl groups in aliphatic alcohols, can form esters or ethers, while those attached to aromatic rings behave like those of phenol. On the other hand, it must be remembered that this generalization is only a useful guide, and in more complex cases one reactive group can to some extent influence another, especially if they are situated near to one another in the molecule.

In aqueous solution, the α-amino acids are ionized to form what is sometimes called an *internal salt* or *zwitterion:*

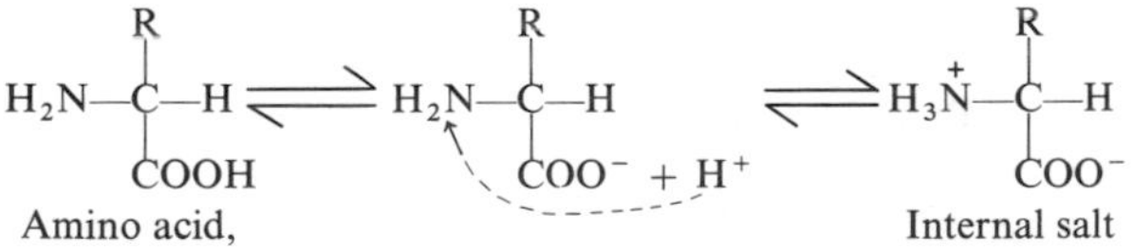

$$\begin{array}{ccccc} \begin{array}{c} R \\ | \\ H_2N{-}C{-}H \\ | \\ COOH \end{array} & \rightleftharpoons & \begin{array}{c} R \\ | \\ H_2N{-}C{-}H \\ | \\ COO^- + H^+ \end{array} & \rightleftharpoons & \begin{array}{c} R \\ | \\ H_3\overset{+}{N}{-}C{-}H \\ | \\ COO^- \end{array} \\ \begin{array}{c}\text{Amino acid,} \\ \text{—covalent form}\end{array} & & & & \text{Internal salt} \end{array}$$

The carboxyl group loses a proton, becoming negatively charged, while the amino group acquires a positive charge by accepting a proton; the close proximity of the carboxyl group to the amino group enhances this ionization process and causes amino acids to have a high dipole moment which makes most of them very soluble in water. Their solubility also depends upon the nature of their side groups (R). Thus the α-amino acid, glutamic acid

($—R = —CH_2CH_2COOH$), is more soluble in water than valine [$—R = —CH(CH_3)_2$] because the former has an additional ionizing centre while valine has a non-polar and therefore hydrophobic side chain.

The properties of amino acids, and indeed of proteins too, are greatly influenced by the hydrogen ion concentration of the solution in which they find themselves. When in acid solution, they tend to take up hydrogen ions, becoming positively charged, but in alkaline solution they lose protons and by doing so they acquire a negative charge:

$$H_3\overset{+}{N}—\underset{COOH}{\overset{R}{C}}—H \underset{-H^+}{\overset{+H^+}{\rightleftharpoons}} H_3\overset{+}{N}—\underset{COO^-}{\overset{R}{C}}—H \underset{-H^+}{\overset{+H^+}{\rightleftharpoons}} H_2N—\underset{COO^-}{\overset{R}{C}}—H$$

High hydrogen ion concentration (low pH)—net positive charge	At the isoelectric point there is no net charge on the molecule	Low hydrogen ion concentration (high pH)—net negative charge

Since amino acids and proteins readily accept or release hydrogen ions—depending on the pH of the solution—they readily act as *buffers*, tending to resist large changes in pH when acid is added or removed from the solution. In cells and in blood plasma, proteins play a vital role in buffering changes in pH which would otherwise occur (Section 18.3). This is important because the activity of many enzymes is greatly affected by pH (Section 6.5).

A *buffer* is typically a solution containing a weak acid, HA, together with a salt of that weak acid which, being highly ionized, provides anions of the acid, A^-. The dissociation of the weak acid can be written in the general form:

$$HA \rightleftharpoons H^+ + A^- \tag{3.1}$$

The hydrogen ion concentration of a buffer solution depends on three quantities, the concentrations of the undissociated acid and of the anion, [HA] and [A^-], and also the dissociation constant of the acid K_a, where

$$K_a = \frac{[H^+][A^-]}{[HA]} \tag{3.2}$$

The pH of the solution is related to these three quantities. By rearranging Eq. (3.2) and taking logarithms:

$$\log H^+ = \log K_a + \log\frac{[HA]}{[A^-]}$$

or (3.3)

$$pH = pK_a + \log\frac{[A^-]}{[HA]}$$

The last equation is known as the Henderson–Hasselbalch equation; in it pK_a is the negative logarithm of K_a and the final term results from the fact that $-\log([HA]/[A^-])$ is equal to $+\log([A^-])/[HA])$.

To appreciate how a buffer system works, imagine that such a system consists of two reservoirs, one of acid HA and the other of its anion A^-. The relative size of the reservoirs depends on the number of moles of acid and salt of the acid used to make the buffer. On adding extraneous hydrogen ions to the buffer, these unite with A^- ions to form HA molecules, thus increasing the size of the HA reservoir very slightly ($+\Delta HA$) and decreasing the reservoir of A^- correspondingly ($-\Delta A^-$). Conversely, on adding extraneous hydroxyl ions, these unite with hydrogen ions to give neutral water molecules, thus disturbing the equilibrium shown in Eq. (3.1) so that the reservoir of HA molecules becomes slightly depleted and that of A^- ions becomes slightly larger. So long as

$$\frac{[A^- - \Delta A^-]}{[HA + \Delta HA]} \quad \text{or} \quad \frac{[A^- + \Delta A^-]}{[HA - \Delta HA]}$$

are not too different from the original quotient $[A^-]/[HA]$, the pH will stay nearly constant.

For a weak carboxylic acid such as acetic acid, the pH at which 50 per cent ionization occurs is equal to the pK_a value of the acid. This is evident from the fact that if $[HA] = [A^-]$ then $\log([A^-]/[HA]) = \log 1$, which is zero. Different acids have different pK_a values, so 50 per cent ionization occurs at different pH values. Buffering is, of course, maximal at pH ranges for which $[A^-]$ and $[HA]$ are nearly equal; i.e., near the pK_a value (Fig. 3.2a).

For amino acids, the same principles apply, but the situation is more complex in that at least two different pK values exist—one corresponding to 50 per cent ionization of the carboxyl group and one for 50 per cent ionization of the amino group. As a result, usually three (sometimes more) molecular species exist at different pH values, as is indicated by the titration curve shown in Fig. 3.2b. In consequence, most amino acids have two pH ranges over which buffering power is significant, but, except for the amino acid histidine, neither is within the pH range 6.2 to 7.8 which is of importance within living cells. Indeed, the isoelectric point, at which buffering is minimal, usually lies within this range. The isoelectric point is the pH at which the zwitterion exists and at which no net charge exists on the molecule. At the isoelectric point the amino acid will not move if subjected to electrophoresis. Conversely, by changing the pH of the buffer at which electrophoresis is occurring, amino acids can be made to move to the positive electrode, to the negative electrode, or not to move at all. The isoelectric point is the pH which is numerically equal to $\frac{1}{2}pK_1 + \frac{1}{2}pK_2$, where pK_1 is the pK value for 50 per cent conversion of an acid to its salt and pK_2 the pK value corresponding to 50 per cent ionization of the amino group.

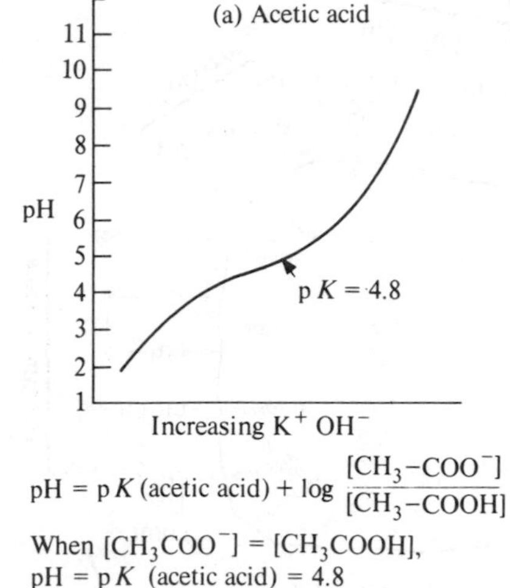

$$pH = pK\,(\text{acetic acid}) + \log \frac{[CH_3{-}COO^-]}{[CH_3{-}COOH]}$$

When $[CH_3COO^-] = [CH_3COOH]$,
$pH = pK$ (acetic acid) = 4.8
This is the pH of maximum buffering, as shown by the fact that there is a relatively slow rise in pH as $[OH^-]$ increases.

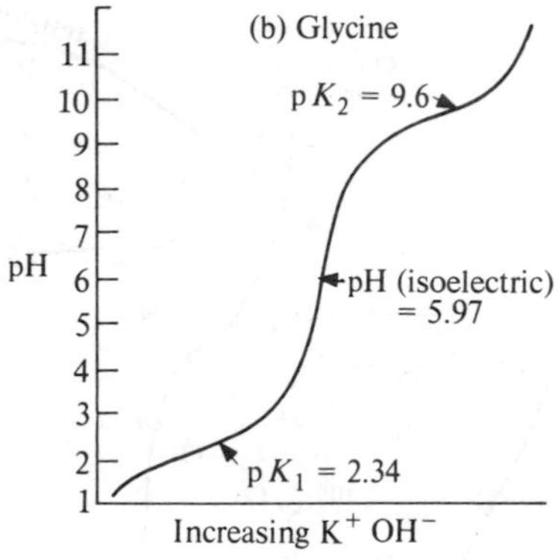

At pK_1, there are equal concentrations of $\overset{+}{N}H_3CH_2COOH$ and $\overset{+}{N}H_3CH_2COO^-$ present. At pK_2, there are equal concentrations of $\overset{+}{N}H_3CH_2COO^-$ and $NH_2CH_2COO^-$ present. Thus there are two regions where buffering is high. At $\frac{1}{2}\,pK_1 + \frac{1}{2}\,pK_2$ (1.17 + 4.8) the only ionic species is the zwitterion, $\overset{+}{N}H_2CH_2COO^-$. Here, at the isoelectric point, buffering is minimal (pH = 5.97).

Fig. 3.2 **Titration curve of (a) acetic acid and (b) glycine against a potassium hydroxide solution**

3.3 Individual amino acid residues

The side groups of the α-amino acids (Fig. 3.3) present in a protein largely determine its three-dimensional structure and its properties. In general, side groups can be divided into five types—although some amino acids have side groups which could be classified in more than one way:

(a) Non-polar
(b) Alcoholic
(c) Those containing sulphur
(d) Acidic and basic
(e) Homo- or heterocyclic

There are, in addition, (f) two *imino* acids present in many proteins.

(a) Amino acids with non-polar side groups

The side groups of the non-polar amino acids (L-leucine, L-isoleucine, L-valine, and L-phenylalanine) are hydrophobic and tend to be attracted to other non-polar compounds. This attractive force helps to maintain the three-dimensional structure of proteins and can result in the formation of

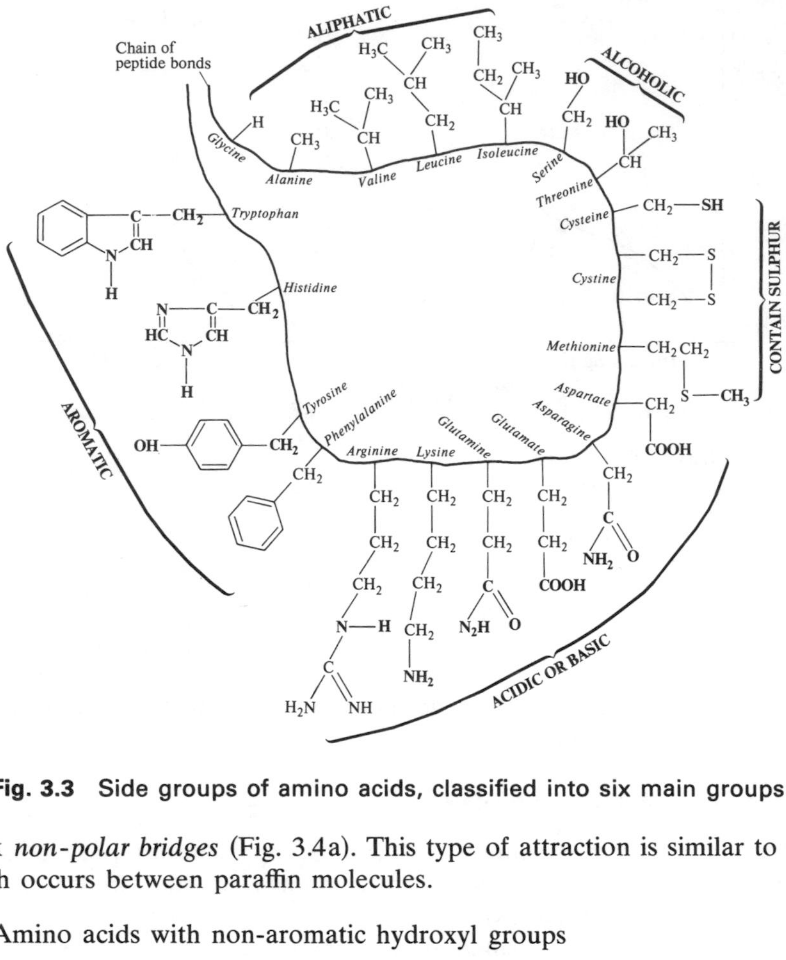

Fig. 3.3 Side groups of amino acids, classified into six main groups

weak *non-polar bridges* (Fig. 3.4a). This type of attraction is similar to that which occurs between paraffin molecules.

(b) Amino acids with non-aromatic hydroxyl groups

The hydroxyl group of the alcoholic α-amino acids, L-serine and L-threonine, is important in many proteins since it is able to take part in ester or ether linkages. Such linkages often attach the protein to a prosthetic group, and in the case of some enzymes, they may bind reactants to the catalytically active part of the enzyme. Sometimes, a phosphate group is attached to the hydroxyl group of serine. This happens, for example, to some of the serine residues present in casein, a protein present in milk and cheese:

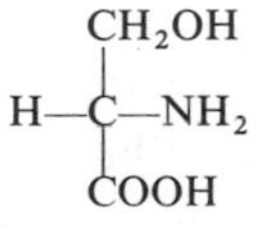

Serine

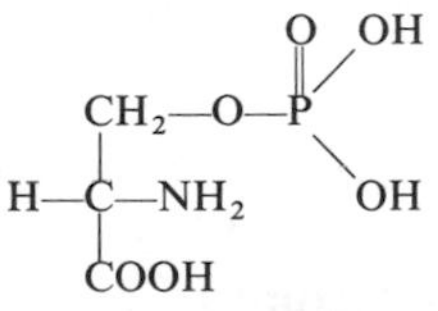

Phosphoserine

(c) Amino acids with side groups containing sulphur

Cysteine, one of the amino acids which contains sulphur, resembles serine except that it possesses a *sulphydryl group* instead of hydroxyl. It often acts as a reducing agent, for the hydrogen atom of the —SH group is more readily lost than that of the hydroxyl group (compare, in inorganic chemistry, the properties of H_2S and H_2O). In a number of enzymes the sulphur atom of cysteine is important because it provides a point of attachment for the reactants—an example is illustrated in Fig. 11.4.

The sulphydryl groups of cysteine residues are frequently involved in maintaining the three-dimensional structure of proteins. Two cysteine residues can link a pair of polypeptide chains together, or form a link between two parts of a single chain. The linkage, known as a *disulphide bridge* (Fig. 3.4b) actually involves an oxidative change which results in the formation of the 'double' or di-α-amino acid, L-cystine:

```
     CH2—S—H              CH2—S—S—CH2
     |                    |       |
  H—C—NH2              H—C—NH2 H—C—NH2
     |                    |       |
     COOH                 COOH    COOH
   Cysteine                 Cystine
```

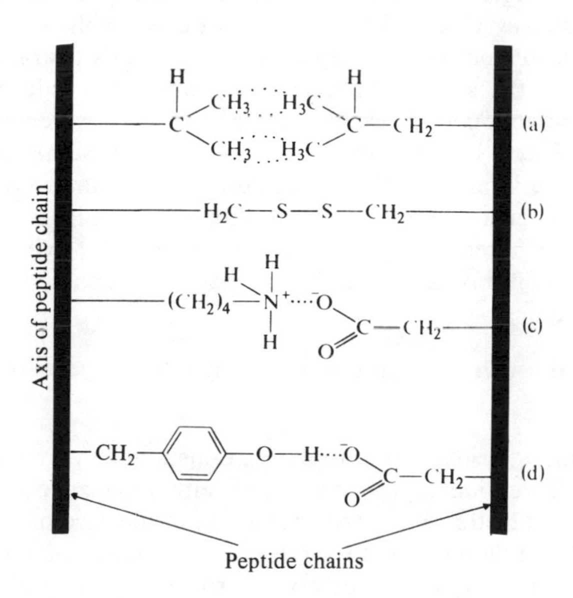

Fig. 3.4 Four types of bridge linking peptide chains together;
(a) non-polar bridge between valine and leucine,
(b) disulphide bridge formed by a trans-chain cystine molecule,
(c) ionic bond between lysine and aspartate,
(d) hydrogen bond between tyrosine and aspartate

(d) Dibasic and diacidic amino acids

The buffering power conferred by the ionizing terminal groups of peptide chains has been mentioned, but most of the buffering power of proteins results from the side groups of the dibasic and diacidic amino acids. A lysine residue, which can take up a proton, is an example of the first, and aspartic acid, which can lose a proton, is an example of the second:

$$\vdash CH_2{-}CH_2{-}CH_2{-}CH_2{-}NH_2 + H^+ \rightleftharpoons \vdash (CH_2)_3{-}CH_2\overset{+}{N}H_3$$

Lysine side group

Axis of peptide chain

$$\vdash CH_2{-}COOH \quad - H^+ \rightleftharpoons \vdash CH_2{-}COO^-$$

Aspartic acid side group

The isoelectric points of dibasic or diacidic amino acids are usually higher or lower than those of most other amino acids. The isoelectric point of free lysine and aspartate is 9.7 and 2.9 respectively.

The side groups of these acidic and basic amino acids enable some enzymes to combine with their substrates. In addition, they are important in maintaining the three-dimensional structure of many proteins by forming *ionic* or *salt bridges* (Fig. 3.4c). An ionic bridge involves an electrostatic attraction between a positively charged and a negatively charged side group.

By contrast to the simple aliphatic amino acids, the side groups of the acidic and basic amino acid residues are attracted to water owing to their ionic nature. There is evidence to suggest that in some proteins (e.g., haemoglobin) the interior of the molecule contains numerous hydrophobic amino acid side groups (derived from simple aliphatic amino acids and phenylalanine), whereas the outer surface, in contact with water under physiological conditions, is surrounded by many hydrophilic acidic and basic amino acid residues.

(e) Amino acids with side groups containing homocyclic or heterocyclic rings

The four amino acids, L-tryptophan, L-histidine, L-tyrosine, and L-phenylalanine have side groups containing either benzene or heterocyclic rings (Fig. 3.3). In histidine and tryptophan, as in the case of the homocyclic compounds, it is not possible to ascribe to the bonds of the rings either double-bond or single-bond characteristics, since each is a *hybrid* of the two types. Other compounds containing these so-called 'aromatic' bonds will be encountered in later chapters—they include several vitamins and the bases of the nucleic acids.

The hydroxyl group of tyrosine is weakly acidic, like that of phenol, and is able to take part in hydrogen bonding, forming links between amino acid side chains (Fig. 3.4d). Furthermore, the 'aromatic' amino acids often fulfil an important function by holding substrates or prosthetic groups in position—this is particularly true of the *imidazole* ring of histidine. In myoglobin, for example, a ferrous ion is held in the centre of a porphyrin ring and, in addition, it is also linked to the nitrogen atom of a histidine residue (page 385).

(f) The imino acids

L-Proline and L-hydroxyproline are present together with true α-amino acids in many proteins and especially in the protein, collagen, a constituent of cartilage. They are *imino* acids, but clearly have a close resemblance to the α-amino acids, with which they form peptide bonds in the normal way:

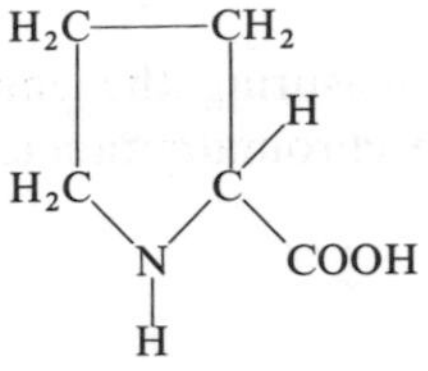

Proline

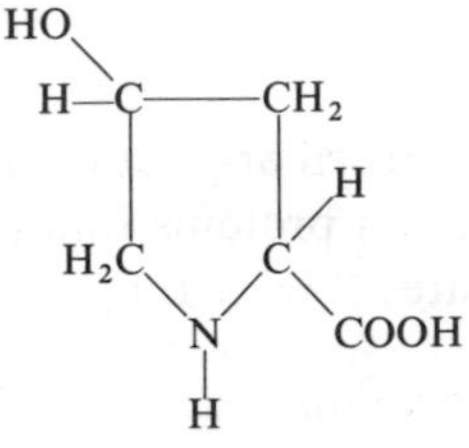

Hydroxyproline

However, because of the imino group, the α-carbon atoms are part of a five-membered heterocyclic ring, and, in consequence, the bond angles of peptide linkages containing them are somewhat different from the values characteristic of the true amino acids. The result is that imino acids tend to produce a somewhat distorted or 'kinked' peptide union, which is one reason why many protein molecules have a complex molecular topography.

3.4 Some important reactions of amino acids

Amino acids undergo several reactions which are important in relation to the qualitative and quantitative investigation of proteins, and tests for proteins nearly always depend ultimately upon the reactions of the amino acid residues they contain.

(a) The Biuret test

Any compound with two or more peptide linkages close together responds to this test, so although proteins give a positive reaction, it is not entirely specific for proteins. When copper sulphate in alkaline solution is added to a protein solution, a violet-purple colour is produced owing to the formation of a copper complex.

(b) The ninhydrin test

This test works best with free amino groups so a protein is usually first hydrolysed by boiling with hydrochloric acid. When a dilute solution of triketohydrindene (ninhydrin) is warmed with amino acids, a purple colour is produced as a result of the following reaction:

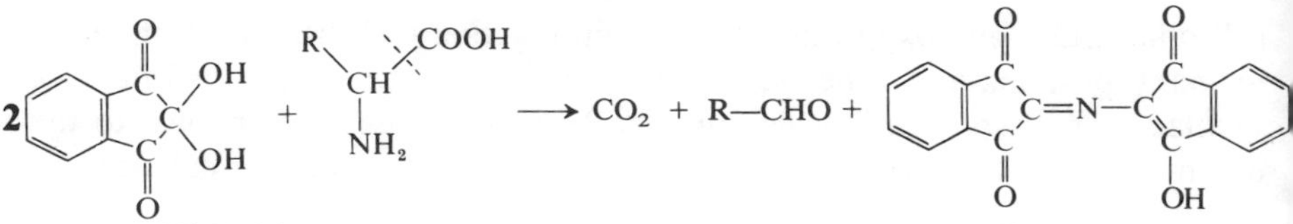

$$2\,\text{Ninhydrin} + R\text{—}CH(NH_2)\text{—}COOH \longrightarrow CO_2 + R\text{—}CHO + \text{Purple-coloured derivative}$$

Ninhydrin

Purple-coloured derivative

This reaction is particularly valuable for measuring the amounts of individual amino acids in proteins following the chromatographic separation of protein hydrolyzates (Fig. 3.1c).

(c) The formol titration

Since amino acids are amphoteric substances they react with acids and with bases to produce salts of different kinds. In consequence, simple titration with either acids or alkalies is not practicable. If, however, the amino group is blocked so that it no longer reacts as a base, titration of the carboxyl group (or groups) with standard sodium hydroxide solution is possible. The amino group can be blocked by addition of an excess of formaldehyde, which undergoes an addition reaction with the hydrogen and nitrogen atoms of the amino group:

$$2HCHO + H_2N\text{—}CH(R)\text{—}COO^- \longrightarrow (HOCH_2)_2N\text{—}CH(R)\text{—}COO^-$$

(d) Sanger's reagent, DNFB

2,4-Dinitrofluorobenzene (DNFB) reacts with the amino groups of amino acids, with the elimination of hydrogen fluoride, to form yellow dinitrophenyl derivatives:

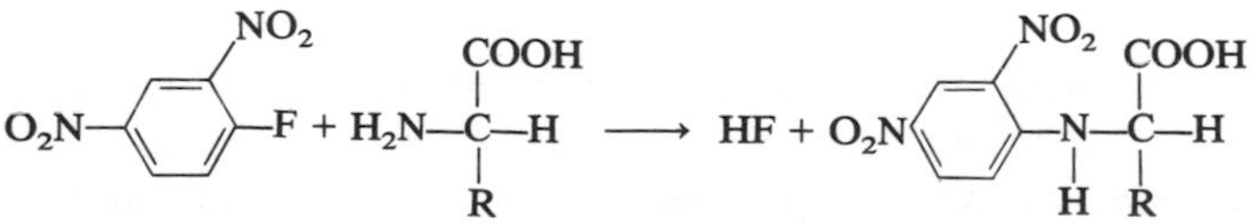

$$O_2N\text{—}C_6H_3(NO_2)\text{—}F + H_2N\text{—}CH(R)\text{—}COOH \longrightarrow HF + O_2N\text{—}C_6H_3(NO_2)\text{—}NH\text{—}CH(R)\text{—}COOH$$

These are fairly stable substances which can be separated and characterized by chromatography.

The reaction occurs even when amino acids are present as residues in a protein or peptide chain, when the amino acid at one end of the chain forms a DNFB derivative. Upon removal by hydrolysis the 'marked' amino acid can be readily identified. This provides a means for identifying the amino acid which originally occupied the terminal position at the 'free amino group' end of the peptide chain. All the other amino acid residues originally present are released as free underivatized amino acids.

(e) Ion exchange chromatography

This technique, unlike paper chromatography, is suited for *quantitative* estimation of the amino acids present in a mixture. A suitable ion exchange resin such as sulphonated polystyrene is placed in a tall glass column filled with dilute acid buffer. The mixture of amino acids is placed at the top of the column in a buffer of pH 3. Solvent is run slowly down the column, the pH *being gradually increased.* The basic amino acids such as lysine, asparagine, and glutamine, being positively charged at low pH, are attracted to negatively charged groups on the resin. Conversely, acidic amino acids have no affinity for the resin and pass fairly rapidly down the column with the solvent. Somewhat later, the non-polar amino acids emerge from the column and finally the basic amino acids are eluted from the resin. As solvent, containing dissolved amino acids, drips from the bottom of the column it is collected in test tubes in fractions of about 1 ml. When the ninhydrin reagent is added to the tubes it is possible to estimate colorimetrically each type of amino acid, and since each passes down the column at a characteristic rate, the total quantity of the individual amino acids in the original mixture can be calculated. This technique has now reached a highly automated stage of development.

3.5 The primary structure of peptides and proteins

It will be recalled (Section 3.1) that the *primary structure* of a protein is concerned with the number and type of amino acids present and their sequence in the polypeptide chain. The percentage of each of the twenty or so amino acids present is usually determined by a quantitative use of ion exchange chromatography, but the determination of the sequence in which they occur is a time-consuming and arduous task.

One method for doing this was pioneered by F. Sanger in the period 1950 to 1960, and depended on the graded hydrolysis of the molecule using enzymes specific for certain points of attack of the polypeptide chains. Four examples are illustrated in Fig. 3.5.

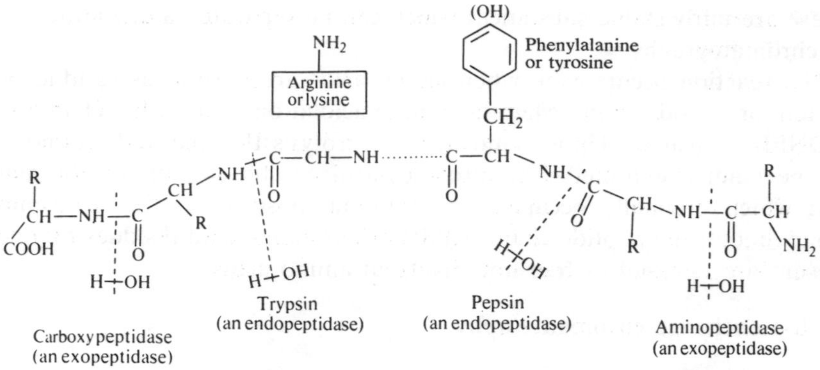

Fig. 3.5 How various peptidases attack proteins. Exopeptidases hydrolyse terminal amino acids; endopeptidases hydrolyse peptide bonds within the peptide chain. Note the specificity of trypsin and pepsin

A second, and rather similar technique, also pioneered by Sanger, involves the partial hydrolysis of polypeptides by acid, the smaller peptides then being separated by chromatography and/or electrophoresis prior to identification. The pieces of information resulting from this approach are then matched together and the original structure worked out in a way reminiscent of the assembly of the pieces of a jigsaw puzzle. This is illustrated by the following example which relates to the small protein, insulin.† The molecule of insulin can be split into several sections by partial acid hydrolysis. One section (now known to be part of the A-chain, Fig. 3.6) can then be isolated from the mixture. This unit, a peptide containing nine amino acid residues, upon careful further hydrolysis, gives a mixture of di-, tri-, and tetrapeptides, including the following:

Ser	Leu		Glu	Leu		Asp	Tyr	
	Leu	Tyr		Leu	Glu		Tyr	$CySO_3H$
		Tyr	Glu		Glu	Asp		
Ser	Leu	Tyr		Leu	Glu	Asp		
					Glu	Asp	Tyr	$CySO_3H$

Sequence: Ser Leu Tyr Glu Leu Glu Asp Tyr $CySO_3H$
(serine, leucine, tyrosine, glutamic acid, leucine, glutamic acid, aspartic acid, tyrosine, oxidized cysteine)

Another method for determining primary structure is to employ a reagent which attacks and dislodges the amino acid at the free amino group end of

† *Acknowledgement.* We acknowledge with thanks permission of the publishers, McGraw-Hill, to quote from *Principles of Biochemistry*, by A. White, P. Handler, and E. Smith.

the peptide chain. One such reagent is phenyl isothiocyanate (Edman's reagent). It reacts with, and removes as a complex, the terminal amino acid, leaving *the rest of the peptide chain intact.* After separation, the process can be repeated so that, one by one, the amino acid residues can be dislodged and identified:

$$C_6H_5{-}N{=}C{=}S + H{-}NH{-}CH(R){-}C({=}O){-}NH{-}CH(R'){-}\text{etc.}$$

$$\longrightarrow C_6H_5{-}NH{-}C({=}S){-}NH{-}CH(R){-}C({=}O){-}NH{-}CH(R'){-}\text{etc.}$$

$$\longrightarrow C_6H_5{-}N\langle C({=}S){-}NH{-}CH(R){-}C({=}O)\rangle + H_2N{-}CH(R'){-}\text{etc.}$$

By these and other similar means the primary structures of many peptides have been worked out, but so far the primary structures of relatively few proteins have been fully elucidated. Insulin is one that has, and its structure is shown in Fig. 3.6.

Insulin, which has a molecule of such a size that it can be regarded as a large polypeptide or a very small protein, is a vertebrate hormone secreted by the pancreas. Its main action is to control animals' carbohydrate metabolism (Section 13.6). It was the first protein to have its primary structure fully worked out. It is now known that the true molecular weight is 5733, although its tendency to dimerize led initially to estimates of about twice this value. Sanger showed that insulin consists of two peptide chains, one containing twenty-one amino acid residues (chain A) and a second with thirty residues (chain B). These are linked by two disulphide bridges (Fig. 3.6), a third cystine bridge occurring within the A-chain itself.

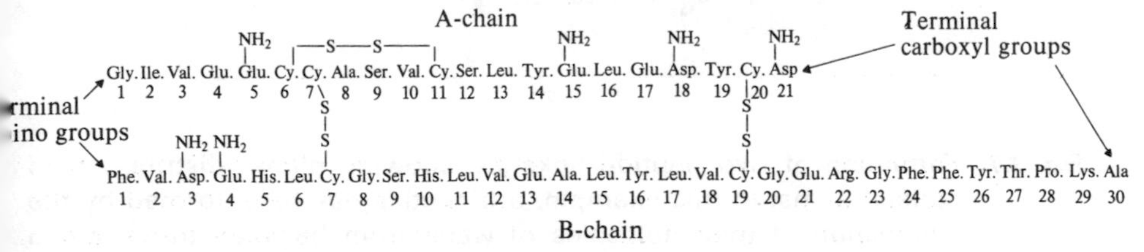

Fig. 3.6 The primary structure of insulin

Insulins from a variety of mammals have now been studied and, in general, have been found to vary only in the amino acid composition of the A-chain at positions 8, 9, and 10. Insulins from man and rabbit are exceptional; in man, the terminal amino acid residue of the B-chain is threonine instead of alanine. Nevertheless—for the treatment of diabetes mellitus—bovine insulin is a very effective hormone in man, a fact which indicates that similar or identical physiological responses can sometimes be evoked by protein-like substances of somewhat different composition.

3.6 The peptide linkage—a planar structure

The amino group of one amino acid can unite with the carboxyl group of another, with the elimination of a molecule of water. The linkage so formed is essentially that of a substituted amide and is called a *peptide linkage*. This is shown in Fig. 3.7.

Since amino acids contain two functional groups (excluding any in the side

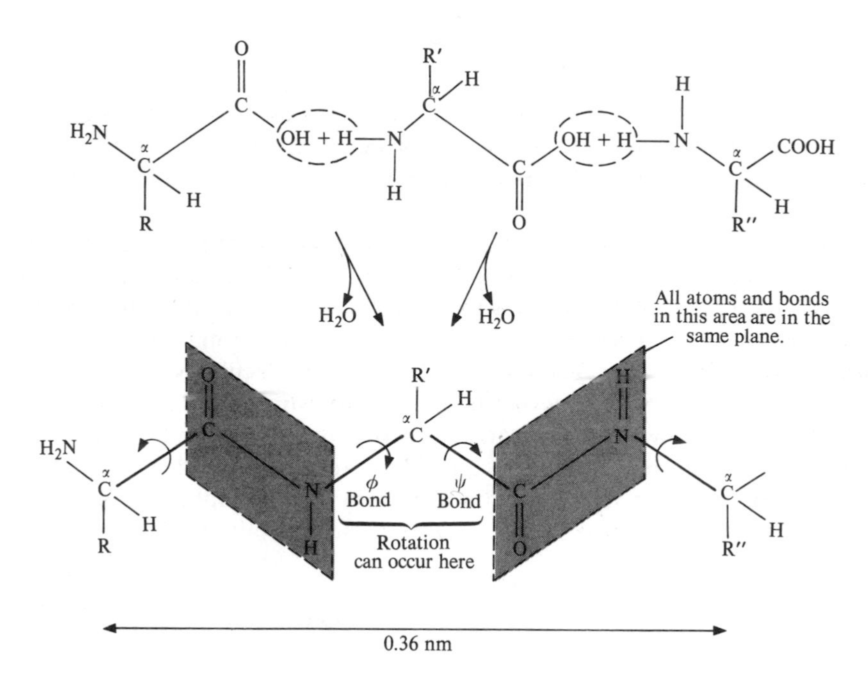

Fig. 3.7 Formation of two peptide linkages, showing planar distribution of certain atoms. In this example, a tripeptide has been formed by the elimination of two molecules of water from between three amino acid residues

groups R), the process can be indefinitely repeated, one end of the peptide still possessing a free amino group and the other a free carboxyl group. In the example in Fig. 3.7, three amino acids have combined with the loss of two water molecules, forming a tripeptide containing two peptide linkages. The prefix *tri*- indicates the presence of three amino acid residues.

The peptide linkage provides a particularly stable bond between the nitrogen atom of one amino acid and the carbon atom of the carboxyl group of the neighbouring amino acid. Its stability can probably be explained by the fact that it is best represented as a hybrid between two forms:

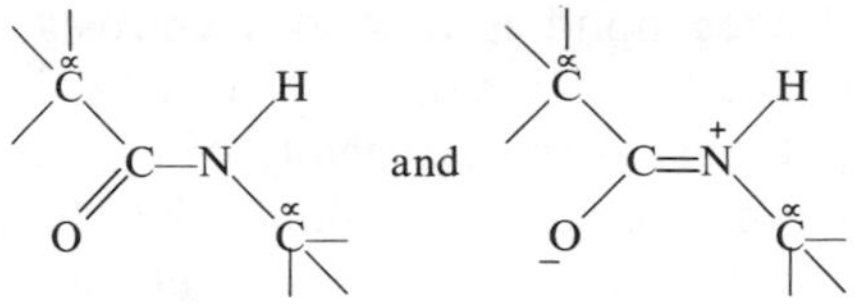

$$\overset{\alpha}{C}-\underset{\underset{O}{\|}}{C}-\overset{H}{N}-\overset{\alpha}{C} \quad \text{and} \quad \overset{\alpha}{C}-\underset{\underset{O^{-}}{|}}{C}=\overset{H}{\overset{+}{N}}-\overset{\alpha}{C}$$

The double-bond character of the linkage between the carbon atom of the carbonyl group and the nitrogen atom makes the peptide linkage coplanar (Fig. 3.7), i.e., all the atoms lie in one plane, just as they do in ethylene. This planar group includes six atoms in all, the α-carbon atoms, with side groups attached, being at the corners of two planes. Both the N—C and the C—C bonds (centre, Fig. 3.7) are free to rotate so as to assume positions of minimum energy—or maximum probability in space. These most probable angles of inclination are known as the ϕ bond and the ψ bond rotations respectively. They are largely determined by the steric and electrical properties of the twenty or so different R groups of the amino acids.

We can thus visualize the rather rigid peptide linkage as imposing a *limited flexibility* to the primary chain of the protein molecule—a series of planes, joined at their corners, and able to *tilt with respect to each other* in a variety of ways, but in ways determined by the requirements of the molecule as a whole. Moreover, if the amino acid side groups are not to interfere with one another—i.e., show steric hindrance—the α-carbon atoms on either side of the peptide bond must normally have these groups arranged *trans* with respect to each other.

The great stability of the peptide bond can be illustrated by the fact that vigorous boiling of proteins in acid (6 *M* hydrochloric acid) or alkali (3 *M* sodium hydroxide) for many hours is often necessary to hydrolyse them completely. Proteolytic enzymes achieve the same result at body temperature and a physiological pH (Fig. 3.5).

3.7 Secondary structure of proteins

A rather elegant consequence of the planarity of peptide linkages and resultant semi-rigidity of the protein chain is that it creates the possibility of *regularly spaced*, repeated *hydrogen bonding*. This confers upon many proteins and peptides a three-dimensional stability which is known as their *secondary structure*. In effect, while an individual hydrogen bond is weak, † if it is repeated hundreds or thousands of times during the length of a long molecule these bonds impart a very considerable strength to the secondary structure.

The stabilizing contribution of the peptide linkage is explained by the fact that there is a tendency for the *carbonyl oxygen atom of one peptide linkage* to share its electrons with the *hydrogen atom attached to the amide nitrogen* of another. The hydrogen bond is a weak electrostatic attraction, usually represented by a dotted line. It can only be effective if the interatomic distances are just right, and so the *possibility* of production in one molecule of thousands of such bonds actually *determines* that the molecule should 'try' to fold or bend to produce such distances—a self-fulfilling 'prophesy', as it were:

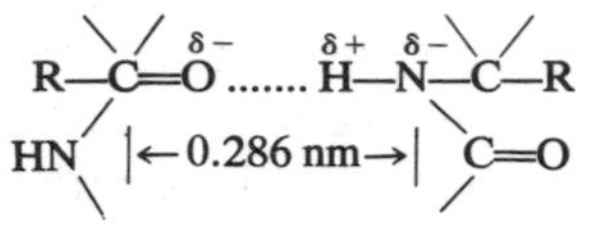

$$\mathrm{R{-}C{=}\overset{\delta-}{O}\cdots\cdots\overset{\delta+}{H}{-}\overset{\delta-}{N}{-}C{-}R}$$

$$\mathrm{HN} \quad |\leftarrow 0.286\ \mathrm{nm} \rightarrow| \quad \mathrm{C{=}O}$$

Within the requirements and constraints imposed above (planar peptide linkage and an ideal hydrogen bond spacing of 0.286 nm between carbon and nitrogen atoms) there are several possible ways in which a primary protein or peptide chain may arrange itself in space. *Fibrous proteins* have been particularly valuable in the study of the *conformation* of polypeptide chains (i.e., the way they fold or twist) because they provide characteristic x-ray diffraction patterns, and the existence of such patterns in itself implies some form of regularity within the molecule. Three such conformations will be briefly described here. The first, the α-helical conformation, is of particular interest, for the much more complicated non-fibrous or *globular proteins* often have a twisting polypeptide chain which, over short lengths, is in the α-helical conformation.

(a) The α-helix

X-ray crystallographic studies have shown that one group of animal fibrous proteins, known as the α-keratins, have a particular repeating pattern which corresponds to a conformation known as the α-helix. The α-keratins occur in ectodermal structures of vertebrates and include hoof, horn, skin, and hair. They are rich in cystine—which contributes additional stability to the fibres as a result of cross-fibre bridging (Fig. 3.4b).

In the α-helix (Fig. 3.8) each peptide bond is planar and each carbonyl group is hydrogen-bonded to the amide group of the third amino acid from

† About 12 to 25 kJ/mol, whereas a covalent bond is about 400 kJ/mol.

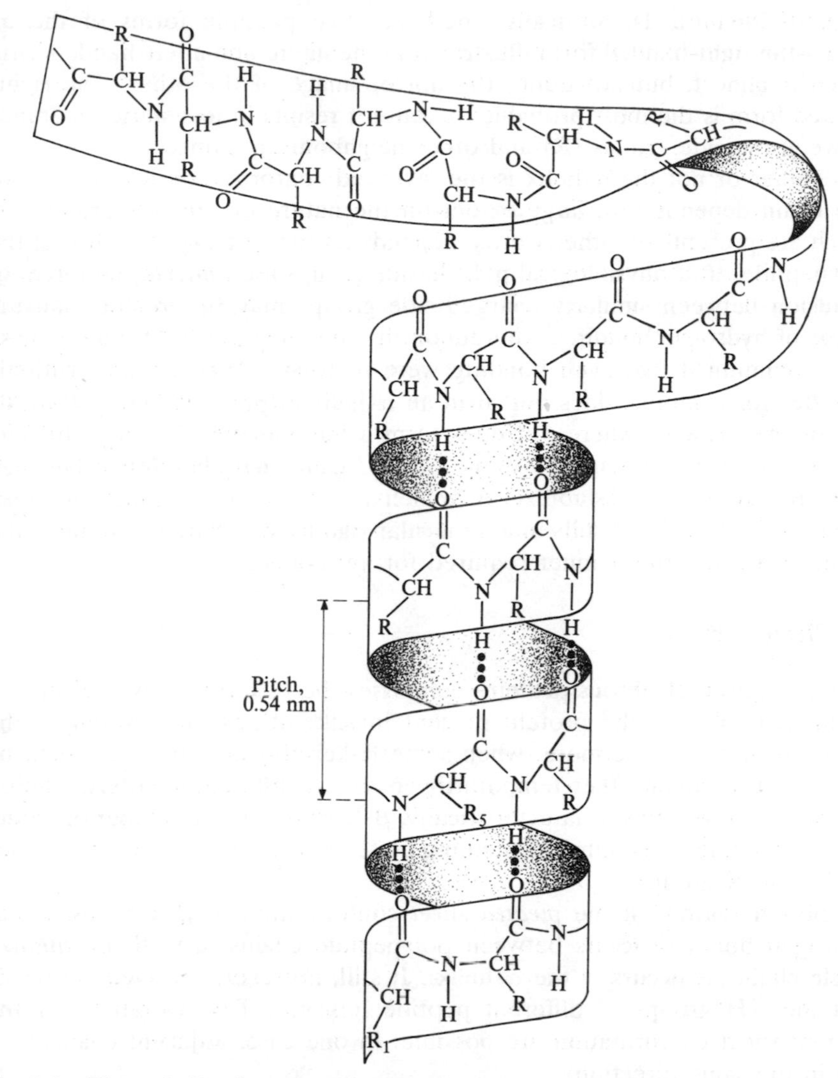

Fig. 3.8 The α-helix. A very common arrangement for amino acid residues in proteins. The configuration of the atoms taking part in the peptide bonds of the right-handed form of the helix are shown, hydrogen bonds being indicated by dotted lines

it along the chain. In Fig. 3.8, the hydrogen bonding is shown by dotted lines. For every ten turns of the helix there are thirty-six amino acid residues with their side groups (R) projecting outwards from the axis of the helix. The hydrogen bonding greatly stabilizes the structure, the distance between the oxygen and nitrogen atoms being almost ideal for a strong hydrogen bond (0.286 nm). Theoretically, there are two possible forms of the α-helix—the right-handed form illustrated in the figure and a left-handed form which is almost, but not quite, the mirror image of the other. The right-handed form is the most probable, because it results in less steric hindrance between the side groups (R) and other neighbouring atoms.

Whether or not the α-helix is the favoured conformation for a polypeptide chain depends to a large extent on the nature of the side groups (R) which are present—in other words, depends on the primary structure of the polypeptide. If at physiological pH the side groups are *ionized*, the forces of repulsion between similarly charged side groups may be greater than the forces of hydrogen bonding, whereupon the free energy would not decrease to a minimum if hydrogen bonding were to occur. A polylysine synthetic peptide, for example, does not form an α-helix at pH 7 (when $—\overset{+}{N}H_3$ side groups are present) whereas it does form a helix at pH 12 (when the side groups are in the unionized $—NH_2$ form). Again, *steric hindrance* between large R groups can destabilize an α-helix. Finally, the presence of *imino acids* such as proline results in a molecular rigidity which does not allow the chain to bend in the manner required for an α-helix.

(b) Pleated sheets

Another group of fibrous proteins comprises the β-keratins. A well-known example is fibroin, the protein present in silk; others are present in the beaks of birds. Furthermore, when some α-keratins are stretched (e.g., by moist heat treatment) they temporarily give x-ray diffraction patterns similar to those of β-keratins. Characteristically, β-keratins lack a number of amino acids, including the sulphur-containing cysteine, but are rich in amino acids with small R groups.

Another feature of the *pleated sheet* conformation of β-keratins is that hydrogen bonding occurs between polypeptide chains instead of *within* a single chain (as occurs in the α-helix). It still, however, involves the C=O and the NH groups of different peptide linkages. Two variations of the pleated sheet conformation are possible; in one case, adjacent chains may run in the same direction:

—C—CO—NH—C— (→)
—NH—C—CO—NH— (→)

whereas, in the other, they face in opposite directions:

—C—CO—NH—C— (→)
—C—NH—CO—C— (←)

Heat-stretched α-keratins are examples of the first, or *parallel*, conformation, whereas fibroin possesses an *antiparallel* conformation.

Fig. 3.9 A diagrammatic representation of tropocollagen molecules within a fibril of collagen

(c) Collagen

Collagen is a fibrous protein present in connective tissue in tendons and skin of vertebrates and may sometimes account for 20 to 30 per cent of total body protein. It is hydrolysed to *gelatine*, a mixture of polypeptides rich in the simple amino acids glycine and alanine and in the imino acids proline and hydroxyproline. Its striking structural characteristic is that the peptide chains have glycine residues in *every third position*. These chains twist to form a left-handed helix, and three such chains intertwine and are hydrogen-bonded together to form a molecule of three intertwined polypeptide strands. This hierarchy of three polypeptide chains is called *tropocollagen*. A (macroscopic) collagen fibril is composed of an array of parallel tropocollagen molecules and appears striated under the electron microscope. This is caused by the fact that the tropocollagen molecules are staggered, with a gap between succeeding molecules in the same fibril, in the manner shown in Fig. 3.9. It is believed that, in the process of bone formation, the hydroxyapatite crystals (a form of calcium phosphate) are deposited in these gaps between tropocollagen molecules in the same row.

3.8 Tertiary structure of proteins

The α-helix only partially describes the structure of the α-keratin type of fibrous proteins. In hair, for example, the basic structure consists of not one α-helix, but a super-helix of three or seven such helices wound around one another like the threads of string constituting a rope. Moreover, the individual strands of the super-helix are held together by cross-strand disulphide bridges formed from cysteine molecules originally present in the strands (resulting, after oxidation, in cross-strand *cystine* molecules). Similarly, tropocollagen comprises a hierarchy of three polypeptide strands, and these are themselves aligned geometrically in a collagen fibril, held in position by extra cross-strand bonding. These additional complexities illustrate what is understood by the *tertiary structure* of proteins.

In the examples above, the tertiary structure is of a relatively simple geometrical type. In the non-fibrous or *globular* proteins, however, the tertiary structure can assume extreme complexity, giving each individual globular protein (and therefore most enzymes) *a unique three-dimensional shape*. One of the most important aspects of the α-helix is that it forms the basis upon which the structure of most globular proteins is patterned. The α-helix can be imagined to twist and turn to provide the overall shape of the molecule—although, in many proteins, parts of the α-helix may be 'unwound' over some part of the total length of the peptide chain. In the crystalline state, for example, about 50 per cent of the insulin molecule is estimated to be in the α-helical form; what form the molecule takes in solution is less certain.

Since the α-helical structure is absent in some regions of many molecules, yet each molecule possesses, under constant environmental conditions, a more or less permanent and unique shape to which its biological characteristics are intimately related, it is apparent that types of bonding, other than hydrogen bonding associated with peptide linkages, must help to stabilize the conformation. Interchain linking via disulphide bridges and the other bridging devices illustrated in Fig. 3.4 are undoubtedly involved in different proteins; in addition, shape is partly determined by the nature and positioning of any prosthetic groups which may be present and by the distorting effect of amino acid linkages.

The complete tertiary structure of proteins can only be determined by elaborate use of x-ray crystallography, a technique largely pioneered for proteins by M. Perutz and J. Kendrew. Their first major achievement was to elucidate the structure of a globular protein, *myoglobin*. This protein occurs in muscle, acting as a temporary oxygen store; it is particularly abundant in marine mammals such as seals and whales, but also occurs in man and other mammals. Myoglobin contains 153 amino acid residues and the chain has been found by Kendrew to have almost the same three-dimensional structure as each of the four peptide chains of haemoglobin. Within its folded structure myoglobin contains the prosthetic group, protohaem, the composition of which is considered later (see Fig. 8.5).

Figure 3.10 provides a somewhat simplified impression of the myoglobin molecule. Some 75 per cent of the structure is arranged in the α-helical form—there are, in fact, eight separate sections of α-helix, separated by regions where the helical structure is absent. Thus the molecule, which contains a protohaem group, has a definite, folded tertiary structure. The amino acids in the interior of the molecule have, for the most part, hydrophobic side groups which play an important role in maintaining the tertiary structure by non-polar bridges. Conversely, hydrophilic amino acid side groups tend to be on the outside of the molecule and interact with the aqueous environment.

It is of particular interest to speculate that, despite the complexity of

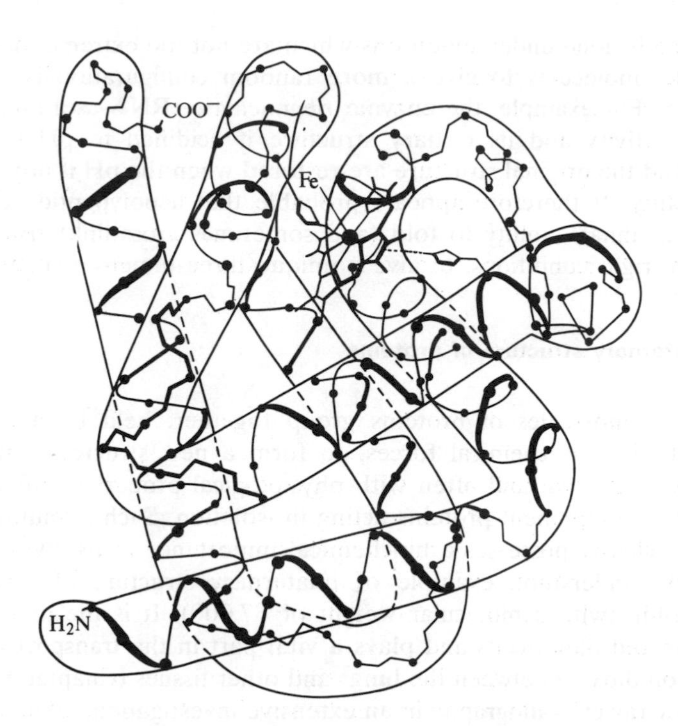

Fig. 3.10 The structure of myoglobin. The ferrous ion is in the centre of a protohaem group (see Fig. 8.5)

protein topography and the likelihood that no two proteins have molecules which are exactly the same shape, nevertheless, the tertiary structure of a particular protein may be an inevitable consequence of the sequence of amino acid residues in its primary structure. If this is so, tertiary structure may be created by successive approximations, but—in the constant environment of the cytosol—with ultimate certainty, at the moment of protein synthesis at the ribosomes (Chapter 16). For, if the ribosome is imagined as moving in a series of jerks along the messenger RNA, attaching one amino acid residue at each jerk, an automatic folding might be anticipated, a folding which, at any instant, will be determined by the attempt of all the *hydrophobic* side chains present at that instant to escape from an aqueous environment by moving to the inside of the molecule, and of the *hydrophilic* groups both to form polar links with one another and to orientate themselves outwards.

When globular proteins are treated with acids or a variety of other chemical reagents they are denatured (Section 3.10). This process implies a distortion or unfolding of the tertiary structure and is usually irreversible, with a loss of any enzymic activity originally present. However, if the denaturing is done under conditions which are not too extreme, the unwinding of the molecules to give a more random configuration is sometimes reversible. For example, the enzyme *ribonuclease* (RNA-ase) loses both its enzymic activity and its tertiary structure if acidified to pH 3, but both activity and the original structure are restored when the pH is adjusted back to neutrality. It therefore appears probable that a polypeptide chain may possess an innate ability to fold (and sometimes cross link) *spontaneously* under the right conditions, to give a unique three-dimensional structure.

3.9 Quaternary structure of proteins

Sometimes, molecules of proteins group together, held by more or less strong physical or chemical forces, to form a new structural unit with a geometry of its own and often with physiological properties different from those of the component proteins acting in isolation. Such a 'multimolecular molecule' clearly possesses a biochemical importance in its own right. The most fully understood example of quaternary structure of a protein is haemoglobin (with a molecular weight of 67 000). It is the major protein present in red blood cells and plays a vital part in the transport of oxygen and carbon dioxide between the lungs and other tissues (Chapter 18). Perutz has used x-ray crystallography in an extensive investigation, taking nearly 30 years, to elucidate the structure of its two main forms, the bright red (arterial) form which is fully oxygenated and the duller red (venous) form which does not contain molecular oxygen.

Both forms of haemoglobin have a roughly spherical structure composed of four subunits, two of which are called α-chains and contain 141 amino acid residues apiece. The other two, the β subunits, have 146 residues. The four subunits fit neatly together in a tetrahedral arrangement, with each subunit retaining a tertiary structure closely resembling myoglobin (Fig. 3.10). Each contains a protohaem group. The means of attachment between the individual subunits is highly complex, for it includes numerous salt and non-polar bridges, together with hydrogen bonding between neighbouring regions of the subunits. Of particular interest is the presence of salt bridges in the deoxygenated form. These occur between the carboxyl end of each α subunit and the terminal amino group of a β subunit. Similarly, there are head-to-tail links between the carboxyl ends of the β subunits and the amino groups terminating the α-chains. Upon oxygenation, these and many other interchain links break and the whole quaternary structure changes slightly to form a different configuration with a number of newly formed interunit links. In other words, the *four subunits tilt differently with respect to one another* in oxygenated and deoxygenated haemoglobin molecules.

J. Monod (1965) coined the term *allosteric* to describe proteins which undergo such changes in shape. In many cases the two forms of the same complex protein have important differences in their properties. In the present instance, the deoxygenated haemoglobin has a much lower affinity for oxygen than has the arterial form. Thus arterial haemoglobin takes up oxygen about 500 times more readily than the venous form. This has great physiological implications which are considered in Chapter 18.

Phosphorylase and *aspartate carbamyl transferase* are two examples of allosteric enzymes with quaternary structures which vary with cellular circumstances and, by so doing, play important regulatory roles. *Phosphorylase,* in its active form, consists of four identical subunits. Each of these subunits contains a phosphorylated serine residue which helps to bind the monomers together. When the phosphate is removed (and cells contain an enzyme which can do this) the tetramer falls into two halves, each of which is a dimer with rather low phosphorylase activity. Since glycogen phosphorylase is the enzyme which helps to mobilize the food reserve glycogen, its activity is crucial to the level of cellular metabolism (Sections 11.4 and 13.6). By modification of its quaternary structure its activity can be greatly altered, and this device offers an elegant method whereby cellular metabolism of the liver and muscles can be regulated.

Aspartate carbamyl transferase is another enzyme with a quaternary structure which is under cellular control. This enzyme is important because it is one of many involved in the synthesis of one of the nucleotides (cytidine triphosphate, CTP) needed to make DNA and RNA. If its activity is suppressed, cell division and protein synthesis slow down. The enzyme is somewhat different in different organisms, but in *Escherichia coli* it contains two catalytic subunits and three subunits with regulatory properties. When the final product of the sequence of enzyme reactions (CTP) is being made too rapidly, CTP accumulates in the cell. Some of it attaches itself to the regulator subunits in the quaternary structure of the carbamyl transferase. By so doing, it causes a conformational change of the total quaternary structure. This greatly diminishes the enzyme's activity and so temporarily switches off the cell supply of new CTP. It is of interest that if the multienzyme is chemically dissociated into its component parts the enzymic *capacity* to produce CTP is maintained but all regulatory *control* of this important process is lost (Section 15.12).

3.10 Native and denatured proteins

It is known that the structure and function of proteins are intimately related, and a simple way to demonstrate the importance of the structure of active, *native* proteins is to *denature* them. There are several ways in which this can be done, but, in all cases, more or less disorganization of their three-dimensional structure occurs, and this leads to an alteration in their physical, chemical, and biological characteristics. This is particularly apparent when the denatured protein is an enzyme, for when enzymes are denatured they lose their ability to catalyse reactions. Similarly, when muscle proteins are denatured, they are unable to contract.

It was seen in Section 3.8 that, for some globular proteins, it is possible for denaturation to be a reversible process, so long as the conditions causing denaturation are not too severe. This demonstrates that uncoiling of polypeptide chains is not necessarily irreversible; indeed, cases are known where broken interchain links such as disulphide bridges spontaneously reform in exactly the right positions. *Irreversible* denaturation sometimes probably results from the formation of new, unnatural covalent bonds, sometimes from modification in amino acid side chains, and sometimes from cleavage of peptide bonds. In some cases, chemical substances causing denaturation combine electrovalently or covalently with groups in the protein. Trichloroacetic acid, for example, unites with cationic sites in the protein molecule.

Denaturing can be brought about by physical or chemical means. Heating is perhaps the most frequently used method for achieving irreversible denaturation. When an egg is boiled, the soluble, almost colourless albumin becomes solid, insoluble, and white. Autoclaving of bacterial cultures and sterilization of medical instruments are carried out for the same reason. Ultraviolet radiation also often leads to irreversible denaturation. Among chemical reagents which cause reversible or irreversible denaturation are detergents, acids and alkalies, urea in high concentration, and water-soluble organic solvents such as acetone. Salts of certain heavy metals such as mercury, copper, and lead are also effective denaturing agents, as are certain anions such as trichloroacetate, tungstate, and tannate. These various chemical reagents act in a variety of ways but, directly or indirectly, they probably alter, *inter alia*, the bonding forces shown in Fig. 3.4 which, in different proteins, help to stabilize secondary, tertiary, or quaternary structure.

Denaturation is often most readily measured or assessed by loss of biological activity. In addition, loss of the normal conformation is often associated with marked alterations in physical chemical characteristics such as the absorption spectrum and optical activity. It is also often found that, after denaturing, globular proteins are more readily attacked by proteolytic enzymes than they are in their original native state.

3.11 Antigens and antibodies

A special protective system exists in vertebrates to protect them from the incursion of foreign macromolecules such as proteins, nucleic acids, and some polysaccharides. Body cells produce *antibodies* which are globulins

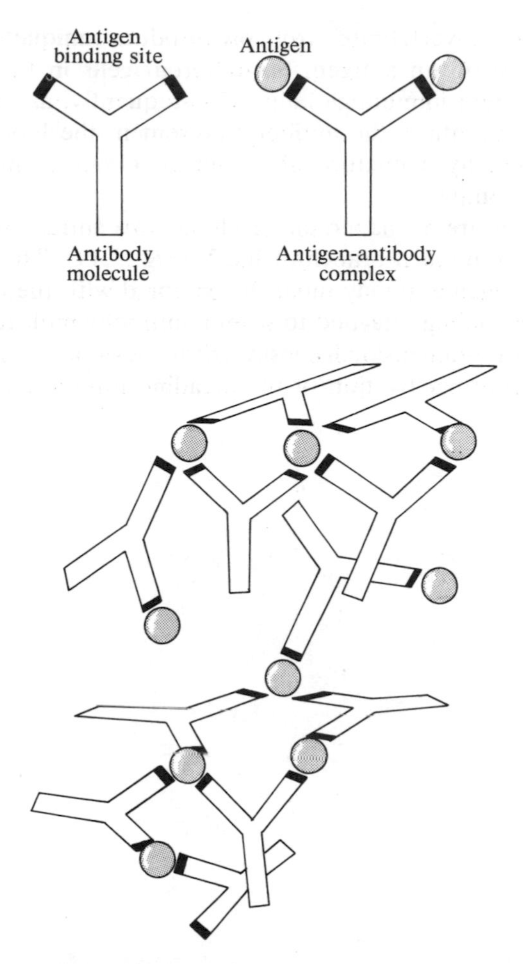

Fig. 3.11 Formation of a three-dimensional lattice when antibodies react with antigens

capable of reacting with the invading molecules, or *antigens.* These *immunoglobulins* pass into the blood stream. By withdrawing some of the blood an *antiserum* specific for a particular antigen can be produced.

Antigen and antibody unite to form an insoluble complex which renders the antigens harmless. The union is very highly specific; if, for example, an antibody has been formed in species A as a result of injecting a little haemoglobin from species B into species A, then haemoglobin from another species, C, will produce little or no antigen-antibody complex when injected into the *immunized* species A. This indicates that, despite their functional similarity, the haemoglobins of species B and C do not possess identical amino acid sequences in their primary structures.

Individual cells in a vertebrate probably produce uniquely structured antibodies in response to an antigen, but different cells in the same body produce slightly different immunoglobulins. Consequently, as can be shown by electrophoretic separation, the antibody present in the blood stream of one individual is actually a mixture of numerous closely similar, but not identical, immunoglobulins.

Antibody molecules are Y-shaped and each has two antigen binding sites, one located at the top of each branch of the Y (Fig. 3.11). The presence of two binding sites on each antibody molecule, coupled with the possibility of antigen molecules becoming attached to several antibody molecules, enables a large, insoluble three-dimensional meshwork to develop, with the consequent precipitation and inactivation of the invading foreign molecules.

4

Carbohydrates

Carbohydrates, lipids, and proteins are the most abundant of the organic compounds present in cells. Carbohydrates are also known as *saccharides* (from a Greek word meaning 'sweetness') since many of those of relatively small molecular weight have a sweet taste, although this is not true of those with large molecules. This chapter will not be concerned primarily with the detailed chemistry of these compounds, for which an organic chemistry text should be consulted, but instead will deal with the range of molecular structure to be found within the group. The diversity of their molecular structure is reflected in a corresponding diversity in biological function. The carbohydrates illustrate the maxim that applies to some other groups of substances as well—namely, that when nature has 'chosen' a particular molecular pattern, it often adapts that pattern so that various members of the group fulfil a variety of biological uses.

The chemical compounds known collectively as carbohydrates include such important materials as sugars, starch, glycogen, and cellulose. Some carbohydrates serve directly or indirectly as *sources of energy* in both plants and animals; others are components of the walls of plant cells or help to *provide rigidity* to the exoskeleton of certain invertebrate animals. They fall into three main groups, the monosaccharides, the disaccharides, and the polysaccharides, and since the last two groups are derivatives of the former, it is to the monosaccharides that attention must first be drawn.

4.1 Classification and naming of carbohydrates

(a) Monosaccharides

Monosaccharides are aldehydes or ketones which also contain primary and (usually) secondary alcohol groups; they are all soluble in water and most of them are sweet. They can be represented by the following group formulae:

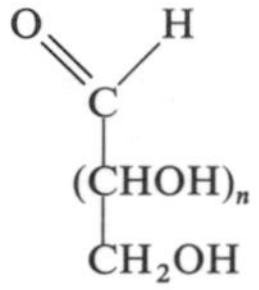

$$\begin{array}{c} O \quad H \\ \diagdown\!\!\diagup \\ C \\ | \\ (CHOH)_n \\ | \\ CH_2OH \end{array}$$

Aldehyde sugar

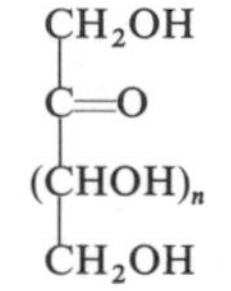

$$\begin{array}{c} CH_2OH \\ | \\ C{=}O \\ | \\ (CHOH)_n \\ | \\ CH_2OH \end{array}$$

Ketone sugar

$C_xH_{2x}O_x$

General formula

Most carbohydrates are distinguished by the suffix *-ose*, the aldehyde sugars being described as *aldoses* and ketone sugars as *ketoses.* In almost all monosaccharides the carbon atoms are joined together in an unbranched chain, and the various members of the group are distinguished from one another by naming according to the number of carbon atoms present in the molecule. The commonest monosaccharides contain six, five, or three carbon atoms in their molecules and are named *hexoses*, *pentoses*, and *trioses*, though sugars with four and seven carbon atoms are occasionally important. Glucose is an aldohexose, fructose a ketohexose, ribose an aldopentose, and glyceraldehyde is an aldotriose.

The optical rotation observed when solutions of sugars are placed in a polarimeter may be dextro (*d*, or +) or laevo (*l*, or −). When isolated from cells the simplest aldose, glyceraldehyde, is always the dextrorotatory isomer, although an unnatural laevo form can, of course, be synthesized. Conventionally—there is no other justification—dextroglyceraldehyde is written on paper with the hydrogen atom of the asymmetric carbon atom on the *left* when the aldehyde group is on *top:*

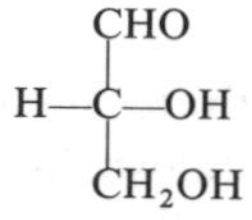

$$\begin{array}{c} CHO \\ | \\ H{-}C{-}OH \\ | \\ CH_2OH \end{array}$$

(+)-Glyceraldehyde, the parent substance for the D family of sugars

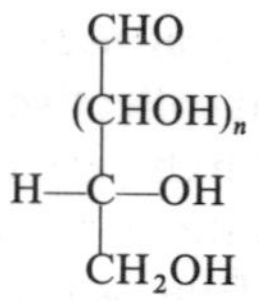

$$\begin{array}{c} CHO \\ | \\ (CHOH)_n \\ | \\ H{-}C{-}OH \\ | \\ CH_2OH \end{array}$$

All natural aldose sugars are of this type, that is, D family

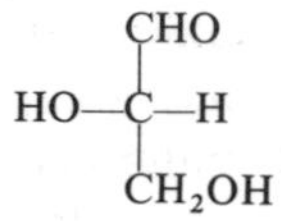

$$\begin{array}{c} CHO \\ | \\ HO{-}C{-}H \\ | \\ CH_2OH \end{array}$$

(−)-Glyceraldehyde, from which unnatural L-family sugars can be made

Virtually all the monosaccharides in nature can be synthesized from D(+)-glyceraldehyde by a series of reactions which elongates the carbon chain. Thus they form a chemical family in the same way that the amino acids present in proteins are all related to L-alanine (Chapter 3). However, the L-amino acids are not all *laevo*rotatory and many monosaccharides differ from D-glyceraldehyde by showing laevo- instead of dextrorotation. Thus D-glucose is dextrorotatory, like D-glyceraldehyde, but D-fructose is laevorotatory. Nevertheless, since all the common aldoses and hexoses belong to the D-glyceraldehyde family, the configuration of the groups attached to the final asymmetric carbon atom of each molecule, as the formulae above indicate, is the same as that of D-glyceraldehyde.

Both aldoses and ketoses contain a carbonyl group and have reducing properties similar to those of aliphatic aldehydes. The apparently anomalous position of the ketoses should be noted, for aliphatic ketones are not reducing agents; it would appear that reducing properties are conferred upon the ketoses by the presence in the molecule not only of a carbonyl group but also of primary and secondary alcohol groups. All monosaccharides consequently reduce Fehling's solution and Barfoed's solution (both of which are changed from a blue cupric compound to red cuprous oxide), and form a silver mirror with ammoniacal silver nitrate.

In general, free monosaccharides play only a minor role in biochemistry. Nearly always it is as derivatives such as phosphate esters that they enter into the metabolic pathways of the cell.

(b) Disaccharides

Disaccharides are sugars formed by elimination of a molecule of water from between two molecules of monosaccharide:

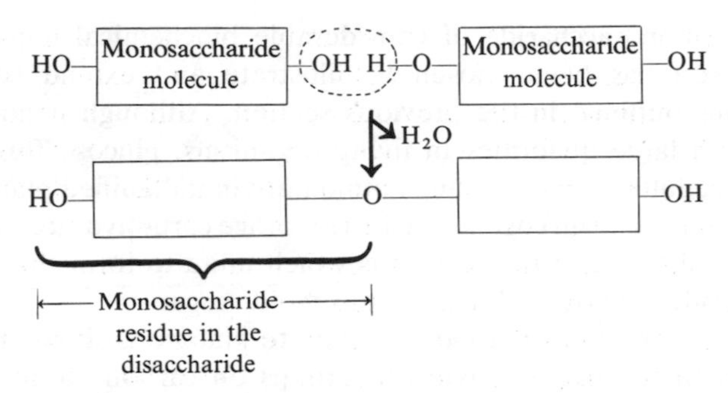

The two monosaccharides which unite to form a disaccharide may be the same or different. Thus two molecules of glucose unite to form maltose but a molecule of glucose unites with one of galactose to form milk sugar, or lactose. Moreover, for reasons outlined later, the monosaccharides can link together in different ways, with the result that some disaccharides are reducing sugars and some are not. Most disaccharides are readily soluble in water.

It is possible for three molecules of monosaccharide to unite with the elimination of two molecules of water, or four to unite with the loss of three molecules of water. Di-, tri-, tetrasaccharides (and so on up to about ten monosaccharide units) are often classified together as oligosaccharides (from a Greek word meaning 'few') but, in practice, the disaccharides are of much more importance than most oligosaccharides of larger molecular weight.

(c) Polysaccharides

These can similarly be regarded as more complex derivatives of monosaccharides comprising a large number, *n*, of monosaccharide units united together by the elimination of $(n - 1)$ molecules of water:

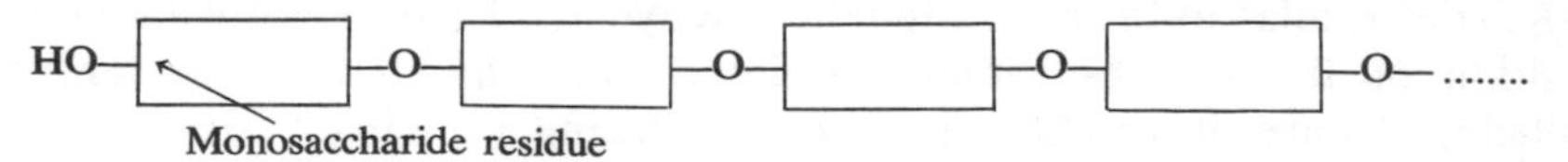

The monosaccharide residues may be so linked together that they form straight or branched chains. Many polysaccharides are insoluble in water or, if they dissolve, form colloidal rather than true solutions. Several of them give characteristic colours with iodine. Polysaccharides of biological importance may have up to half a million monosaccharide residues in their molecules. Sometimes these residues are identical (homopolysaccharides) and sometimes different (heteropolysaccharides), while in more complex cases some or all of the residues are not true monosaccharides but derivatives of monosaccharides.

4.2 Glucose—an aldohexose sugar

Glucose is a monosaccharide of considerable biochemical importance and for this reason has been chosen to illustrate and extend some of the considerations outlined in the previous section. Although it does not very often occur in large quantities in living organisms, glucose forms the unit structure of cellulose, which occurs abundantly in unlignified plant cell walls, as well as of starch and glycogen, which are storage carbohydrates in plants and animals; it is also one of the two units which unite to form the molecule of the disaccharide, sucrose, which is 'household' sugar.

The simplest structure that can be given to glucose is shown in Fig. 4.1a, which shows that it has five hydroxyl groups on carbon atoms 2 to 6, the aldehyde carbon atom being conventionally designated as carbon atom 1. Since the valency bonds of carbon are tetrahedrally distributed and four of these carbon atoms have four different monovalent groups attached to them, it follows that 2^4 different isomers of aldohexose exist. (If the reader is unfamiliar with the phenomenon of stereoisomerism, he is strongly advised to construct models of sugars, remembering that the test of identity is superimposability and that pairs of isomers are related as an object is to its mirror image.)

It was explained in the previous section that almost all naturally occurring monosaccharides, irrespective of their optical rotation, are members of the D series of sugars, the configuration around the penultimate carbon atom (number 5 in this case) being the same as that in D-glyceraldehyde. The configuration around the other asymmetric carbon atoms can likewise be represented in two dimensions by writing the hydrogen atom either on the left or on the right; the arrangement in the case of glucose (left, right, left, left) is shown in the figure.

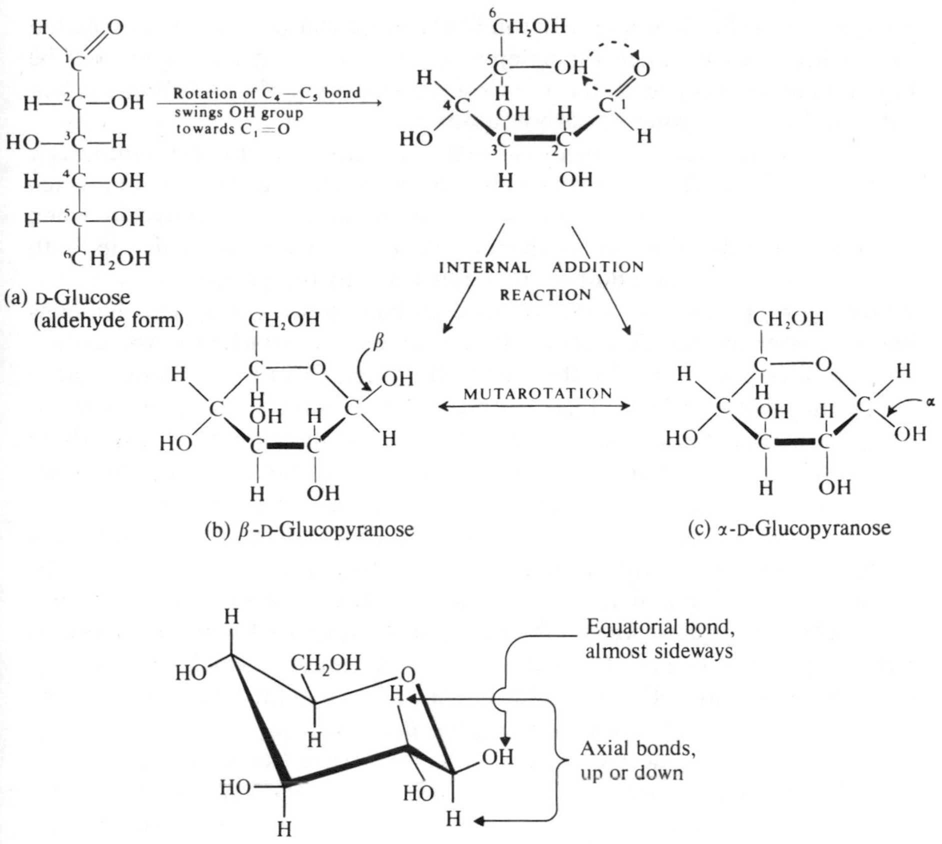

(a) D-Glucose (aldehyde form)

(b) β-D-Glucopyranose

(c) α-D-Glucopyranose

(d) β-D-Glucopyranose (in chair-shaped representation)

Fig. 4.1 The structure of glucose

Like other aldehydes, glucose is a reducing agent, although in many reactions it is weaker in this respect than might have been anticipated. This diminished reducing power is probably a consequence of a complication which occurs when solid glucose and other hexoses are dissolved in water—namely, that a tautomeric change occurs in which a large proportion of the molecules take on a cyclic form. This tautomeric change can be regarded as an internal addition reaction, in which the hydrogen atom of the hydroxyl group attached to carbon atom 5 attaches itself to the oxygen atom of the carbonyl group on carbon atom 1, so as to form a new hydroxyl group (Fig. 4.1). As a result of the opening of the carbonyl double bond in this manner, the free valency of the carbonyl group completes a heterocyclic, six-membered *pyranose* ring by union with the oxygen atom attached to carbon atom 5. This ostensibly somewhat unlikely event can be readily appreciated by making a model, when the valency angle between carbon atoms will be found quite naturally to result in a close juxtaposition of the carbonyl group with the hydroxyl group of carbon atom 5.

This cyclization leads in turn to a further complication, for two tautomeric ring forms are possible. The reason for this is exactly the same as that which leads to, say, the formation of two forms of aldehyde cyanhydrin when hydrogen cyanide adds on to the double bond of acetaldehyde; in both addition processes, the addition reaction leads to the production of asymmetry where, previously, a symmetrical carbon atom existed. This happens because, after the addition, four different groups are attached to the carbon atom of what was originally the carbonyl group. This carbon atom is often called the *anomeric* carbon atom and the two isomers produced when it becomes asymmetric are said to be anomers. In the case of glucose there are, at the moment of ring formation, theoretically equal chances of either of the two double bonds of the carbonyl group rupturing, and, according to which of the two breaks, two different sequences of the four monovalent groups around this carbon atom are possible. The two forms can be represented, as in Fig. 4.1, by directing the newly formed hydroxyl groups either upwards or downwards, the plane of the ring itself being imagined as *projecting at right angles* out of the paper. The form for which the hydroxyl group is shown directed downwards is known as α-D-glucopyranose, the one in which it is directed upwards being the β form.

The β form of D-glucopyranose is somewhat more stable in solution than its α counterpart, probably because, in the β form, the hydroxyl groups on carbon atoms 1 and 2 are farther apart. Consequently, even though, theoretically, the carbonyl double bonds may have equal chances of opening, a 50:50 mixture of α and β forms is not achieved in practice owing to the intervention of a final complication—mutarotation. This process leads to the interconversion of α and β forms, until an equilibrium mixture is established. Thus, commencing with the pure α isomer, upon contact with water, more and more of the β form is produced until equilibrium is reached. For glucose, the equilibrium mixture contains 64 per cent of the β form, about 36 per cent of the α form, and a trace of the non-cyclic aldehydic reducing form; starting with the pure β form, exactly the same equilibrium mixture is achieved. The process of mutarotation is accelerated by the enzyme *aldose mutarotase.*

The difference between α- and β-D-glucose has been stressed because, when this and other monosaccharides unite to give disaccharides and polysaccharides, the linkage is sometimes of one type and sometimes the other and, very often, the physical properties of the product are intimately bound up with the nature of the linkage. Cellulose, for example, is constructed from β-glucose groups, and comprises zig-zag chains of monosaccharide units, whereas starch and glycogen, which contain α-glucose units, have molecules where the chains tend to wind round one another.

The cyclic forms of sugars used here and elsewhere in this book were introduced by W. Haworth, a pioneer in the structural chemistry of sugars. They have the advantage of being relatively easy to draw. Strictly speaking, however, like many other saturated ring structures (e.g., cyclohexane), the rings are not quite planar. One of the two non-planar representations is chair-shaped, and in Fig. 4.1d the formula of β-D-glucopyranose is shown in this form.

4.3 Other aldoses and ketoses

In addition to glucose, there are two other aldohexoses of considerable biological importance. These are D-galactose and D-mannose, and their structures are shown in Fig. 4.2. They resemble glucose except that the configuration of the groups around certain of the asymmetric carbon atoms

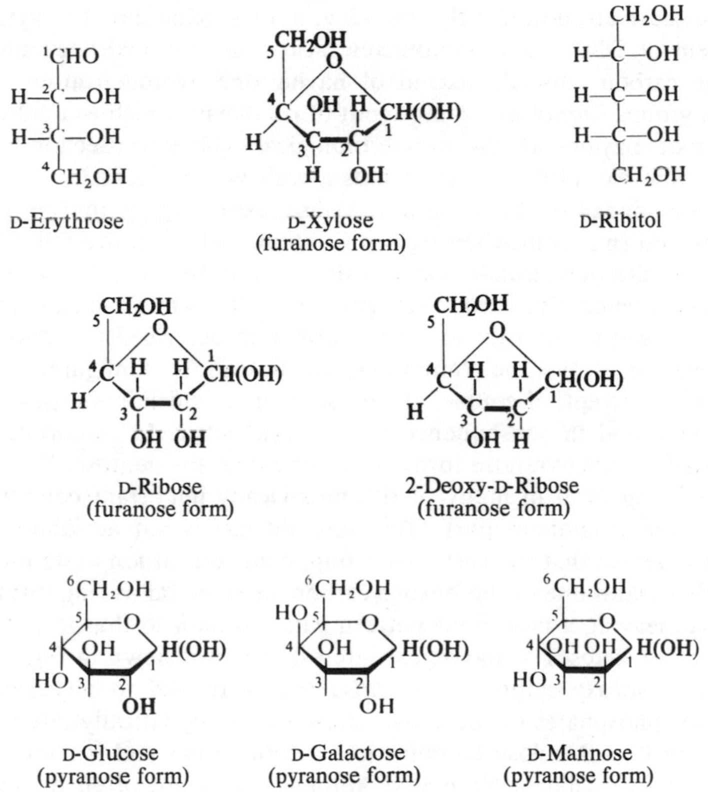

D-Erythrose

D-Xylose (furanose form)

D-Ribitol

D-Ribose (furanose form)

2-Deoxy-D-Ribose (furanose form)

D-Glucose (pyranose form)

D-Galactose (pyranose form)

D-Mannose (pyranose form)

Fig. 4.2 Some important aldoses and related compounds

is different—galactose differs at carbon atom 4 and mannose at carbon atom 2. Each has an α and β form analogous to the two forms of glucose—a comparison of models will show that α and β forms of a sugar are technically quite different substances, for they neither superimpose nor are mirror images of one another.

D-Glyceraldehyde (Section 4.1) is an aldotriose formed *in vivo* when glycerol is oxidized. Its phosphorylated derivatives are of great importance in the glycolytic pathway, a reaction sequence which results in the breakdown of hexose sugars. D-Erythrose (Fig. 4.2) is an example of a tetrose; its phosphate is involved in the photosynthetic carbon cycle and in a route by which sugars are oxidized known as the pentose phosphate pathway. Aromatic amino acids cannot be made by all organisms but, in those which can, D-erythrose phosphate frequently participates in the biosynthesis.

There are two important aldoses with five carbon atoms. These are D-ribose and D-xylose. In addition, there are two other materials, 2-deoxy-D-ribose and the ribose reduction product, ribitol, which have functional (as distinct from energetic) uses of great significance. Though related to the pentoses, these are not, strictly speaking, monosaccharides. Deoxyribose, a component of DNA (deoxyribonucleic acid), has two hydrogen atoms attached to carbon atom 2, instead of having one hydrogen atom and one hydroxyl group. Ribitol is a component of riboflavin, which is a cofactor for a number of enzymes which catalyse biological oxidations (Section 5.4). The configurations for all these compounds are shown in Fig. 4.2.

D-Ribose, deoxy-D-ribose, and D-xylose exist largely in the *pyranose* six-membered ring form when free in solution. But when present as part of larger molecules they usually occur in the form of five-membered rings (Fig. 4.2). This *furanose* ring structure arises from the straight-chain form in a comparable way to the pyranose ring formation described for glucose.

Fructose (Fig. 4.3) is the commonest ketohexose. Its configuration resembles glucose—except, of course, at carbon atom 2, which in ketoses carries a double bond and in consequence is not asymmetric. In solution, fructose exists largely in the pyranose form but it resembles the pentoses by assuming a furanose ring form in many of the biologically important compounds of which it is a component part. This resemblance is not accidental, but is related to the fact that the carbonyl group, being on carbon atom number 2, effectively isolates one of the hexose carbon atoms so far as ring formation is concerned, leaving a molecular pattern which is akin to that in pentoses.

Some other ketoses of biological interest are also shown in Fig. 4.3. The simplest, dihydroxyacetone, is a triose and is related to glyceraldehyde; indeed, the phosphates of these two trioses are enzymatically interconvertible in the cell. D-Ribulose contains five carbon atoms and is configurationally related to D-ribose. Note that ketonic sugars are often distinguished from their aldehydic counterparts by the ending *-ulose*. D-Sedoheptulose is a

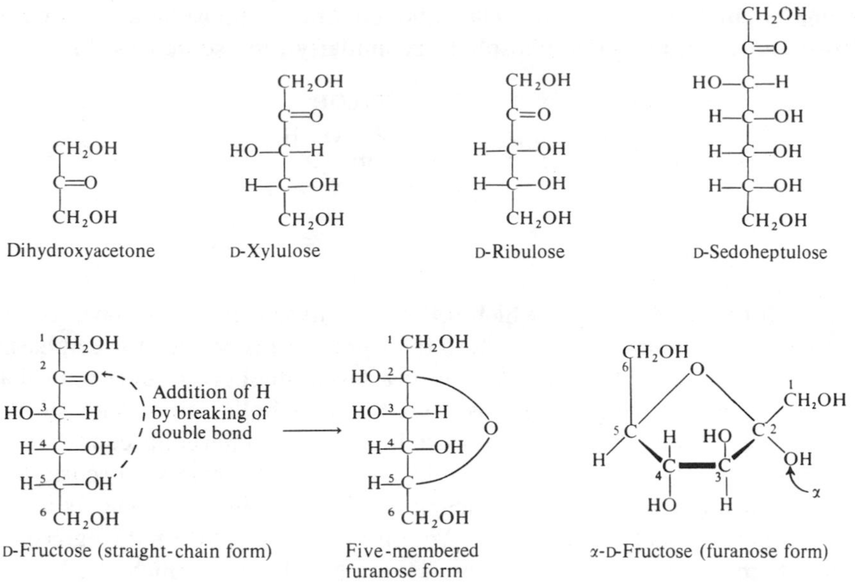

Fig. 4.3 Some important ketoses

sugar containing seven carbon atoms; its phosphate, like those of ribulose and xylulose, plays an important role in the photosynthetic carbon cycle. Readers are, however, advised that it is unnecessary at this stage to commit all these names and formulae to memory. Such a process is tedious and only an overall impression of nomenclature is necessary until individual compounds are encountered in relation to particular aspects of cell metabolism.

4.4 Derivatives of the monosaccharides

Simple monosaccharides are seldom directly involved in biochemical reactions, but are first transformed to esters, to ether-like compounds, or to oxidation products. Occasionally, reduction products or amino sugars are of importance as well. Taking glucose as the principal example, the structure and nomenclature of some of these derivatives of sugars will now be illustrated.

The hydroxyl group on carbon atom 1 (number 2 in ketoses) is manifestly rather different from those on the other carbon atoms, for it originates from the carbonyl group of the straight-chain form. It is not unusual to find that it is, in certain circumstances, more reactive than the other hydroxyl groups. Phosphoric acid, for example, readily esterifies glucose, *in vitro*, the product being D-glucose-1-P (in formulae and equations, phosphate is often abbreviated to P; inorganic phosphate is similarly represented as P_i):

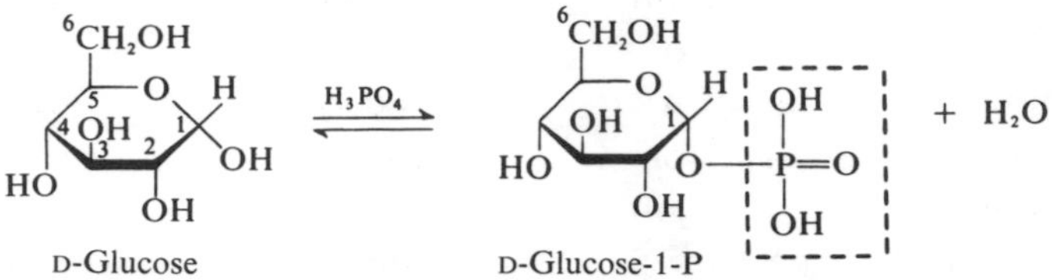

D-Glucose
(α form)

$\xrightarrow{H_3PO_4}$

D-Glucose-1-P
(α form)

+ H_2O

In the formulae for sugars which follow, the hydrogen atoms have, conventionally, been omitted except where their presence is needed for emphasis or clarity. Glucose-1-phosphate is one of a large number of sugar phosphates to be found in living organisms, and the metabolic importance of these derivatives cannot be overemphasized. Although energetic considerations are undoubtedly involved (Section 7.9), the almost universal occurrence in biochemical reactions of sugar phosphates rather than simple sugars is a fact not always readily explained. Another glucose phosphate, glucose-6-phosphate, is also well known, as is a diphosphate in which the hydroxyl groups on carbon atom 1 and carbon atom 6 are both esterified. This diphosphate and its very important fructose counterpart have the following formulae:

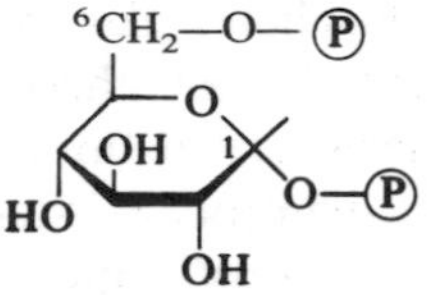

D-Glucose-1,6-di-P
(a pyranose compound)

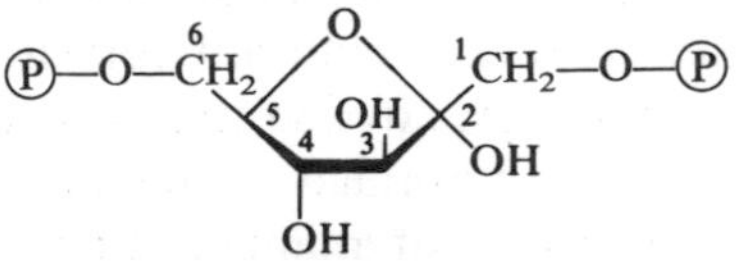

D-Fructose-1,6-di-P
(a furanose compound)

Under suitable conditions, alcohols react with the anomeric (Section 4.2) hydroxyl group of monosaccharides to yield ether-like substances termed *glycosides*. The term 'ether-like' is used, for sugars are not typical alcohols and the resulting compounds, although structurally like ethers, are far less stable to hydrolysis than ordinary ethers, being in many respects more like esters:

$$R-O-H + H-O-R \longrightarrow R-O-R + H_2O$$

Two molecules of an alcohol

An ether

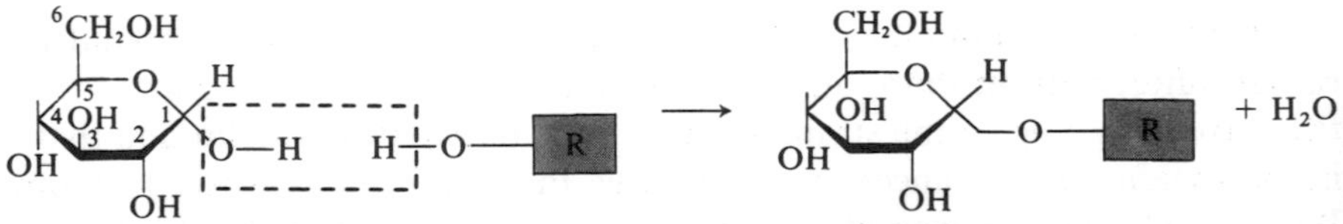

One molecule of glucose
(α form)

$\longrightarrow$

R-α-glucoside, or
1-R-α-D-glucose

+ H_2O

Glycoside is a general name which can be used whenever it is unnecessary to specify the participation of any one particular sugar—it can be compared to such general names as alkyl, aryl, and acyl. Glycosides exist, of course, in α and β forms since the anomeric hydroxyl group is involved in their formation.

For the alcohol which helps to form the glycoside linkage a second sugar molecule may be substituted—sugars contain alcoholic hydroxyl groups—and when this occurs a disaccharide is formed. Thus two α-D-glucose residues unite together in the following manner to give maltose, a sugar which is, essentially, a glucose-α-glycoside. That is to say, in the molecule of maltose, the sugar of the 'glycoside' part of the molecule is also glucose:

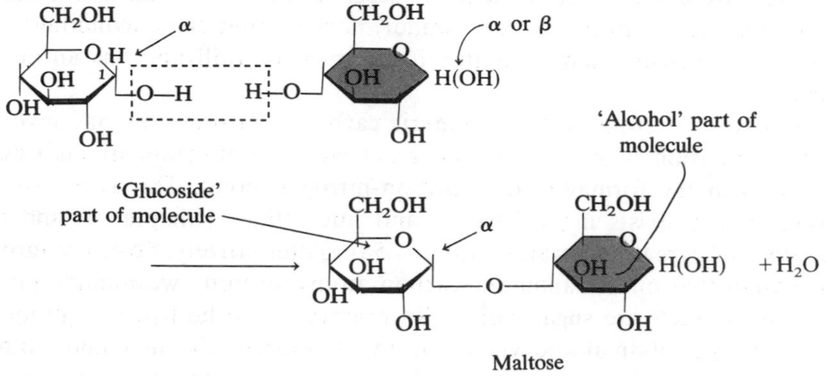

'Glucoside' part of molecule

'Alcohol' part of molecule

$+H_2O$

Maltose

The repetition of this process in such a way that many monosaccharide units are held together by α- or β-glycoside linkages leads to the formation of polysaccharides; conversely, the hydrolysis of appropriate polysaccharides results in the liberation of the monosaccharide units from which they are constructed.

Several other monosaccharide derivatives play important but more restricted roles in cells or in the tissues of particular groups of organisms. They include oxidation and reduction products of monosaccharides as well as 2-amino and substituted 2-amino sugars. Some of these, with figure references, are included in Fig. 4.4.

Of the many possible oxidation products of monosaccharides two kinds are of particular interest. If the aldehydic 'head' of an (aldo)glycose (to use the general name for sugars) is oxidized to carboxyl, the hydroxy acid so formed is called a *glyconic* acid (Fig. 4.4). Conversely, the primary alcohol group at the 'tail' of the molecule may be oxidized to carboxyl, when the product is known as a *glycuronic acid* (whether accidentally or otherwise, 'uronic' recalls 'urea' and 'uric acid' and this is a useful mnemonic to recall that this name refers to the 'tail' of the molecule!). The best-known example of the first of these is 6-*phosphogluconic* acid, which is produced in an important oxidative step of the pentose phosphate pathway (Fig. 14.9).

Galacturonic acid and *glucuronic* acid are examples of the second possibility; they are present, with or without additional modification, in certain complex polysaccharides (Section 4.6). Ascorbic acid, or vitamin C, is an oxidation product more distantly related to hexose sugars. It was deficiency of this vitamin, brought on by inadequate amounts of fresh fruit and vegetables on board ship during long voyages, which led to serious outbreaks of scurvy among sailors during the great days of sea exploration.

Reduction of aldehydes and ketones produces polyhydroxy alcohols. Two important examples are *ribitol,* formed by the reduction of ribose, and *glycerol,* formed by the reduction of glyceraldehyde. Ribitol is a component of the important electron carriers FMN and FAD, whose structures and functions are described in the next chapter. Glycerol is the alcohol liberated when fats are hydrolysed and, under normal metabolic conditions, its oxidation to glyceraldehyde is the first step in its utilization as an energy source.

The hydroxyl group on the anomeric carbon atom (i.e., the one involved in the formation of glycosides) can react with various aliphatic and cyclic amines with the formation of a carbon-nitrogen bond. This is the sort of linkage which exists in nucleosides and nucleotides (Chapter 5) and it is therefore of great importance (Fig. 4.5a). Alternatively, hydroxyl groups other than that on the anomeric carbon atom can unite with amino groups and the products are sugar amines. In practice, it is the hydroxyl group on carbon atom 2 of an aldose which is normally replaced by an amino group in compounds of biological interest (Fig. 4.5b). Glucosamine and galactosamine are two such sugar amines which participate, with or without further modification, as units in the structure of various more complex polysaccharides (Section 4.6).

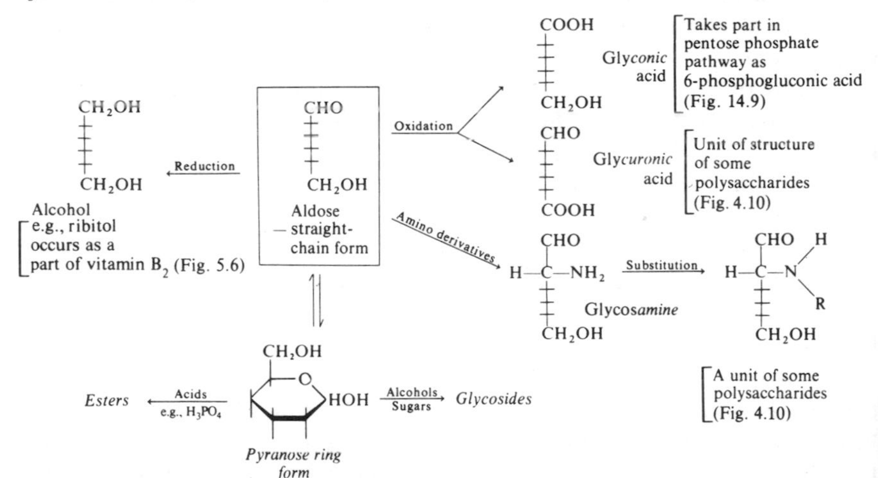

Fig. 4.4 Some derivatives of monosaccharides of biological importance

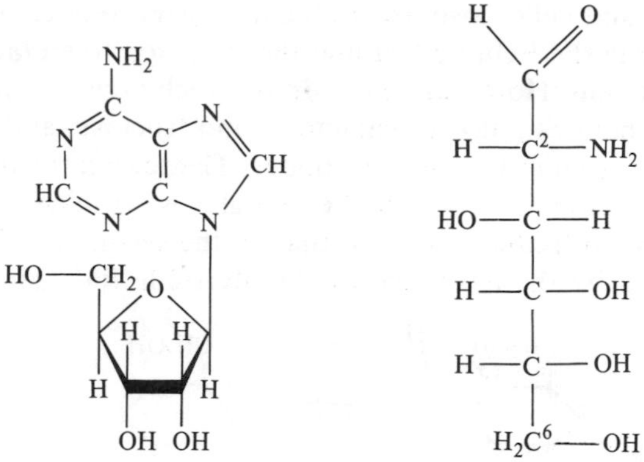

(a) Adenine nucleoside (b) D-Glucosamine

Fig. 4.5 Sugar derivatives containing the carbon-nitrogen linkage

4.5 Disaccharides

Maltose is a disaccharide in which *two* glucose units are joined through an α-1,4-glycoside linkage (Fig. 4.6). It is readily attacked by dilute acids and by *maltase*, an enzyme which specifically hydrolyses α-glycoside linkages and which has no action upon β-glycosides. Consequently, *cellobiose*, which is a disaccharide comprising glucose units joined by a β-glycoside linkage, is not attacked by maltase, although it is hydrolysed by β-glycosidases and by dilute acids. *In vivo*, maltose is formed by the partial hydrolysis of starch; cellobiose is a breakdown product of cellulose.

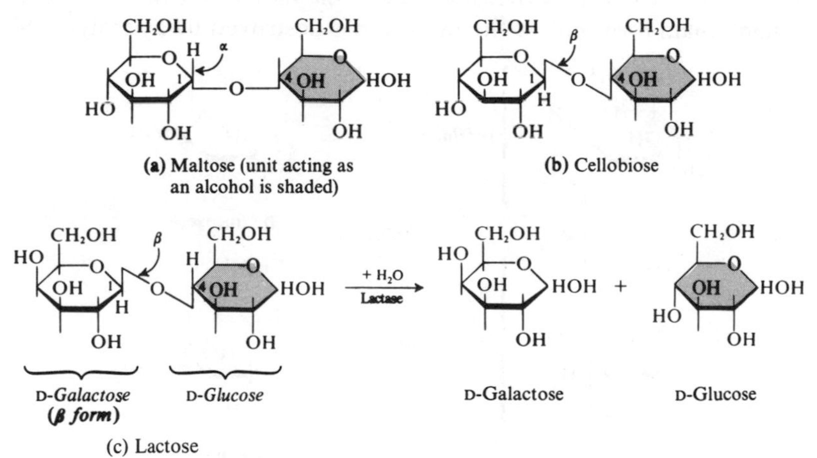

Fig. 4.6 Three reducing disaccharides

Both maltose and cellobiose are reducing sugars, reducing cupric copper to cuprous oxide in the Fehling test and the silver ion to metallic silver in the silver mirror reaction. However, they are not such powerful reducing agents as the monosaccharides glucose, galactose, and fructose, and do not readily reduce Barfoed's reagent to cuprous copper. The reason for this difference is that while carbon atom 1 of one of the two glucose residues is free to revert back to the carbonyl (reducing) form, that on the second residue is stabilized in the cyclic form by the existence of the glycoside linkage:

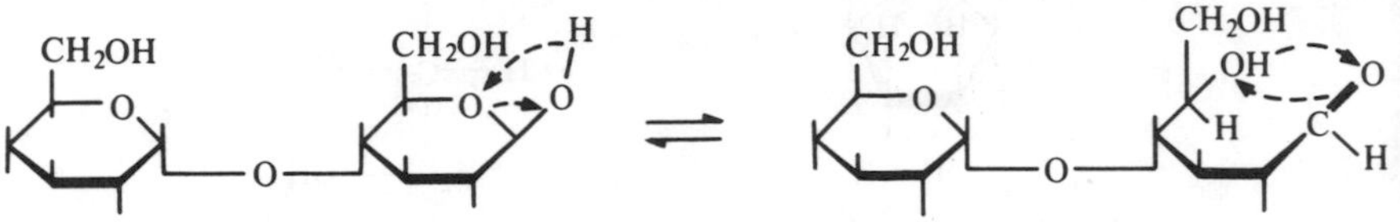

Maltose—outline of double cyclic form

Maltose—outline of aldehyde form

Some of the disaccharides which occur naturally are constructed from two different monosaccharide units, important examples being D-lactose, which is the main carbohydrate in milk, and sucrose, which is especially abundant in sugar cane and sugar beet. D-Lactose consists of D-galactose in its β form, linked to carbon atom 4 of D-glucose (Fig. 4.6c). It is hydrolysed to these two monosaccharides by *lactase*, a mammalian digestive enzyme.

Sucrose comprises a molecule of D-glucose linked through its carbon atom 1 to carbon atom 2 of fructose (Fig. 4.7). Here, however, not only is the linkage α with respect to the glucose and β with respect to the fructose, but the glucose is present in its six-membered pyranose form while the fructose is in the five-membered furanose form. Since the two carbon atoms participating in the glycoside linkage are the two carbonyl-carbon atoms of the individual monosaccharides, neither half of the molecule can revert back to the straight-chain form unless the molecule is destroyed by hydrolysis. Since

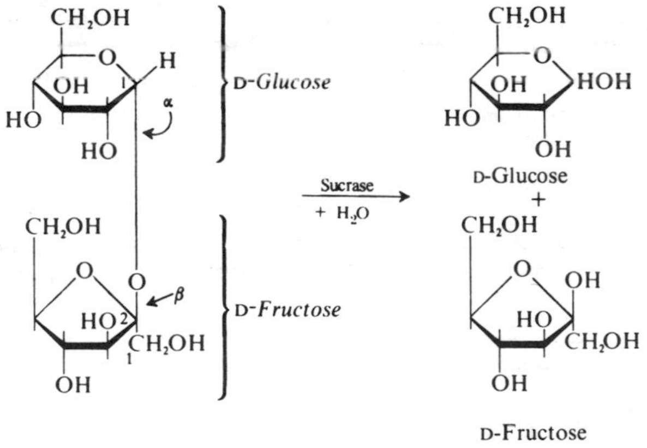

D-Glucose

D-Fructose

Sucrase + H_2O →

D-Glucose + D-Fructose

Fig. 4.7 Sucrose, a non-reducing disaccharide, and its hydrolysis

there is no potential carbonyl group anywhere in the molecule, sucrose is a *non-reducing disaccharide.* An enzyme, *sucrase*, capable of hydrolysing sucrose, is present in plant cells and in the absorptive cells of the small intestine of mammals. The enzyme is also called invertase from the fact that, whereas sucrose is dextrorotatory, the equimolar mixture of glucose and fructose obtained from it is laevorotatory; the 'inversion' of rotation occurs because fructose ('laevulose') is more laevorotatory than an equimolar quantity of glucose ('dextrose') is dextrorotatory.

4.6 Polysaccharides

Polysaccharides are constructed from monosaccharide units by elimination of water from a number of such units. *Cellulose* is the most abundant polysaccharide in nature, since it is a major component of unlignified plant cell walls (Chapter 1). It has great tensile strength and is resistant to chemical attack. It can only be hydrolysed by the most vigorous acid treatment, and few, if any, animals have enzymes capable of destroying it. Its breakdown by soil bacteria is, however, of great practical importance, as is its metabolism by bacteria in the rumen of such herbivores as cattle.

Starch is not a single substance but a mixture of at least two components. The simplest, amylose, is a straight-chain 'condensate' of glucose residues and thus resembles cellulose except that, whereas the glycoside linkages in cellulose are β linkages, those in starch are of the α form (Fig. 4.8). Starch is much more readily hydrolysed than cellulose, hydrolysis being possible either by means of dilute acid or by the action of enzymes termed *amylases.* The amylases hydrolyse amylose with the formation of maltose; the disaccharide is then normally further degraded by maltase. Amylose reacts with

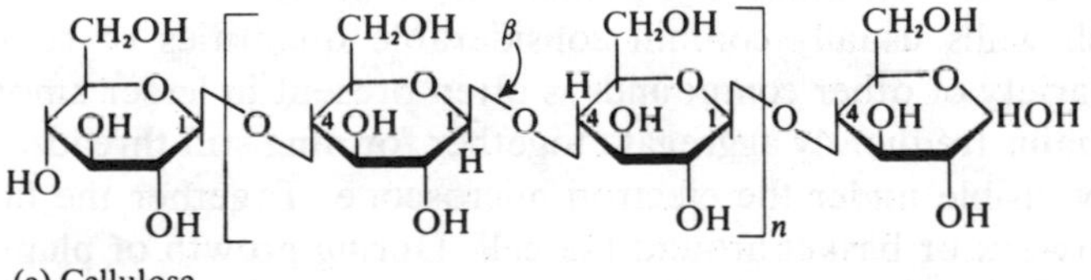

(a) Cellulose

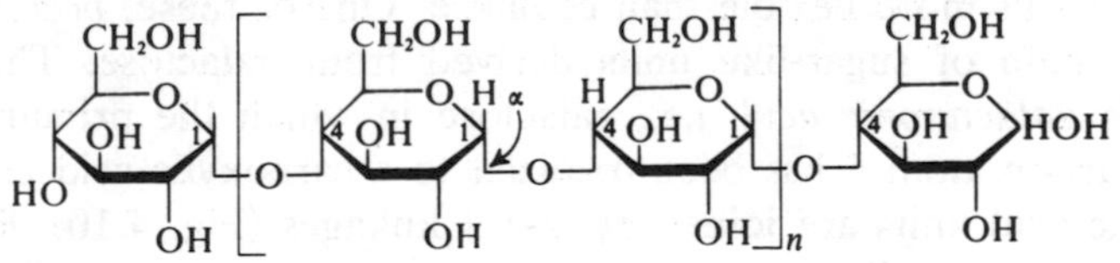

(b) Amylose

Fig. 4.8 The structures of cellulose and amylose

iodine to give the well-known dark blue colour which disappears on warming and reappears on cooling. In common with other polysaccharides, starch is a non-reducing carbohydrate since carbonyl groups of all units (except one of the two terminal ones) participate in the glycoside linkages.

The two polysaccharides considered so far, amylose and cellulose, are linear polymers of glucose residues linked by 1,4-glycoside linkages. But some polysaccharides have a branched structure. For example, besides amylose, starch contains a compound known as amylopectin. On digestion with pancreatic amylase, this yields not only maltose but also a second disaccharide, isomaltose, in which two glucose residues are united by a 1,6-α-glycoside linkage:

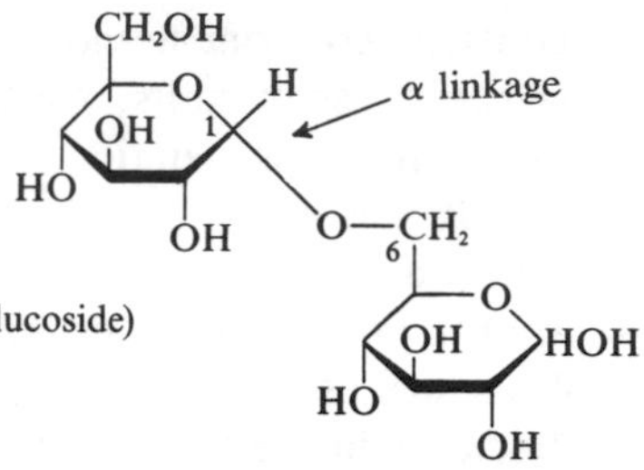

Isomaltose
(glucose 1,6-α-glucoside)

This suggests that amylopectin consists of chains of glucose residues resembling amylose (α-1,4-glycoside linkages) but having secondary branches which are united to the main chains by 1,6-glycoside linkages. The structure of amylopectin is indicated in Fig. 4.9. The average chain contains 20 to 25 glucose units and the number of glucose residues between branching points within a chain is 5 to 8. Glycogen, the major carbohydrate reserve of animals, resembles amylopectin in structure but is more highly branched, the average chain length being 8 to 14 glucose residues and the number of glucose units between branching points only 3 or 4.

Plant cell walls usually contain considerable quantities of cellulose, although a variety of other compounds is often present in lesser amounts. The cellulose chains frequently aggregate together forming stiff threads (microfibrils) readily visible under the electron microscope. Together the fibres form a tough network or basket around the cell. During growth of plant cells the cellulose wall develops gradually, and very young plant cell walls contain several materials more flexible than cellulose. One of these, *pectin*, consists of a long chain of sugar-like units derived from galactose. The unit of structure is *galacturonic acid*, i.e., galactose in which the primary alcohol group of carbon atom 6 has been oxidized to a carboxylic acid group. The galacturonic acid units are joined by α-1,4 linkages (Fig. 4.10). Pectin has jelly-like properties allowing easy expansion of the growing cell. The carboxyl groups are sometimes esterified with methanol and are sometimes in the form of their calcium salts. It has been suggested that the proportion of

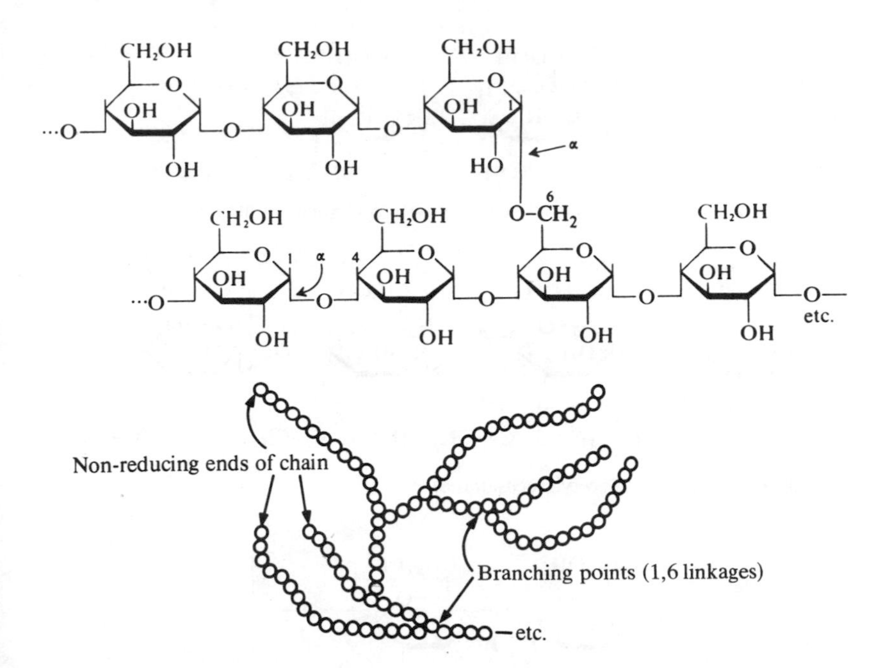

Fig. 4.9 Amylopectin. (Above: the interchain link. Below: a small fragment of amylopectin showing branching. Each circle represents a glucose unit)

pectin in each of these two forms determines the degree of elasticity in the cell wall of the growing plant cell. The jellifying properties of pectin are frequently employed to make jams 'set'. Other components of plant cell walls are *hemicellulose* and *lignin.* Hemicelluloses are polysaccharides but they are not related to cellulose, comprising principally residues of the pentose xylose linked by β-1,4 linkages. Lignin, which accounts for about 20 per cent of the dry weight of *wood,* is *not a polysaccharide* but is a complex condensate of phenolic acids.

In animals, too, polysaccharides play an important structural role. The cuticle of insects, for example, contains *chitin,* a polymer of a derivative of glucose in which the hydroxyl group attached to carbon atom 2 is replaced by amino groups which are in turn acetylated. Chitin consists of residues of *N-acetylglucosamine* joined by β-1,4 links (Fig. 4.10). The same structural unit is present in *hyaluronic acid.* This polysaccharide-like substance occurs as a component of several gelatinous tissues of vertebrates, including the vitreous humor of the eye and the synovial fluid which lubricates the joints between bones. Hyaluronic acid is composed of alternating units of *N*-acetylglucosamine and glucuronic acid (the glucose equivalent of galacturonic acid). Hyaluronic acid is of interest in that its molecule contains, in

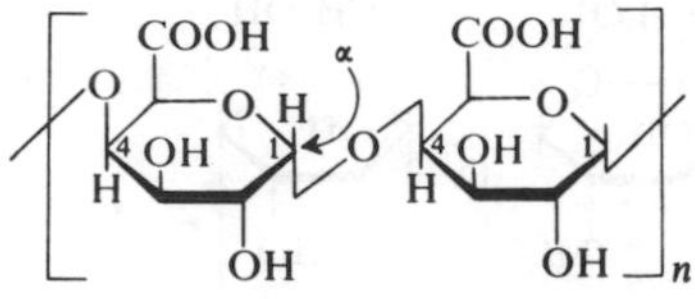

Repeating unit of pectin (residues of galacturonic acid)

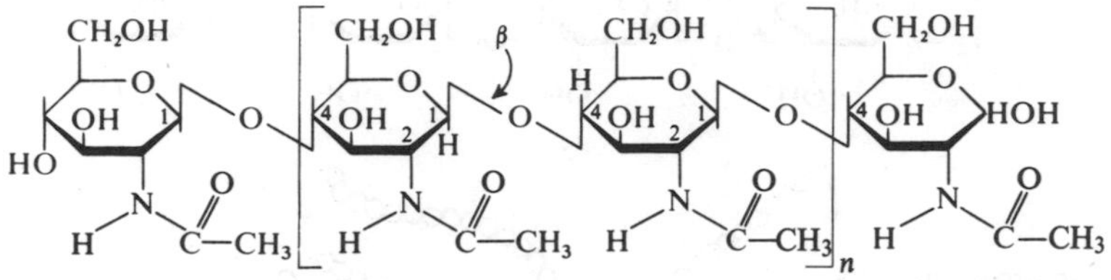

Chitin (residues of *N*-acetylglucosamine)

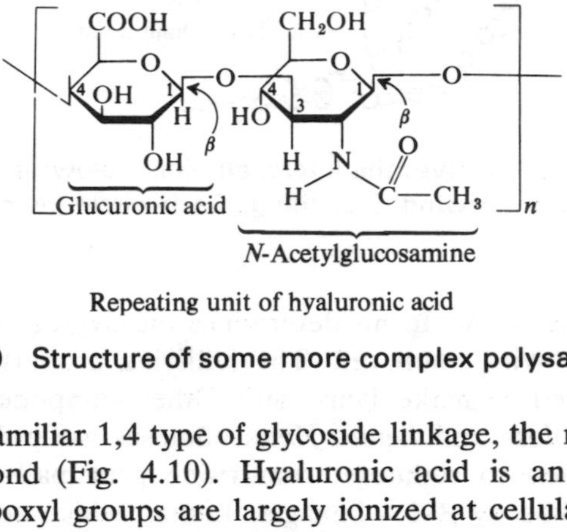

Glucuronic acid

N-Acetylglucosamine

Repeating unit of hyaluronic acid

Fig. 4.10 Structure of some more complex polysaccharides

addition to the familiar 1,4 type of glycoside linkage, the much less common 1,3-glycoside bond (Fig. 4.10). Hyaluronic acid is an acidic substance, because the carboxyl groups are largely ionized at cellular pH.

Chondroitin is another acidic polysaccharide, but comprises units of *N*-acetylgalactosamine united to glucuronic acid residues by β-1,3 links. It occurs, in the form of a sulphate, in skeletal tissues such as cartilage and bone. *Heparin*, which occurs in the walls of blood vessels, prevents the clotting of blood. It is a sulphated condensate of glucosamine and glucuronic acid.

4.7 Role of carbohydrates inside and outside cells

The bulk of the carbohydrate in cells occurs as polysaccharide. This may be of structural importance as in the case of cellulose and the polysaccharide-like substances chitin and pectin (Fig. 4.10) or it may be a storage material such as starch in plants and glycogen in animals. The polysaccharides have molecular weights ranging from 10^3 to about 10^7, and, unlike the mono- and disaccharides, they are not readily soluble in water (some, such as starch, do however form colloidal solutions). In the cell, therefore, monosaccharide units, once incorporated into polysaccharides, are effectively prevented from exerting any substantial osmotic pressure. Mono- and disaccharides tend to be the forms of carbohydrate which are transported between cells, e.g., glucose is carried about the mammal in the blood and sucrose is transferred in plants from one organ to another in the phloem. Once inside the cell, however, monosaccharides are usually converted back into polysaccharides except for the small amount which is required immediately for metabolic purposes; the latter is converted instead to monosaccharide phosphates.

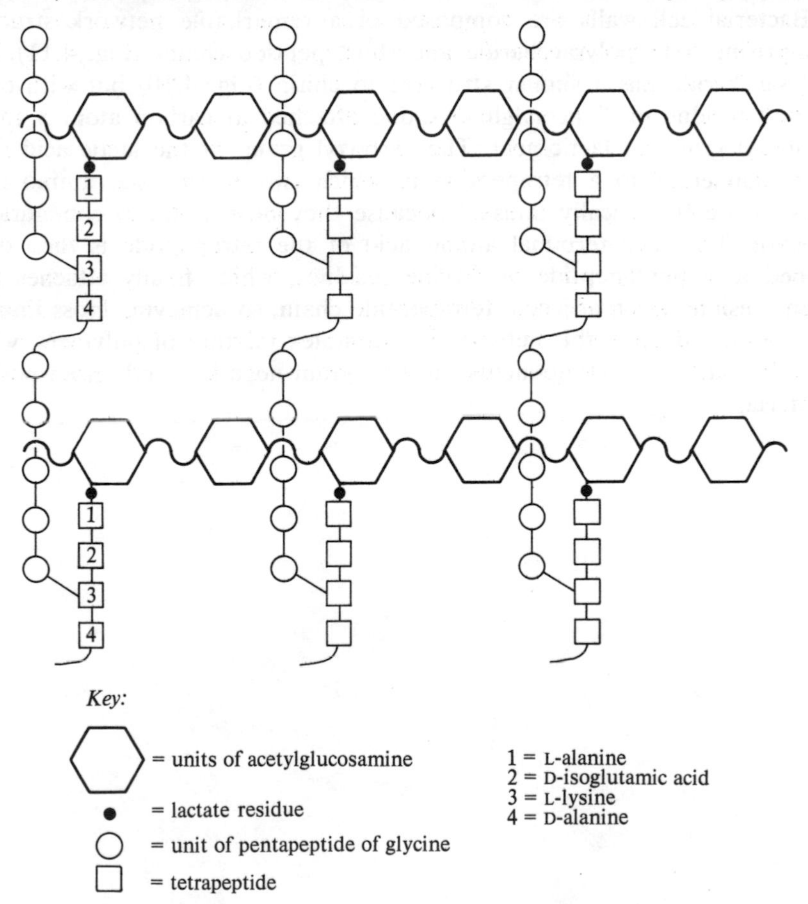

Fig. 4.11 Diagram of the giant macromolecule providing the basic structure of bacterial cell walls. Upon it different macromolecules are superimposed; these are not the same in gram-positive and gram-negative bacteria

Mucoproteins secreted from cells are conjugate proteins containing oligosaccharide side branches, each with a relatively few monosaccharide residues. Very often, the final residue in the oligosaccharide chain is *N*-acetylneuraminic acid (an acetylated mannosamine derivative with pyruvic acid forming an addition compound with the aldehyde group). Many of the proteins on the outside of cell walls are *glycoproteins* and they probably play an important role in cell recognition and in antigen-antibody reactions. Finally, human beings possess different blood groups, because of slightly different proteins in their blood. These proteins are also glycoproteins, and it is the different oligosaccharide side chains attached to the proteins that are responsible for the existence of these different blood groups.

Bacterial cell walls are composed of a remarkable network structure comprising both polysaccharide and short peptide chains (Fig. 4.11). The polysaccharide has a similar structure to chitin (Fig. 4.10) but with every second residue of *N*-acetylglucosamine attached at carbon atom 3 to the hydroxyl group of lactic acid. The carboxyl group of the lactic acid is, in turn, connected to a tetrapeptide in which two of the four amino acids present are biologically unusual because they possess the D configuration (Section 3.2). The terminal amino acid of the tetrapeptide is then often joined to a pentapeptide of glycine residues, which finally attaches to a lysine residue in an adjacent tetrapeptide chain, so achieving cross linkage. Superimposed upon this network is a complex mixture of polymers, which contains different components in the gram-negative and gram-positive bacteria.

5

Some biological carriers

In many of the cellular reactions described later in this book, an electron or a molecular unit is transferred from one molecule to another *via* an intermediate carrier molecule. Sometimes the carrier is an independent entity. In such a case, if the enzyme reaction is carried out *in vitro*, the carrier can be added in known and variable quantity to the system (for example, NAD^+, which accepts electrons from suitable substrates, and ATP, which carries high-energy phosphate). Many enzymes will only function with a particular carrier, even when other closely similar carriers are available. Thus lactate dehydrogenase only operates in the presence of NAD^+, whereas glucose-6-phosphate dehydrogenase only catalyses transport of electrons from glucose-6-phosphate to $NADP^+$. Sometimes the situation is more complex, for the carrier may be the prosthetic group of an enzyme, as, for example, is the hydrogen carrier FAD which is a part of the molecule of succinate dehydrogenase. Or the carrier may be an integral part of a biological membrane, an example being cytochrome **b**, which is firmly attached to the inner mitochondrial membrane.

This chapter will deal with the structure and general significance of a number of carriers. They are considered in the following sections:

5.1 Mononucleotides (for example, AMP, TMP)
5.2 Dinucleotides (NAD, NADP)
5.3 Polynucleotides (RNA, DNA)
5.4 Substances related to mono- and dinucleotides but which contain the flavin base dimethylisoalloxazine (FAD, FMN)
5.5 A dinucleotide-like substance containing the base phosphopantotheine (coenzyme A)

This list is not exhaustive, for many important carriers such as the cytochromes, quinone derivatives, thiamine pyrophosphate, and lipoic acid are more conveniently considered elsewhere. All the carriers in the list, however, despite their multifarious uses, have a certain structural similarity. In particular, they all contain one or more nitrogenous bases and they usually contain one or more sugar residues and one or more phosphate groups.

5.1 Mononucleotides and their pyrophosphates

These compounds are of major biological importance in their own right since their phosphates are *carriers of energy*. One compound, ATP, is outstanding in this respect, although several similar compounds play subordinate roles in cellular biochemistry. In addition, however, the mononucleotides can be regarded as the building bricks from which dinucleotides (Section 5.2) and polynucleotides (Section 5.3) are constructed, rather as monosaccharides can be considered the structural units of di- and polysaccharides (Chapter 4).

Mononucleotides comprise three essential parts: a *heterocyclic base* linked to carbon atom 1 of a *ribose* (or deoxyribose) molecule which, in turn, has a *phosphate group* attached to it. Almost all the nucleotides discussed in this book have the phosphate group linked to carbon atom 5 of the pentose and they will be represented thus:

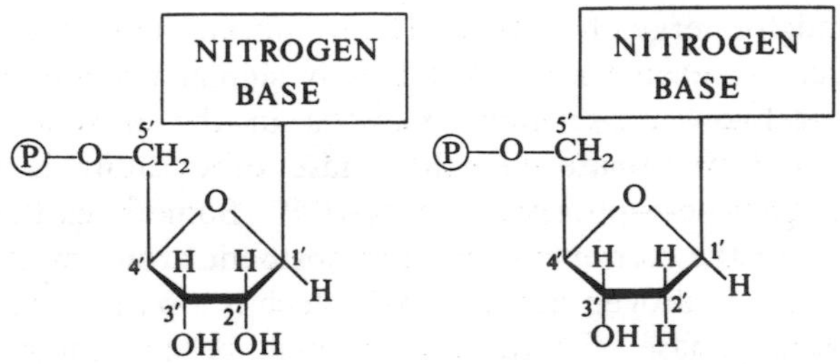

A ribose nucleotide

A deoxyribose nucleotide

The name of the heterocyclic nucleotide base (Table 5.1) will be inserted each time in the rectangle. In most cases the base is a derivative of *purine* or *pyrimidine*, though one important base, nicotinamide, is a *pyridine* derivative (Fig. 5.4). The commonest purine nucleotides (Fig. 5.1) are adenosine monophosphate (AMP) and guanosine monophosphate (GMP).

The structures of *purine* nucleotides and the method of numbering the atoms in the heterocyclic base are shown in Fig. 5.1. To differentiate the carbon atoms of the pentose from atoms in the base it is often convenient to refer to them as 1′, 2′, 3′, etc. Thus the linkage between sugar and base in AMP involves carbon atom 1′ and nitrogen atom 9.

Two methods of numbering the atoms of the *pyrimidine* ring are currently employed. That adopted here is, for simplicity, the one which resembles that used for the six-membered ring within the purine structure. Figure 5.2 shows two frequently encountered pyrimidine nucleotides, namely uridine monophosphate (UMP) and cytidine monophosphate (CMP). In addition, Fig. 5.2 shows thymidine deoxyribose nucleotide (dTMP), which is present in nucleic acids (Section 5.3). CMP can exist as both CMP and dCMP but

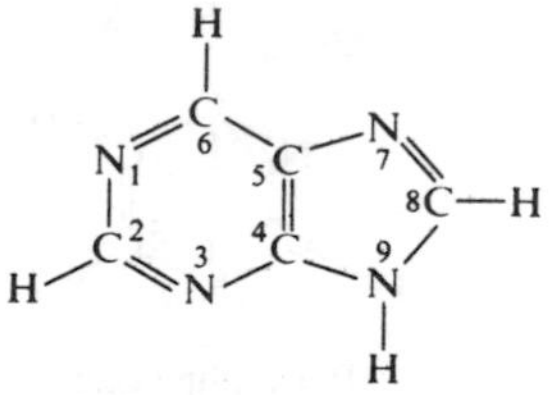

Purine ring system

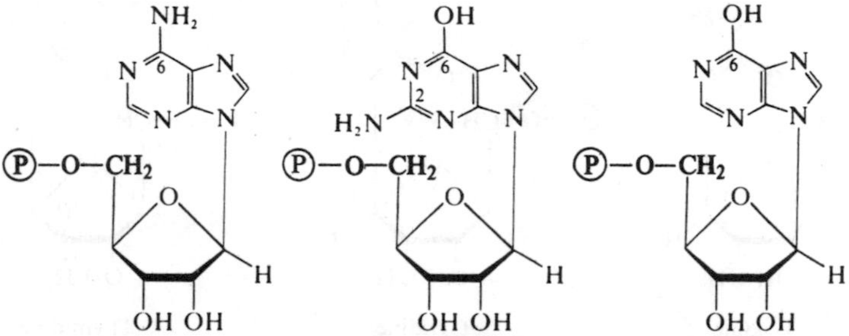

Adenosine monophosphate (AMP) Guanosine monophosphate (GMP) Inosine monophosphate (IMP)

Fig. 5.1 Mononucleotides containing the purine ring. (AMP and GMP also exist as the deoxyribose forms dAMP and dGMP)

the deoxyribose form of UMP is only rarely found in nature. It will be observed that thymine is 5-methyluracil, so unmethylated uracil is generally present only in ribose nucleotides and methyluracil only in deoxyribose nucleotides. It will be seen that, in the pyrimidine nucleotides, as in the purine nucleotides, the base is attached to the ribose anomeric carbon atom by a β-*N*-glycosidic bond:

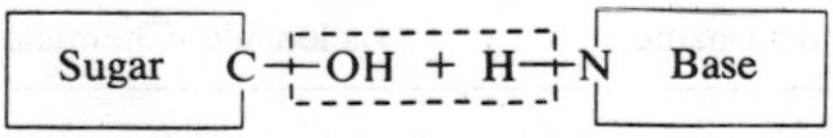

Sugar C—OH + H—N Base

For the present, only the nucleotides containing D-ribose will be considered, for, generally, the deoxyribose nucleotides occur only in a polynucleotide, deoxyribonucleic acid (DNA), the structure of which is discussed in Section 5.3.

Nucleosides are similar to the nucleotides except that they *lack the phosphate group.* For example, adenosine is a nucleoside corresponding to AMP (Fig. 5.1) but the hydroxyl group attached to carbon atom 5′ of the ribose is not esterified by attachment to phosphate. Nucleosides, as such, only very rarely take part in cell metabolism whereas nucleotides participate directly

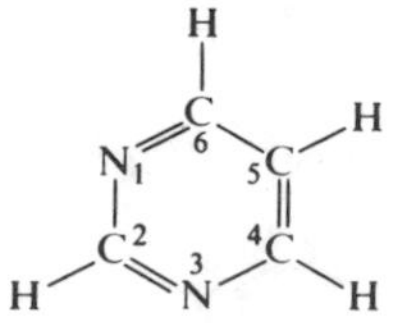

Pyrimidine ring

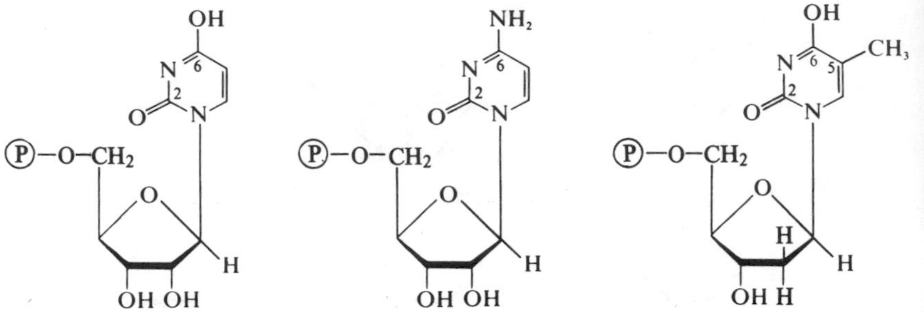

Uridine monophosphate (UMP)

Cytidine monophosphate (CMP)

Thymidine monophosphate (dTMP)

Fig. 5.2 Mononucleotides containing the pyrimidine ring. UMP exists only in the form shown, with ribose as the sugar residue. Thymidine mononucleotide, dTMP, exists only in the deoxyribose form. CMP can also exist as dCMP; i.e., with deoxyribose instead of ribose

Table 5.1 Nucleosides and nucleotides containing purine or pyrimidine bases united to ribose

Base	Chemical name	Nucleoside	Chemical name of nucleotide
1. Purines			
Adenine	6-amino purine	Adenosine	Adenosine monophosphate (AMP)
Guanine	2-amino-6-oxypurine	Guanosine	Guanosine monophosphate (GMP)
Hypoxanthine	6-oxypurine	Inosine	Inosine monophosphate (IMP)
2. Pyrimidines			
Uracil	2,6-dioxypyrimidine	Uridine	Uridine monophosphate (UMP)
Cytosine	2-oxy-6-amino pyrimidine	Cytidine	Cytidine monophosphate (CMP)

Note: Inosine monophosphate is normally only important in intermediate metabolism of purines. Uridine monophosphate exists mostly as a ribose nucleotide. All the other nucleotides also exist with deoxyribose instead of ribose (see Section 5.3). In addition, the base *thymine*, 5-methyluracil, forms a deoxyribose nucleoside called thymidine and a nucleotide called thymidine monophosphate (dTMP).

in biochemical reactions. Table 5.1 lists the names of the principal purine and pyrimidine bases and their corresponding nucleosides and nucleotides. The reader is recommended first to get to know the names of the nucleotides, and then, at a later stage, the names and formulae of the bases.

The formulae of the nucleotide bases in Figs 5.1 and 5.2 are the simplest which can be ascribed to them. In reality, they are resonance hybrids of several structures and, moreover, the hydrogen atoms of hydroxyl groups undergo tautomeric changes. For example, the base guanine can be represented by either of the two formulae:

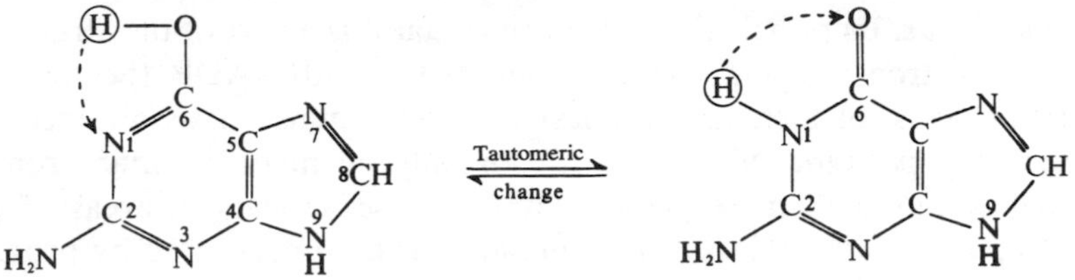

Tautomeric change

The hydrogen atom on nitrogen atom 9 is the one which is lost when the base unites with ribose or deoxyribose. It is also worthy of note that the hydrogen atoms attached to the nitrogen and oxygen atoms of the nucleotide bases readily form hydrogen bonds which are vital in maintaining the structure of DNA.

Adenosine monophosphate (AMP) is a very important nucleotide for not only is it one of the components of DNA and RNA, substances which can be regarded as *carriers of information*, but it is also part of an *electron carrier* called NAD (nicotinamide adenine dinucleotide). Moreover, it plays a vital role as the organic component of the ubiquitous *energy carrier*, ATP (adenosine triphosphate). When energy of an appropriate form is available, inorganic phosphate (P_i) will combine with AMP to form, first, adenosine *di*phosphate (ADP) and then adenosine *tri*phosphate (ATP). The —P—O—P— linkages of these last two compounds resemble those of inorganic pyrophosphate. Rather a large amount of energy is associated with these linkages— or, more precisely, is made available when they are broken. At cellular pH, the phosphate hydrogen atoms are ionized and AMP, ADP, and ATP exist in the cell partly in the form of complexes with divalent magnesium ions (Fig. 5.3).

The function of ATP in energy transfer was mentioned in Chapter 1 and will be discussed in relation to bioenergetics (Chapter 7). Briefly, however, when the terminal phosphate group of ATP is removed, energy is released which, in the presence of appropriate machinery, is not squandered as heat (as it is when hydrolysis takes place), but can be utilized to cause a muscle to contract or to drive reactions which only occur if a supply of energy is available.

The other nucleotides shown in Figs. 5.1 and 5.2 also form di- and triphosphates but are rather less frequently concerned in energy transfer than are their adenosine counterparts. Nevertheless, they have specialist uses in this connection; in particular, they will be encountered in relation to the biosynthesis of di- and polysaccharides (Chapter 12) and of the nucleic acids (Chapter 16).

Two other mononucleotides of considerable importance do not fit into the groups already discussed (there are others of local importance in the molecules of certain nucleic acids). One of these contains the pyridine ring instead of a purine or pyrimidine (Fig. 5.4a); it is called *nicotinamide mononucleotide.* Its principal significance is that it is a part of the structure of the two electron-carrying dinucleotides NAD and NADP (Section 5.2). Nicotinamide is, in fact, the simplest of the vitamins; in its absence, man may develop *pellagra.* Nicotinamide can only be made by man from the essential amino acid, tryptophan, so the disease appears in areas of poor nutrition, when protein intake is minimal. It can be prevented by providing meat or milk to those at risk.

The second important nucleotide is *cyclic AMP*—more properly, adenosine-3′,5′-cyclic phosphate (Fig. 5.4b). An enzyme, *adenyl cyclase,* present on the inner side of the plasma membrane of cells, catalyses the removal of a molecule of pyrophosphate and water from ATP in such a way as to form an internal ester between the phosphate on carbon atom 5′ and the hydroxyl group on carbon atom 3′ of the ribose. The enzyme carries out this conversion in animal cells when certain hormones which circulate in the blood reach the cell. Cyclic AMP plays an important part in the regulation of cellular metabolism, in particular by helping to activate phosphorylase (Chapter 13), which in turn converts stored glycogen to glucose-1-phosphate.

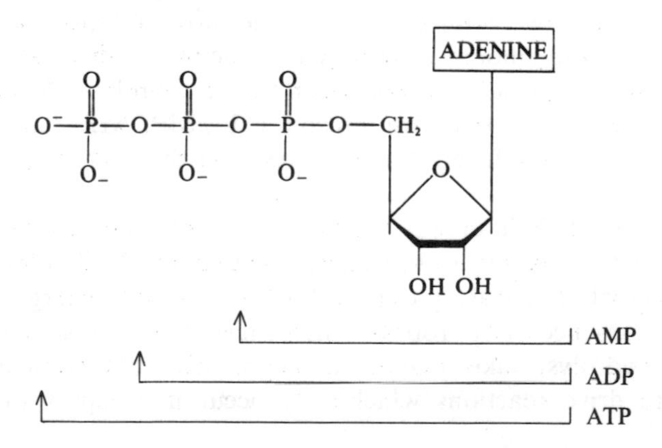

Fig. 5.3 Structures of AMP, ADP, and ATP (ionized forms)

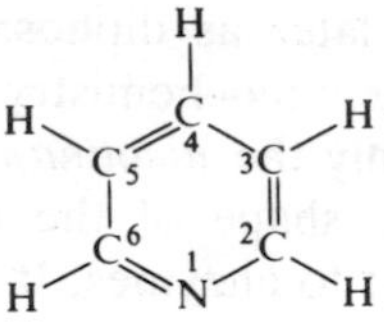

Pyridine ring

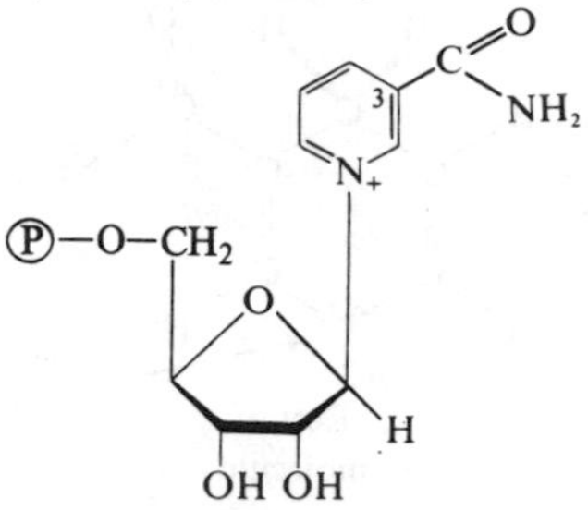

Nicotinamide
mononucleotide
(NMN)
(a)

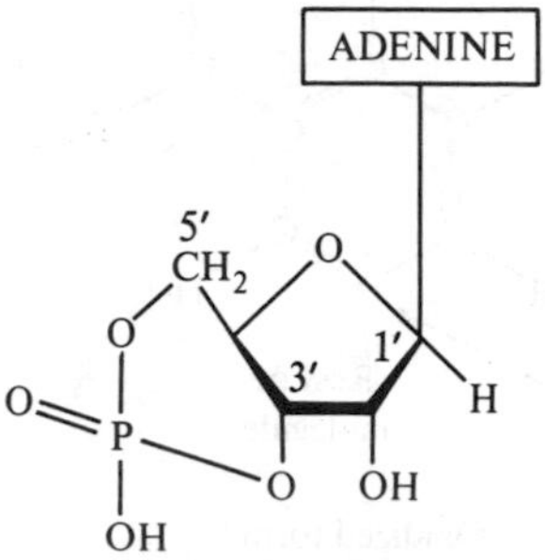

Cyclic AMP
(b)

Fig. 5.4 Structures of (a) nicotinamide mononucleotide and (b) cyclic AMP

5.2 Dinucleotides

Rather as two monosaccharides can be linked to form a disaccharide, two mononucleotides can be linked together by their phosphate groups to give biologically important compounds called dinucleotides:

BASE-RIBOSE—Ⓟ—Ⓟ—RIBOSE-BASE

First mononucleotide | Second mononucleotide

The most frequently encountered dinucleotide consists of the (mono) nucleotide, AMP, linked to the nucleotide which has nicotinamide as the heterocyclic base (Fig. 5.4). It is known as *nicotinamide adenine dinucleotide* (NAD) and it can be represented in the following way:

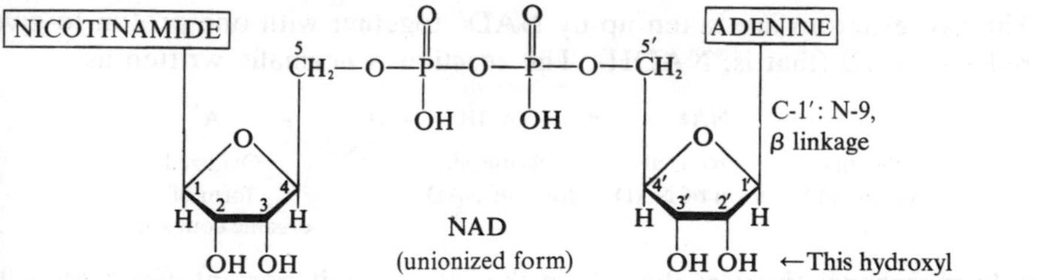

NAD
(unionized form)

NAD, originally known as coenzyme I and later as diphosphopyridine nucleotide (DPN), is of great importance in cell biochemistry because it participates in numerous oxidation reactions. Only the *nicotinamide* part of the molecule is directly involved, although the shape of the molecule is important in relation to stereospecific attachment to enzymes. It can exist in oxidized and reduced forms with the following structures:

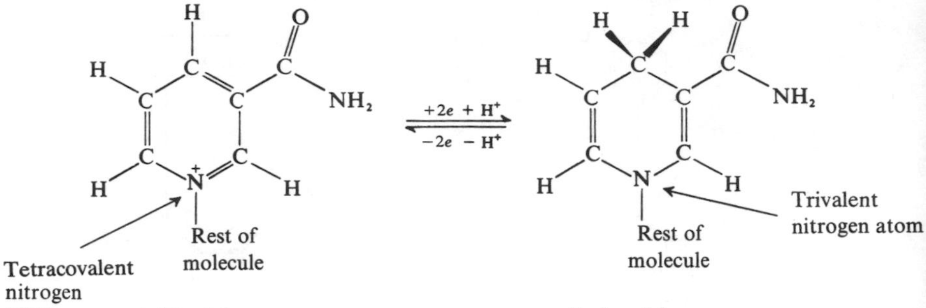

Tetracovalent nitrogen atom — Rest of molecule — Oxidized form

$\underset{-2e\ -\ H^+}{\overset{+2e\ +\ H^+}{\rightleftharpoons}}$

Rest of molecule — Trivalent nitrogen atom — Reduced form

Oxidized NAD is often written as NAD^+ because it has a positive charge on its pyridine tetracovalent nitrogen atom. (A somewhat similar relationship is shown by that of NH_4^+ to NH_3.) In the reduced form, NAD possesses two *electrons* and one *proton* more than the oxidized form and is often written as NADH:

$$NAD^+ + 2\ \text{electrons} + H^+ \rightarrow NADH$$

Numerous oxidation reactions occur in the cell and many of these require the cooperation of NAD. The mechanism can be explained briefly by considering a hypothetical organic compound AH_2. Its oxidation involves the removal of the two hydrogen atoms (i.e., two protons together with two electrons) and can be represented thus:

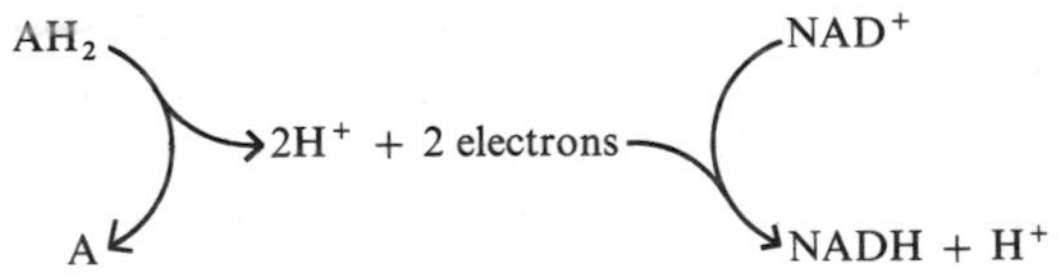

$AH_2 \rightarrow A$; $2H^+ + 2$ electrons; $NAD^+ \rightarrow NADH + H^+$

The two electrons are taken up by NAD^+ together with one proton to give reduced NAD (that is, NADH). The reaction is normally written as

AH_2	+	NAD^+	$\rightleftharpoons$	NADH	$+ H^+$	+	A
Organic compound		Oxidized form of NAD		Reduced form of NAD			Oxidized form of organic compound

In subsequent chapters the role in the cell of this important *coenzyme* will become evident. It is noteworthy that there is a similar dinucleotide which differs from NAD only in having the hydroxyl group on carbon atom 2′ of the ribose in the adenosine part of the molecule esterified by an additional phosphate group. It is known as nicotinamide adenine dinucleotide phosphate, or NADP. Like NAD, it is an electron carrier, and cooperates with enzymes responsible for oxidation and reduction of substrates.

The enzymes which have these dinucleotides as cofactors are in most cases specific for only one of them. The NAD-linked enzymes are very frequently those taking part in catabolic pathways. In most instances the NAD is reduced to NADH which then passes the two electrons associated with its reduction to the cytochrome chain in the mitochondria, so providing the energy for synthesizing ATP from ADP. The NADH is thus reoxidized to NAD^+ allowing a further oxidation of substrate AH_2. Several important examples of the catabolic dehydrogenases linked to NAD are found in the main pathway for catabolizing glucose (glycolysis), in the TCA cycle, and in the pathway for oxidizing fatty acids.

By contrast, it is often found that the enzymes which have $NADP^+$/NADPH as cofactor catalyse steps in pathways which have an anabolic function. For example, reducing power for the synthesis of long-chain fatty acid molecules is derived from NADPH, and the reduction of carbon dioxide in photosynthesis requires NADPH. There are two main ways in which NADPH is produced from NADP. One is part of the 'light' reaction of photosynthesis and the other is a special pathway for catabolizing glucose, known as the pentose phosphate pathway (Chapter 14).

Although there are several important exceptions to the generalization that NAD is the coenzyme of catabolism and NADP that of anabolism, it does appear that the two coenzymes have distinctly different roles. This is probably the basis for some enzymes being linked to one of the dinucleotides while others are specific for the second.

5.3 Structure of polynucleotides

Nucleoproteins are conjugated proteins (Section 3.1) comprising a basic protein united to a non-protein compound called a *nucleic acid*. The acid is loosely bound to the protein by electrovalent acid-base forces. Nucleic acids are large molecules formed by the union of tens, hundreds, or even thousands of mononucleotides (Section 5.1), the phosphate group of one nucleotide being attached to the sugar unit of the next.

Two types of nucleic acid exist, distinguished by the nature of the sugar units they contain and by a minor difference in the structure of one of the four nucleotide bases characteristically present. In addition, the two nucleic acids are differently located within the cell, DNA being largely in the nucleus whereas RNA, although present in the nucleus, is for the most part outside it. It should be added that extranuclear DNA also exists—it occurs, for example, in mitochondria and chloroplasts.

Although unusual nucleotides are occasionally present, each of the two nucleic acids typically hydrolyses to a mixture of four sorts of nucleotides, two of which contain purine bases and two pyrimidines. In both nucleic acids, the purine bases are adenine and guanine, but whereas RNA contains the pyrimidine bases cytosine and *uracil*, DNA contains cytosine and *thymine*. Formulae of the five bases are given in Figs. 5.1 and 5.2; it will be observed that thymine is, in fact, 5-methyluracil.

The second structural difference between DNA and RNA relates to the nature of the 'sugar' part of the mononucleotides from which they are constructed, for the mononucleotide units of RNA (*ribo*nucleic acid) contain the furanose form of ribose, whereas those of DNA (*deoxy*ribonucleic acid) contain deoxyribose (Section 5.1). The latter possesses two hydrogen atoms and has no hydroxyl group on carbon atom 2′ of the furanose ring (Fig. 5.5). In addition to these differences, structural differences of a secondary nature also exist, for whereas RNA is often single-stranded, DNA consists typically of two chains of DNA linked together to form a double helix.

It will be recalled that the dinucleotide NAD comprises two mononucleotide residues united by a pyrophosphate linkage (Section 5.2). Clearly, such a method of union cannot result in the formation of a molecule larger than a dinucleotide—the 'monomeric' units of a polynucleotide, like those of a polysaccharide, must unite in such a way that an indefinite repetition of the linkage is possible. In polynucleotides, the phosphate of one mononucleotide unites *with the* 3′-*hydroxyl group* of the sugar unit of the next mononucleotide in the sequence. The result is a phosphate 'bridge' leading from carbon atom 5′ of one nucleotide to carbon atom 3′ of the next (Fig. 5.5; nomenclature is discussed in Section 5.1).

Finally, it should be mentioned that, while adenine, guanine, cytosine, and thymine are by far the most common bases present in DNA, a small proportion of certain others may also be present. Perhaps the most important of these is 5-methylcytosine, which may account for 20 per cent of the nucleotide bases of plant DNA. Traces of this base are occasionally present in animal DNA. 5-Hydroxymethylcytosine is present in a virus, the T-2 bacteriophage, which attacks *Escherichia coli*. These methyl and hydroxymethyl groups are added after the synthesis of the DNA and they help to protect the DNA from enzymic degradation.

The purpose of polynucleotides is to act as carriers of *information*. The sequence of bases in a strand of chromosomal DNA ultimately determines the nature and quantity of the proteins, and therefore the enzymes, present in the cell. Thus, indirectly, DNA probably controls the composition of each of the components of the cell and thereby determines almost all of the activities of the cell from the moment it is formed until its death. In

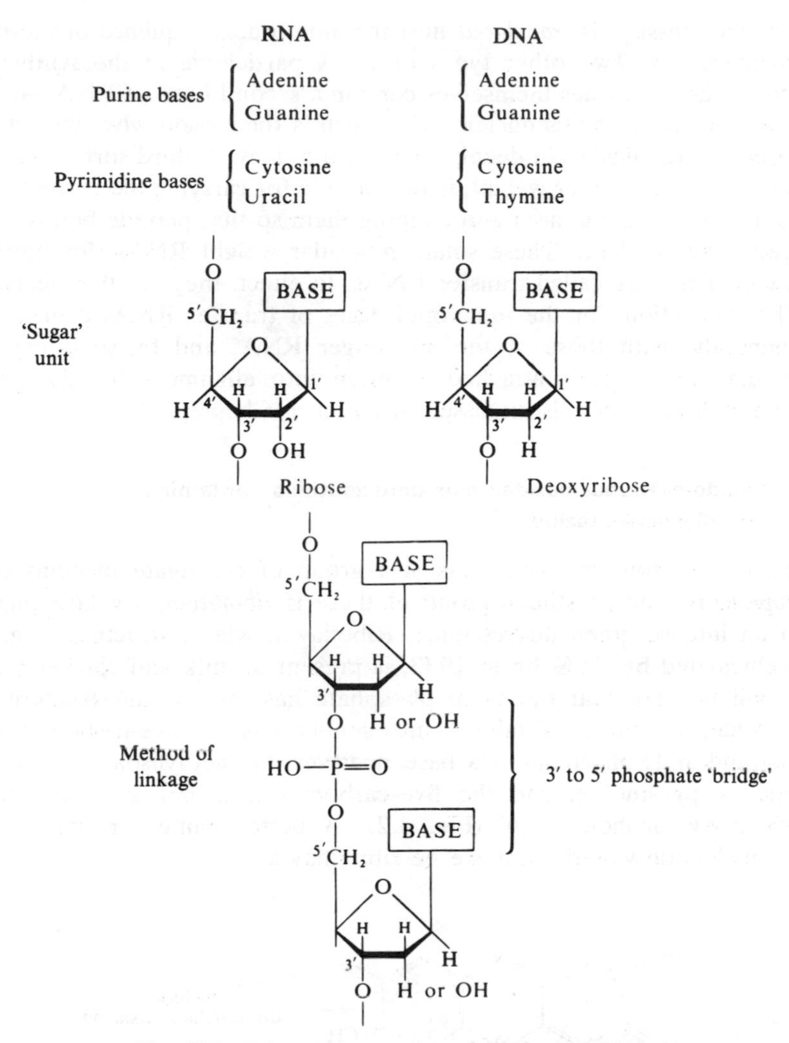

Fig. 5.5 Primary structure of RNA and DNA.

addition, of course, through the processes of mitosis or meiosis, the DNA determines the perpetuation of the characteristics of the species, without or with minor variation, from generation to generation.

DNA does not itself leave the cell nucleus of eukaryotic cells, yet most of the cellular protein is synthesized outside the nucleus. One sort of RNA acts as the connecting link. Machinery exists which *transcribes* the message contained within the molecule of DNA, imprinting it upon a molecule of messenger RNA. The latter moves out of the nucleus to the ribosomes, where the message is *translated* into the amino acid sequence of the newly formed protein. Two other types of RNA participate in the synthesis of protein. The ribosomes themselves contain a second form of RNA—indeed, it is the presence of this nucleic acid which is the reason why ribosomes or 'basophilic granules' stain deeply with certain dyes. A third sort of RNA, of much smaller molecular weight, is responsible for carrying the correct amino acids to the ribosomal area and aligning them so that peptide bonds can be forged between them. These small molecular weight RNAs (for there are many of them) are called transfer RNAs. In effect, they are the mechanism of the translation, for the functional bases of transfer RNAs complement, sequentially, with those of the messenger RNA, and by so doing draw particular amino acids attached to them into alignment for the protein synthesis. The process is discussed in detail in Chapter 16.

5.4 Pseudo-mononucleotides and dinucleotides containing dimethylisoalloxazine

Almost ubiquitous in living cells is a group of conjugate proteins called *flavoproteins.* The prosthetic group of these is *riboflavin*, a yellow pigment with an intense green fluorescence. Riboflavin, whose structure (Fig. 5.6) was elucidated by R. Kuhn in 1933, is present in milk and chicken's eggs.

It will be seen that riboflavin phosphate has the overall structure of a nucleotide, in that it contains a nitrogenous base, a five-carbon hydroxy compound, and a phosphate. The base, however, is dimethylisoalloxazine, not a purine or pyrimidine, and the five-carbon unit is not a sugar but the polyhydroxy alcohol, ribitol (Fig. 4.2). A better name for this pseudo-mononucleotide would therefore be ribitylflavin.

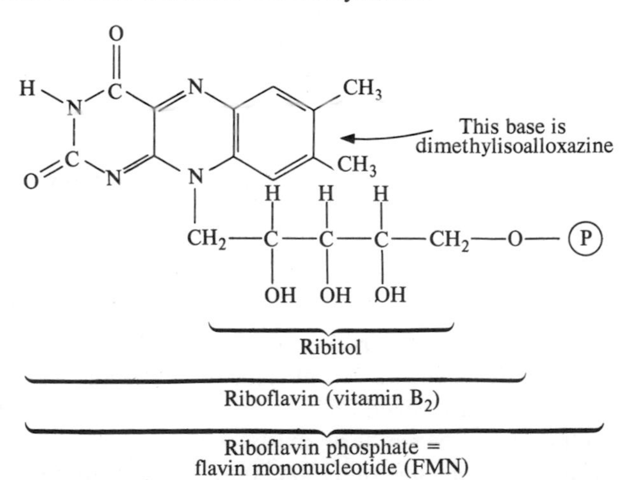

Fig. 5.6 Structure of flavin mononucleotide (FMN)

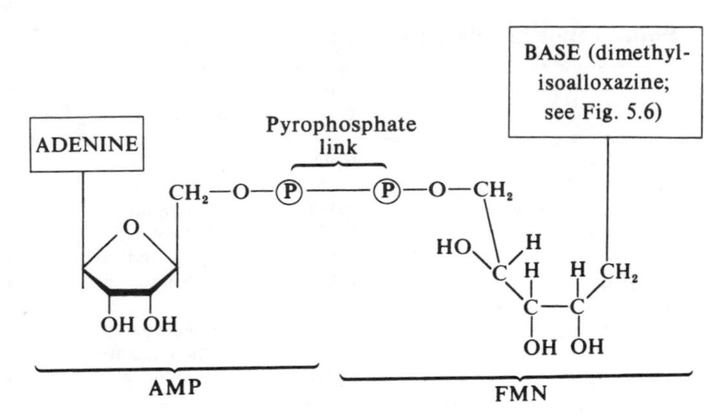

Fig. 5.7 Structure of flavin adenine dinucleotide (FAD)

The dimethylisoalloxazine base can be made by plants but not by animals. This means that riboflavin is an essential growth factor in animals, for dimethylisoalloxazine does not occur in any other common biological substance. Riboflavin is water soluble and was originally known as vitamin B_2. Like nicotinamide (Section 5.2) it undergoes reversible oxidation and reduction. Since it is not used up in this process, only a small intake is necessary to replace wastage.

Flavoproteins can be divided into two groups, according to the exact nature of the prosthetic group. In one group the riboflavin is in the form of its phosphate, i.e., the prosthetic group is FMN. On mild acid treatment, the FMN is more or less easily removed from the protein with which it is associated, indicating that the binding is probably not covalent. The second group of flavoproteins contains a pseudo-dinucleotide as the prosthetic group. This dinucleotide, flavin adenine dinucleotide, contains FMN united to adenosine monophosphate (Fig. 5.7).

A very important enzyme with a prosthetic group of this type is succinate dehydrogenase. In this enzyme, but not in all FAD-containing proteins, the prosthetic group is covalently attached to the protein. Some flavoproteins, including succinate dehydrogenase, contain, or work in close association with, heavy metal ions, particularly those of iron or molybdenum. Some important flavoproteins are listed in Table 5.2, which also shows the nature of the prosthetic group, the location or source of the enzyme, and refers to the chapters where the functions of many of them are further discussed.

Flavoproteins function as electron carriers, and, whether in FMN or FAD, the isoalloxazine is the part of the molecule which undergoes reversible oxidation and reduction. Like NAD (Section 5.2), the base exists in oxidized and reduced forms, the reduced isoalloxazine having two more electrons and two more protons—i.e., two more hydrogen atoms—than the oxidized form.

Table 5.2 Some important flavoproteins

Enzyme	Prosthetic group	Source	Chapter reference
NADH dehydrogenase†	FMN	Mitochondria	9
Succinate dehydrogenase†	FAD	Mitochondria	9
NADPH-cytochrome P_{450} reductase	FAD	Microsomes	14
NADPH-ferredoxin reductase	FAD	Chloroplasts	14
Lipoyl dehydrogenase	FAD	Mitochondria	11
Saturated fatty acyl CoA dehydrogenase	FAD	Mitochondria	10
L-Amino acid oxidase	FAD	Snake venom	
Glucose oxidase	FAD	*Aspergillus niger*	
L-Lactate dehydrogenase†	FMN	Yeast	11

† These enzymes also contain iron.

The difference between the two forms is shown in the shaded area of Fig. 5.8. It will be seen that reduction results in a decrease in the number of double bonds present (compare, for example, the hydrogenation of ethene to ethane or acetaldehyde to ethanol). The principle of the reaction is that the oxidized form of isoalloxazine accepts electrons from a more powerful reducing agent, AH_2, so oxidizing AH_2 to A. It then passes them on to an acceptor B, which is a stronger oxidizing agent than reduced isoalloxazine. The stronger oxidizing agent is thus reduced to BH_2 and reduced isoalloxazine is reoxidized. The prosthetic group of flavoprotein enzymes thus provides a 'shuttle service' of electrons from a more reduced (i.e., more oxidizable) substance to a more oxidized (i.e., more reducible) substance.

5.5 Coenzyme A

Coenzyme A carries *acyl groups*, and, by so doing, activates them. It plays a most important role in biochemistry, for it is a necessary cofactor both in fatty acid breakdown and (indirectly) in their synthesis (Sections 10.2 to 10.4).

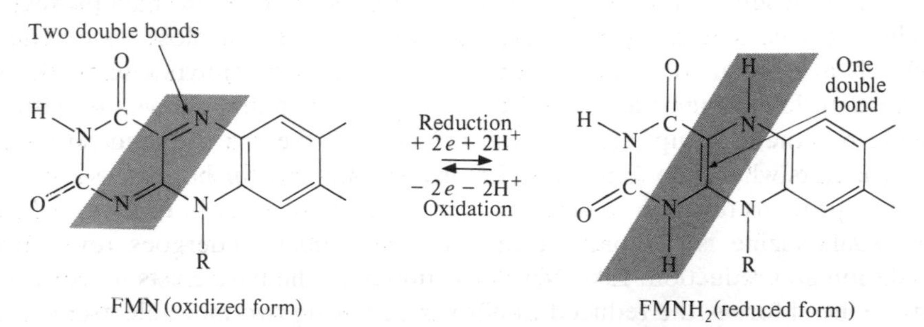

Fig. 5.8 Oxidized and reduced forms of dimethylisoalloxazine

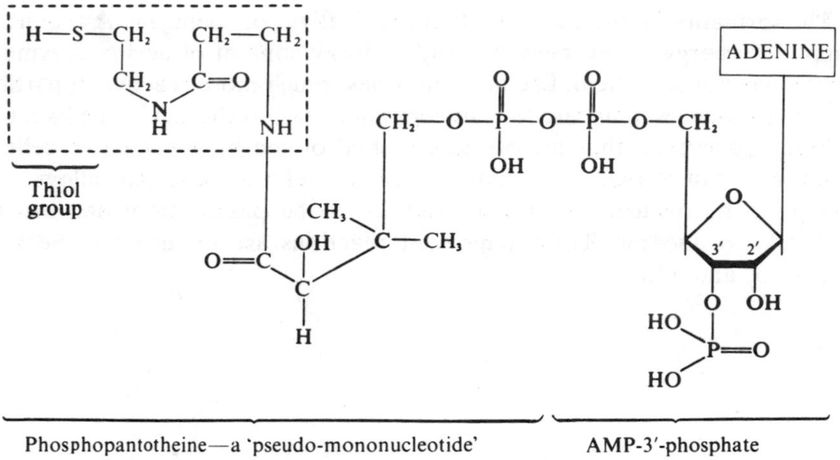

Phosphopantotheine—a 'pseudo-mononucleotide' AMP-3′-phosphate

Fig. 5.9 Structure of coenzyme A

It is involved at a vital phase of carbohydrate breakdown, enabling the product of aerobic oxidation of carbohydrates to enter the TCA cycle (Section 11.8). It also participates in one reaction of the TCA cycle itself (Section 9.6). Its structure is shown in Fig. 5.9. It invites the title of 'pseudo-dinucleotide', though more dubiously so than FAD. Half of the molecule is the mononucleotide AMP, except that an additional phosphate group is attached to ribose carbon atom 3′. The second half of the molecule consists of a nitrogenous base, but the central six-carbon moiety is not discernibly a ribose derivative.

Phosphopantotheine actually subdivides into three parts—phosphate, thioethylamine ($HSCH_2CH_2NH_2$), and a central portion called pantothenic acid. Animals cannot synthesize the pantothenic acid, so it has to be provided in the diet. It is needed only in small quantities and is therefore another example of a vitamin. Given pantothenic acid, animals can synthesize the rest of the coenzyme A molecule for themselves. 'Active acetate' was recognized to exist long before its structure was known, for free acetic acid does not readily participate in acetylation reactions. It is, in fact, *acetyl* coenzyme A. It is now known that many acyl groups, and not just acetyl, are carried in active form by coenzyme A.

The function of coenzyme A, then, is to unite with (low-energy) carboxylic acids to form 'high-energy' acyl derivatives which are analogous in certain ways to acyl chlorides:

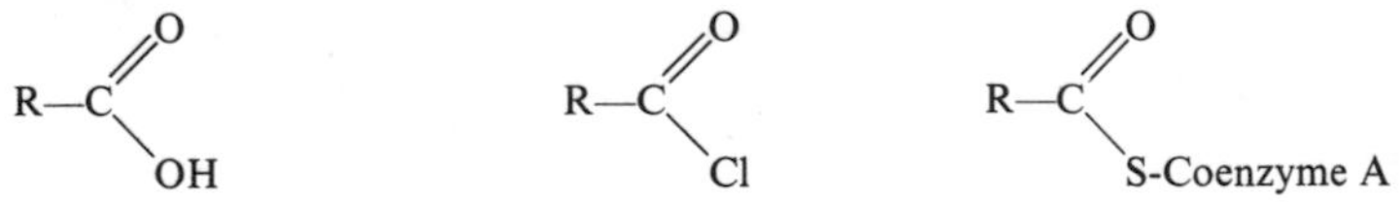

$R-C(=O)-OH$	$R-C(=O)-Cl$	$R-C(=O)-$S-Coenzyme A
Carboxyl group	Acyl chloride	Acyl coenzyme A

The formation in the cell of acylated derivatives of coenzyme A requires a supply of energy. Conversely, the high-energy content of acyl coenzyme A derivatives renders them, like acyl chlorides, much more reactive in particular circumstances than are the uncombined carboxylic acids. Unlike acyl chlorides, however, they are not readily hydrolysed by water under cellular conditions, but change only under the control of enzymes. This allows acyl groups to be oxidized in various ways or to be passed from one organic molecule to another. These important reactions are discussed in Sections 9.6, 10.3, and 11.8.

6

Enzymes

6.1 Enzymes as catalysts

Enzymes are catalysts formed by cells but capable of functioning *in vitro* if the conditions are appropriate. In the simplest case the reaction will occur if a solution of the enzyme is added to its *substrate* (the reactant), held at a suitable temperature and pH. Sometimes, however, as in other catalytic reactions, cofactors of various kinds may be essential, and, early in the history of biochemistry, it was discovered that the purification of an enzyme was frequently associated with diminution of enzymic activity owing to loss of the cofactors. For living cells enzymes are of vital importance for they accelerate a very large number of essential biochemical reactions. The cell must perform chemical reactions within a relatively narrow range of physical conditions—high temperature and extremes of pH are incompatible with the existence of the living cell. Enzymes thus enable reactions to occur under physiological conditions which would otherwise be unacceptably slow.

All enzymes so far isolated are proteins or polypeptides and are therefore of large molecular weight and do not dialyse. In common with other proteins, they are denatured by various agents and when denatured they lose their activity. They are often adversely affected by heat although a few can tolerate damp heat near the boiling point—heat lability is one important characteristic of enzymes which distinguishes them from most inorganic catalysts. A second difference is the *specificity* enzymes usually show for the *substrate* they act upon—by contrast, inorganic catalysts often accelerate several quite different reactions. Platinum, for example, catalyses numerous oxidation reactions including the contact process, the catalytic oxidation of ammonia, and the combustion of hydrogen in oxygen. Enzymes are often named from the substrate they act upon, the ending *-ase* being added to the root of the substrate's name (e.g., RNA-ase, lipase). Sometimes the name also incorporates the *type* of reaction involved (e.g., succinate dehydrogenase, cytochrome oxidase).

Enzymes can often be prepared to a high degree of purity; indeed, some 250 enzymes have been produced in pure crystalline form since J. Sumner first crystallized *urease* in 1926. For these enzymes, it is possible to express enzyme activity in terms of the weight of protein which converts 1 μmol of the natural substrate in 1 min under the most favourable conditions. This is known as the unit of enzyme activity. The *specific activity* of an enzyme preparation (pure or impure) is then defined as the number of units, with respect to a particular substrate transformation, per milligram of *total* protein present. Clearly, at each stage of purification of an enzyme, the specific activity will rise as extraneous protein is removed. Assuming any necessary cofactors are added to the monitoring system, the activity will rise to a maximum when all the protein present in the preparation is the protein of the enzyme. The *turnover number*, or *molecular activity*, of an enzyme is the number of substrate molecules changed per enzyme molecule each minute when excess substrate and any necessary cofactors are present. For example, a molecule of carbonic anhydrase converts 36×10^6 molecules of carbon dioxide to carbonic acid per minute. Similarly, the molecular activity of catalase (hydrogen peroxide to water and oxygen) is 5.6×10^6 and that of succinate dehydrogenase is 1.1×10^3.

Many enzymes are simple proteins but others are conjugate, having a non-protein group more or less closely associated with the protein *apoenzyme*; the total complex is termed the *holoenzyme*. When the non-protein part of the complex is firmly bound, it can be regarded as an integral part of the structure and is then called a *prosthetic* group, whereas, if it is very loosely bound, it can more properly be considered as a separate entity, termed a *coenzyme*. In practice, however, there are intermediate examples, and no sharp dividing line exists. Many coenzymes are of relatively low molecular weight and can be dialysed away from the protein apoenzyme, whereas true prosthetic groups remain attached to the apoenzyme.

In addition, many enzymes require inorganic anions or cations to be present. The enzyme amylase, present in saliva, is activated by chloride ions, the *kinase* enzymes usually require the presence of magnesium ions, and *alcohol dehydrogenase* requires zinc ions. Other examples are to be found in Section 6.10.

6.2 Extraction, concentration, and purification of enzymes

Some enzymes are firmly bound to mitochondrial, chloroplastic, or endoplasmic reticular membranes and, in consequence, are often difficult to separate. The majority, however, form colloidal solutions in water and can be more or less readily separated and concentrated by the same methods as are used for other proteins. Originally, the process of separation was often long and arduous, based upon differential solubility, in various solutions, of the enzyme under study and the numerous inert proteins present as contaminants. In the last two or three decades, however, chromatographic and electrophoretic techniques have greatly simplified purification, and some 1500 enzymes have been prepared in pure or nearly pure form.

The first step in the extraction of a potentially soluble enzyme is often to homogenize the tissue containing it and to centrifuge to remove insoluble matter. In some cases, brief *heat treatment* (50°C) of the resulting colloidal solution coagulates heat-labile unwanted proteins and so results in further concentration. A historically important process, still often valuable, is to *salt out* either impurities or the enzyme by addition of an appropriate amount of a salt containing a divalent anion (e.g., ammonium sulphate) or a divalent cation (e.g., magnesium chloride). This process causes precipitation because the ions of the electrolyte become hydrated, thus decreasing the amount of solvent available to dissolve the colloids present. Alteration of the pH of a mixture of proteins also causes *differential precipitation*, for the solubility of each protein component of the mixture is least at its isoelectric point (Section 3.2). A third general method often used to achieve initial purification is to make use of the different solubilities of different proteins in aqueous solutions to which varying amounts of such water-miscible organic solvents as ethanol or acetone have been added.

Following initial purification, various forms of chromatography and electrophoretic systems can often prove valuable. An example of the separation of proteins by *electrophoresis* has already been given in Section 3.1, and since enzymes are proteins the technique is equally applicable to the purification of enzymes. The method of *ion exchange chromatography*, described in Section 3.4 in relation to the separation of amino acids, can also be adapted to separate proteins. For example, acidic proteins can be chromatographed on basic ion exchange materials such as diethylaminoethyl cellulose (DEAEC) and basic proteins can similarly be chromatographed on columns with a negatively charged stationary phase such as carboxymethyl cellulose.

Another technique is called *gel filtration.* It can be used not only to separate proteins of different molecular weights but also, after calibration of a column (using substances of known molecular weight), can provide an estimate of the apparent molecular weight of an enzyme of unknown molecular constitution. Common packings in columns used for gel filtration are cross-linked dextran gels (e.g., 'Sephadex®') and polyacrylamide gels. The tiny granules, or beads, of the packing behave rather like dialysis membranes (page 41) in that water and smaller molecules can penetrate the granules but large molecules cannot. Moreover, 'small' and 'large' are relative terms, in that granules can be manufactured with different porosity or '*molecular sieve*' characteristics. The result is that whereas water outside *and* inside granules is effectively available to smaller molecules only the water outside granules is available to larger ones. In consequence, *larger* molecules move down a 'Sephadex' column faster than smaller ones and so

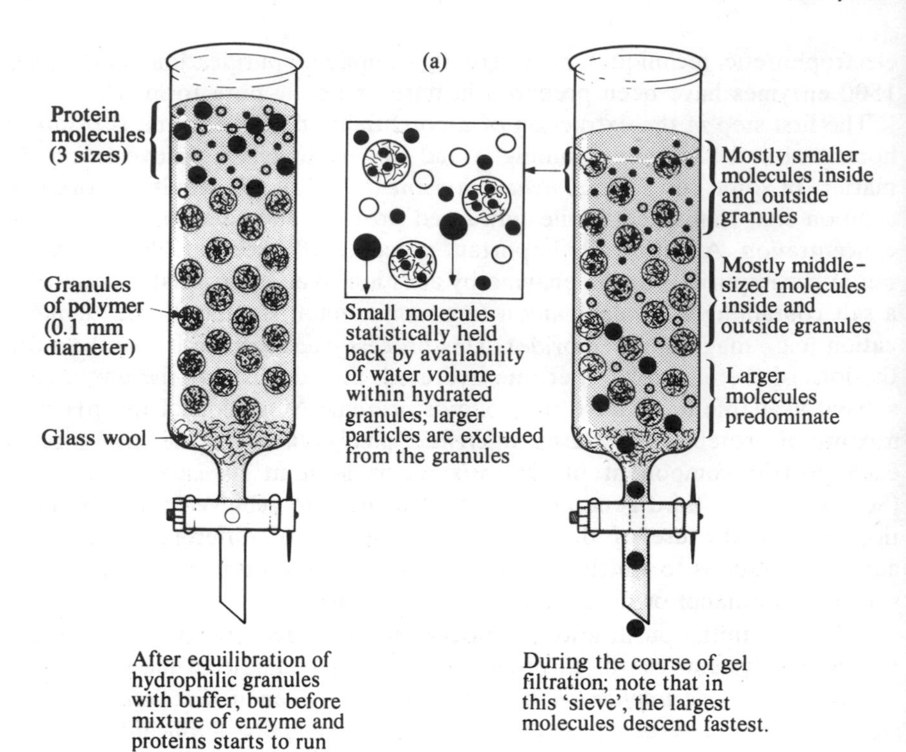

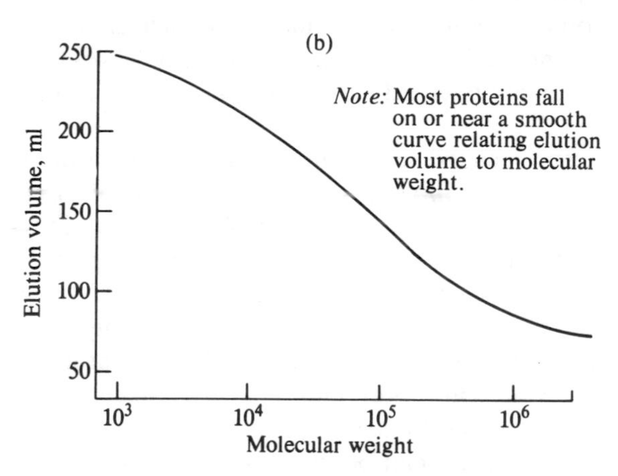

Fig. 6.1 (a) The gel filtration method for separating proteins and purifying enzymes. (b) Relation of elution volume to molecular weight

emerge from the base of the column first (Fig. 6.1). When, as in enzyme purification procedures, several proteins are present, molecules of lower and lower molecular weight emerge in succession as elution volume increases. Using a fraction separator and an appropriate assay technique for the enzyme in question, a considerable purification is often possible.

6.3 Classification of enzymes

Over the years, the naming of enzymes developed in a somewhat haphazard manner, but usually in accordance with the principle mentioned in Section 6.1, namely, that the substrate for the reaction provides, directly or indirectly, the stem of the name of the enzyme. Enzymes can often be grouped into families, united by the common factor of the type of substrate they attack; there are, for example, numerous proteinases (e.g., pepsin), numerous esterases, and numerous carbohydrases (e.g., sucrase and maltase). Rather similarly, the dehydrogenases all remove electrons and protons (often apparently in the form of hydrogen atoms) from an oxidizable substrate and pass them on to NAD or some other carrier, while phosphatases remove phosphate groups from phosphate esters. Yet others, such as the aminotransferases and acyl transferases, form families named according to the nature of the groups they transfer.

As more and more enzymes were isolated and an increasing insight was obtained into the mechanism of enzyme action, it became apparent that many of the names given to the earlier-discovered enzymes were in various ways inadequate. Nevertheless, well-established names die hard, especially when they have the advantage of being reasonably short; amylase, decarboxylase, hexokinase, even succinate dehydrogenase, are names which are

Table 6.1 Classification of enzymes into six main groups

1. *Oxidoreductases*	These add or subtract electrons, oxygen, or hydrogen; e.g., oxidases, dehydrogenases.
2. *Transferases*	These transfer a group from one organic molecule to another; e.g., methyl group transferases, acyl group transferases, phosphotransferases, aminotransferases.
3. *Hydrolases*	These split a molecule in two by the action of water; e.g., those acting on ester bonds, on polysaccharide linkages, on peptide linkages.
4. *Lyases*	These remove groups non-hydrolytically, leaving a double bond, or, in reverse, add groups to double bonds; e.g., decarboxylases, citrate synthase, aldolase.
5. *Isomerases*	These bring about a redistribution of atoms, or groups of atoms, within a molecule; e.g., epimerases, mutases.
6. *Ligases*	These link together two molecules, always at the expense of a high-energy compound, usually ATP; e.g., the enzymes which unite amino acids to their specific s-RNAs, the enzymes leading to acyl coenzyme A by uniting the free acid and coenzyme A.

Table 6.2 Systematic classification of enzymes, illustrating, for a few representativ examples, the unique numbering system

Number	Trivial name	Reaction	Notes
1 OXIDOREDUCTASES			
1.1.1.27	Lactate dehydrogenase	Lactate + NAD^+ = pyruvate + NADH + H^+	Cofactor: Zn
1.1.1.49	Glucose-6-phosphate dehydrogenase	Glucose-6-P + NADP = glucono-δ-lactone-6-P + NADPH + H^+	Acts in pentose-P pathway
1.2.4.1	Pyruvate dehydrogenase	Pyruvate + oxidized lipoate = acetylhydrolipoate + CO_2	Requires thiamine pyrophosphate
1.4.1.3	Glutamate dehydrogenase	Glutamate + NAD^+ + H_2O = α-ketoglutarate + NH_3 + NADH + H^+	Cofactor may be NAD or NADP
2 TRANSFERASES			
2.2.1.1	Transketolase	Sedoheptulose-7-P + glyceraldehyde-3-P = ribose-5-P + xylulose-5-P	Contains thiamine pyrophosphate and Mg^{++}
2.2.1.2	Transaldolase	Sedoheptulose-7-P + glyceraldehyde-3-P = erythrose-4-P + fructose-6-P	This and transketolase are involved in pentose-P cycle
2.4.1.11	Glycogen synthase	UDP-glucose + $(glycogen)_n$ = UDP + $(glycogen)_{n+1}$	Involved in glycogen synthesis
2.6.1.1	Aspartate aminotransferase	Aspartate + α-ketoglutarate = oxaloacetate + glutamate	Contains pyridoxal-P
2.7.1.1	Hexokinase	ATP + glucose = glucose-6-P + ADP	Contains Mg^{++}, takes part in glycolysis
3 HYDROLASES			
3.1.1.3	Lipase	Triglyceride + H_2O = diglyceride + fatty acid	Also hydrolyses di- and monoglyceride
3.1.1.7	Acetylcholinesterase	Acetylcholine + H_2O = choline + acetate	Plays important part in neuromuscular transmission
3.2.1.1	α-Amylase	Hydrolyses starch to maltose	Contains Ca^{++}
3.5.1.5	Urease	Urea + H_2O = CO_2 + $2NH_3$	Not involved in the *synthesis* of urea
4 LYASES			
4.1.1.32	Phosphoenolpyruvate carboxylase	GTP + oxaloacetate = GDP + phosphoenol pyruvate + CO_2	Important in gluconeogenesis
4.2.1.2	Fumarate hydratase	L-Malate = fumarate + H_2O	Takes part in the TCA cycle
5 ISOMERASES			
5.1.3.1	Ribulosephosphate 3-epimerase	Ribulose-5-P = xylulose-5-P	Takes part in pentose-P cycle
5.1.3.2	UDP-glucose epimerase	UDP-glucose = UDP-galactose	Contains NAD
6 LIGASES			
6.2.1.2	Acyl-CoA synthetase (fatty acid thiokinase)	ATP + fatty acid + CoA = AMP + pyrophosphate + acyl CoA	Activation of fatty acids
6.4.1.1	Pyruvate carboxylase	ATP + pyruvate + CO_2 + H_2O = ADP + P_i + oxaloacetate	Contains biotin; important in gluconeogenesis

fairly easily memorized. Short names are, however, sometimes misleading, especially when, in addition to enzyme and substrate, some obligatory third substance actively participates in the reaction. For this reason, the Enzyme Commission of the International Union of Biochemistry, in 1964, devised a scheme of classification which brought considerable order to the rather chaotic situation then existing.

The Enzyme Commission allocated a unique four-part number and a systematic name to each enzyme. To begin with, all enzymes belong to one or other of six main groups, so the four-part number starts with one of the numbers 1 to 6. The principal characteristics of the enzymes in these six groups are shown in Table 6.1.

Each of these six groups is then subdivided according to the nature of the linkage being attacked or the group being transferred. These subgroups are then further subdivided. Individual enzymes are then designated by a fourth number.

This system is illustrated by reference to an enzyme in group 3 of Table 6.1. Acetylcholinesterase is an enzyme which splits acetylcholine into the base, choline, and acetic acid. It is therefore a hydrolase (group 3). But hydrolases are themselves of different kinds: some hydrolyse ester linkages, some hydrolyse glycoside linkages (Chapter 4), others hydrolyse peptide linkages (Chapter 3), and so on. Acetylcholinesterase is in subgroup 3.1, which contains enzymes which act only on *ester* linkages. Ester linkages are themselves of several types: carboxylic acid esters, phosphate esters, etc. Acetylcholinesterase splits the carboxylic ester linkage of *acetyl* choline [$CH_3COOCH_2CH_2N^+(CH_3)_3$] and so is assigned to the subgroup 3.1.1. Each enzyme capable of splitting various carboxylic acid esters is allotted a different number—the fourth number of the four-tier numbering system—so that each is uniquely defined. The full designation for acetylcholinesterase is 3.1.1.7, and its systematic name is acetylcholine acetylhydrolase. Table 6.2 provides a few examples of enzymes classified according to the Enzyme Commission's numbering system, together with the trivial (common) name of the enzyme, the reaction it catalyses, and cofactors needed for the reaction to proceed.

6.4 The specificity of enzymes

It is possible to subdivide enzymes into groups according to the degree of specificity they exhibit. The most striking example of specificity is demonstrated by enzymes exhibiting an *absolute specificity* for one substrate. *Urease*, for example, catalyses the reaction:

$$\underset{\text{Urea}}{H_2O + NH_2CONH_2} \rightarrow CO_2 + 2NH_3$$

Sometimes, and particularly when the reaction involves substrate oxidation, it is found that an absolute dependence also exists for a *second* substance, such as one which accepts electrons from the substrate. The acceptor is, in a sense, a secondary substrate, since its molecules are aligned on the enzyme surface, just as are the molecules of the substrate.

One such example is *glucose-6-phosphate dehydrogenase.* This enzyme removes two electrons and two protons from glucose-6-P to give phosphogluconic acid lactone. The reaction only occurs, however, if $NADP^+$ is present as an electron acceptor:

$$\text{Glucose-6-P} \xrightarrow[\text{NADP}^+ \rightarrow \text{NADPH} + \text{H}^+]{\text{Glucose-6-phosphate dehydrogenase}} \text{6-Phosphogluconolactone}$$

Under biological conditions this reaction (Section 14.6) is virtually irreversible, but similar reactions exist which can proceed, according to the conditions, in either direction. *Lactate dehydrogenase,* for example, removes two electrons and two protons from lactate to form pyruvate, or alternatively can reduce pyruvate to lactate. The coenzyme is $NAD^+/NADH$:

$$CH_3-CH(OH)-COOH \underset{\text{NAD}^+ \;\; \text{NADH} + \text{H}^+}{\overset{\text{Lactate dehydrogenase}}{\longleftrightarrow}} CH_3-C(=O)-COOH$$

The manner in which $NADP^+$ and NAD^+ are able to accept electrons was described in the previous chapter.

In the example above, the double-headed arrows imply that the reaction is reversible. Such reversibility can lead to a semantic difficulty, for, if the enzymic reduction of pyruvate is encountered in isolation, it might appear as though the enzyme responsible, lactate dehydrogenase, had been named incorrectly by reference to the product rather than to the reactant.

Lactate dehydrogenase of animal muscle is one of many enzymes which exhibit *stereochemical specificity,* for it only oxidizes L(+)-lactic acid. More strikingly, perhaps, when acting in reverse, it reduces pyruvic acid only to L(+)-lactic acid, whereas chemical synthesis by addition of hydrogen to the carbonyl double bond always gives the racemic mixture. Similarly, it is usual for only one of a pair of geometrical isomers to be formed by enzymic action. *Succinate dehydrogenase,* for example, converts fumarate to succinate, but has no action on the corresponding *cis*-isomer, maleic acid:

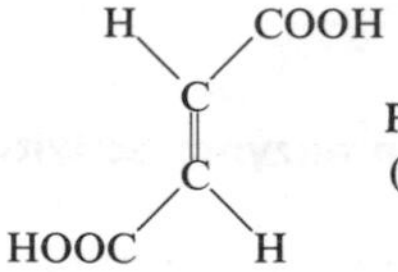

H(COOH)C=C(H)HOOC — Fumaric acid (*trans*-isomer)

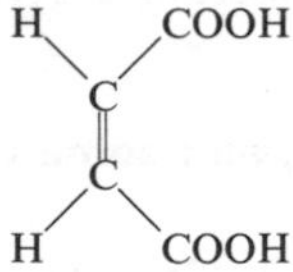

H(COOH)C=C(H)COOH — Maleic acid (*cis*-isomer)

$$\text{FAD} + \begin{matrix}CH_2\text{—COOH}\\ |\\ CH_2\text{—COOH}\end{matrix} \underset{\text{dehydrogenase}}{\overset{\text{Succinate}}{\rightleftharpoons}} FADH_2 + \text{HOOC—CH=CH—COOH}$$

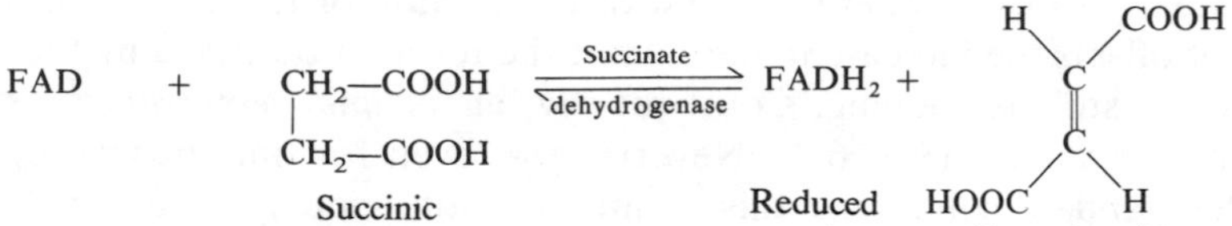

Succinic acid; Reduced carrier; Fumaric acid

A further type of specificity shown by some enzymes is known as *group specificity*. As the name implies, the enzymes catalyse reactions involving a series of substrates which have in common one identical group but differ in some other way. For example, *α-glucosidase* hydrolyses several α-glucosides:

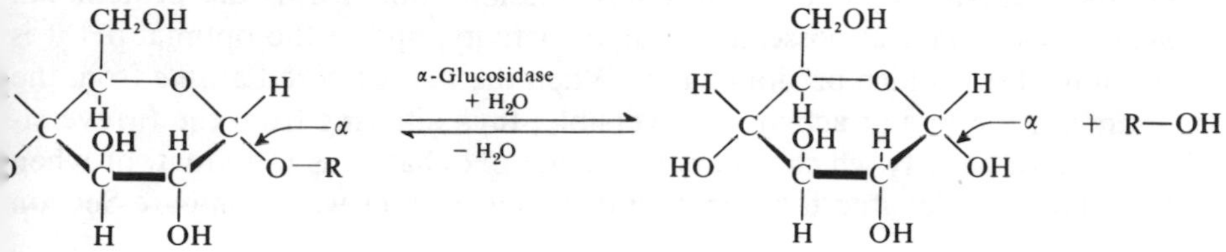

$$\text{R-}\alpha\text{-Glucoside} \underset{-H_2O}{\overset{\alpha\text{-Glucosidase} + H_2O}{\rightleftharpoons}} \alpha\text{-Glucose} + \text{R—OH}$$

R-α-Glucoside; α-Glucose; Alcohol or sugar

In each case a hydroxyl compound (R—OH) such as methanol, ethanol, or glucose is removed from the α-glucoside, the other product in each case being glucose. The enzyme maltasc, which splits maltose into two molecules of glucose, is an example of an α-glucosidase, for maltose is an α-glucoside (Section 4.5). But in addition to hydrolysing maltose, maltase will attack a range of α-glucosides, though it attacks each at a different rate, indicating that the affinity of the enzyme with each substrate is different. On the other hand, maltase does not hydrolyse cellobiose because this sugar is a β-glucoside.

A final group of enzymes comprises members of *low specificity*; here, the only requirement from the substrate is that is should possess a particular linkage. Thus certain esterases will attack a wide range of compounds of the type R—COO—R′, where both R and R′ can be varied considerably. Lipase, for instance, will hydrolyse (at different rates) most of the natural fats—if this were not so, there would have to be a specific lipase for each of the several dozen different glycerol–carboxylic acid linkages which exist in dietary fats and oils (Section 2.1).

6.5 Effect of heat, pH, and some chemical agents on enzyme activity

Enzyme-catalysed reactions, like all others, proceed more rapidly as the temperature is raised, but since enzymes are denatured by heat, a point is eventually reached where the rate of destruction of the enzyme is so great that it offsets the increase in the rate of the reaction catalysed by the enzyme which is still functioning. Consequently, an *optimal temperature* exists for enzyme reactions (Fig. 6.2). Nevertheless, even for one enzyme operating under standard conditions, this optimum is not uniquely fixed, for the factor of time must also be taken into account. In reactions lasting only a few seconds, for example, the optimum might be 20 to 30°C higher than when the same enzyme-catalysed reaction is studied over a longer period of time.

The catalytic efficiency of any enzyme is markedly affected by the pH of the surroundings, the activity rising rapidly to a maximal value at the *optimal pH* and then falling off again as the pH is increased further (Fig. 6.2). This occurs because only one of the many possible ionic forms the protein can assume (Section 3.2) possesses catalytic activity, and at the optimal pH it is this ionic form which predominates. When the pH is altered a *little* from the optimum, the loss of activity is reversible, thus differing from the (irreversible) loss of activity which occurs when the pH change is too great, or when the enzyme is denatured by heating or in some other way (compare Section 3.10).

Since enzymes are proteins they may be inactivated by numerous chemical reagents which coagulate, precipitate, and denature proteins. There is,

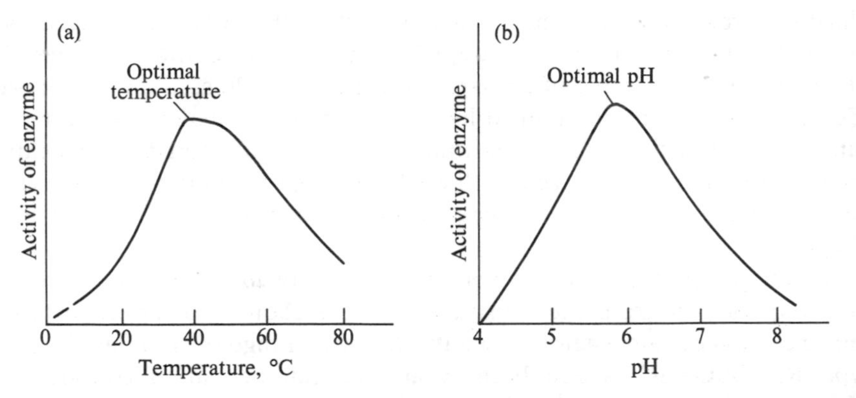

Fig. 6.2 The effect (a) of temperature and (b) of pH on the activity of an enzyme. The actual optimal values vary with the enzyme concerned

Table 6.3 Some enzyme inhibitors

Substance	Site of attack	Example of enzyme
Arsenic compounds	Phosphate groups	Phosphorylating enzymes
Azide	Ferric porphyrins	Cytochrome oxidase
Carbon monoxide	Ferrous porphyrins; Cu enzymes	Cytochrome P_{450}
Cyanide	Ferric porphyrins; Cu enzymes	Peroxidase
Diisopropylfluoro-phosphonate (DFP)	OH group of serine	Cholinesterase trypsin, chymotrypsin
Dithiocarbamates	Cu-dependent enzymes	
EDTA	Metallic ion cofactors	Arginase; Amylase
Fluoride	Mg- and Ca-dependent enzymes	Enolase
Heavy metallic ions (e.g., lead)	SH groups	δ-Aminolaevulinic acid dehydratase
Iodoacetate	SH groups, imidazole	Several dehydrogenases
Organomercury	SH groups	Several dehydrogenases
Oxalate	Mg- and Ca-dependent enzymes	Pyruvate carboxylase
Sulphide (e.g., H_2S)	Ferrous porphyrins	Cytochrome oxidase

however, a number of more subtle ways by which enzymes may be *inhibited*—that is to say, partially or completely put out of action at least temporarily, and sometimes permanently.

An irreversible inhibitor may unite with the site on the enzyme surface normally occupied by the natural substrate. Alternatively, it may act elsewhere than this site, but in such a way as to alter the properties of the protein to such an extent that it no longer functions as an enzyme. An example of the former is *iodoacetate,* which unites with the crucially important thiol groups (—SH) which form part of the active sites of many dehydrogenases (ESH):

$$ESH + ICH_2COOH \longrightarrow ESCH_2COOH + HI$$

In those cases where attachment occurs away from the active site it usually causes a change in the topography of the enzyme molecule of such a nature as to hinder or prevent enzyme-substrate complex formation. Some important examples of enzyme inhibitors are listed in Table 6.3. It will be seen that often a thiol (sulphydryl) group is attacked. In other cases metallic ions which are present in many enzymes, and which are essential for their activity, form more or less stable complexes with various inhibitors. For example, *fluoride* forms a complex with magnesium ions and in consequence inactivates enolase (Section 11.7). Similarly, carbon monoxide, by attacking iron in some haem groups, can inactivate mitochondrial cytochrome oxidase (Section 8.7) and cytochrome P_{450} on the endoplasmic reticulum (Section

14.10). It should be noted that unlike general protein precipitation reagents (Section 3.10), the inhibitors just described are often highly specific with regard to the enzymes that they attack.

Reversible inhibition of enzyme action is also well known and can take several forms. One of these is termed *competitive inhibition.* It sometimes happens that a compound exists which is sufficiently like the natural substrate in structure for it to be able to take the place of that substrate on the enzyme surface. By simply occupying the active site it prevents the natural substrate molecules from attaching themselves. It does not damage the enzyme, and only forms a *loose* complex with it. Consequently, if a large amount of natural substrate is added, mass action considerations ensure that the inhibitor can be replaced. Thus the level of competitive inhibition varies with the concentration of natural substrate present.

The most famous example of competitive inhibition is the effect of malonate on the activity of succinate dehydrogenase, the enzyme which catalyses the formation of fumarate from succinate. Malonic acid (HOOC-CH_2COOH), like succinic acid, is a dicarboxylic acid, but is one methylene group smaller; it becomes attached to the enzyme, blocking sites normally occupied by succinate. Its effect is illustrated by the following equations, where S stands for succinate, M for malonate and A and B are the products of the reaction catalysed by the enzyme, E:

$$E+S \rightleftharpoons ES \rightarrow A+B+E$$
$$E+M \rightleftharpoons EM \text{ (inactive; no further action)}$$

Since S and M are clearly competing for a limited amount of enzyme, the concentration of ES and hence the rate of formation of A and B depends on the dissociation constants of ES and EM and upon the relative concentrations of S and M. A and B in this instance are $FADH_2$ and fumaric acid. In fact, the dissociation constants of ES and EM are such that the enzyme is 50 per cent inhibited when the concentration of malonate is about 2 per cent that of succinate.

In contrast to the competitive inhibition just discussed, other forms of reversible inhibition (of which at least two are known), as well as irreversible inhibition, are independent of the concentration of the enzyme's natural substrate. For these, the level of inhibition depends only on the concentration of inhibitor present. The subject is discussed further in Section 6.8.

6.6 How enzymes accelerate reactions

The function of enzymes is to hasten attainment of the equilibrium state. Without them, many cellular reactions would occur too slowly to support life. For a reaction to be possible, it is necessary for reacting molecules to possess a certain minimal energy, E_A, the activation energy, which is an attribute of a particular chemical reaction occurring by a particular reaction mechanism. When molecules of potential reactants which possess less than this minimal energy are brought together, they fail to react; indeed, for most reactions, only a very small proportion of the total number of collisions between molecules of reactants results in a chemical change.

Possession of energy in excess of E_A is a fundamental requirement if molecules are to have a *chance* of reacting, but, even then, reaction may not occur, for it is also necessary that the molecules of each reacting species should be appropriately orientated with respect to the other when they come together. Enzymes probably affect the rate of biochemical reactions both by enabling more molecules to overcome the energy barrier for the reaction and by increasing the probability of correct orientation at the moment of 'collision'.

Mathematically, activation energy is related to the rate of reaction by the Arrhenius equation:

$$\text{Rate} = PZe^{-E_A/RT} \qquad \text{or} \qquad \log_{10}\text{rate} = \log_{10}PZ - \frac{E_A}{2.303RT}$$

where e is 2.72 (the base of natural logarithms), R is the gas constant (8.3 J/deg mol), T the absolute temperature, Z the frequency of molecular collision, and P is the probability of successful collision (maximal value, 1.0). When orientation improves, P will tend to rise; Z is related, among other things, to the concentration of the reactants. The exponential term shows that a small decrease in E_A has a large effect on reaction rate. For example, if E_A falls from 42 to 21 kJ (P and Z factors remaining unchanged), it is clear that the reaction will accelerate by approximately a hundredfold.

Enzymes enable reactions to bypass high activation energy barriers by *re-routing* reactions; re-routed reactions are essentially different reactions, with their own activation energies, even though both the substances disappearing and the substances being formed are the same as when no enzyme is present. This is usually achieved by reactant molecules forming an intermediate complex with the enzyme so that they are suitably placed and in an appropriate state of electron activation for the reaction to occur.

An analogy which illustrates some aspects of activation may be helpful so long as it is recognized that no analogy is perfect. Water will not flow spontaneously from one beaker to another placed beneath it, because the walls of the upper vessel act as a barrier. However, the barrier can be overcome by several means; an obvious one would be to put marbles into the beaker so as to raise the water level. Such a method affects all the water just as *heating* a reaction mixture raises the average kinetic energy of the molecules in the system. A more efficient way to move the water is to use a siphon; this causes water to flow without the necessity of raising the level

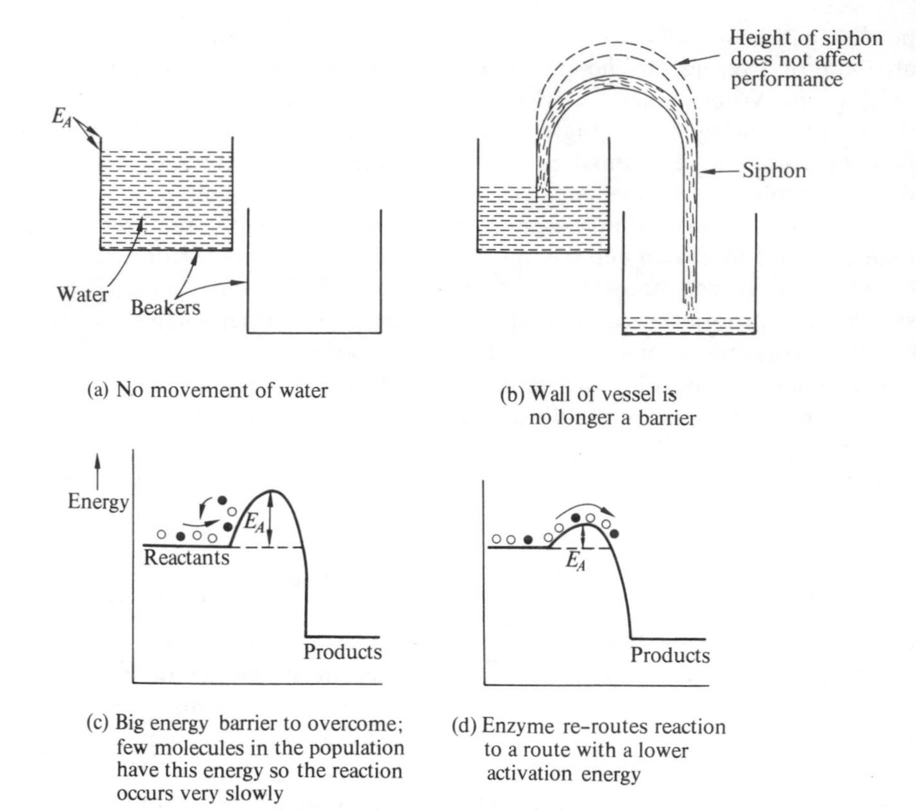

Fig. 6.3 **An analogy illustrating the meaning of potential energy and 'activation' in a physical and a chemical system**

(i.e., the potential energy) of the whole mass of water, i.e., only the water in the siphon is 'activated'.

An enzyme no more affects the energy change occurring when reactants form products (Fig. 6.3) than the siphon affects the relative height of the two beakers—*the enzyme, like the siphon, merely organizes a transitional stage.* Water stops flowing through the siphon as soon as the level of water is the same in the two beakers. Similarly, a reaction catalysed by an enzyme effectively ceases when chemical equilibrium is reached. The analogy can be taken a step further, for if water is constantly removed from the lower beaker to ensure that the water levels do not equalize, water will continue to flow through the siphon. So, too, an enzyme continues to catalyse a reaction if the products of the reaction are constantly removed by the intervention of some additional biochemical system.

It should be noted that the magnitude of the activation energy has nothing to do with the difference in energy between reactants and products (a subject discussed in the next chapter). Moreover, energy put into molecules of reactants to activate them is released again as they surmount the energy barrier. Similarly, in the analogy above, the height of the bend of the siphon tube above the wall of the beaker does not affect the way it enables water to escape from the higher to the lower reservoir.

Catalysts re-route reactions by forming intermediate complexes with one or more reactants, and often with cofactors as well. The classic work of D. Keilin still offers one of the most elegant and simple démonstrations of enzyme-substrate complex formation. The enzyme peroxidase, which catalyses the reaction

$$H_2O_2 + X \rightarrow H_2O + XO$$

is a conjugate protein containing a haem group. This haem group possesses a very distinctive absorption spectrum with four prominent lines, but when hydrogen peroxide is added to the enzyme, these four absorption bands disappear and two new ones appear. Such a change in the pattern of light absorption is an indication that a change has occurred in the molecular structure of the haem prosthetic group. If the enzyme-substrate (i.e., peroxidase–H_2O_2) complex so formed is separated and a substance is added capable of accepting oxygen from that complex, it is found that the four-banded spectrum of the peroxidase reappears and simultaneously the added substance, X, is oxidized:

HAEM—APOPROTEIN (with H_2O_2 attached to HAEM) $+ X \longrightarrow XO + H_2O +$ HAEM—APOPROTEIN

(Two absorption bands) (Four absorption bands)

The highly specific nature of many enzyme-substrate interactions and the loss of enzyme activity which accompanies denaturation leave little doubt about the fundamental importance of the three-dimensional structure of proteins in relation to enzyme action. Furthermore, three-dimensional structure is itself an intricate function of the attractive and repulsive forces inherent in the hydrophilic and hydrophobic qualities of the side chains of constituent amino acids (page 48). An interesting theory of enzyme action is based upon these facts. If a groove or crevice in the enzyme topography known as the 'active centre' allows the substrate molecule to fit neatly into it, new forces will inevitably come into operation whenever enzyme molecule and substrate molecule come into close proximity. This, in turn, implies that the balance of forces which originally led to the enzyme's tertiary structure becomes 'upset' as *the substrate molecule fits into position*, with the result that the enzyme molecule will *change its shape.* It can be imagined, in particular, that, in the immediate vicinity of the substrate

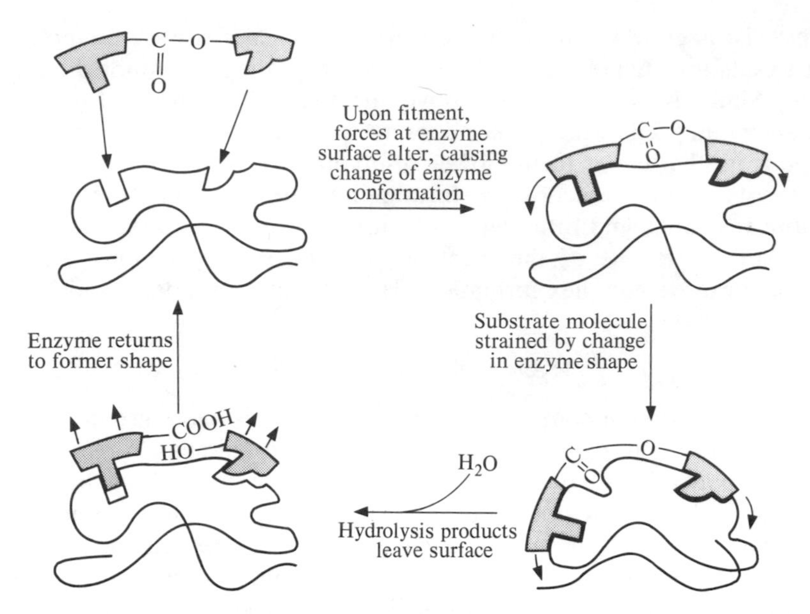

Fig. 6.4 Diagrammatic representation of the distortion theory of the mechanism of enzyme action

molecule, a profound redistribution of forces is likely, resulting in both a stretching and a contraction of parts of the enzyme structure, distorting the substrate molecule as it does so (Fig. 6.4).

More recently a modified, and probably preferable, form of the enzyme-substrate fitment theory has been advanced. Known as the *induced-fit* model, it suggests that the unstrained, or relaxed, form of the enzyme fits the *intermediate complex* rather than either the substrate or the products of an enzymic reaction. The advantage of this model is that it better explains enzyme activity for *reversible* reactions. Otherwise, comments made in the preceding paragraph (about the way the local balance of forces is probably upset when enzyme and substrate interact) are probably equally applicable.

6.7 Enzyme-substrate interaction and the Michaelis constant

Evidence such as that provided by Keilin's work leads to the conclusion that the substrate and enzyme combine together—and therefore that the enzyme can be regarded as a *temporary* reactant which, unlike the substrate, S, is not used up during the course of the reaction. Since the rate of every reaction, whether catalysed or not, is governed, *inter alia*, by the Law of Mass Action, let us consider a situation in which the concentration of enzyme is fixed and only the concentration of the substrate varies. In such a case, the initial reaction velocity will be expected to increase with increasing levels of substrate concentration until a limiting initial reaction velocity is reached (Fig. 6.5). This will happen when the substrate occupies all the active sites of all the enzyme molecules. Similarly, if the reaction is repeated with half as much enzyme, the limiting value of the initial reaction velocity will be half its former value. Note that the term 'initial reaction velocity' is used, for we are *not* discussing here the progress of a reaction with the passage of time.

Such observations are explained by the application of classical equilibrium concepts and provide a useful constant, the Michaelis constant. This constant gives a measure of the relative activity of an enzyme in different circumstances. In particular, it is of interest in relation to the activity of enzymes of rather low specificity—i.e., of enzymes capable of catalysing several related reactions, but each to an extent determined by the exact nature of the substrate. The Michaelis constants of certain allosteric enzymes are also of importance in relation to some mechanisms regulating cellular metabolism (Section 13.2).

The derivation of the Michaelis constant in the simplest possible case assumes that one molecule of substrate S forms an intermediate complex,

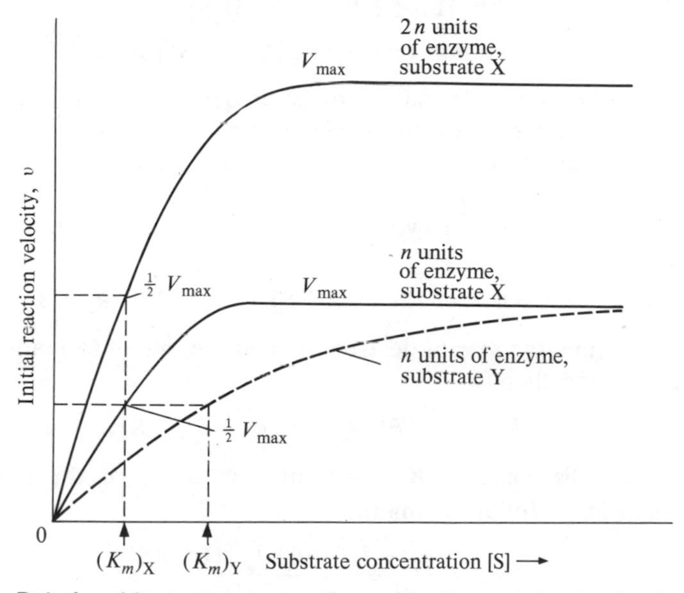

Fig. 6.5 Relationship between reaction velocity and level of substrate. The continuous lines show V_{max} when n and $2n$ units of enzyme are present. The dashed and lower continuous lines show the effect on K_m of equal amounts (n units) of enzyme on two related substrates, X and Y; the affinity of enzyme for substrate X is greater than for substrate Y

ES, with one molecule of the enzyme E, and that the products A and B do not readily form a complex with the enzyme:

$$E + S \underset{k_2}{\overset{k_1}{\rightleftharpoons}} ES \xrightarrow{k_3} A + B \tag{6.1}$$

In order that these reactions can be considered from a kinetic viewpoint, the concentrations of the substances participating can be represented by square brackets, thus: free enzyme [Free E], enzyme-substrate complex [ES], and substrate [S]. k_1, k_2, and k_3 are the velocity constants of the reactions in the directions indicated by the arrows. Using these symbols and applying the ideas of mass action:

(a) At fixed temperature, and at any moment of time,

$$\text{Rate of formation of ES} = k_1[\text{Free E}][\text{S}] \tag{6.2}$$

The *total* enzyme concentration [E] consists of two fractions, the enzyme which is free at any instant and that which is bound to the substrate

Therefore,

$$[\text{Free E}] = [\text{E}] - [\text{ES}]$$

and $\quad$ Rate of *formation* of ES $= k_1([\text{E}]-[\text{ES}])[\text{S}]$ $\quad$ from (6.2)

Note that the rate of formation of ES is dependent upon time, since the concentration of the substrate [S] will decrease as time passes—for isolated enzyme systems of the sort now being considered.

(b) The disappearance of ES can be considered in similar terms, except that Eq. (6.1) shows that two different reactions lead to its disappearance:

$$\text{Rate of } \textit{destruction} \text{ of ES} = k_2[\text{ES}] + k_3[\text{ES}] = (k_2 + k_3)[\text{ES}]$$

(c) At equilibrium, the rate of destruction and the rate of formation of ES are equal, and therefore

$$k_1([\text{E}] - [\text{ES}])[\text{S}] = (k_2 + k_3)[\text{ES}]$$

The Michaelis constant K_m is defined in terms of the three rate constants in the following manner:

$$K_m = \frac{k_2 + k_3}{k_1} = \frac{([\text{E}] - [\text{ES}])[\text{S}]}{[\text{ES}]} \tag{6.3}$$

Equation (6.3) is rarely suitable for estimating K_m since the concentrations of the enzyme and of the enzyme-substrate complex are usually difficult to measure. However,

(d) A simple and rather ingenious modification of the equation leads to quantities readily determined by experiment. The rate of formation of the products A and B from E is, by definition, the overall velocity v of the multiple reaction represented by Eq. (6.1). Therefore,

$$v = k_3[\text{ES}] \tag{6.4}$$

In the presence of a *large excess of substrate*, almost all the enzyme becomes bound to the substrate to form ES. Therefore, when the system is 'overwhelmed' with substrate there is hardly any difference between [ES] and [E], and, inevitably, the enzyme is working at maximal velocity V_{max}. In other words, the maximal value of the concentration of enzyme-substrate complex, $[\text{ES}_{\text{max}}]$, approaches, but cannot exceed, the concentration of enzyme E originally present:

$$V_{\text{max}} = k_3[\text{ES}_{\text{max}}] \doteq k_3[E] \tag{6.5}$$

Therefore, from Eqs. (6.5) and (6.4):

$$\frac{V_{\text{max}}}{v} = \frac{k_3[\text{E}]}{k_3[\text{ES}]} = \frac{[\text{E}]}{[\text{ES}]} \tag{6.6}$$

From (6.3),

$$K_m = \left(\frac{[\text{E}]}{[\text{ES}]} - 1\right)[\text{S}]$$

Substituting Eq. (6.6) into this expression:

$$K_m = \left(\frac{V_{\text{max}}}{v} - 1\right)[S] \tag{6.7}$$

If the observed initial velocity of the reaction, v, is measured under otherwise standard conditions but at varying levels of substrate concentration [S], right up to complete saturation with substrate, K_m can be calculated. In fact, as substitution of $2v = V_{\text{max}}$ in Eq. (6.7) demonstrates, K_m is numerically equal to the substrate concentration when the reaction is working at half its maximal rate.

An important consequence of the Michaelis relationship is shown in Fig. 6.5. Cholinesterases, for example, show a varying affinity for different choline esters. One of these enzymes, in addition to hydrolysing acetylcholine, is able to attack propionylcholine and butyrylcholine, but the K_m value is different in all three cases. As the figure illustrates, the more active an enzyme is in yielding products from a particular substrate, the smaller is the value of the Michaelis constant.†

H. Lineweaver and D. Burk rearranged Eq. (6.7) to provide an alternative

† K_m expresses the affinity of an enzyme for its substrate, but only provided that Michaelis kinetics or conditions operate.

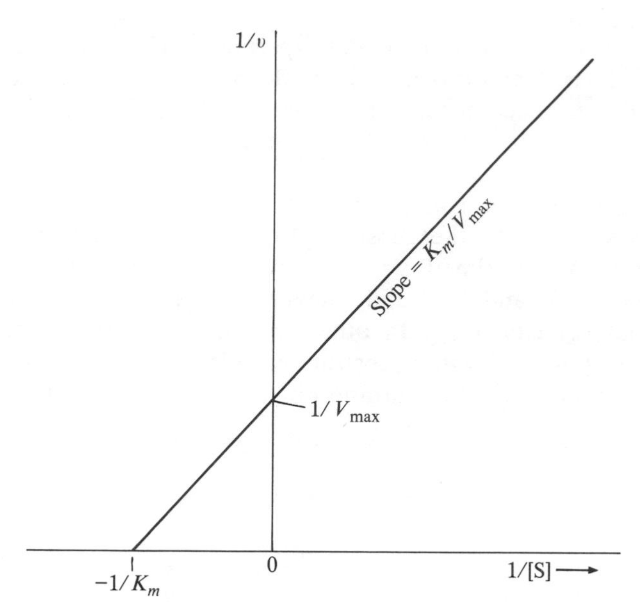

Fig. 6.6 Lineweaver–Burk plot of the reciprocal of the initial velocity against the reciprocal of the substrate concentration

and often more convenient graphical method of expressing experimental data for determining K_m:

$$\frac{K_m}{[S]} = \frac{V_{max}}{v} - 1 \qquad \frac{V_{max}}{v} = \frac{K_m}{[S]} + 1 \tag{6.8}$$

Therefore,

$$\frac{1}{v} = \frac{K_m}{V_{max}}\frac{1}{[S]} + \frac{1}{V_{max}} \tag{6.8a}$$

Upon comparison of this equation with that for a straight line

$$y = mx + c$$

the following relations are apparent:

(a) When the reciprocal of initial reaction velocity is plotted against reciprocal of the substrate concentration which resulted in that velocity, a *straight line* is to be expected. This happens because, in the Lineweaver–Burk equation, $y = 1/v$ and $x = 1/[S]$.
(b) The *slope* of this line is K_m/V_{max} since this is the term in Eq. (6.8) corresponding to m in Eq. (6.8a).
(c) In the general equation (6.8a), when $x = 0$, $y = c$. For the Lineweaver–Burk equation, when $1/[S] = 0$, the intercept on the $1/v$ axis is numerically equal to $1/V_{max}$.

(d) Similarly, in the general equation, when $y = 0$, $x = -c/m$. In the present instance, when $1/v = 0$, in Eq. (6.8):

$$\frac{1}{[S]}\frac{K_m}{V_{max}} = \frac{-1}{V_{max}} \qquad \text{or} \qquad \frac{1}{[S]} = \frac{-1}{K_m}$$

It is apparent from these relations that K_m and V_{max} can be readily estimated in simple cases by measuring the velocity of the reaction at several levels of substrate concentration (Fig. 6.6).

Two aspects of the Michaelis constant and its applications are particularly worthy of note: (a) the Michaelis constant has the *same units as the substrate concentration*, and K_m is consequently expressed as moles per litre (actual values are often within the range 10^{-2} to 10^{-6} mol/litre). (b) Since the value of K_m is estimated by measuring initial reaction velocity at various substrate concentrations, but without reference to the quantity of *enzyme* present, it is possible to obtain estimates of K_m using very impure preparations of enzymes. Most other measures of enzyme activity require a knowledge of the exact amount of enzyme present and sometimes even of its molecular weight—e.g., the *turnover number* of an enzyme is the number of moles of reactant converted to product per minute by one mole of enzyme (Section 6.1).

6.8 Reversible inhibition and the Lineweaver–Burk plot

It was mentioned on page 114 that the inhibition of succinate dehydrogenase by malonate was an example of *competitive inhibition*. When the rate of the (uninhibited) reaction is determined using a fixed amount of enzyme but different levels of substrate, a Lineweaver–Burk plot resembling Fig. 6.6 is obtained. If the experiment is then repeated in the presence of a *fixed* concentration of malonate, a linear relationship is still evident (Fig. 6.7), but the *slope of the line is greater*. Moreover, the intercept on the $1/v$ axis is *the same* as that of the line obtained when succinate is studied alone.

The fixed intercept implies that when the system is overloaded with substrate ($[S] \to \infty$, therefore $1/[S] \to 0$), the maximal rate of reaction, V_{max}, is the same no matter what concentration of malonate was employed. In other words, under these limiting conditions, all the enzyme molecules are available for reaction and each works with unimpaired efficiency. The higher slope of the line obtained when inhibitor is present shows that, for the reaction to proceed at any *given rate* (below V_{max}), the concentration of the substrate must be higher when inhibitor is present than when it is not. That 'given rate' could be half the maximal speed if we so desired. But the substrate concentration [S] at which the rate is half V_{max} is, by definition, K_m. It follows that the *apparent* K_m value in the presence of the inhibitor

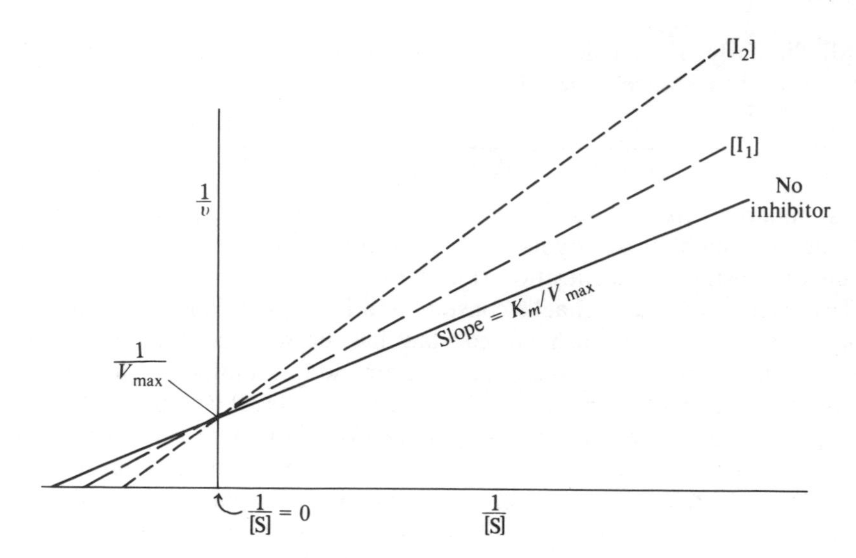

Fig. 6.7 Lineweaver–Burk plots showing the effect of increasing concentrations, [I], of a competitive inhibitor. Not all enzyme molecules are immediately available to the substrate since active sites of some of them are blocked by inert molecules. Slopes of lines are given by $(K_m/V_{max}) \times (1+[I]/K_i)$. For the solid line, $[I] = 0$, giving a slope of K_m/V_{max} as in Fig. 6.6. Upon adding a large excess of substrate, $[S] \to \infty$ or $1/[S] \to 0$; the graph shows that in this circumstance the inhibitor does not affect the rate of reaction ($1/V_{max}$ is the same for $[I_1]$, $[I_2]$, etc.), implying that all enzyme molecules are then available for reaction and functioning normally

must be greater than the actual value in its absence. The difference is, in fact, a function of the chosen concentration [I] of the inhibitor and the dissociation constant K_i of the enzyme-inhibitor complex:

$$\text{Slope (inhibited reaction)} = \text{slope (no inhibitor)} \times \left(1 + \frac{[I]}{K_i}\right) \qquad (6.9)$$

Two other forms of reversible inhibition are common, known respectively as *uncompetitive* and *non-competitive* inhibition. *Uncompetitive* inhibitors do not bind with the enzyme but unite instead with the enzyme-substrate complex, forming a triple complex, ESI, which *does not break down to give the products A, B of the reaction*:

$$E + S \rightleftharpoons ES \longrightarrow A + B \qquad \text{compare equation (6.1)}$$

$$ES + I \rightleftharpoons ESI$$

$$ESI \not\longrightarrow A + B$$

This puts a proportion of the enzyme temporarily out of action. Inhibitors of this type give Lineweaver–Burk plots for which the lines obtained in the presence of the inhibitor are parallel to that obtained when no inhibitor is present. This implies that the K_m value for molecules not incapacitated is the same as in the absence of the inhibitor, but since fewer active enzyme molecules are present, the rate of formation of products A, B decreases as the concentration of the inhibitor rises.

Non-competitive inhibitors bind in such a way that either the enzyme-substrate complex is formed more slowly or, alternatively, once formed, it breaks up to give products less rapidly:

$$\mathrm{E + S} \underset{k_2}{\overset{k_1}{\rightleftharpoons}} \mathrm{ES} \xrightarrow{k_3} \mathrm{A + B}$$

(k_1 and/or k_3 lower when inhibitor present)

Usually, the binding is not at the same site on the enzyme as that to which the substrate binds. The Lineweaver–Burk regression lines in the presence of the inhibitor are of higher slope than the line for the uninhibited reaction.

It can be shown that, *in their simplest form, all reversible reactions involving inhibitors* can be represented by the equation:

$$\frac{1}{v} = \frac{K_m}{V_{\max}}(1 + \mathrm{A})\frac{1}{[\mathrm{S}]} + \frac{1}{V_{\max}}(1 + \mathrm{B}) \qquad (6.10)$$

where A and B can be zero or $[\mathrm{I}]/K_i$.

(a) When A and B are both zero, the equation reverts to the simple form of Eq. (6.8) (if A, B are both zero, clearly no inhibitor is represented in the equation).
(b) When A is $[\mathrm{I}]/K_i$ and B is zero we get the equation for *competitive* inhibition, the slope of the Lineweaver–Burk line being greater than that in the absence of the inhibitor by the factor $(1+[\mathrm{I}]/K_i)$
(c) For *uncompetitive* inhibition A is zero and B is $[\mathrm{I}]/K_i$ whereas
(d) for *non-competitive* inhibition both A and B are $[\mathrm{I}]/K_i$.

Clearly, experiments of the sort which enable Lineweaver–Burk plots to be produced can also provide evidence about the probable way a particular inhibitor affects an enzyme system. It should, however, be stressed that numerous complications can occur and the simplified account given here is far from rigorous. *Apparent* Michaelis constants (as determined from the position where $[\mathrm{S}] = \frac{1}{2}V_{\max}$ for inhibited reactions) can be compared with *actual* K_m values as obtained in reactions where no inhibitor is present. Comparisons of slopes and intercepts of these various regression lines allow K_m, $V_{\max}$, and K_i to be determined.

6.9 Steady states

Reversible reactions catalysed by isolated enzymes, like other reversible chemical reactions, tend always to move to a state of equilibrium. However, in the living cell, unlike most reactions *in vitro*, the overall set of chemical reactions tends to run at a nearly constant rate, since it is necessary for the cell to keep a number of integrated processes ticking over throughout its life. Performance of work is, however, inconsistent with the establishment of a general state of equilibrium, for at equilibrium no useful work is done and there is no net formation of products from reactants.

What actually occurs *in vivo* is that many (but not all) enzymic reactions are associated in chains in such a way that one or more of the products of the first reaction become the reactants of the next. Clearly, the removal of the products of the first reaction prevents the establishment of equilibrium even though the concentrations of reactants and products in the mixture often depart little from what the equilibrium concentrations would have been under the same conditions. The same is true for all other reactions in the chain. Such a *steady-state* system resembles certain industrial assembly lines; each step operates at a rate governed by the provision of the appropriate reactants and by the removal of the products through the intervention of the next reaction in the sequence. This interdependence of the reactions ensures that the whole chain or cycle settles down to a constant level of performance unless additional or external factors disturb it in some way (see Chapter 13).

The velocity with which reactants are converted to products will be different at each step of the sequence and will depend not only upon the concentrations of reactants but also upon the *quantity and effectiveness* of the participating enzymes. Thus if, in the sequence:

$$\mathrm{A} \xrightarrow{\text{Enzyme } E_1} \mathrm{B} \xrightarrow{\text{Enzyme } E_2} \mathrm{C} \xrightarrow{\text{Enzyme } E_3} \mathrm{D}$$

one enzyme, let us say, for example, enzyme E_2, is either slower acting or in shorter supply than neighbouring enzymes it controls the rate at which the whole sequence works. In this situation enzyme E_1 would supply so much B to enzyme E_2 that enzyme E_2 would be converted almost entirely to the E_2-B complex. Enzyme E_2 would therefore work at maximum rate (page 121). No matter how fast enzyme E_3 is able to function *in vitro*, it cannot, when part of such a sequence, act faster than enzyme E_2 provides it with substrate C. Consequently, enzyme E_2 will act as a brake, or *pacemaker*, determining the rate at which D is formed from A.

One of the chief characteristics of living cells is their ability to regulate their activities according to the circumstances. Suppose, for example, the cell has need of a certain amount of a particular metabolite D. Clearly, for optimal efficiency, it must have neither an excess nor a deficiency of D and, in practice, there are several ways in which the level of D can be controlled. Sometimes a later enzyme in the sequence (for example, E_3) is dependent on a cofactor also required by enzyme E_1. Competition for a limited amount of cofactor can restrict the rate of formation of D. Such cofactors as NAD, NADP, and ATP are frequently rate limiting.

Another way in which the level of D can be controlled introduces a new concept—that of *allosteric* inhibition or activation of an enzyme. The principle of allosteric action is illustrated in Fig. 6.8. An inhibiting substance forms a loose combination with the enzyme at a site other than that occupied by the natural substrate. This loose attachment causes a change in the shape, or *conformation,* of the enzyme which makes it difficult for the natural substrate to fit into the active site. When, as a result of general metabolic activity, the concentration of the inhibiting substance falls, the loosely bound molecules of the inhibitor or *modulator* (I in Fig. 6.8) tend to vacate the enzyme surface, thus enabling the enzyme to return to a shape more suitable for reaction with the natural substrate. Thus, if a product, D, of a sequence of reactions is being formed faster than it is needed it tends to accumulate. As it accumulates, more and more molecules of an enzyme somewhere in the sequence which leads to its formation are rendered less active or inactive. In consequence, *feedback inhibition* cuts off the supply of D, the concentration of which then falls. Examples of this important self-regulatory process are considered in Sections 15.13 and 15.14.

In practice, allosteric enzymes are normally more complex than the example shown in Fig. 6.8, for usually they comprise two or more subunits. It is believed that these multimolecular complexes have several binding sites, one or more being specific for the substrate and others for the modulator substances. The latter either activate or inhibit the enzyme complex by changing the shape of the principal active centres.

Another method whereby the quantity of D can influence the rate at which the whole sequence of reactions operates is known as *enzyme repression.* In this case, D does not affect the speed at which an enzyme functions but the speed at which one of the enzymes in the sequence is *synthesized.* A final possibility, of which several examples are known, is that a *deficiency* of D can lead indirectly to the synthesis of more D by initiating the formation *de novo* of one or more of the enzymes involved in its synthesis. When this occurs the process is known as *enzyme induction* or adaptive enzyme synthesis.

6.10 Inorganic ions and small organic molecules as enzyme cofactors

Cofactors of various kinds play an indispensable role in numerous enzyme-catalysed reactions. Their existence was first recognized in 1905 when

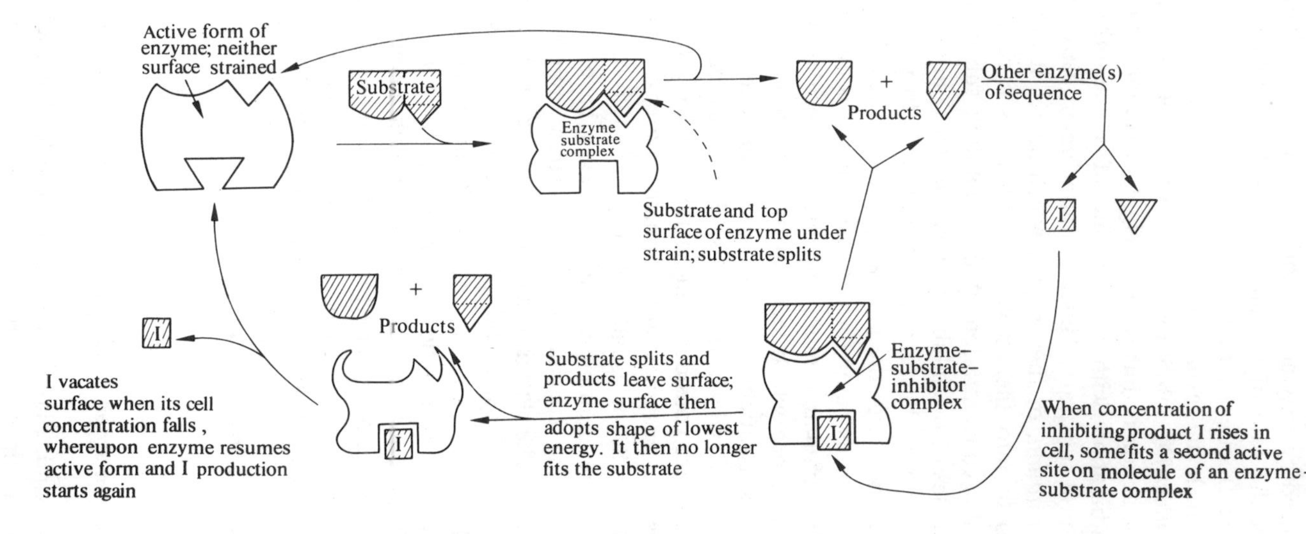

Fig. 6.8 One possible interpretation of allosteric inhibition of an enzyme by a product of the reaction sequence

A. Harden and W. Young showed that dialysed yeast juice did not ferment sugars, but that undialysed juice, or dialysed juice to which some of the dialysate was subsequently added, fermented sugars actively. These cofactors, removed from the enzymes by dialysis, were found to be of relatively small molecular weight; in the case of yeast juice, NAD later proved to be one of them and magnesium ions another. It is now known that many enzyme-catalysed reactions require the participation of inorganic ions or of organic molecules which, compared to enzymes, are of relatively small molecular weight.

Numerous enzymes have metal cations as cofactors: magnesium, iron, calcium, manganese, copper, and zinc are all important examples. When some of these ions are absent, deficiency diseases occur in all types of organisms but can often be remedied by addition of the deficient cation to the diet or growth medium. *Trace elements* are elements which, for the reason above, are only needed in minute quantities.

Inorganic ions can participate in enzymic reactions in a number of different ways. Sometimes a *cation* may be an integral part of the enzyme. Thus ascorbic acid oxidase and cytochrome oxidase (Section 8.7) are two examples of a group of enzymes which are *copper* proteins. Other enzymes are proteins containing *iron* bound within a haem group (e.g., peroxidase, Section 6.6) and yet others are non-haem iron proteins. Nitrogenase and nitrate reductase are two enzymes which contain *molybdenum* (Section 15.1).

A second possibility is that the cation is an important part of the structure of a *coenzyme*. An important example is *cobalt*, which is present within the molecule of vitamin B_{12} (see below). A third role of metal ions is to help bind enzyme to coenzyme, substrate to enzyme, or substrate to coenzyme. This binding ensures that the various components of an enzymic reaction are correctly orientated with respect to one another. The role of *zinc* in the alcohol dehydrogenase reaction provides an example (Fig. 6.9). The zinc ions are apparently involved in the attachment of the NAD^+ to the enzyme surface, probably by bonding which involves the sulphydryl groups of cysteine residues.

A fourth way in which metals assist enzymic action is by helping to stabilize the three-dimensional shape of the apoenzyme. Amylase produced by the salivary glands of vertebrates offers an example. Salivary amylase is a metalloprotein, each molecule of the enzyme containing one ion of *calcium*. If this calcium is removed by chelating agents, the enzyme's activity is not directly impaired, but it is much more readily inactivated by proteolytic enzymes. Amylase is also an example of an enzyme which requires an *anion* for effective action; if *chloride* ions are removed by dialysis, amylase no longer hydrolyses glycogen and starch to maltose.

Coenzymes are relatively low molecular weight organic substances which

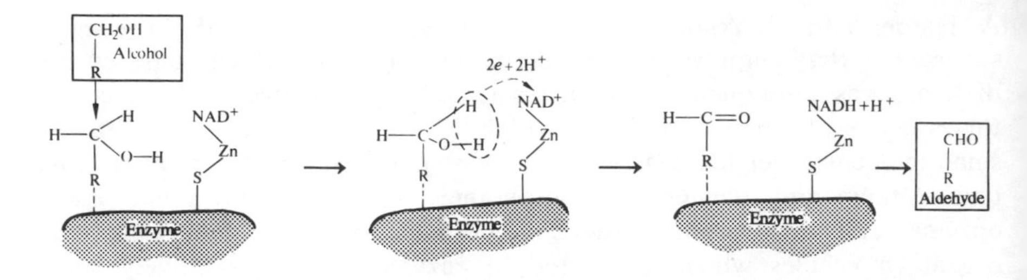

Fig. 6.9 The action of alcohol dehydrogenase

Table 6.4 Some vitamins, their sources, deficiency symptoms, and coenzymes or other biological systems involving them

Vitamin	Source or comment	Disease or symptom	Coenzyme or function
Nicotinamide, Niacin	Liver, meat	Pellagra; dermatitis; diarrhoea	NAD, NADP
Thiamine (B_1)	Wholemeal bread	Beriberi; muscular degeneration; polyneuritis	Thiamine pyrophospha
Pyridoxine (B_6)	Liver, kidney	Growth failure in rats; anaemia	Pyridoxal phosphate an pyridoxamine phosphate
Cobalamine (B_{12})	Made only by micro-organisms; present in sewage sludge	Pernicious anaemia	C_1 metabolism
Riboflavin (B_2)	Liver, milk, eggs	Magenta-coloured tongue	FMN, FAD
Pantothenate	Yeast, eggs, milk	Growth failure in rats	Coenzyme A; acyl-carrier protein
Lipoate	Liver		Lipoamide
Folic acid	Spinach leaf; made by gut flora	Growth failure and anaemia	Tetrahydrofolic acid
L-Ascorbic acid (vitamin C)	Citrus fruit, potatoes	Scurvy; sore gums; collagen synthesis impaired (only a vitamin in primates and guinea pig)	Not clear; redox reactions involved
Vitamin A	Carotenoids in green leaf vegetables	Night blindness; poor growth; sterility in male	Visual process, chondroitin sulphate synthesis
Vitamin D	Fish liver oils	Rickets; poor bone formation	Regulation of calcium metabolism
Vitamin E	Plant seed oils	Infertility in rats; muscular dystrophy in rabbits	Not clear; metabolism of unsaturated fatty acids?
Vitamin K	Made by gut flora	Haemorrhage	Plays a part in blood clotting, but also present in plants
Biotin	Egg yolk, yeast	Dermatitis and paralysis in rats	Carboxylation

are essential for enzyme action. As was seen in Section 6.1, there is no sharp dividing line between cofactors which are firmly bound to enzymes as their prosthetic groups and loosely bound coenzymes which can usually be removed by dialysis. In either case, the organic cofactor, or a part of it, is sometimes not synthesized by an animal species (some species differences are known) and hence must be provided in the diet. Such compounds are termed *vitamins.* Examples are thiamine and riboflavin. Plants must, of necessity, be able to synthesize all the organic cofactors they require, but some microorganisms do need to be provided with exogenous cofactors.

Historically, a pellagra-preventive factor was recognized and classified as one of the water-soluble B group of vitamins. It is now known that the anti-pellagra vitamin is *nicotinamide,* a part of the molecule of NAD. Rather similarly, vitamin B_1, the anti-beriberi vitamin, proved to be *thiamine,* an essential part of thiamine pyrophosphate, an essential cofactor in the process of oxidative decarboxylation (Sections 9.6 and 11.8). Vitamin B_2 turned out to be *riboflavin,* a prosthetic group of many enzymes involved in oxidation-reduction reactions (Section 5.4). These and some other common vitamins are listed in Table 6.4. The table also shows coenzymes which contain them and diseases or symptoms which can arise in animals of various kinds if insufficient of the vitamin is present in the diet or insufficient is produced by gut microorganisms.

6.11 The absorption spectrum of NAD and its use in the study of some enzymic reactions

It will be recalled (Sections 5.2, 6.4, and 6.10) that NAD and NADP are essential cofactors in a number of important enzyme-catalysed redox reactions. In the absence of the appropriate pyridine nucleotide, the reaction does not proceed to a finite degree. In isolated *in vitro* systems where AH_2 is being oxidized to A, it is therefore possible to add a known amount of NAD^+ and to determine how rapidly it is reduced to NADH. Conversely, of course, in reactions where B is being reduced to BH_2, it is possible to add NADH and determine the rate of its oxidation to NAD^+. The same comments apply to $NADP^+$/NADPH in reactions where enzymes are linked to this nucleotide.

Both NAD and NADP have absorption spectra which are different for the oxidized and reduced forms. This difference offers an excellent method for following the time course of some pyridine nucleotide-linked reactions. The absorption occurs in the ultraviolet region of the spectrum within the range 230 to 400 nm, and so is monitored using an ultraviolet spectrophotometer with cuvettes of quartz. Figure 6.10 shows the absorption spectra of NAD^+ and NADH, with absorbance plotted against wavelength in nanometres. It is evident that absorption maxima occur at approximately 260 and 340 nm.

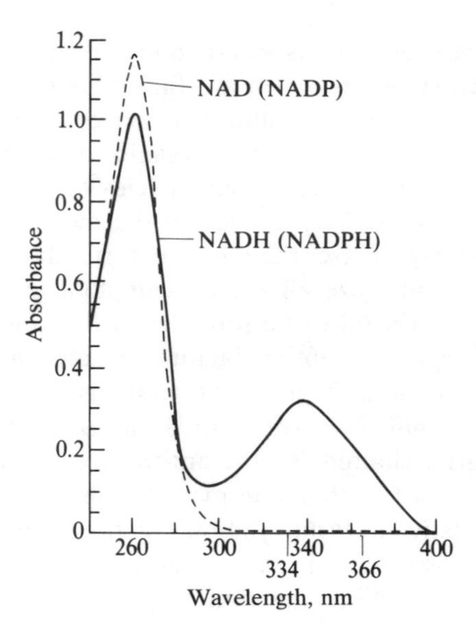

Fig. 6.10 Ultraviolet absorption spectra of oxidized and reduced NAD and NADP

The 260-nm peak is caused by the absorption of the adenine ring system. This ring structure is the same in oxidized and reduced forms of the nucleotide and its absorption changes little when NAD^+ is converted to NADH. However, the absorption at 340 nm is due to the pyridine ring system which is responsible for electron transport (Section 5.2). The absorption at this point rapidly increases as NAD^+ is changed to NADH. Consequently, the change in optical density at 340 nm reflects the extent to which NAD^+ has been reduced during a reaction such as

$$\text{Lactate} + NAD^+ \rightleftharpoons \text{Pyruvate} + NADH$$

or

$$\text{Malate} + NAD^+ \rightleftharpoons \text{Oxaloacetate} + NADH$$

Reactions catalysed by NADH, $NADP^+$, or NADPH can be studied in a similar manner.

7

Bioenergetics

7.1 Energy for the living cell

An athlete, on completing a sprint, may say, rather colloquially, that he has 'used up a lot of energy'. In fact, he has done nothing of the sort; his muscles have converted certain chemical compounds into a variety of products by reactions which liberate energy. This energy is then employed to do work; it makes his muscle fibres contract and expand and so propels him along. During the exercise, he becomes 'hot'; he also stirs air as he moves along, and his feet produce frictional heat where they press against the ground. No energy has actually been used up in the sense that it has been 'destroyed', but energy which was an integral part of certain molecules has been converted into a variety of other forms of energy.

Before exercise, the athlete needs to supply his muscles with energy-supplying materials, and his appetite is likely to be good. An energy source is also needed to enable the tissues of his body to carry out numerous other processes which include growth, tissue repair, and the pumping of certain substances across cell membranes.

Food, very indirectly, is the fuel which drives a complex and efficient machine. Subsequent chapters describe how useful energy is derived by living cells from chemical substances such as sugars and fats and how this energy is harnessed to do work of many kinds. Together, these many forms of work occurring within the cell (or organism) constitute what is known as 'living' or 'being alive'. The discipline of *thermodynamics* deals with such things as energy, heat, and work, and consequently provides explanations for many of the energetic processes occurring in cells. In this book, the approach is deliberately *non-mathematical*, even at the expense of some loss of precision. Having acquired a 'feeling' for the qualitative aspects of energetics, readers requiring a quantitative, mathematical approach will be in a position to proceed to more rigorous treatments.

7.2 Interconversion of energy

The sprinter was described as converting chemical energy to the mechanical energy of muscular contraction. The conversion of energy from one form to another occurs in the inanimate world and in all living systems. Examples can be found involving several types of energy. Chemical energy can be changed to heat, electrical, or light energy; it can also be converted to osmotic energy. Light energy may be converted into electrical or chemical energy. Similarly, electrical energy may be transformed into chemical energy. Whenever the interconversion of two forms of energy occurs, it takes place according to the requirement of the First Law of Thermodynamics—that *energy can neither be created nor destroyed.* The law implies that, when a decrease in the quantity of one form of energy occurs this is *exactly compensated* by an equivalent increase in the quantity of other forms. The principle of conservation of energy was first formulated by V. Mayer in 1841 as a result of a study of energy transformation in the inanimate world, but it is equally applicable to living systems.

One of the most vital of all energy transformations occurs when light energy reaches the leaf of a plant. This is the process of *photosynthesis* whereby carbon dioxide is reduced with the initial production of a sugar. It has been estimated that the amount of carbon dioxide reduced is in excess of 10^{11} tons per year (only about 10 per cent of it by land-based plants) and the light energy thus transformed to chemical energy represents between one-millionth and one-hundred-thousandth of the total solar energy impinging on the earth. In fact, only about one-fiftieth of the light energy falling on most plants actually excites electrons in chlorophyll—the process which leads eventually to the reduction of carbon dioxide (Chapter 14). The rest of the incident light is either reflected or, if absorbed by the leaf, is converted to heat.

In due course, the carbohydrate made by the process of photosynthesis is catabolized by the plant or by a foraging animal. This *loss of potential energy* by carbohydrate as it is oxidized can be harnessed in such a way that some of it *increases the potential energy of some other compound.* The way this harnessing is effected is considered later (Section 7.4). In addition, some of the energy of oxidation is squandered as *heat* while some is lost as energy associated with the random movement of the carbon dioxide and water molecules formed by the oxidation process (energy of *entropy*).

Since energy transformations occur with no overall gain or loss of energy, there is an *exact equivalence* between different forms of energy. This means that there is a precise—and constant—'exchange rate' between the different 'currencies' of energy. For example, the heat produced by an electric current of I amperes at a voltage of V volts for 1 second is equivalent to IV joules, and 4.18 J are equivalent to 1 cal.

When one form of energy disappears it is often replaced by two or more forms of energy, as was seen in the example in a previous paragraph. It is therefore convenient to express all these energy forms in terms of the same energy unit. In biochemistry the unit often chosen is the joule (J) or kilojoule (kJ), the latter being equal to 1000 J. This policy will be followed here, although until recently most biologists favoured the use of the calorie (cal). In practice, those more familiar with the calorie can, of course, convert data from joules to calories by dividing by 4.18. Those preferring the calorie must watch out for one serious pitfall. If it is stated that a change in chemical energy of x calories occurs during a chemical reaction it does *not* mean that *heat*—a form of energy directly capable of causing change in temperature—is necessarily involved, but only that energy *equivalent* to x calories of heat has been gained or lost. An advantage of using the joule rather than the calorie as the unit of energy is that this pitfall is avoided.

7.3 Potential energy in physical and chemical systems

A factor of considerable importance in relation to the energetics of cellular processes is the direction in which chemical reactions tend to proceed, and a familiar situation from the inanimate world—the waterfall—will provide a clue to the principles involved. Static water cannot drive a water wheel but, if the water is allowed to flow downhill, the energy of the moving water can be harnessed so that it does the work of rotating a wheel at the foot of the fall. The greater the distance through which the water falls the more the energy is harnessed by the water wheel. Thus the energy that unit quantity of water can provide to do work depends on the initial height of the water, relative to a wheel situated at the foot of the waterfall. Clearly, there is a spontaneous movement of water downhill; it does not move *spontaneously* in the opposite direction, though suitable machinery, driven by an external energy supply, can be used to pump water uphill. The fall of the water is an energy-yielding process, for the water loses *potential energy* as it moves from a higher to a lower level. Any process, whether it is physical or chemical, which results in the surroundings gaining energy is termed an *exergonic* process, i.e., the system or substance becomes poorer in energy. Energy provided by the exergonic process can be harnessed to drive an energy-consuming or *endergonic* process, so long as appropriate machinery is available, although the harnessing is never 100 per cent efficient. In the absence of this machinery the energy is not harnessed but is just squandered, usually manifesting itself as heat.

Precisely the same argument can be applied to chemical processes as to physical ones. When any *chemical reaction* occurs, the potential energy of reactants is replaced by the potential energy of products, and the difference between these two energy levels represents the energy liberated (or absorbed) in the reaction. Thus when glucose is oxidized to carbon dioxide and water, energy is released from the glucose molecules. The chemical energy

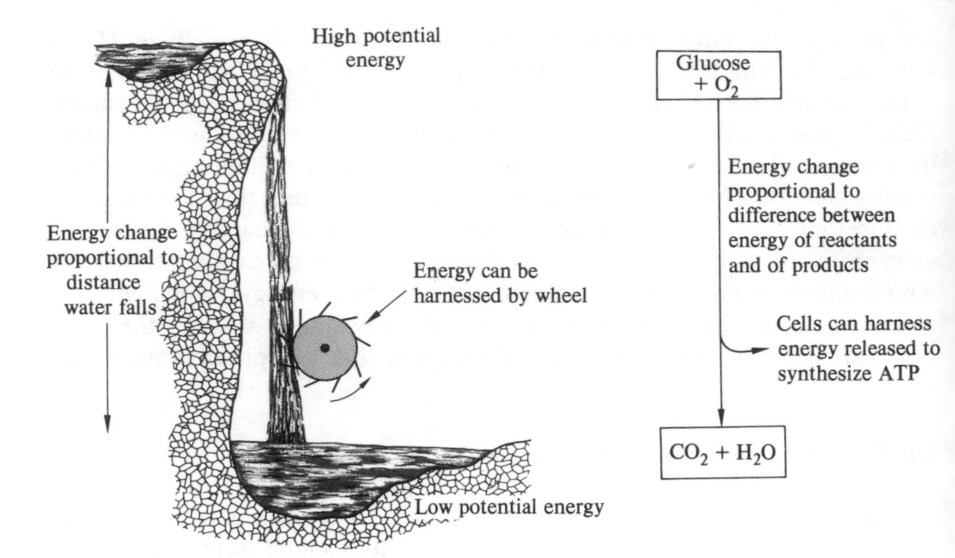

Fig. 7.1 **The potential energy lost as water descends a waterfall (left) can be harnessed to do work just as energy released in a chemical reaction can be used by cells to drive endergonic processes**

inherent in the glucose molecule is a *potential* energy, and it is greater than the potential energy of the oxidation products. Glucose oxidation is thus analogous to the fall of water from a higher to a lower level—energy is released which can be harnessed to do work if appropriate machinery is available (Fig. 7.1).

The change in chemical potential energy occurring when reactants give rise to products in an exergonic reaction is of great importance from a bioenergetic viewpoint, because it is a measure of the maximal quantity of energy made available by the reaction for driving endergonic processes in the cell. Similarly, when the change is endergonic, the energy difference measures the energy that must be provided to force the reaction to proceed. The *free energy change* (ΔG) is thus defined as the difference between the (summed) free energy of the various reactants, $\sum G_R$, and the (summed) free energy of the various products, $\sum G_P$:

$$\Delta G = \sum G_P - \sum G_R \tag{7.1}$$

In Eq. (7.1), the symbol $\sum$ simply means 'the sum of', so $\sum G_R$ for a reaction involving the combustion of sugar in oxygen would just mean 'the free energy of the sugar undergoing reaction plus the free energy of the oxygen undergoing reaction'. If there is only one reactant, $\sum G_R$ and G_R have, of course, the same meaning. In practice the absolute values of $\sum G_R$

and $\sum G_P$ are of less importance than ΔG, the free energy released or absorbed as the reaction occurs. Just as a reaction can only proceed spontaneously if it is exergonic (ΔG is negative, since $\sum G_P$ is smaller than $\sum G_R$) so, if it is endergonic (ΔG positive), it will only occur if external energy is provided.

In considering the energy change accompanying a chemical reaction it is necessary to take account of the *mass of the reactants* in addition to the *change in potential energy*. This is analogous to the waterfall, where the energy change is proportional both to the distance the water falls and to the quantity of water flowing over the top. The capacity factor of chemical reactions is taken into consideration by the use of a quantity known as the *standard free energy change*, ΔG^0. This corresponds to the free energy change for a *molar* reaction. Thus, for the complete oxidation of one mole of glucose (180 g) to carbon dioxide and water, the standard free energy change (ΔG^0) is −2870 kJ, and therefore if 360 g of glucose were to be oxidized the *actual* free energy change, ΔG, would be −5740 kJ.

The overall equation for photosynthesis is the reverse of the one for glucose oxidation:

$$C_6H_{12}O_6 + 6O_2 \underset{\text{Photosynthesis}}{\overset{\text{Oxidation}}{\rightleftharpoons}} 6CO_2 + 6H_2O + \text{Energy}$$

In the forward direction, free energy is released by the oxidation, whereas, in photosynthesis, light energy indirectly raises the energy level of carbon dioxide and water to the higher free energy of glucose plus oxygen. To compare the free energy change of the two reactions it is *not necessary to know details of the individual pathways* (which are, in fact, quite different) because the Law of Conservation of Energy requires that the free energy intake in the one direction is inevitably the same as the free energy released in the other. This is shown in the following scheme:

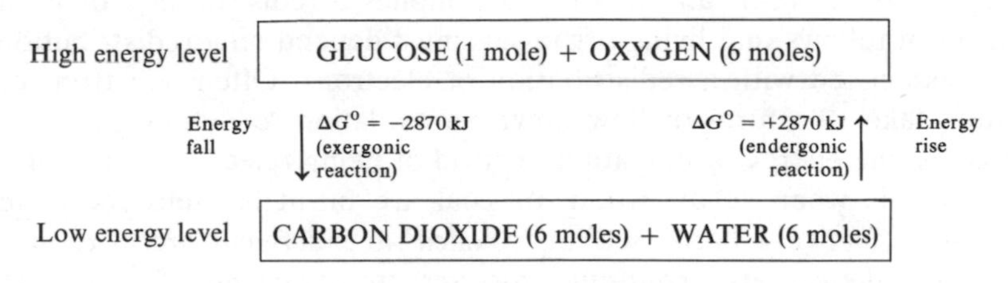

It should be noted that although the oxidation of glucose is spontaneous—the direction of the chemical change is always towards the formation of carbon dioxide and water—nevertheless, as is well known, glucose remains stable almost indefinitely in the presence of oxygen at room temperature. This is because oxidation only occurs at finite speed if either (a) energy is supplied in sufficient amount to ensure that enough molecules overcome the barrier presented by the activation energy E_A or (b) if the value of E_A is effectively reduced by the presence of a catalyst or enzyme (see Section 6.6). Note that the magnitude of E_A in no way affects the value of ΔG, since energy supplied to drive reacting molecules over the 'energy hump' is regained in full as the barrier is surmounted.

7.4 The driving force of the living machine (re-routing of endergonic reactions)

Present knowledge of living processes and of the way that classical thermodynamic considerations can be applied to them has shown that there is no difference in principle between energetic aspects of life and the energetic processes which drive inanimate machines. In particular, no 'vital force' need be invoked to account for how endergonic processes occur in living systems.

Energy changes are often conspicuous in certain reactions because *heat* is evolved. Early on in our scientific training most of us became aware of the existence of 'exo*thermic*' and 'endo*thermic*' reactions, perhaps without fully realizing that gain or loss of energy by chemical systems does not necessarily take the form of gain or loss of *heat*. In biochemistry, however, 'energy change' and 'heat change' seldom even approximately correspond, for heat, it must be remembered, is like the water of a waterfall, in that it can only do work when it changes potential, i.e., when if flows from a hotter to a colder body.

Clearly, since the living machine operates at an almost constant temperature, it is not driven by the movement of *heat* energy down a temperature gradient. Instead, it depends on *gradients of chemical potential energy.* Since changes in chemical (potential) energy require that a change in molecular structure should occur, and this in turn implies a redistribution of valency electrons, it follows that living organisms must depend on a redistribution of energy associated with a redistribution of electrons. Often, electron redistribution takes the form of flow down an oxidation-reduction gradient. In such cases, the energy of oxidation, instead of being released in the form of heat—as it is when sugar, wood, or coal are burnt as fuels for a *heat*-engine—is effectively 'harnessed' or '*coupled' to endergonic processes in such a way that these energy-requiring processes are driven uphill.* This sort of 'coupling' is in no way unique to biochemistry: physical energy is similarly harnessed whenever a car-jack is employed, and a decrease in chemical potential energy is translated into an increase in physical potential energy whenever burning petrol drives a car uphill.

One of the reasons why energy transactions are handled efficiently by living cells is that potential energy changes occur in small discrete steps, and the cell has an elaborate organization both at the molecular level and above it, which ensures effective energy coupling. For example, the oxidation of glucose in a calorimeter results in a one-step heat change, whereas the cell splits glucose up, step by step, so that *at no one stage is there a massive release of energy*. Moreover, when the cell degrades glucose in steps, only a fraction of the energy so made available is manifest as heat, for the quantity of energy released at some of those steps is such that it can be passed on at once to molecules of some other substances nearby, thereby raising *their* potential. This 'trapped' energy is frequently used to drive *formally endergonic* reactions.

The phrase 'formally endergonic' has been used because at this point a difficulty is encountered which relates more to the use of words than to any biochemical problem. Suppose we have a reaction $A \rightarrow B$ in which B is richer in energy than A. The formation of B from A is undoubtedly *endergonic* and requires an input of external energy if it is to occur. But suppose that an exergonic reaction $C \rightarrow D$, instead of occurring in a simple or 'direct' manner, takes place in such a way that it involves the participation of substance A of the endergonic reaction $A \rightarrow B$. So long as the energy change $C \rightarrow D$ is greater than that of $A \rightarrow B$, a new spontaneous exergonic reaction becomes feasible:

$$A + C \rightarrow B + D \qquad (\Delta G, \text{negative}) \qquad (7.2)$$

In such a 'double' reaction, A does not change to B in quite the sense that the simple reaction $A \rightarrow B$ implies, yet B is still formed from A. The resolution of this paradox is quite simple; the reaction $A \rightarrow B$ has, in fact, been *re-routed*. Since we often need to take such 'coupled' reactions apart (it should be remembered that the reaction $C \rightarrow D$ can occur without A being present, just as falling water need not drive a water wheel) it is convenient to think of the *biologically impossible* 'direct' reaction $A \rightarrow B$ as 'formally' endergonic. The standard free energy change (ΔG^0) of the formally endergonic reaction is clearly of importance for it tells us the minimal amount of energy which must be supplied by the exergonic reaction which is coupled to it.

The formation of the important substance adenosine triphosphate, or ATP, from the energy-poorer diphosphate ADP, is an apparently endergonic process which proceeds as a result of 'coupling' with an exergonic change. Figure 7.2 shows that ADP combines with inorganic phosphate to form ATP in a reaction requiring an input of about 33 kJ/mol. Thus ADP and phosphate together correspond to 'A' in the coupled example above, and ATP corresponds to 'B'. Except when the energy for the reaction originates from sunlight, as it does in photosynthesis, it is usually derived from one of several exergonic reactions $C \rightarrow D$ which are steps in the process of respiration.

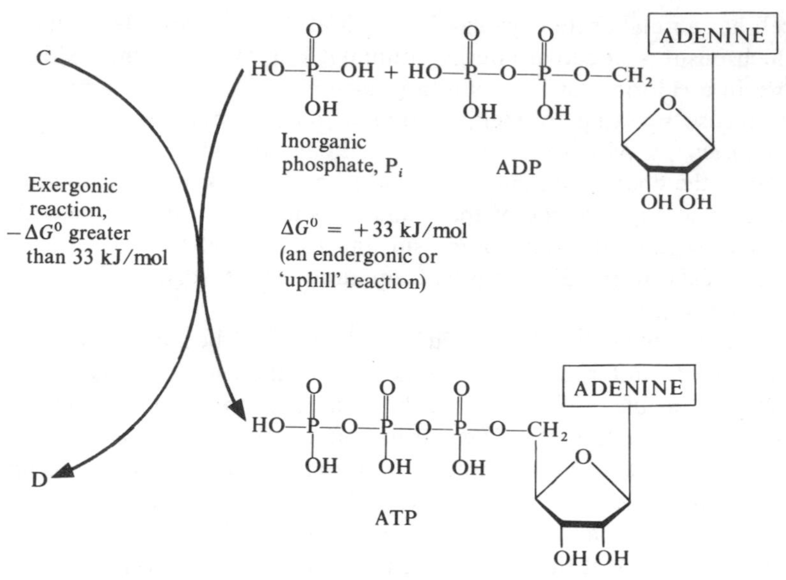

Fig. 7.2 The exergonic reaction C → D 'drives' the change ADP + P_i → ATP

7.5 Free energy, enthalpy, and entropy

In Section 7.4 it was seen that the capacity of a chemical system to do useful work often depends on the 'downhill' flow of electrons. Just as the potential difference in the case of falling water depends on the height through which the water descends, so in a chemical system the change in *free energy*, ΔG, is the difference between the sum of the free energy of all the reactants (R) and that of all the products (P):

$$\Delta G = \sum G_P - \sum G_R$$

Free energy G is a property of a system just as are volume and mass; once the size and certain environmental conditions have been fixed, G is accurately defined. Neither the previous history of the system nor the route whereby a chemical system of energy $\sum G_R$ approaches another one, of energy $\sum G_P$, plays any part in determining the magnitude of ΔG.

The free energy change, ΔG, is the thermodynamic quantity of most significance to the biochemist, for the following reasons:

(a) Its sign indicates whether a reaction is theoretically able to proceed spontaneously. If it is negative, the reaction can proceed spontaneously and if it is positive, it cannot. However, thermodynamics do not pronounce upon how *fast* a reaction occurs (the aerobic oxidation of protein and of cellulose are strongly exergonic, but both the reader and his book have hopefully survived so far).

(b) For exergonic reactions, the magnitude of $-\Delta G$ indicates the maximal energy available for useful work.

(c) For reversible reactions, the 'driving force', $-\Delta G$, declines as the mixture of reactants and products moves towards equilibrium. At equilibrium, $\Delta G = 0$; this indicates that no further net change, and no useful work, is then possible (Section 7.6).

A second thermodynamic quantity is the internal energy of a system. When the pressure is kept constant, this is known as the *enthalpy* and rather unfortunately, perhaps, is given the symbol H. Like volume and free energy (G), H is a state function; that is to say, its magnitude depends only on the condition and mass of the system before and after change. The change in enthalpy when a reaction takes place is therefore independent of the route, and is called ΔH.

The absolute enthalpy H is the sum total of numerous atomic and subatomic energy factors.† On the other hand, ΔH is a relatively simple and straightforward experimental quantity; it is the *heat of reaction as measured in a constant pressure calorimeter.* A word of warning is, however, necessary, for ΔH is only the *heat* of reaction if the reaction remains uncoupled to other changes. If coupling occurs, the heat measured in the calorimeter is, of course, the heat evolved in the overall coupled reaction. Thus when the breakdown of glucose *in vivo* is coupled with endergonic reactions, only part of the enthalpy change is manifest as a heat change, and it is in this sense that the use of the symbol ΔH is somewhat misleading.

Most exothermic reactions (ΔH, negative) are also exergonic, and the oxidation of glucose is a case in point ($\Delta G^0 = -2870$ kJ/mol; $\Delta H^0 = -2810$ kJ/mol). Nevertheless, the heat of reaction is not always a good guide as to whether a reaction is exergonic and therefore theoretically spontaneous. It may seem strange that the complete oxidation of glucose to carbon dioxide and water yields more energy capable of doing useful work (ΔG) than the difference of the internal energy (ΔH) between the system oxygen/glucose and the oxidation products carbon dioxide/water. Additional energy to the extent of $\Delta G^0 - \Delta H^0$, that is, 2870 − 2810 kJ, becomes available as a result of the oxidation. This additional energy is known as energy associated with *entropy* and is given the symbol ΔS. The entropy, s, of a system is a state property, just like G and H, and the quantitative relation between changes in these three forms of energy is given by the equation

$$\Delta G^0 = \Delta H^0 - T\Delta S^0 \tag{7.3}$$

† Absolute values of H, G, and S are unknown.

where T is the absolute temperature. Applying this equation at 20°C to the total oxidation of one mole of glucose, we get:

$$-2870 = -2810 - (293\,\Delta S^0)$$

$$-60 = -293\,\Delta S^0$$

Thus the entropy change in this reaction, ΔS^0, is $+60/293$ kJ per degree per mole.

The quantities S and ΔS have unnecessarily worried many biologists. The entropy of a substance or system represents the freedom of movement of molecules within that substance or system, and of the atoms which constitute each molecular particle. The more the freedom of movement, the more the chaos and the higher the energy associated with entropy. Thus, as the temperature is raised, the energy of entropy increases; the increased chaos of gaseous molecules at higher temperatures is well known. Similarly, if a regular crystal lattice is destroyed by melting, or an ordered large covalent molecule is split up by oxidation, ΔS will be a positive quantity.

Glucose has a highly organized molecule with atoms arranged in an ordered structure and the molecules of solid glucose are arranged in a crystal lattice, whereas the products are less complex molecules moving about haphazardly. Thus, when solid glucose is oxidized, the total entropy of the products is greater than the total entropy of reactants and ΔS is consequently positive. The molar figure is, in fact, 60/293 kJ per degree, as was seen above, when measured at 20°C.

Equation (7.3) is a representation of the Second Law of Thermodynamics. Unlike the first law, the second law indicates the *direction* of chemical change, for it states that all reactions will occur so that free energy of the system under study tends to decrease and the entropy of the universe (the system plus its surroundings) tends to rise. It is noteworthy that, in a particular reaction, the entropy of the system may increase or decrease. In the example above, the entropy of the system increased when glucose was burnt. When a mixture of hydrogen and oxygen is ignited, however, the highly random movement of the molecules of two gases is replaced by the relatively restricted movement of the particles of liquid water, so the entropy of the *system* decreases. Nevertheless, the heat evolved imparts greater kinetic energy to molecules of the surrounding air, so the entropy of the *universe* increases.

7.6 Free energy, equilibria, and the equilibrium constant

Most readers will be familiar with the basic ideas of chemical equilibrium, having already encountered such systems as the esterification of acetic acid. Commencing with several vessels containing varying amounts of acetic acid and ethyl alcohol, the (mathematical) product of the concentrations of the products of the reaction divided by the (mathematical) product of the concentrations of *residual* reactants is eventually the same for all the systems, at one particular temperature, so long as sufficient time has elapsed for *equilibrium* to be reached. Using square brackets to represent molar concentrations at equilibrium, the equilibrium constant K is thus given by the expression

$$K_{(T)} = \frac{[\text{ethyl acetate}][\text{water}]}{[\text{ethanol}][\text{acetic acid}]} \tag{7.4}$$

Values of K have been determined for numerous biochemical reactions and are of considerable value to the biochemist. A reaction is said to be *reversible* if it does not proceed so near to completion, under the conditions of the experiment, that the denominator of equations such as that above tends to zero due to near-total depletion of one or both reactants.

A reaction of importance in glycolysis is the interconversion of dihydroxyacetone phosphate and glyceraldehyde-3-phosphate by the enzyme triose phosphate isomerase:

$$\text{Dihydroxyacetone phosphate} \rightleftharpoons \text{glyceraldehyde-3-phosphate}$$

This reaction will now be employed to illustrate the use of an equilibrium constant in simple calculations. Analysis shows that the equilibrium mixture contains approximately 95.7 per cent dihydroxyacetone phosphate and 4.3 per cent glyceraldehyde-3-phosphate. What is the value of the equilibrium constant for the reaction?

Since

$$K = \frac{[\text{product}]}{[\text{reactant}]} \tag{7.5}$$

and concentration is proportional to the amounts present, then

$$K = \frac{4.3}{95.7} = 0.045$$

Conversely, if we knew the equilibrium constant to be 0.045 for the reaction as written, how could we calculate the proportions of the two substances in the equilibrium mixture? Suppose we start with one unit of dihydroxyacetone phosphate before adding enzyme, and that, by equilibrium, a proportion x has been converted to glyceraldehyde-3-phosphate. Then, at equilibrium $1 - x$ units of dihydroxyacetone phosphate remain and x units of glyceraldehyde-3-phosphate have been formed. Therefore, since

$$K = \frac{[\text{product}]}{[\text{reactant}]}, \qquad \text{and} \qquad 0.045 = \frac{x}{1 - x}, \qquad \text{then} \qquad x = 0.043$$

$$100(1 - x) = \text{percentage of dihydroxyacetone phosphate} = 95.7$$

All spontaneous reactions are exergonic, so when pure A is converted to an equilibrium mixture of A + B, free energy is liberated. Similarly, the approach to the equilibrium mixture, starting from pure B, is spontaneous and exergonic. At equilibrium, no further change in concentrations occurs; the system is static. A 'static' chemical system, like water in a motionless reservoir, is incapable of doing any useful work and, from the moment of equilibrium onwards, $\Delta G = 0$. This is summarized in the following scheme:

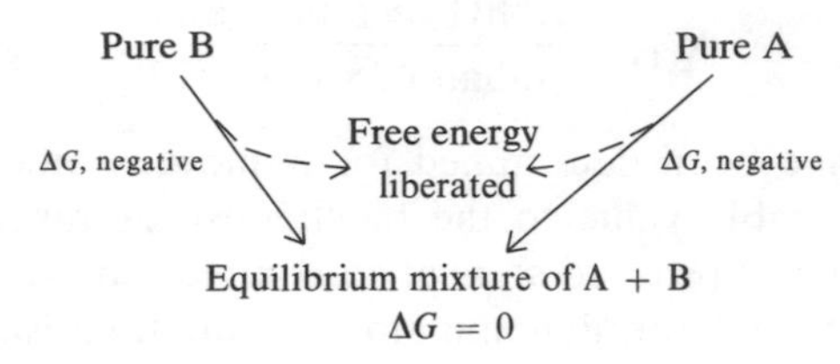

Note carefully that when a freely reversible reaction is being considered (that is, K is not too far from unity under the experimental conditions), it is only formally true to write

$$A \underset{-\Delta G\,=\,x\text{J}}{\overset{+\Delta G\,=\,x\text{J}}{\rightleftharpoons}} B$$

for the 'uphill' part of the reaction is from the *point of minimal potential energy*, namely, from the equilibrium mixture of A and B.

This is illustrated in Fig. 7.3 where the humps represent activation energies. These do not concern the present discussion because the value of the activation energy does not affect the overall energy change in a chemical

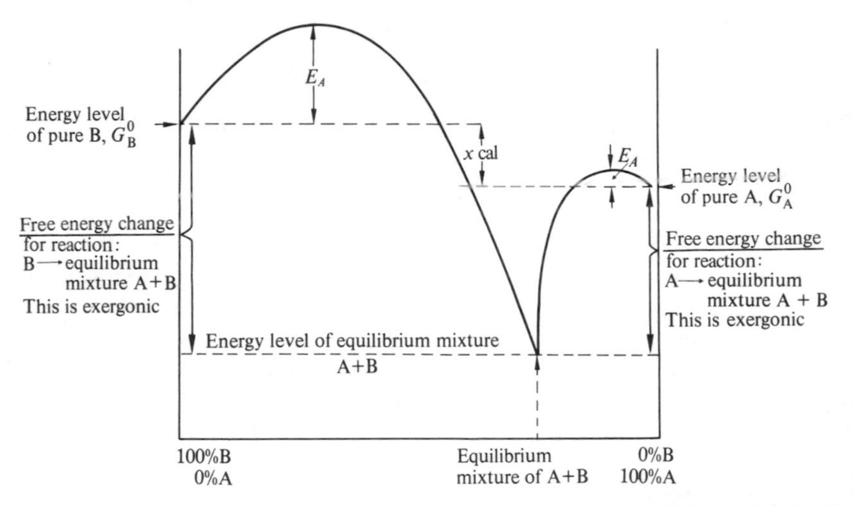

Fig. 7.3 Energy relationship for two substances A and B for which $K = [B]/[A] = 0.33$. Both the diagram and the value of the equilibrium constant show that the equilibrium mixture is richer in A than B

reaction. A calculation of the free energy change which would occur if pure A were to be converted to pure B is not directly relevant because, whatever the initial composition of a mixture of A and B, the composition after equilibrium is reached is always the same. The theoretical energy change when pure A is converted to pure B is $G_B - G_A$ and, when pure B is converted to A, it is $G_A - G_B$, but to move from the *equilibrium mixture* towards *either* pure A or pure B requires the input of energy. What tends to happen in real situations in the cell is that, for substances A, B, *linked by a readily reversible reaction*, the lower-energy compound A can, in practice, be converted continuously to B provided that B is constantly *removed* from the equilibrium mixture. In other words, although the proportion of B in the equilibrium mixture may be exceedingly small, the reaction is pulled over towards the higher-energy form B either by transfer of B to another part of the cell or by the conversion of B to other compounds.

When the equilibrium constant K for a reaction is measured under standard conditions (25°C; 1 mol/litre), the value of the free energy change in terms of joules *per mole* of reactant can be obtained from the expression

$$\Delta G^0 = -RT \ln K \tag{7.6}$$

The logarithm used in this equation is a natural logarithm (ln), not a logarithm to base 10. Arithmetically, it is often more convenient to use $\log_{10} K$; this is related to $\ln K$ by a factor of 2.303:

$$2.303 \log_{10} K = \ln K \quad \text{or} \quad \log_{10} K = 0.4343 \ln K$$

As an example, let us calculate the *standard free energy change* for the triose phosphate reaction occurring at pH 7 at 25°C under conditions where the concentrations of both reactants and products are maintained at 1 M by addition of the reactant and extraction of the product. Then

$$\begin{aligned}\Delta G^0 &= -8.314 \times 298 \times 2.303 \log_{10} 0.045 \\ &= -5706 \log_{10} 0.045\end{aligned}$$

where $\log_{10} 0.045 = \bar{2}.6532 = -1.3468$, and $R = 8.314$ J/deg · mol
Therefore,

$$\begin{aligned}\Delta G^0 &= -5706 \times (-1.3468) \\ &= 7690 \text{ J/mol}\end{aligned}$$

Equation (7.6) shows that the standard free energy change is an inevitable consequence of the value of K; standard conditions must apply, but the nature of the reactants and products are irrelevant in themselves. Consequently, it is possible to tabulate ΔG^0 and K values, and this has been done in Table 7.1. It will be seen that a K value of 0.045 corresponds to a ΔG^0 value of a little less than +8 kJ/mol.

Table 7.1 Relationship between the standard free energy change, ΔG^0, and the equilibrium constant K at 25°C

ΔG^0 (kJ/mol)	K† (25°C, pH 7)	Log K	ΔG^0 (kJ/mol)	K† (25°C, pH 7)	Log K
−24	16180.0	4.209	+24	0.0000618	$\bar{5}.791$
−20	3221.0	3.508	+20	0.000310	$\bar{4}.492$
−16	639.7	2.806	+16	0.00156	$\bar{3}.194$
−12	127.4	2.105	+12	0.00785	$\bar{3}.895$
−8	25.3	1.403	+8	0.0395	$\bar{2}.597$
−6	11.3	1.052	+6	0.0887	$\bar{2}.948$
−4	5.04	0.702	+4	0.199	$\bar{1}.293$
−2	2.24	0.351	+2	0.446	$\bar{1}.649$
0	1.00	0.0	0	1.00	0.0

$$\dagger\; K = \frac{\text{products of the concentrations of products of reaction}}{\text{products of concentrations of reactants of reaction}};$$

thus, the greater the magnitude of K, the more 'complete' the reaction is, when approached from the direction the equation is written. Note that $1/16180 = 0.0000618$, and so on.

7.7 Free energy changes of reversible reactions under non-standard conditions

In the previous section we saw how standard free energy changes can be calculated. These are the figures quoted in reference tables such as Table 7.2, for it would be impossible to record free energy change values for all the different concentration levels at which reactions actually occur. In practice, however, cellular concentrations are much less than 1.0 M (many soluble metabolites having a concentration of the order of 0.1 to 10 mM) and the cellular environment may ensure that ratios of the concentrations of reactants to products are far removed from those required by the equation for a particular chemical reaction. In such circumstances, the *actual* ΔG value—i.e., the value which determines the 'driving force' of a reaction—can be very different from the ΔG^0 value for the reaction. This will be illustrated for the following reaction, catalysed by aldolase (Section 11.5):

$$\text{Fructose-1,6-di-P} \rightleftharpoons \text{dihydroxyacetone P} + \text{glyceraldehyde-3-P}$$

For the aldolase reaction, $\Delta G^0 = 23.95$ kJ/mol, so the reaction is highly endergonic under standard conditions. The equilibrium constant for the reaction is 0.000063 (compare Table 7.1). Suppose, however, that the initial concentration of fructose-1,6-diphosphate is only 0.5 mM and that the environmental concentration of dihydroxyacetone phosphate is 0.24 mM and of glyceraldehyde-3-phosphate is 0.0105 mM. What would be the actual free energy change in these circumstances if the reaction occurs at 25°C?

The equation used to calculate actual free energy changes under non-standard conditions is

$$\Delta G = \Delta G^0 + 2.303RT \log_{10} \frac{\text{product of concentrations of products}}{\text{product of concentrations of reactants}}$$

So, for the reaction when the initial concentrations are as stated above,

$$\Delta G^0 = +23\,950 \text{ J/mol}; \qquad R = 8.314 \text{ J/deg.mol}$$

$$T = 298 \text{ K}; \qquad 2.303RT = 5706$$

and

$$\frac{[\text{Dihydroxyacetone P}][\text{glyceraldehyde-3-P}]}{[\text{Fructose-1,6-di-P}]} = \frac{(0.24 \times 10^{-3})(0.0105 \times 10^{-3})}{(0.5 \times 10^{-3})}$$

$$= 5 \times 10^{-6}$$

Therefore,

$$\Delta G = +23950 + 5706 \log_{10} 5 \times 10^{-6}$$

where

$$\log 5 \times 10^{-6} = \bar{6}.699, \qquad \text{that is,} \qquad -5.301$$

Therefore,

$$\Delta G = +23950 + (5706 \times -5.301) = -6274 \text{ J}, \quad \text{or} \quad -6.27 \text{ kJ}$$

Table 7.2 Standard free energy changes of some common enzyme-catalysed reactions

Reaction	Enzyme	ΔG^0(kJ/mol)
Dihydroxyacetone P $\rightleftharpoons$ glyceraldehyde P	Triose P isomerase	+7.69
Fructose-1,6-di-P $\rightleftharpoons$ dihydroxyacetone P + glyceraldehyde P	Aldolase	+24.0
Glucose-6-P $\rightleftharpoons$ fructose-6-P	Glucose phosphate isomerase	+1.67
Glucose-6-P + H_2O $\rightleftharpoons$ glucose + P_i	Glucose-6-phosphatase	−13.80
Glucose + ATP $\rightleftharpoons$ glucose-6-P + ADP	Hexokinase	−16.7
3-P-glycerate $\rightleftharpoons$ 2-P-glycerate	Phosphoglyceromutase	+4.43
2-P-glycerate $\rightleftharpoons$ phosphoenol pyruvate + H_2O	Enolase	+18.40
Pyruvate + NADH + H^+ $\rightleftharpoons$ lactate + NAD^+	Lactate dehydrogenase	−25.10
Fumarate + H_2O $\rightleftharpoons$ malate	Fumarate hydratase	−3.14
Pyruvate + NAD^+ + CoA $\rightleftharpoons$ acetyl CoA + NADH + H^+ + CO_2	Pyruvate dehydrogenase	−33.44
Acetyl CoA + oxaloacetate + H_2O $\rightleftharpoons$ citrate + CoA	Citrate synthase	−32.20
Malate + NAD^+ $\rightleftharpoons$ oxaloacetate + NADH + H^+	Malate dehydrogenase	+28.00

Thus under these particular non-standard conditions, the reaction is exergonic even though the standard free energy change is positive.

7.8 Electron flow; relation of free energy to electrode potential

The following equations represent the combustion of magnesium in oxygen and in chlorine:

$$Mg + \tfrac{1}{2}O_2 \rightarrow Mg^{++}O^{2-} \qquad x \text{ J liberated per g atom Mg}$$

$$Mg + Cl_2 \rightarrow Mg^{++}2Cl^{-} \qquad y \text{ J liberated per g atom Mg}$$

The fundamental similarity of these two changes is obvious; in both, each atom of magnesium *loses two electrons* and an atom of a non-metal receives them. Since the electrons originate in both reactions from magnesium, they start at the same potential, and hence the difference in the energies of reaction (x and y J) must be attributed to the fact that the electron acceptors, oxygen and chlorine, lie at different levels 'downhill'.

The energy change in an oxidation-reduction, or *redox*, process is as inevitably tied up with the height of the 'energy-fall' as is the change of potential of water with the height of the 'waterfall' (Fig. 7.1). The capacity factor, needless to say, is the mass of the reactant. One gram atom or one mole always consists of about 6×10^{23} particles (Avogadro's number), and therefore, in the equations above, the standard energy changes x, y J represent the energy changes associated with the loss of 12×10^{23} electrons, for magnesium is divalent.

In reactions such as those above, the energy changes are usually measured spectroscopically and expressed in kilojoules. Redox reactions are also very common in aqueous solution. Such reactions often involve inorganic ions and relevant standard electrode (or redox) potentials are to be found in textbooks of physical chemistry. However, organic compounds can participate in oxidation-reduction reactions, and redox changes of this type are of fundamental importance in the aqueous environment of the cell.

Redox reactions in aqueous solution can be measured electrically, the energy change being initially expressed in electronvolts. Since, however, energy units are interconvertible, such voltage differences can be reexpressed as joules, the interconversion factor, for the transfer of *one* electron per molecule, being 96 500, since the charge on 6×10^{23} electrons is 96 500 coulombs (the Faraday):

$$-\Delta G_{\text{joules}} = n \times 96\,500 \times \Delta E \qquad (7.7)$$

where n is the number of electrons transferred from, or to, an ion or molecule and ΔE is the measured energy change in electronvolts.

The tendency for substances to undergo redox reactions differs considerably and depends on the relative ease with which they accept or give up electrons. This tendency can be measured, and, for comparative purposes, can be referred to the readiness with which hydrogen gas is oxidized to hydrogen ions:

$$H_2 \rightleftharpoons 2H^+ + 2e$$

Compounds in the upper part of Table 7.3 lose electrons more readily than hydrogen gas, while those in the lower part have a lesser tendency to become oxidized. For each substance the tendency of its reduced form, under standard conditions, to give up electrons and so become oxidized, is known as the standard redox potential E_0'.

Standard conditions need to be specified, for the actual redox potential varies with such things as concentration and pH; E_0' refers to a molar solution containing equal quantities of the oxidized and reduced forms of the compound at pH 7.0. When E_0' is *negative* the substance tends to become oxidized more readily than hydrogen gas, i.e., a high negative E_0' indicates a high *electron-donating*, or reducing, capacity. Conversely, substances with positive values of E_0' tend to become reduced because they readily accept electrons and so act as oxidizing agents.

Table 7.3 shows the values of E_0' for a number of biochemically important donors and acceptors of electrons. By definition, the E_0' value implies the

Table 7.3 Redox potentials of some important cellular reactions

Redox reaction (oxidized form → reduced form)	E_0' volts (25 or 30°C; pH 7.0)
Succinate → α-ketoglutarate	−0.67
CO_2 + α-ketoglutarate → isocitrate	−0.48
Ferredoxin (ox.) → (red.)	−0.43
NAD^+ → NADH	−0.32†
Riboflavin → reduced riboflavin	−0.21‡
Acetaldehyde → ethanol	−0.20
Pyruvate → lactate	−0.18
Oxaloacetate → malate	−0.17
Fumarate → succinate	+0.03
Cyt. **b** $Fe^{3+} \rightarrow Fe^{2+}$	+0.07
Ubiquinone (CoQ) → $CoQH_2$	+0.10
Cyt. **c** $Fe^{3+} \rightarrow Fe^{++}$	+0.25
Cyt. **a** $Fe^{3+} \rightarrow Fe^{++}$	+0.29
$Cu^{++} \rightarrow Cu^+$	+0.35
Oxygen, $\frac{1}{2}O_2 \rightarrow H_2O$	+0.82

† NADP has the same redox potential as NAD.
‡ Riboflavin as part of FAD or FMN in flavoproteins has a wide range of redox potentials.

presence of a 50:50 mixture of the oxidized and reduced forms of each redox pair shown on the left-hand side of the table. Many of these substances are electron carriers which play an important role in the process of mitochondrial electron transport described in Chapter 8. Others (themselves often derived from fatty acids and carbohydrates) provide the electrons which, when they move down that transport system, give rise to ATP (compare Section 7.4 and Fig. 7.2).

The use of Eq. (7.7) in the calculation of energy changes associated with the movement of electrons down a redox gradient can be illustrated by reference to the E_0' values of NAD and oxygen in Table 7.3. When electrons 'fall' from NADH to oxygen (so as to give NAD^+ and water under standard conditions), the distance they move down the redox scale is from −0.32 V to +0.82 V. The potential difference $\Delta E_0'$ is consequently 1.14 V. Applying Eq. (7.7) and remembering that $n = 2$, since two electrons are transferred per molecule:

$$-\Delta G^0 = \frac{2 \times 96\,500 \times 1.14}{1000}$$
$$= 220 \text{ kJ/mol}$$

Since the reaction is exergonic, some 220 kJ/mol of energy is made available when one mole of NADH transfers its electrons to oxygen.

A second important example is succinate oxidation to fumarate, oxygen being again the final electron acceptor:

$$\Delta E_0' = 0.82 - 0.03 \qquad \text{or} \qquad 0.79 \text{ V}$$

The number of electrons involved in the equation is again two, and consequently,

$$-\Delta G^0 = \frac{2 \times 96\,500 \times 0.79}{1000}$$
$$= 153 \text{ kJ/mol}$$

The reaction is exergonic but less energy is released than when NADH is the electron donor. It might therefore be correctly anticipated that fewer molecules of ATP are produced when succinate is oxidized than when a substrate first donates electrons to NAD (Section 8.9).

7.9 Adenosine triphosphate and other 'energy-rich' compounds

Since energy is indestructible, it follows that when the energy-rich phosphorus bond of ATP is ruptured, 33 kJ of energy must become available per mole ATP. If the rupturing is by hydrolysis, the energy appears as heat.

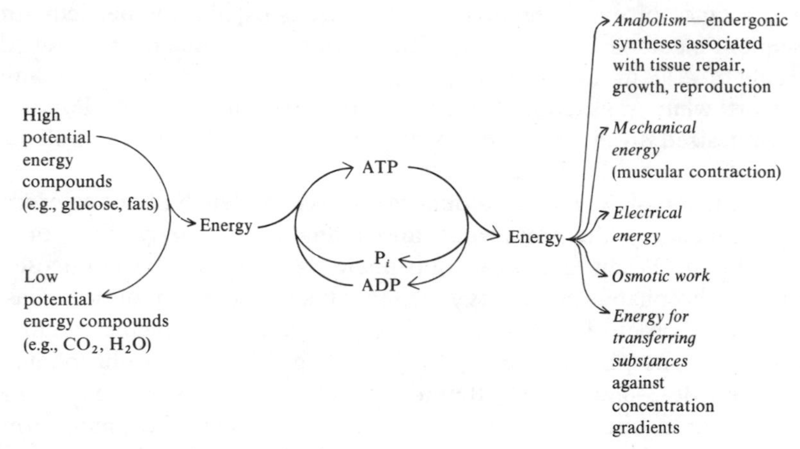

Fig. 7.4 Central role of ATP in cellular energetics

Under biological conditions, however, and in the presence of enzymes, much of the energy is converted to forms other than heat. In particular, it can, as we have seen, be harnessed so that it can increase the potential energy of another chemical system. In later chapters it will be shown that the energy can also be made to do mechanical, osmotic, or other forms of work.

In Fig. 7.4, the ADP/ATP cycle is presented as a coupling device, linking energy released from a catabolic change with absorption of energy by endergonic processes. Often, the terminal phosphate of ATP is not released as inorganic phosphate ions, but instead becomes combined *covalently* with a compound whose chemical potential energy is thereby raised. For example, the initial reaction in the metabolism of glucose is the formation of glucose-6-phosphate, the catalyst being *hexokinase:*

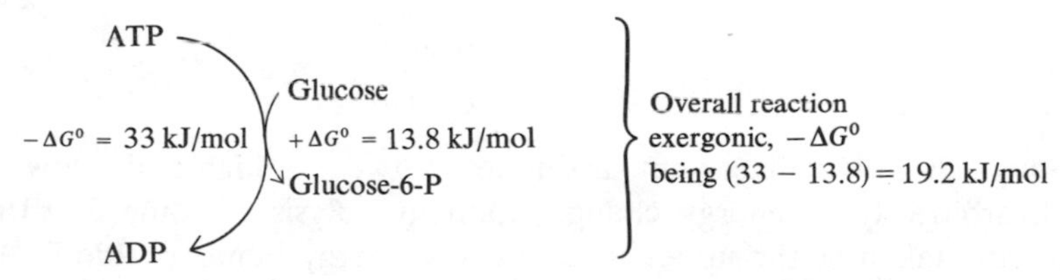

The reaction above is an example of a *priming reaction.* In a priming reaction a compound (in this case, glucose) is converted to a derivative of higher potential energy by the attachment of some additional group to it. Priming is by no means peculiar to biochemistry; if acetic acid is converted to acetyl chloride by the action of phosphorus pentachloride, the product, acetyl chloride, is far more reactive than the original acetic acid. So too is its biological counterpart, acetyl coenzyme A (page 101). In the present instance, glucose-6-phosphate, unlike unphosphorylated glucose, is converted to a substance of even higher energy which is capable of participating in subsequent metabolic conversions. The utilization of some of the energy of a molecule of ATP in the formation of glucose-6-phosphate is a biochemically most worthwhile transaction, for eventually more than thirty ATP molecules are synthesized from thirty of ADP as the primed glucose molecule is oxidized to carbon dioxide and water.

No discussion of ATP or of bioenergetics is complete without reference to what is meant by the somewhat misleading term '*energy-rich*' or '*high energy*' bond. When glucose-6-phosphate is hydrolysed to glucose and inorganic phosphate, the energy change takes the form of a release of energy of the order of 13.8 kJ/mol. Comparable energy changes occur when other sugar phosphates are hydrolysed (e.g., fructose-6-phosphate and glyceraldehyde-3-phosphate). It will be noted that this energy change is small compared with that occurring when the terminal phosphate group of ATP is removed by hydrolysis to form ADP. Compounds such as ATP which undergo reactions yielding a large change in free energy are often called *high energy compounds*, and the particular bond cleaved in the highly exergonic reaction is often described as a *high energy bond* and represented thus: ~. In fact, ATP has two such bonds for, when the terminal and middle phosphate groups are removed by hydrolysis, in each case some 33 kJ/mol are released. On the other hand, the removal of the phosphate group from AMP, adenosine *mono*phosphate, only results in the liberation of some 12 kJ/mol. To indicate the position of the high energy bonds ATP can be written as follows:

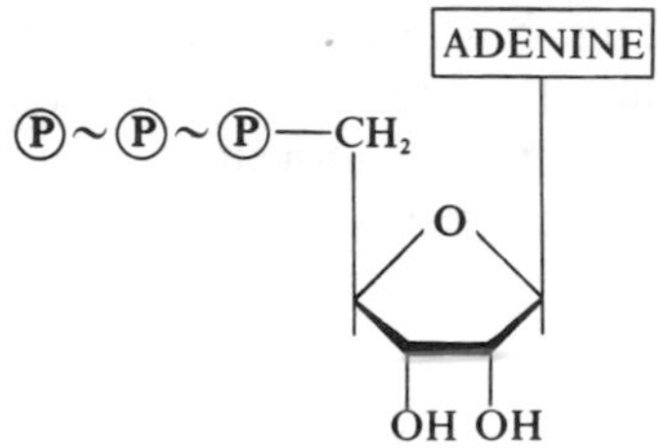

ADENINE

Ⓟ~Ⓟ~Ⓟ—CH_2

O

OH OH

Although there is no clear-cut distinction between a high and a low energy bond, arbitrarily an energy change upon hydrolysis of some 24 kJ/mol is frequently taken as the upper limit for low energy bonds (Table 7.4).

The total energy of ATP and of other high energy compounds is best regarded as being distributed *throughout* the molecule rather than being associated with a particular bond, though at the *moment of cleavage* the bond being cleaved is effectively associated with a disproportionately high level of energy. Upon hydrolysis of ATP, the high energy change occurs because the pyrophosphate (anhydride-type) linkage is replaced by more stable phosphate ions or covalent phosphates—and a 'stable' chemical sub stance is, of course, one which is relatively low in energy. Hence energy is

Table 7.4 Standard free energy changes for the hydrolysis of some biologically important compounds

Compound	$-\Delta G^0$ (kJ/mol) (pH 7, 25°C)
Glycerol-1-P	9.2
Glucose-6-P	13.8
Fructose-6-P	16
Glucose-1-P	20
UDP-glucose (to UDP + glucose)	31
ATP (to ADP + P_i)	33†
ATP (to AMP + pyrophosphate)	36
Acetyl CoA	35
Succinyl CoA	43
Creatine phosphate	44
Carbamyl phosphate	51
Phosphoenol pyruvate	57

† Readers may find that other textbooks quote a different figure. Various books use values ranging between 30 and 40. The accuracy of measurement is probably no greater than 90 per cent, and, additionally, the presence of certain inorganic ions (for example, Mg^{++}) can significantly affect the result. The value of 33 kJ/mol is a useful average figure and is employed throughout this book.

released not because of any unique property of a particular bond, but because of the *overall chemical structure* of the ATP molecule. Nevertheless, the designation 'high energy bond' provides a *convenient shorthand way* of signifying at what locations in molecules particularly important energy changes occur when the compounds participate in biochemical reactions.

An alternative—but not distinct—way of approaching the phenomenon of 'high energy' is to say that the equilibrium constants for the hydrolysis of low energy phosphate esters are nearer to unity than are those for the hydrolysis of high energy phosphate esters. This, however, follows inevitably from the relationship between the equilibrium constant and the standard free energy represented in Eq. (7.6) and Table 7.1. If the gap between the standard free energy of reactant and product is large, the reaction will inevitably be more nearly irreversible than if it is small. For ATP hydrolysis, $-\Delta G^0 = 33$ kJ/mol, so $\log_{10} K$ is evidently greatly in excess of 5.0 (Table 7.1). For the hydrolysis of glucose-6-phosphate, on the other hand, $-\Delta G^0 =$ 13.8 kJ/mol (Table 7.2), so $\log_{10} K$ is just a little more than 2.2.

One important product of glucose metabolism is known as *phosphoenol pyruvate* (PEP). It is a high energy compound bearing a phosphate group. When PEP undergoes simple hydrolysis pyruvate and inorganic orthophosphate are formed, the reaction being highly exergonic ($-\Delta G^0 = 57$ kJ/mol). However, within the cell, an enzyme, *pyruvate kinase*, normally catalyses the transfer of the phosphate from PEP to ADP, much of the energy of cleavage accompanying the phosphate, in the sense that it provides the energy needed

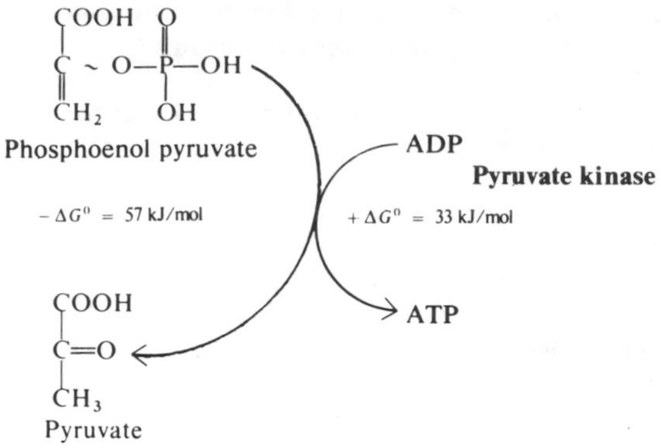

COOH O
C ~ O—P—OH
CH$_2$ OH
Phosphoenol pyruvate

ADP
Pyruvate kinase

$-\Delta G^0 = 57$ kJ/mol

$+\Delta G^0 = 33$ kJ/mol

COOH
C=O
CH$_3$
Pyruvate

ATP

(a) Transfer of high energy phosphate from phosphoenol pyruvate to ADP (ΔG^0 for coupled reaction = $-57 + 33$ or -24 kJ/mol)

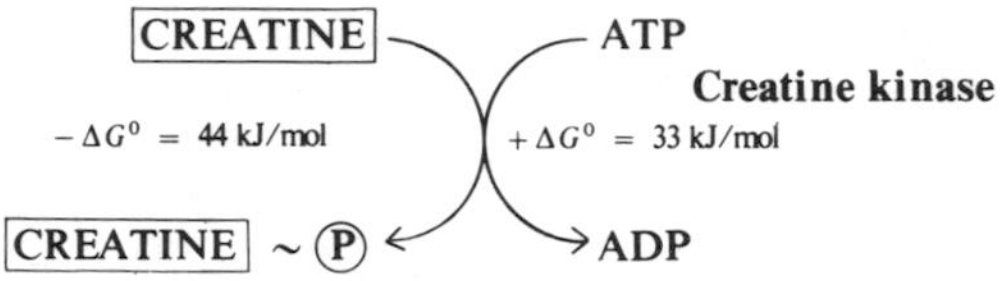

CREATINE

ATP
Creatine kinase

$-\Delta G^0 = 44$ kJ/mol

$+\Delta G^0 = 33$ kJ/mol

CREATINE ~ Ⓟ

ADP

(b) The 'Lohmann' reaction ($\Delta G^0 = -44 + 33 = -11$ kJ/mol). For the formula of creatine, see page 416.

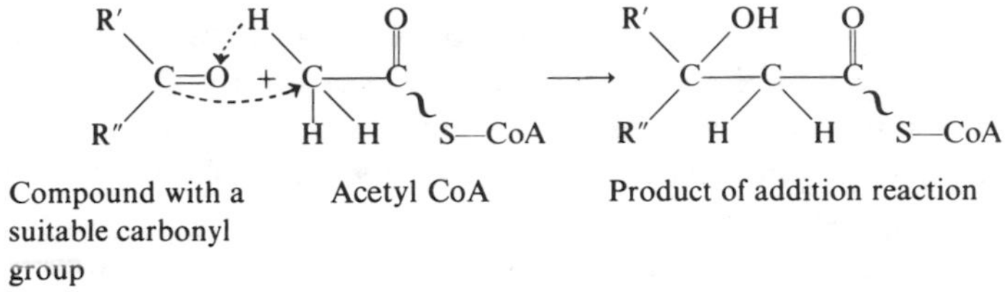

R′ C=O R″ + H–C(H)(H)–C(=O)–S—CoA ⟶ R′ R″ C(OH)–C(H)(H)–C(=O)–S—CoA

Compound with a suitable carbonyl group

Acetyl CoA

Product of addition reaction

(c) A typical reaction of high energy acetate (acetyl coenzyme A)

Fig. 7.5 Three reactions involving high energy compounds

to attach the phosphate to the ADP and so to raise the energy level of the latter to the higher level of ATP (Fig. 7.5a). This is one way in which ATP is synthesized in the cell. PEP is clearly an example of compound C in Fig. 7.2. Pyruvate kinase is one of a family of *kinases* or phosphoryl transferases, each of which either transfers organic phosphate to ADP (so converting it to ATP) or, as in the case of hexokinase (Section 11.4), passes phosphate from ATP to an organic compound, thereby converting it to a phosphate ester.

A rather similar reaction is known as the Lohmann reaction but it differs from the preceding examples in that it is *reversible*. In vertebrate muscle, ATP reacts with creatine to form a high energy compound, creatine phosphate (Fig. 7.5b). The reaction, catalysed by creatine kinase, is of particular importance because ATP never accumulates in cells. When the muscle cell needs more ATP than it can synthesize in the time available, creatine phosphate donates its phosphate group to ADP and so regenerates ATP. For this reason, creatine phosphate acts as an energy store and is called a *phosphagen*. The creatine kinase reaction is reversible because the free energy change is relatively small. By contrast, the reaction catalysed by pyruvate kinase is not reversible for the coupled reaction involves too high an energy change for pyruvate to be directly phosphorylated by ATP.

Acetyl coenzyme A is yet another example of a high energy compound. Figure 7.5c shows one sort of reaction in which this compound can participate—it is an addition reaction which resembles the aldol condensation of classical organic chemistry. The 'acetyl' part of acetyl coenzyme A has a higher energy than free acetate (Section 5.5) and, unlike the latter, is able to 'condense' with suitable carbonyl groups. Several examples of this special kind of addition reaction will be given in later chapters.

It is not possible to present or to acquire a very meaningful picture of what biochemistry is all about without giving consideration to the energy changes which accompany biochemical reactions, for all reactions which occur in cells must of necessity be both *mechanistically* and *energetically feasible*. This should be constantly kept in mind as, in later chapters, the various biochemical pathways are described. Interesting though the machinery of the living engine frequently proves to be, it is nevertheless the energetic function that the machinery fulfils which is worthy of special consideration, for this function is the *raison d'être* for its existence.

8

Mitochondrial electron transport and oxidative phosphorylation

8.1 The purpose of electron transport

In Chapter 1, the mitochondrion was described as the 'power house of the cell' and it is proposed now to show why it merits this name. To do this, it will be necessary to make use of certain energetic considerations outlined in the previous chapter, and especially the idea of electron loss as an essential factor in oxidation. To show the subject matter of this chapter in perspective, the overall process of mitochondrial oxidation and electron transport will be briefly described first; then follows a more detailed consideration of the mechanisms of electron transfer and of an important process in which the energy lost by the electrons leads to the formation of ATP.

The oxidation of carbohydrates and of fats results in the production of *acetyl coenzyme A* (Section 1.8), the acetyl part of which undergoes controlled oxidation with the eventual production of ATP, carbon dioxide, and water. The controlled oxidation is brought about by an elegant mechanism called the tricarboxylic acid cycle, which is a catalytic device involving a cyclic series of reactions. The nature of the reactions is not our present concern; they represent the *machinery* of the oxidation whereas this chapter is concerned with the *energetic purpose* for which that machinery exists. All that needs to be stressed now is that, by means of this machinery, the breakdown of the acetyl unit is subdivided into four oxidative stages, and that the electrons released at each of these four stages are passed down a redox gradient until in each case they reduce oxygen to water. Thus, if the machinery of oxidation is represented as the circumference of a wheel (Fig. 8.1) the electron transport redox systems can be represented as four spokes directed to a central hub of oxygen.

The oxidizable substrates which are the electron donors for the four 'spokes' are *isocitrate*, *α-ketoglutarate*, *succinate*, and *malate*. Each of these, when oxidized, loses two electrons together with two hydrogen ions, the electrons passing down the energy gradient of the transport system. When

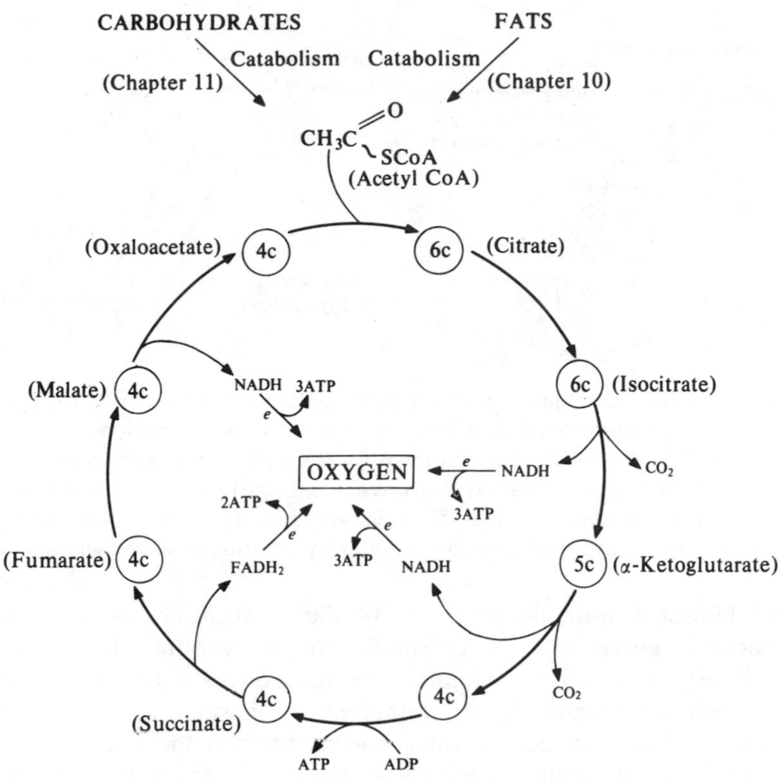

Fig. 8.1 A simple outline of the tricarboxylic acid cycle. (Acetyl coenzyme A, resulting from carbohydrate or fatty acid catabolism, enters the cycle by combining with a four-carbon compound. One turn of the cycle produces 3NADH, $1FADH_2$, and $2CO_2$. The reduced carriers are the source of energy for ATP synthesis by oxidative phosphorylation, $\xrightarrow[ATP]{e}$ indicates the passage of electrons from carriers to oxygen)

electrons originate from succinate, the transport system is in certain small respects different from that which operates when the electrons come from the other three substrates, but for the moment any one of the four substrates can be represented by AH_2:

$$AH_2 \xrightarrow{AH_2\text{ dehydrogenase}} A + 2H^+ + 2e^-$$

The electrons are eventually accepted in turn by each member of a series of electron carriers neatly arranged on the inner membranes (cristae) of the mitochondrion; they pass along the chain of carriers, each of which has a

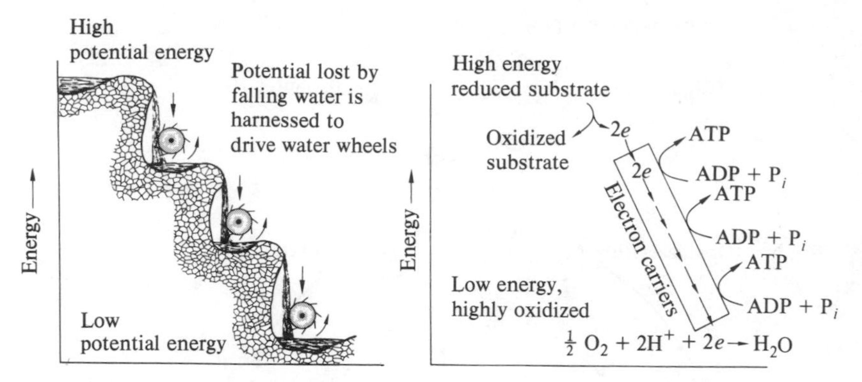

Fig. 8.2 The flow of electrons along a series of electron carriers from a high energy level to a lower level is used to drive the endergonic reaction ADP + $P_i \rightarrow$ ATP. The diagram on the right represents one 'spoke' of the previous figure. At the lowest energy level, the electrons finally reduce oxygen to water. The system is energetically analogous to the energy capture of cascading water by a series of water wheels (left)

potential lower than the previous one, till the electrons finally reach oxygen; the process is, energetically, a 'downhill' process, even though it is 'along' a chain of carriers. The total system can be likened to a waterfall divided into several separate cascades by the intervention of water wheels situated at different heights. Just as the water wheels harness the energy from the falling water so the mitochondrion uses the energy of oxidation to drive a concomitant 'uphill' reaction in which inorganic phosphate and ADP react to give the high energy compound ATP (Fig. 8.2). This latter process is termed *oxidative phosphorylation* and contrasts with reactions such as the one catalysed by pyruvate kinase (Fig. 7.5a) in which high energy phosphate is transferred to ADP without the intervention of oxygen. Reactions of this latter type are called *substrate level phosphorylations.*

8.2 NAD, the first electron acceptor

The coenzyme NAD is the initial electron acceptor in the oxidation of three of the substrates mentioned in Section 8.1, namely, isocitrate, α-ketoglutarate, and malate. NAD is therefore the first of the series of electron carriers so important in oxidative phosphorylation.

The mechanism of the reversible oxidation-reduction of NAD was described in Chapter 5: the reduction of NAD^+ to NADH involves the addition of *two electrons* (and one proton) to the nicotinamide part of the molecule. The nicotinamide moiety is thereby converted from a heterocyclic ring with *three* double bonds to a ring with *two* double bonds (Section 5.2).

This process is similar to numerous redox reactions involving organic compounds; reduction involves the gain of two electrons and the loss of a double bond, while conversely, the loss of two electrons and the formation of a double bond is frequently observed when organic compounds are oxidized.

A specific example will illustrate how NAD^+ links with substrate oxidation. By referring to Fig. 8.1 it will be seen that the conversion of malate to oxaloacetate is one of the reactions of the tricarboxylic acid (TCA) cycle. One of the 'spokes' of the wheel is located here; in other words, when the enzyme *malate dehydrogenase* oxidizes malate to oxaloacetate, the specific cofactor of the enzyme, NAD^+, is reduced to NADH. The oxidation and reduction phases of the reaction are closely coupled but can be described in two parts. First, two electrons and two protons are removed from the malate, yielding oxaloacetate. Second, the two electrons and a proton combine with NAD^+ producing the reduced form, NADH.

Malic acid: $HOOC{-}CH_2{-}CH(OH){-}COOH$ —① → $2e + 2H^+$ —② → $NADH + H^+$ (with NAD^+); $H^+ \xrightarrow{H_2O} H_3O^+$

Oxaloacetic acid: $HOOC{-}CH_2{-}C(=O){-}COOH$

Only one of the two protons removed from the malate is transferred to NAD^+; the other combines with water, tending to make the (buffered) solution slightly acidic. Usually, this reaction is written in the simpler form:

$$HOOC{-}CH_2{-}CH(OH){-}COOH + NAD^+ \xrightarrow[\text{dehydrogenase}]{\text{Malate}} HOOC{-}CH_2{-}C(=O){-}COOH + NADH + H^+$$

For notes on how to remember the formulae of malate and oxaloacetate, see Section 9.1.

8.3 Flavin mononucleotide (FMN)

The NADH produced by the malate dehydrogenase reaction and by certain other mitochondrial enzymes is reoxidized to NAD^+ by passing the two electrons to the prosthetic group of a flavoprotein. This protein, like many other components of the electron transport chain, forms an integral part of the inner mitochondrial membrane. The prosthetic group, flavin mononucleotide (FMN), provides the mechanism for electron transport by this, the second electron transport carrier of the chain. The protein of this carrier catalyses the removal of electrons from NADH and so is also an enzyme. Indeed, this example illustrates the small difference in meaning which sometimes (but not always) exists between 'carrier' and 'enzyme'. The apoenzyme passes the electrons to the tightly bound FMN, so the whole structure (which also contains iron and sulphur) is *both* the second carrier in the sequence *and* an enzyme, *NADH dehydrogenase.*

The molecular changes which occur as NADH is oxidized and FMN is reduced are described in Sections 5.2 and 5.4, but their essential features are shown in Fig. 8.3. A minor complication is that when NADH is oxidized, two electrons but only one proton are lost, whereas when FMN is reduced, two electrons and two protons enter the molecule. The extra hydrogen ion is taken up from the surrounding medium. Hydrogen ions are responsible for the acidity of the medium but, since cellular fluids are buffered, only a small fall in pH in the liquid outside the inner mitochondrial membrane has, in fact, been observed.

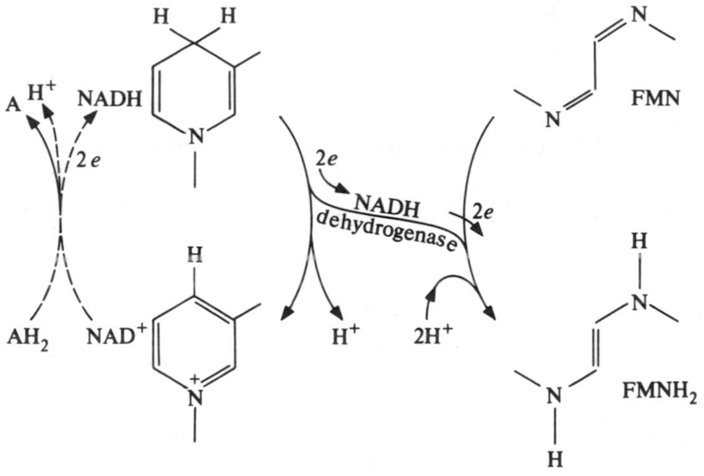

Fig. 8.3 Disposition of protons and electrons in the NADH dehydrogenase reaction (partial formula). One proton enters the environment for each molecule of substrate, AH_2, oxidized. Another enters when NADH is reoxidized by coupling to FMN, but FMN requires two environmental protons to accompany the two electrons originating from NADH

An important question may now be asked. Is the difference between the redox potentials of the $NADH/NAD^+$ and FMNH/FMN systems sufficient to allow a molecule of ATP to be synthesized from ADP and P_i? The energy change under standard conditions could theoretically be obtained by inserting the appropriate E_0' values into Eq. (7.7). Unfortunately, while E_0' for the $NADH/NAD^+$ system is known (−0.32 V), the precise value for FMN is not. It is probably a little more positive than the value for the riboflavin system shown in Table 7.3. Taking a value of −0.15 V, a value of about 33 kJ/mol is obtained. This is just sufficient to allow the coupled reaction

$$P_i + ADP \rightleftharpoons ATP$$

to be freely reversible. Experimental evidence demonstrates, in fact, that ATP *is* synthesized at this stage (page 172).

8.4 Succinate dehydrogenase and FAD

Of the four oxidizable substrates derived from acetyl coenzyme A, succinic acid is exceptional in not passing electrons to NAD. There is instead a lipoflavoprotein which specifically catalyses the oxidation of succinate to fumarate and is known as succinate dehydrogenase.

Flavoproteins are conjugate proteins containing a derivative of riboflavin as prosthetic group. The protein which oxidizes NADH (Section 8.3) is an example of one in which the prosthetic group is flavin *mono*nucleotide (FMN). By contrast, succinate dehydrogenase contains *flavin adenine dinucleotide* (FAD). FAD consists of the nucleotides AMP and FMN linked together by a pyrophosphate bond to form a dinucleotide.

The structure of FAD is described in Section 5.4. It should be compared with that of the dinucleotide NAD (Section 5.2). The riboflavin part of the FAD is reduced by electrons derived directly from succinate, *without the intervention of NAD*:

Succinic acid: COOH–CH_2–CH_2–COOH → $2e + 2H^+$ → $FADH_2$ (from FAD)

Fumaric acid: H(COOH)C=C(H)HOOC

It is noteworthy that succinate is oxidized to the *trans* unsaturated dicarboxylic acid, fumarate, and never to the *cis* isomer.

Once reduced in this manner, the $FADH_2$ is reoxidized in its turn by passing the electrons to ubiquinone in exactly the same way as is described in relation to the reoxidation of $FMNH_2$ in the next section.

8.5 Ubiquinone

Mitochondria contain considerable quantities of a quinone known as ubiquinone, or coenzyme Q. Its reduction involves the addition of two electrons together with two protons, resulting in a decrease in the number of double bounds from four to three (Fig. 8.4). $FMNH_2$ is oxidized by mitochondrial ubiquinone to FMN, the ubiquinone being thereby reduced to a diphenol:

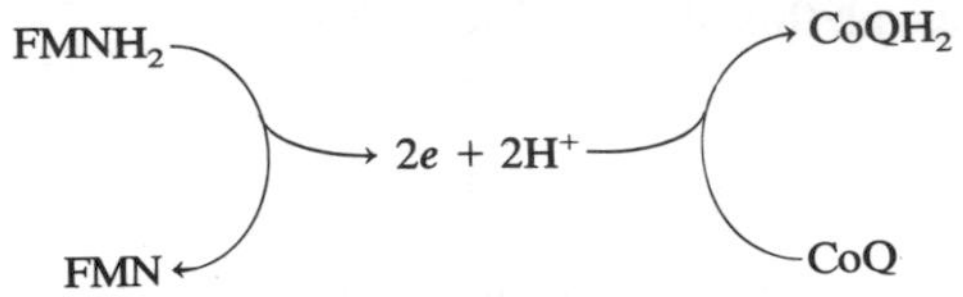

$$FMNH_2 \rightarrow FMN, \quad 2e + 2H^+, \quad CoQ \rightarrow CoQH_2$$

As was mentioned in Section 8.4, the reoxidation of $FADH_2$, the prosthetic group of succinate dehydrogenase, occurs in an identical manner.

The long unsaturated side chain of coenzyme Q renders the molecule readily soluble in the lipids of the inner membrane of the mitochondrion. It is not firmly attached to a protein and can in consequence diffuse short

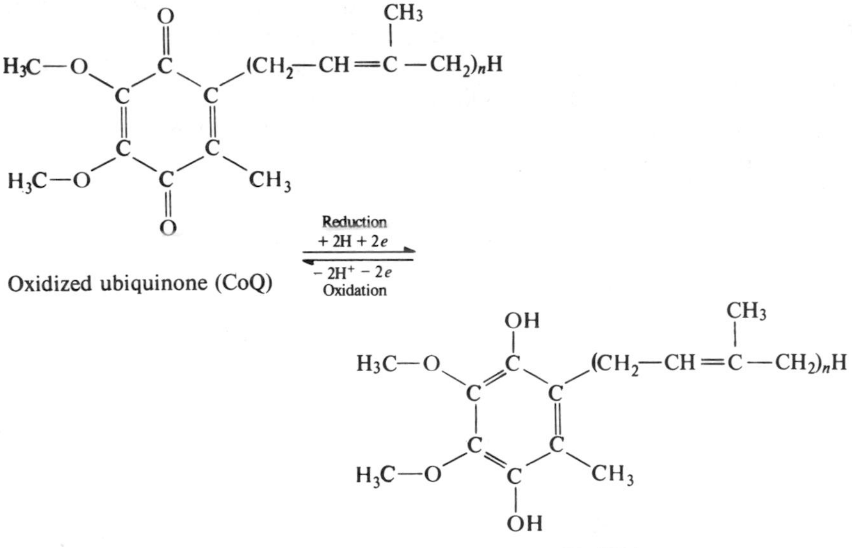

$H_3C—O$, $H_3C—O$, $(CH_2—CH{=}C(CH_3)—CH_2)_nH$, CH_3, O, O

Oxidized ubiquinone (CoQ)

Reduction $+2H + 2e$ ⇌ $-2H^+ - 2e$ Oxidation

$H_3C—O$, $H_3C—O$, OH, OH, $(CH_2—CH{=}C(CH_3)—CH_2)_nH$, CH_3

Reduced ubiquinone ($CoQH_2$)

Fig. 8.4 The structure of oxidized and reduced ubiquinone (coenzyme Q). (*n* is 10 in mammalian mitochondria and 6 in most microorganisms)

distances within the membrane. It is thus a 'carrier' in the strictest sense, unlike FMN or FAD. We shall see later (Section 8.8) that the electron transport system is actually spatially divided into separate 'particles', and the function of coenzyme Q is to remove electrons from one particle (number I) and convey them to another (number III).

8.6 Cytochrome b

Reduced mitochondrial ubiquinone ($CoQH_2$) must be reoxidized if it is to continue to function as an electron carrier capable of accepting electrons from $FMNH_2$. The two electrons are transferred to a lipoprotein, cytochrome **b**, which, like all other cytochromes, contains an iron ion which undergoes reversible oxidation and reduction between the ferrous and ferric states.

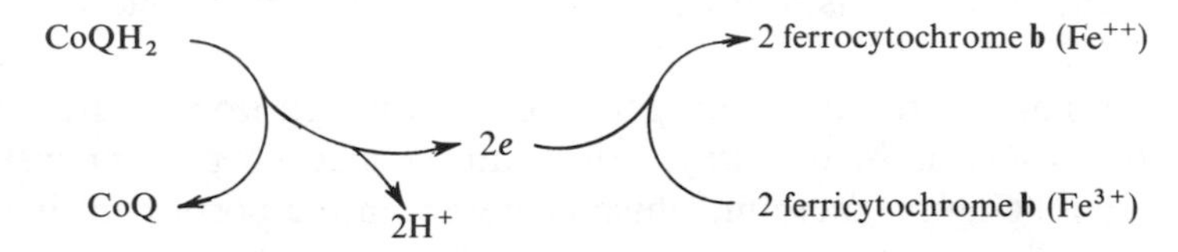

Since the reaction ($Fe^{3+} + e \rightleftharpoons Fe^{++}$) requires the transfer of only *one* electron, two molecules of cytochrome **b** are involved in the oxidation of one molecule of reduced ubiquinone. The two protons released in the reaction are temporarily accommodated in the buffered solution nearby, being (statistically) accounted for at the cytochrome oxidase stage described later.

Cytochrome **b**, like several other cytochromes participating in the mitochondrial electron transport system, is of interest because it possesses an iron-containing haem prosthetic group. The group present in cytochrome **b** is called *protohaem*; the same prosthetic group is an essential constituent of haemoglobin and myoglobin (Chapter 17). Protohaem belongs to a family of compounds known as *porphyrins*, many of which are important biological pigments. Like other porphyrins, protohaem has, as its nucleus, four pyrrole rings (Fig. 8.5). This nucleus is highly 'aromatic', each bond having some single- and some double-bond characteristics, so that, as in the case of benzene or the purine and pyrimidine bases, it is not possible to draw a formula which accurately represents the position of the two 'different' types of bond. In the various porphyrins, different side groups are often attached to the nucleus at the eight outer carbon atoms. For protohaem (Fig. 8.5) the side groups, numbered 1 to 8, are indicated in light type. The porphyrin nucleus is planar, and frequently, as in the case of protohaem in haemoglobin or cytochrome **b**, there is a central iron ion which is held by four

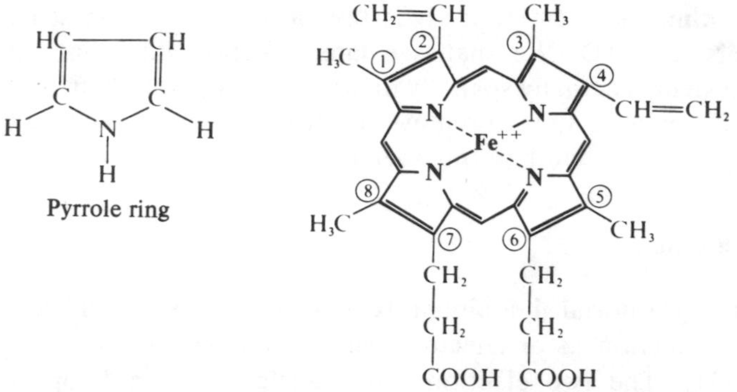

Pyrrole ring

Fig. 8.5 Protohaem—the prosthetic group of cytochrome **b**, haemoglobin, and myoglobin. The porphyrin nucleus (heavy type) has eight positions for side groups; those for protohaem are indicated in light type. The iron ion has four coordinate links with nitrogen atoms

coordinate links to the four nitrogen atoms. Although iron is often coordinated to porphyrins in this way, other cations may be present instead, a notable case being magnesium, which is present in the porphyrin-like group of chlorophyll. In different flavins and cytochromes, the prosthetic groups are also, of course, attached to different *proteins*.

8.7 The other cytochromes

Reduced cytochrome **b** is reoxidized by passing an electron on to a second porphyrin lipoprotein, cytochrome $\mathbf{c}_1$, the iron of which is thereby converted from the ferric to the ferrous state:

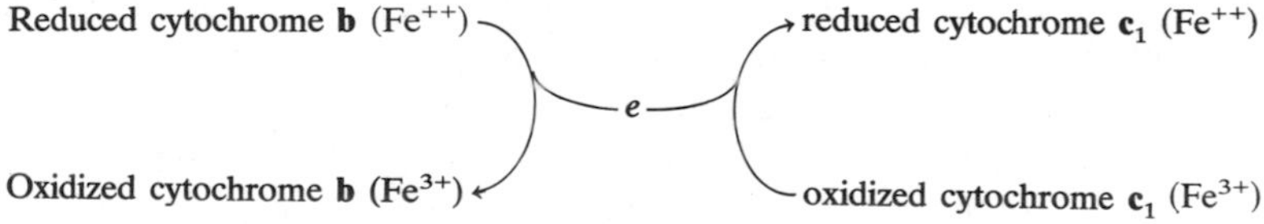

Reduced cytochrome **b** (Fe^{++}) → *e* → reduced cytochrome $\mathbf{c}_1$ (Fe^{++})

Oxidized cytochrome **b** (Fe^{3+}) ← oxidized cytochrome $\mathbf{c}_1$ (Fe^{3+})

Thus electrons derived from an enzyme reaction such as that catalysed by malate dehydrogenase reach cytochrome $\mathbf{c}_1$ by the following route:

$$\text{Malate} \rightarrow \text{NAD} \rightarrow \text{FMN} \rightarrow \text{ubiquinone} \rightarrow \text{cyt. } \mathbf{b} \rightarrow \text{cyt. } \mathbf{c}_1$$

Their final destination is oxygen and they reach it from cytochrome $\mathbf{c}_1$ via at least two additional cytochromes, known by the letters **c** and **a**. Like cytochrome **b**, these contain a porphyrin ring with iron at the centre. The transfer of electrons is associated with a ferric to ferrous valency change on the part of the acceptor cytochrome and with a valency change in the other direction on the part of the donor. Each change is comparable to the reduction of cytochrome $\mathbf{c}_1$ by reduced cytochrome **b**.

The porphyrins of cytochrome $\mathbf{c}_1$ and **c** are identical, being similar to the protohaem of cytochrome **b** except that the vinyl groups of side chains 2 and 4 (Fig. 8.5) are linked to two cysteine residues of the lipoprotein molecule. The porphyrin of cytochrome **a** resembles protohaem except that the methyl group at position 8 is replaced by —CHO, and there is a long branched aliphatic side chain at position 2 instead of the vinyl group. All but one of the cytochromes are insoluble, being firmly bound to the inner mitochondrial membrane. The exception is cytochrome **c**, which is soluble in water. It fulfils a role similar to that of ubiquinone, in that it acts as a diffusible carrier between fixed cytochromes in two different electron transport particles situated a short distance apart (Section 8.8). Some of the differences between the mitochondrial cytochromes are summarized in Table 8.1. It will be seen that the difference between appropriate standard redox potentials is such that ATP can be generated as electrons pass from reduced cytochrome **b** to oxidized cytochrome $\mathbf{c}_1$, but that it cannot be formed as electrons pass from cytochrome $\mathbf{c}_1$ to **c** or from cytochrome **c** to **a**.

The terminal link in the electron transport chain is a lipoprotein complex known as *cytochrome oxidase.* It is given this name because the electrons it receives from reduced cytochrome **c** are passed on to molecular oxygen. Its precise structure is uncertain, but it contains two porphyrins and two ions of copper which give the purified cytochrome a greenish colour. The two porphyrins are identical (Table 8.1) and the conjugate proteins containing

Table 8.1 Physical and chemical properties of cytochromes

Cytochrome	Location†	Membrane bound	E_0' volts	Molecular weight	Notes on chemical structure (position numbers as in Fig. 8.5)
b	ETP III	Yes	+0.07	25 000	Vinyl ($—CH{=}CH_2$) groups in positions 2 and 4
$\mathbf{c}_1$	ETP III	Yes	+0.22	37 000	Thioether [$—CH(CH_3)—S—$protein] group in positions 2 and 4
c	Mobile carrier from ETP III to IV	No	+0.25	12 500	As $\mathbf{c}_1$
$\mathbf{a} + \mathbf{a}_3$ (cytochrome oxidase)	ETP IV	Yes	(+0.29)	200 000	A complex containing two porphyrins and two copper ions Each haem has a chain containing seventeen carbon atoms in position 2, a vinyl group in position 4, and a —CHO group in position 8

† See Section 8.8 (ETP = electron transport particle).

them are designated cytochrome **a** and $\mathbf{a}_3$ respectively. They can be distinguished by their different sensitivities to certain poisons: in particular, cytochrome $\mathbf{a}_3$ is inactivated by carbon monoxide. The standard redox potential of the cytochrome $\mathbf{a}_3$ moiety is approximately +0.38, and it appears that a molecule of ATP is synthesized as electrons pass from the porphyrin of cytochrome **a** to that of $\mathbf{a}_3$.

The mechanism by which reduced cytochrome $\mathbf{a}_3$ transfers its electrons to molecular oxygen is unknown, but the overall process can be represented as follows:

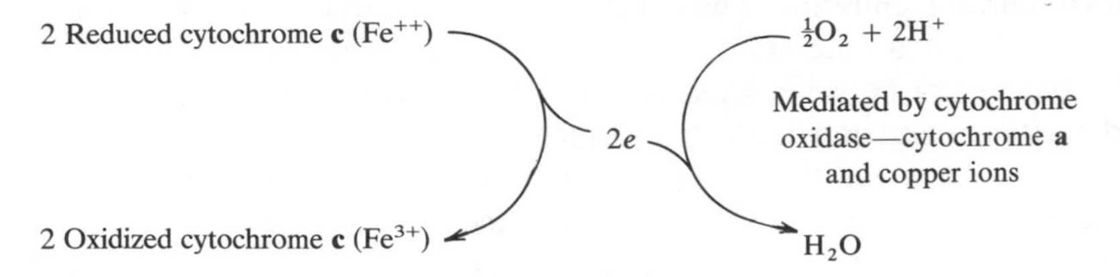

Note that, because two electrons and two protons are required to convert $\frac{1}{2}O_2$ to H_2O, it is necessary for stoichiometric reasons to indicate the participation of two molecules of reduced cytochrome **c**. The copper associated with cytochrome $\mathbf{a}_3$ plays a part in the reaction which leads to the reduction of oxygen; it probably undergoes a cuprous-cupric valency change in the process.

8.8 The site of oxidative phosphorylation in the cell

In Chapter 1 it was seen that the mitochondrion consists of a folded inner membrane and a separate outer membrane. There is a body of evidence to suggest that the electron transport components from NAD to cytochrome oxidase are located in small particles (electron transport particles) forming an integral part of the inner membrane. By means of fractionation techniques it has been possible to break up these particles and thus to show that they consist of four subparticles closely bound together. They are usually numbered I, II, III, and IV.

Particle I contains FMN and is responsible for passing electrons from NADH to ubiquinone (Fig. 8.6). Particle II is succinate dehydrogenase, particle III contains several catalysts, including cytochromes **b** and $\mathbf{c}_1$, and particle IV is cytochrome oxidase. Ubiquinone and cytochrome **c** are apparently not bound to any particle but act as mobile electron carriers between the particles. Oxidative phosphorylation occurs as a result of electron transport in particles I, III, and IV but not in II, in accordance with the observation that succinate oxidation produces one less molecule of ATP than does the oxidation of malate, isocitrate, or α-ketoglutarate, all of which

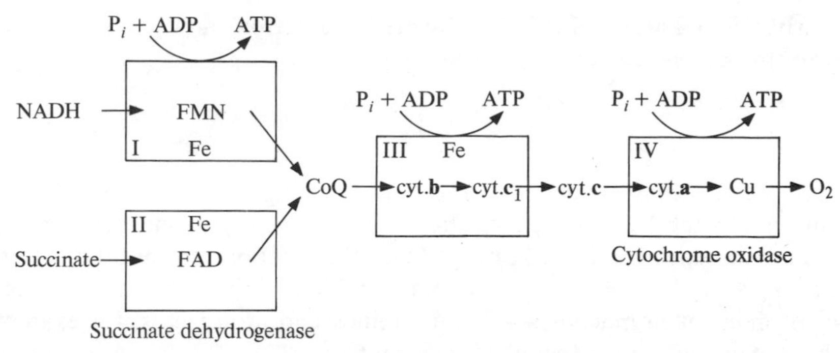

Fig. 8.6 Outline of the electron transport assembly showing the four particles. (Although drawn separately, they are closely linked together. Coenzyme Q and cytochrome **c** are mobile and shuttle the electrons between the particles at the points indicated. Note that ATP synthesis only occurs in particles I, III, and IV. Fe indicates non-haem iron protein)

pass their electrons, via NAD, to particle I. Each particle apart from IV also contains considerable quantities of a non-haem iron protein.

8.9 Oxidative phosphorylation

The major function of the mitochondrial electron transport system is the synthesis of ATP. The quantity of inorganic phosphate, P_i, which joins with ADP to form ATP can be measured, as can the quantity of oxygen converted to water. This has been achieved using both intact isolated mitochondria and inner membrane particles derived from mitochondria by their partial disintegration. The ratio of the number of atoms of inorganic phosphorus incorporated into ATP to the number of atoms of oxygen consumed is known as the P:O ratio. The measured P:O ratio for NADH oxidation is approximately 3.0, while for succinate oxidation it is 2.0. This is in agreement with the prediction in Section 7.8, for it means that less ATP is synthesized when succinate is oxidized than when the substrate is malate, isocitrate, or α-ketoglutarate.

In the previous chapter, it was explained that ATP synthesis was a formally endergonic process involving an input of some 33 kJ/mol. So, if 3 mol of ATP are synthesized from ADP and P_i, 3×33 or 99 kJ of energy are captured. The oxidation of NADH was shown in Section 7.8 to be exergonic and to release some 220 kJ/mol. Hence, from the point of view of energy capture, the efficiency with which electrons are transported from NADH along the electron transport chain is

$$\frac{99}{220} \times 100 = 45\%$$

Similarly, when 2 moles of ATP are formed as electrons flow from $FADH_2$ to oxygen, the efficiency is

$$\frac{2 \times 33}{152} \times 100 = 43.4\%$$

These calculations assume that the energy changes occur under standard conditions. In the living cell, the efficiency may well be somewhat different since non-standard conditions prevail (e.g., the solutions are not molar). It is interesting to note that the figures compare favourably with those for many types of man-made machines—steam engines and motor cars, for example, work at efficiencies well below 15 per cent.

It is found that mitochondria deprived of oxygen fail to show significant oxidative phosphorylation. Under anaerobic conditions, each component of the electron transport chain becomes and remains in the reduced state and hence both the flow of electrons along the chain and the synthesis of ATP ceases. Similarly, poisons such as carbon monoxide and cyanide are able to react with certain members of the chain (e.g., cytochrome oxidase) so as to stop electron transport, thus preventing oxidative phosphorylation.

As was seen in the previous section, the three molecules of ATP produced when the electrons from NADH pass down the electron transport chain are thought to be synthesized at three separate points in the chain. The exact mechanism by which this occurs is, however, not yet understood. There is some evidence that when the reduced form of a carrier C passes on its electrons, the energy of the oxidation is conserved by the simultaneous formation of a high energy compound, which can be represented, for example, as $C^+ \sim X$. The decomposition of this compound is thought to be coupled directly or indirectly with the ADP → ATP system, so that the reaction, in its simplest form becomes

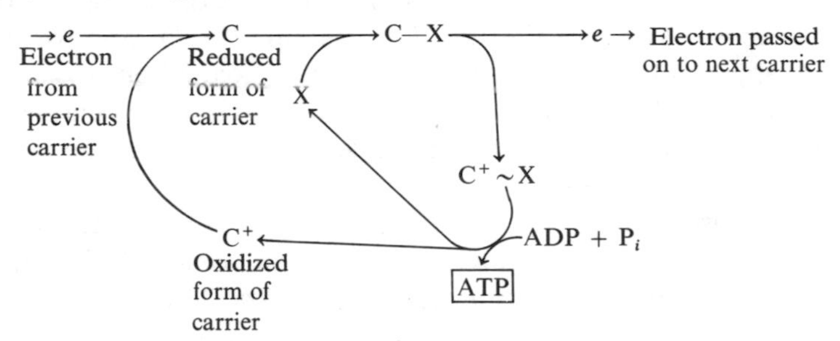

Certain poisons known as 'uncoupling agents' (e.g., 2,4-dinitrophenol and its homologues) cause an increase in the rate of electron flow down the chain and hence greater oxygen consumption, but ATP synthesis nevertheless ceases. The reason is thought to be that they cause the link between C and X to break ($C^+ \sim X \rightarrow C^+ + X +$ heat), so making C^+ available more

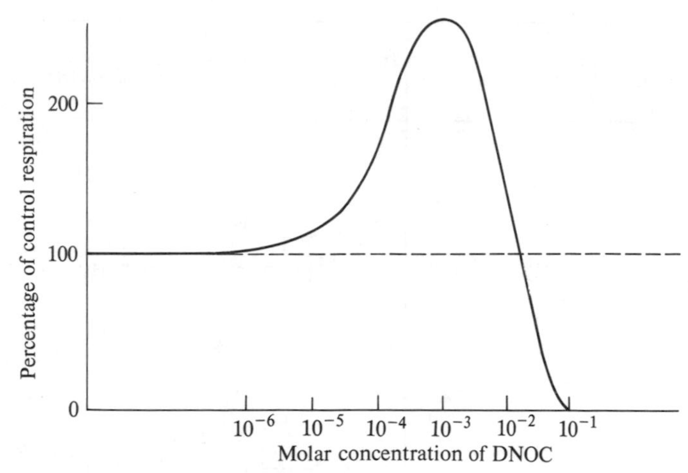

Fig. 8.7 Effect of dinitroorthocresol on the respiration of starved *Chlorella vulgaris* (pH 6.4, 30°C, in the dark)

quickly for further electron transport. In other words, the phosphorylation which accompanies oxidation is normally the pacemaker, acting as a friction brake on the whole process. An analogy is offered by the case of a motor car being driven with a slipping clutch; the more the clutch is allowed to slip, the more the accelerator must be depressed to maintain constant speed, until eventually no amount of 'gas' will compensate for the fully depressed clutch, and motion ceases. Figure 8.7 illustrates how the herbicide 2,4-dinitro-6-methylphenol (DNOC) stimulates oxygen uptake over a certain concentration range; above about 10^{-3} *M*, death of the plant ensues.

A major objection to the high energy *intermediate complex* theory (of which that described above is but one variant) is that no such complex has ever been isolated. Moreover, the theory does not explain why mitochondrial membrane integrity is indispensible for phosphorylation to occur. Other theories have therefore been proposed.

One of these theories postulated that an intermediate exists but that it is an integral macromolecular component of the membrane. The suggestion is that this component undergoes a *conformational* change (compare Section 3.9) to a high energy form as a result of electron transport in its vicinity. The high energy form then relinquishes its energy in such a way as to lead to ATP synthesis. This theory is reminiscent of that proposed to account for muscular contraction (Section 20.4), during which process ATP is hydrolysed to ADP and inorganic phosphate.

The currently most popular proposition is known as the *chemiosmotic* theory, first proposed by H. Lundegardh and later developed by P. Mitchell. This postulates that the carriers responsible for electron transport are so

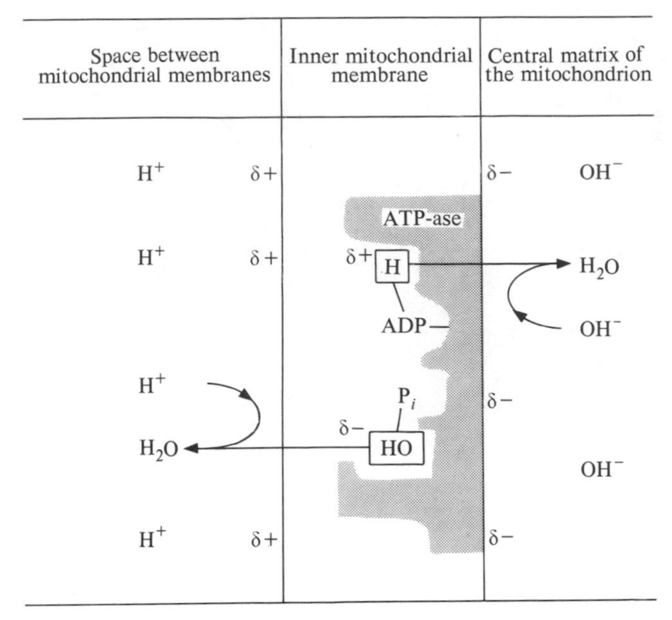

Fig. 8.8 Diagram to illustrate the chemiosmotic theory of oxidative phosphorylation

arranged in the body of the inner mitochondrial membrane that the combined effects of their activity is the pumping of hydrogen ions to the outside of the membrane and of hydroxyl ions to the inside. This would result in the creation of a potential difference across the membrane, the outer side (i.e., that nearest to the outer membrane) becoming positive with respect to the inside. This gradient is then postulated to result in hydroxyl ions being 'sucked' from P_i ions attached to a membrane-bound ATP-ase to neutralize hydrogen ions on the outside of the membrane. Hydrogen ions are postulated to be similarly displaced from a nearby active site where the ADP is bound; they then move in such a way as to neutralize hydroxyl ions on the inside of the membrane (Fig. 8.8). Thus, according to this hypothesis, the result would be a *dehydration* reaction—i.e., the ATP-ase would work in reverse, leading to ATP synthesis:

$$\text{ADP} + \text{P}_i \xrightarrow[\text{Minus OH}^-]{\text{Minus H}^+} \text{ATP} + \text{H}_2\text{O}$$

The theory has certain advantages. It has been shown that the mitochondrial membrane is impermeable to hydrogen ions; consequently, an active pumping system makes sense. Moreover, the mitochondrial membrane has indeed a trans-membrane potential when electron transport is taking place, the outside being positive with respect to the inside. Again, membrane-bound ATP-ase occurs in many membranes, and is associated sometimes with ion movements (compare, for example, the sodium pump, Section 17.6). Finally, uncouplers of oxidative phosphorylation render the membrane 'leaky' with respect to hydrogen ions and so would tend to reduce the trans-membrane charge postulated to be essential for the dehydration of ATP-ase bound ADP and inorganic phosphate (Fig. 8.8). In this figure the enzyme is shown associated with the inner side of the inner membrane. There is some evidence that it may, in fact, be located on the knob-like particles which electron micrographs show are attached to 'stalks' which project into the inner matrix.

8.10 Measurement of oxidative phosphorylation

The study of mitochondrial function has utilized some very sophisticated physical techniques, most of which are beyond the scope of this book. Some mention is, however, desirable of methods used to measure oxygen uptake. Readers may have encountered the special manometer developed by O. Warburg in the 1920s. This is capable of measuring the uptake of oxygen by mitochondria by following changes in gas pressure. A more convenient device now widely used is called the *oxygen electrode*. Figure 8.9 indicates its

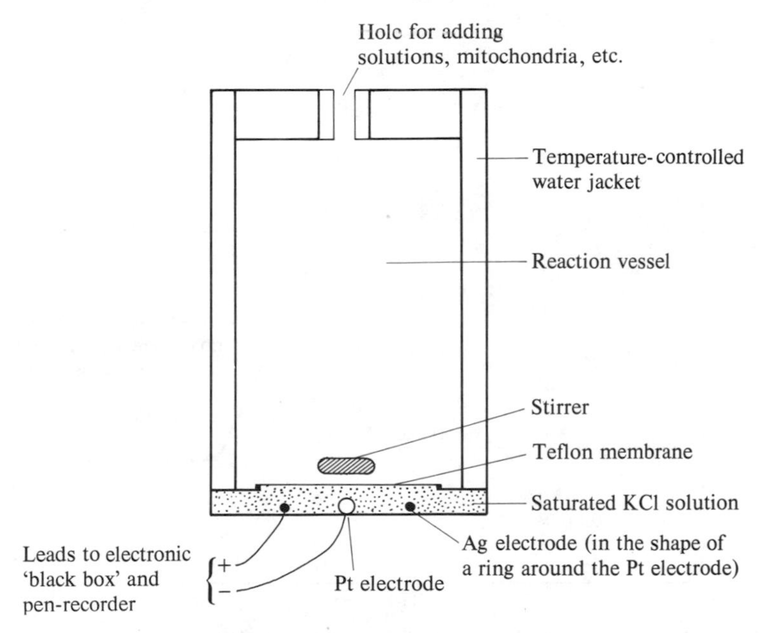

Fig. 8.9 The oxygen electrode

main features. Essentially, the system consists of a temperature-controlled reaction vessel whose contents are stirred by a magnetic stirrer. The base of the vessel is separated from the electrodes by a membrane made of Teflon which is permeable to oxygen. Below this a saturated solution of potassium chloride bathes the two electrodes, one of which is silver and the other platinum. A potential difference of 0.6 V is applied between these electrodes. Under these conditions, oxygen in solution is reduced at the platinum electrode:

$$O_2 + 4e \rightarrow 2O^{--}$$

This causes a small current to flow between the electrodes, its magnitude being proportional to the oxygen concentration. In practice, the instrument is calibrated with solutions of known oxygen concentrations and readings are plotted automatically by a pen-recorder. The apparatus is no more difficult to use than a glass electrode for measuring pH.

An example of a specific experiment will illustrate some of the many applications of the oxygen electrode. Initially a buffered substrate solution (malate) is placed in the reaction vessel maintained at 37°C. When carefully prepared rat mitochondria (Section 1.5) are added, a slow decrease in the

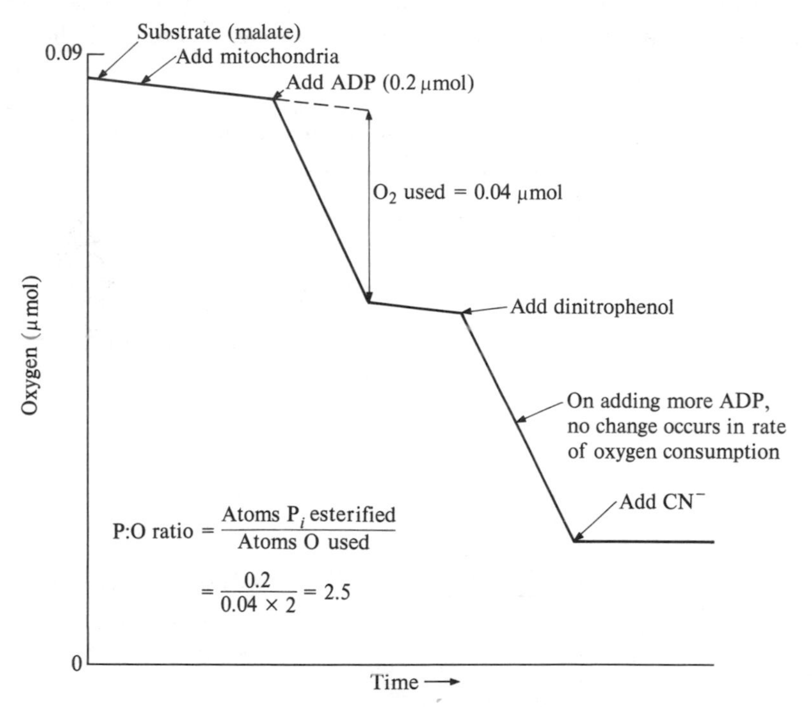

Fig. 8.10 Example of results obtained with the oxygen electrode

oxygen concentration occurs. On adding a solution of ADP the consumption of oxygen increases (Fig. 8.10) since this stimulates oxidative phosphorylation. When all the ADP has been converted to ATP the respiration of the mitochondria decreases again. Knowing the quantity of ADP used and the oxygen consumed it is easy to calculate the P:O ratio. In this experiment a ratio of 2.5 was obtained. This is lower than the value of 3.0 expected for malate oxidation, a probable reason being slight damage to the mitochondria during their preparation. Similarly, if succinate had been the substrate used, a P:O ratio of 2.0 or slightly less would have been expected.

The experimental results shown in Fig. 8.10 indicate a rise in oxygen consumption on adding dinitrophenol. This uncouples oxidative phosphorylation (compare Fig. 8.7) so that when ADP is subsequently added no further increase in oxygen uptake occurs. Finally, adding cyanide halts oxygen consumption since this poison is a strong inhibitor of cytochrome oxidase.

9

The tricarboxylic acid cycle

Much of the NADH essential for the process of oxidative phosphorylation (Chapter 8) is produced by a series of reactions known as the tricarboxylic acid (TCA) cycle. This cycle is often called the Kreb's cycle after Sir Hans Krebs, whose work helped to elucidate its reactions. In addition to its energetic function, the cycle is important as a route allowing the interconversion of carbohydrates and certain amino acids (Chapter 15).

The TCA cycle is an elegant catalytic device which brings about the combustion of the acetyl group of acetyl coenzyme A which, as has been seen (Fig. 8.1), lies on the catabolic pathway of both fats and carbohydrates. The routes from the primary food materials—fats and carbohydrates—will be considered in the next two chapters, but the essential point to recognize now is that most of the energy produced as a result of the catabolism of these two types of materials actually *arises from the breakdown of acetyl coenzyme A*, and *not* from the various preliminary reactions which lead to the formation of acetyl coenzyme A.

The reactions which together form the machinery of the TCA cycle are represented by the circumference of the wheel of Fig. 8.1. The machinery is self-perpetuating; acetyl coenzyme A is fed in at the beginning of the cycle and, as the reaction sequence undergoes one cycle, two carbon atoms are eliminated as (respiratory) carbon dioxide. At the same time, reducing power, in the form of three molecules of NADH and one of $FADH_2$, is created; it is the *production of this reducing power*—from which *ATP is generated* by oxidative phosphorylation—which is the principal function of the cycle. Clearly, there is a close association, both functionally and locationally, between the reactions at the circumference of the wheel, which will shortly be described, and the 'spokes' of the wheel which formed the subject of the last chapter.

9.1 Formulae and names of some important intermediates

Before the individual steps of the TCA cycle are considered in detail, it may be helpful to those not already familiar with such compounds a oxaloacetate, citrate, and α-ketoglutarate if we indicate simple ways of remembering the structures of some compounds of biochemical importance and the interconnections between them.

(a) Fumarate, succinate, and malate

Readers may have encountered the unsaturated dicarboxylic acid, fumaric acid, in relation to geometrical (*cis–trans*) isomerism. Fumaric acid is the *trans* acid. If fumaric acid is saturated by addition of hydrogen, succinic acid is obtained, and if it is saturated by addition of the elements of water, malic acid results. The latter has an asymmetric carbon atom and therefore exists in L and D forms. However, in the cell, only L-malic acid is involved in metabolism in most situations:

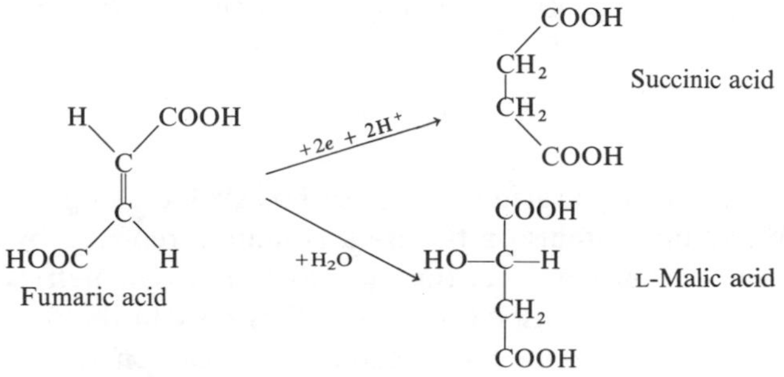

$$\underset{\text{Fumaric acid}}{\text{HOOC–CH=CH–COOH (trans)}} \xrightarrow{+2e\,+\,2H^+} \underset{\text{Succinic acid}}{\text{HOOC–CH}_2\text{–CH}_2\text{–COOH}}$$

$$\underset{\text{Fumaric acid}}{\text{HOOC–CH=CH–COOH (trans)}} \xrightarrow{+H_2O} \underset{\text{L-Malic acid}}{\text{HOOC–C(H)(OH)–CH}_2\text{–COOH}}$$

(b) Acetoacetate, oxaloacetate, and oxalosuccinate

Each of these can be regarded as a composite structure resulting (theoretically) from the reaction of the hydroxyl group of the acid forming the initial part of the name with an '*alkyl*' *hydrogen atom* of the acid indicated by the last part of the name:

$$\underset{\text{Acetic acid}}{H_3C\text{—CO}\boxed{\text{OH H}}} \underset{\text{Acetic acid}}{CH_2\text{—COOH}} \rightarrow \underset{\text{Acetoacetic acid}}{CH_3\text{—CO—}CH_2\text{—COOH}}$$

$$\underset{\text{Oxalic acid}}{\text{HOOC—CO}\boxed{\text{OH H}}}\underset{\text{Acetic acid}}{\text{—}CH_2\text{—COOH}} \rightarrow \underset{\text{Oxaloacetic acid}}{\text{HOOC—CO—}CH_2\text{—COOH}}$$

$$\underset{\text{Oxalic acid}}{\text{HOOC—CO}\boxed{\text{OH H}}}\underset{\text{Succinic acid}}{\text{—CH(}CH_2\text{—COOH)—COOH}} \rightarrow \underset{\text{Oxalosuccinic acid}}{\text{HOOC—CO—CH(}CH_2\text{—COOH)—COOH}}$$

(c) Oxalate, malonate, succinate, glutarate, and adipate

These acids have the general formula $HOOC(CH_2)_nCOOH$, where n is zero in oxalic acid, one in malonic, two in succinic, three in glutaric, and four in adipic acid. The carbon atoms can be lettered or numbered in formulae to distinguish them from one another. Both of the following two conventions are frequently employed—their use is illustrated by reference to glutaric acid:

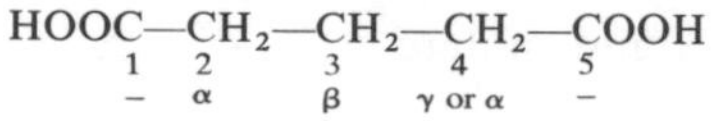

$$\underset{\underset{-}{1}}{HOOC}-\underset{\underset{\alpha}{2}}{CH_2}-\underset{\underset{\beta}{3}}{CH_2}-\underset{\underset{\gamma \text{ or } \alpha}{4}}{CH_2}-\underset{\underset{-}{5}}{COOH}$$

Thus α-hydroxyglutaric acid is:

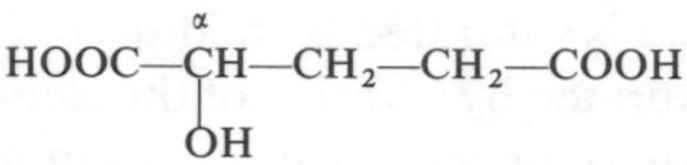

$$HOOC-\overset{\alpha}{\underset{OH}{CH}}-CH_2-CH_2-COOH$$

The oxidation of this secondary alcohol yields α-ketoglutaric acid:

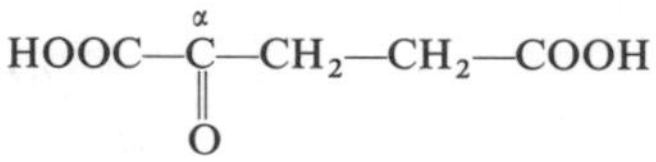

$$HOOC-\overset{\alpha}{\underset{O}{\overset{\|}{C}}}-CH_2-CH_2-COOH$$

(d) Citrate and isocitrate

Citric acid is a tricarboxylic acid with the carbon skeleton of glutaric acid but with the two hydrogen atoms on the β-carbon atom replaced by a hydroxyl group and a carboxyl group. Note that the structure is symmetrical about the β-carbon atom. Its isomer, isocitric acid, differs only in the position of the hydroxyl group, for it is on the α- instead of the β-carbon atom:

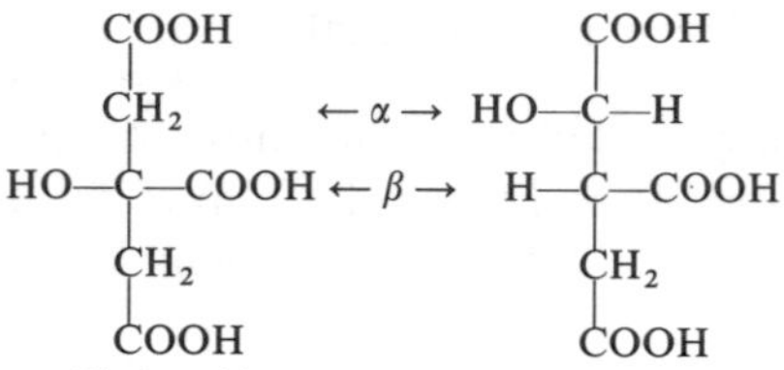

$$\begin{array}{ccc} COOH & & COOH \\ | & & | \\ CH_2 & \leftarrow \alpha \rightarrow & HO-C-H \\ | & & | \\ HO-C-COOH & \leftarrow \beta \rightarrow & H-C-COOH \\ | & & | \\ CH_2 & & CH_2 \\ | & & | \\ COOH & & COOH \\ \text{Citric acid} & & \text{Isocitric acid} \end{array}$$

(e) The naming of weak acids and their salts

In previous sections, it will no doubt have been observed that the words 'fumarate' and 'fumaric acid', etc., have apparently been regarded as interchangeable, the most convenient form of the word-pair having been employed in headings and in text. The reason is that the organic acids considered in this section, while capable of losing one hydrogen ion for each carboxylic acid group present, are nevertheless only partially dissociated at physiological pH. In other words, they are examples of *weak acids*, in contrast to *strong* or nearly completely dissociated acids such as sulphuric acid. For an acid such as succinic acid, both the undissociated acid and its conjugate base, the succinate ion, are present in cells in equilibrium with one another.

Therefore, although a given enzyme may conceivably act only upon the conjugate base or upon the acid, it is not necessary to indicate the specific form since the two are related by a self-adjusting equilibrium. The ratio of the two forms of any such substance depends only on the dissociation constant of the acid and on the pH of the environment (Section 3.2).

(f) A common sequence of reactions

Finally, before considering the tricarboxylic acid cycle in detail it may help readers if attention is drawn to the following reaction sequence:

$$\begin{array}{c}R\\|\\H-C-H\\|\\H-C-H\\|\\R'\end{array}\underset{2e\,+\,2H^+}{\overset{\text{Dehydrogenase}}{\rightleftharpoons}}\begin{array}{c}R\\|\\C-H\\\|\\C-H\\|\\R'\end{array}\underset{\text{Hydratase}}{\overset{H_2O}{\rightleftharpoons}}\begin{array}{c}R\\|\\H-C-OH\\|\\H-C-H\\|\\R'\end{array}\underset{2e\,+\,2H^+}{\overset{\text{Dehydrogenase}}{\rightleftharpoons}}\begin{array}{c}R\\|\\C=O\\|\\H-C-H\\|\\R'\end{array}$$

This sequence occurs not only in the tricarboxylic acid cycle but in other important biochemical processes as well (Sections 10.2 and 10.4). The first step (from left to right as written above) begins with the oxidation of a pair of carbon atoms ($-CH_2-CH_2-$) with the formation of a double bond. The process involves the removal of two protons and two electrons and is often catalysed by a *flavoprotein* enzyme, the FAD prosthetic group of which is reduced to $FADH_2$. In the second step a water molecule is added to the unsaturated compound so producing the hydroxylated derivative ($-CH_2-CHOH-$). The enzymes catalysing this type of hydration are known as *hydratases*. In the final step of this reaction sequence the hydroxyl compound is oxidized to a keto compound. Again, two protons and two electrons are removed, the electrons reducing NAD^+ to NADH.

9.2 Acetyl coenzyme A and its entry into the TCA cycle

Acetyl coenzyme A can be regarded as the 'fuel' which drives the TCA cycle. It is itself formed from fats by the process of β-oxidation of fatty acids (Chapter 10) and from carbohydrates *via* the glycolytic pathway followed by oxidative decarboxylation of pyruvate (Chapter 11). In addition, acetyl coenzyme A is formed either directly or indirectly from the carbon skeletons of unwanted amino acids (Section 9.11). The TCA cycle thus plays a vital role in the energy-providing catabolism of all three main organic constituents of the animal diet.

Acetyl coenzyme A is a thioester, in which the acetyl group is attached through a sulphur bond to coenzyme A:

$$H-S-CoA \qquad CH_3-\underset{\underset{O}{\|}}{C}\sim S-CoA$$

Coenzyme A   Acetyl coenzyme A

The thioester is a high energy compound, the carbon–sulphur link being an example of a high energy bond (page 155). This bond is of great importance, for the attachment of coenzyme A in this manner has the effect of making the acetyl group more reactive than it is in either the free acid or in ordinary esters such as $CH_3—C(=O)—O—R$. The reactivity of the acetyl group of acetyl coenzyme A is, as we have already said, more comparable to that of the acetyl group of acetyl chloride—except that, whereas the latter is reactive in the presence of water and many other substances, the reactivity of acetyl coenzyme A is usually only evident in the presence of the appropriate enzymes. As in other 'high energy' compounds, the energy is not confined to a single bond but is spread over the whole acetyl group. A consequence of this, in the present instance, is that the hydrogen atoms of the methyl group are often found to be more reactive than they would be if the coenzyme were not attached to the acetyl group.

In elementary chemistry, a well-known reaction of acetaldehyde is known as the aldol condensation. In this reaction, which is really an addition reaction, a labile hydrogen atom is transferred from the methyl group of one acetaldehyde molecule to the carbonyl oxygen atom of the other:

$$\begin{matrix} O{=}CH{-}CH_2{-}H \\ H{-}C(=O){-}CH_3 \end{matrix} \xrightarrow[\text{—an addition reaction}]{\text{Aldol 'condensation'}} O{=}CH{-}CH_2{-}CH(OH){-}CH_3$$

Two molecules acetaldehyde — One molecule aldol

A mitochondrial enzyme known as citrate synthase catalyses the comparable 'condensation' of oxaloacetate and acetyl coenzyme A. The high energy associated with acetyl coenzyme A acts as a driving force ensuring that the reaction proceeds almost to completion. One of the carbon-hydrogen bonds of the methyl group of acetyl coenzyme A is particularly active (Section 7.9) and 'condenses' with the keto group of oxaloacetic acid with the formation of a carbon-carbon bond. The keto group becomes a hydroxyl group, just as though a typical addition reaction were occurring, although the precise mechanism of the reaction under biological conditions is not fully understood. The reaction, an example of the one shown in Fig. 7.5c, is completed by the hydrolytic removal of coenzyme A.

Citric acid has six carbon atoms; a complete turn of the cycle results in the elimination of two carbon atoms as carbon dioxide, and oxaloacetate with four carbon atoms is again produced. This can then condense with more acetyl coenzyme A to reform citrate. Hence with each turn of the cycle, one of the initial reactants, the oxaloacetate, is regenerated, making the cycle in this sense self-perpetuating:

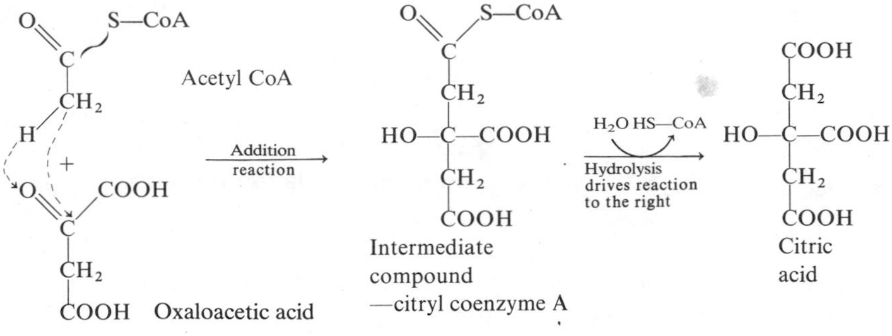

Acetyl CoA (CH_3—CO—S—CoA) + Oxaloacetic acid (HOOC—CH_2—CO—COOH) —Addition reaction→ Intermediate compound—citryl coenzyme A (CoA—S—CO—CH_2—C(OH)(COOH)—CH_2—COOH) —H_2O, HS—CoA; Hydrolysis drives reaction to the right→ Citric acid (COOH—CH_2—C(OH)(COOH)—CH_2—COOH)

9.3 Intramolecular rearrangement of citrate to isocitrate

Citric acid, produced by the first reaction, is next converted to its isomer, isocitric acid. The reaction is catalysed by aconitate hydratase (aconitase), an enzyme dependent on ferrous ions; removal of these ions by dialysis inactivates the enzyme. In effect (though probably not mechanistically), this hydratase catalyses the removal of the elements of water from the citrate, to produce an unsaturated intermediate, *cis*-aconitic acid, and isocitrate results from the subsequent hydration of the aconitate, the orientation of the entering water molecule being the opposite to that of the one initially removed. At equilibrium, there is a mixture of 90 per cent citrate and roughly equal quantities of aconitate and isocitrate:

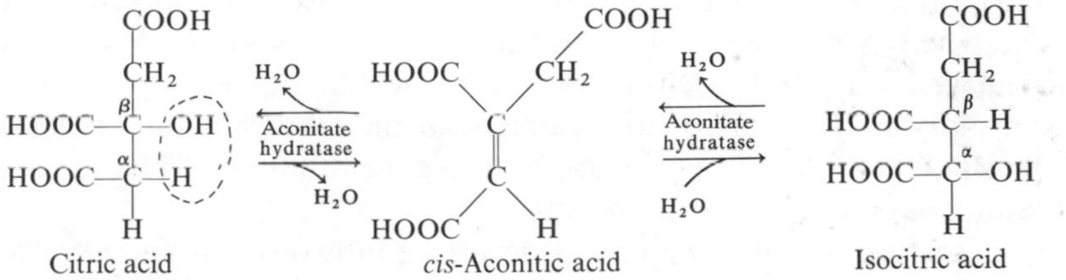

Citric acid (COOH—CH_2—C_β(COOH)(OH)—C_α(COOH)H—H) ⇌ (H_2O; Aconitate hydratase) ⇌ *cis*-Aconitic acid (HOOC—C(CH_2COOH)=C(H)—COOH) ⇌ (H_2O; Aconitate hydratase) ⇌ Isocitric acid (COOH—CH_2—C_β(COOH)H—C_α(COOH)(OH)—H)

The need for this isomeric change will be apparent if it is realized that here, as so often in biochemical oxidations, oxidation of a secondary alcohol group on an α-carbon atom is shortly to lead to the production of an α-keto acid.

9.4 Two types of decarboxylation reactions

The next two steps of the TCA cycle, described in Sections 9.5 and 9.6, involve two different kinds of decarboxylation reactions. These are known

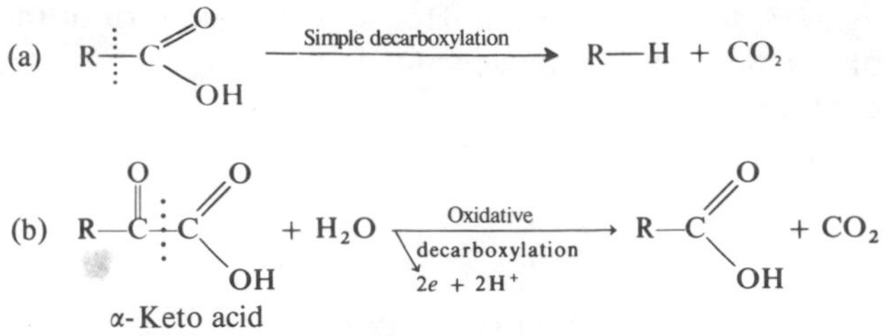

$$\text{(a)}\quad R\!-\!C\!\begin{matrix}\nearrow O\\ \searrow OH\end{matrix} \xrightarrow{\text{Simple decarboxylation}} R\!-\!H + CO_2$$

$$\text{(b)}\quad R\!-\!\overset{O}{\overset{\|}{C}}\!-\!C\!\begin{matrix}\nearrow O\\ \searrow OH\end{matrix} + H_2O \xrightarrow[\text{decarboxylation}\;\searrow 2e + 2H^+]{\text{Oxidative}} R\!-\!C\!\begin{matrix}\nearrow O\\ \searrow OH\end{matrix} + CO_2$$

α-Keto acid

Fig. 9.1 Two types of decarboxylation reaction

respectively as (simple) decarboxylation and oxidative decarboxylation and they are illustrated diagrammatically in Fig. 9.1.

Simple decarboxylation reactions are analogous to the conversion of benzoate to benzene in organic chemistry. The enzymes promoting simple decarboxylation usually require no cofactor other than manganous ions. The conversion of oxalosuccinate to α-ketoglutarate is a reaction of this sort.

In contrast, enzymes catalysing the oxidative decarboxylation of α-ketoglutarate (Section 9.6) and of pyruvate (Chapter 11) comprise three closely bound enzymes cooperating together and require the participation of no less than five organic cofactors.

9.5 Conversion of isocitrate to α-ketoglutarate

This, the second reaction of the TCA cycle, is a dehydrogenation *followed by* a decarboxylation. It must not be confused with oxidative decarboxylation, where dehydrogenation and decarboxylation are intimately linked together.

The whole conversion, from isocitrate to α-ketoglutarate, is catalysed by one enzyme, isocitrate dehydrogenase, which has a molecular weight of about 380 000. It is one of the reactions of the cycle which is coupled with the reduction of NAD and leads to the generation of ATP by oxidative phosphorylation.

The reaction is best explained as occurring in two steps. The isocitric acid first undergoes an oxidation reaction whereby the secondary alcohol group is converted to an α-keto group. In this reaction, two electrons and two protons are removed from the isocitrate (compare Section 8.2), the electrons being transferred, directly or indirectly, to NAD.† The intermediate, oxalosuccinate, still firmly attached to the enzyme, is then decarboxylated on

† Most organisms appear to have two mitochondrial isocitrate dehydrogenases—one specific for NAD and the other for NADP. In rat skeletal muscle and brain the NAD-linked enzyme predominates but in heart, kidney, and liver the NADP-coupled enzyme is present in greater amounts.

the *central* or β-carbon atom, a reaction which leads to the formation of α-ketoglutarate. Magnesium ions are necessary for the NAD-linked first step of the reaction and manganous ions are essential for the decarboxylation in the second step.

$$
\begin{array}{c}
COOH \\
{}^{\gamma}CH_2 \\
HOOC-{}^{\beta}C-H \\
H-{}^{\alpha}C-OH \\
COOH
\end{array}
\xrightarrow{NAD^+ \quad NADH + H^+}
\begin{array}{c}
COOH \\
{}^{\gamma}CH_2 \\
HOOC-C-H \\
C=O \\
COOH
\end{array}
\xrightarrow[Mn^{++}]{CO_2}
\begin{array}{c}
COOH \\
{}^{\gamma}CH_2 \\
{}^{\beta}CH_2 \\
{}^{\alpha}C=O \\
COOH
\end{array}
$$

Isocitrate — Oxalosuccinate (six carbon atoms) — α-Ketoglutarate (five carbon atoms)

9.6 The oxidative decarboxylation of α-ketoglutarate

The decarboxylation mentioned in the previous section accounts statistically (see Section 9.10) for one of the two carbon atoms which enter the TCA cycle in the acetyl part of acetyl coenzyme A. We now come to a second decarboxylation which again leads to the production of carbon dioxide, accounting (statistically) for the second of the two carbon atoms of acetyl coenzyme A. The overall reaction can be represented as follows:

$$
\begin{array}{c}
COOH \\
CH_2 \\
CH_2 \\
{}^{\alpha}C=O \\
COOH
\end{array}
\xrightarrow[2e + 2H^+]{H_2O \qquad CO_2}
\begin{array}{c}
COOH \\
CH_2 \\
CH_2 \\
COOH
\end{array}
$$

α-Ketoglutarate — Succinate

The decarboxylation which occurs with α-ketoglutarate is quite different from that described for isocitrate, for it is an example of the second type of decarboxylation illustrated in Fig. 9.1. The reaction is oxidative as well as being a decarboxylation and results in carbon dioxide, two protons, and two electrons being eliminated in a reaction which is virtually *irreversible.* The mechanism is far more complicated than is indicated in the overall scheme above, because an ingenious device exists for capturing some of the energy involved in the cleavage of the carbon-carbon bond.

Since readers may find this multi-step process somewhat involved, it is intended to present its main features in this section so that the nature of the underlying chemical reactions is evident. In Section 11.8 a second opportunity will exist to assimilate the significance of the individual reactions when these are discussed in a different biochemical setting (the oxidative decarboxylation of pyruvate to acetyl coenzyme A).

Oxidative decarboxylation of α-ketoglutarate is brought about by a multienzyme complex comprising three sorts of enzyme. The whole complex is called the *α-ketoglutarate dehydrogenase complex.* The first enzyme contains thiamine pyrophosphate (TPP). Thiamine is well known as vitamin B_1, a deficiency of which causes polyneuritis in birds and beriberi in human beings. The second enzyme contains lipoic acid, a vitamin which is a *readily reducible cyclic disulphide.* The role of the third enzyme relates to the subsequent fate of the lipoic acid.

Catalysed by the *first* enzyme of the complex, α-ketoglutarate unites with thiamine pyrophosphate, which is here conveniently expressed as TPPH, since the hydrogen atom is mobile:

$$\text{HOOC—CH}_2\text{—CH}_2\text{—C(=O)(COOH)} + \text{H—TPP} \xrightarrow[1]{\text{Enzyme}} \text{HOOC—CH}_2\text{—CH}_2\text{—CH}_2\text{—C(OH)(TPP)}\boxed{\text{COO}}\text{H}$$

The complex then loses carbon dioxide from the position indicated by the box, forming an *aldehyde* complex.

The dehydrogenation reaction (which gives α-ketoglutarate *dehydrogenase* its name) occurs as the non-TPP part of the aldehyde complex is passed on to the prosthetic group of the *second* enzyme. One hydrogen atom from the aldehyde complex reduces one disulphide sulphur atom of lipoic acid to a sulphydryl group. The other hydrogen atom ionizes, leaving its electron behind, and eventually returns to the (ionized) TPP. By ionizing off in this way, the hydrogen leaves behind a C=O group in a newly-formed succinyl radicle, $HOOCCH_2CH_2CO$—. The latter then becomes attached to the second sulphur atom of reduced lipoic acid, thus forming a *thioester:*

$$\text{CH}_2\text{—CH}_2\text{—C(·H)(O:—H)(TPP)} \text{ (COOH)} + \text{S—S (lipoic)} \xrightarrow[\text{1 and 2}]{\text{Enzymes}} \text{TPP(H)} + \text{(H)—S ... CH}_2\text{—CH}_2\text{—C(=O)—S—} \text{ (COOH)}$$

The aldehyde complex undergoes one homolytic and one heterolytic fission

Outline of structure of α-lipoic acid

Succinyl derivative of reduced lipoic acid

In due course, the second enzyme of the complex enables the succinyl group to be passed on to coenzyme A:

$$\text{Succinyl lipoic acid} + \text{CoA} \xrightarrow[2]{\text{Enzyme}} \text{reduced lipoic acid} + \text{succinyl CoA}$$

For this reason it can be called a *trans-succinylase.*

The reduced form of lipoic acid so liberated has to be converted back to the oxidized disulphide form, and this is the task of the *third* enzyme of the complex, *dihydrolipoyl dehydrogenase.*

More detailed formulae will be given when the closely analogous pyruvate dehydrogenase complex is considered in Section 11.8.

9.7 Conversion of succinyl coenzyme A to succinic acid

When α-ketoglutarate is the substance undergoing oxidative decarboxylation, the product is *succinyl coenzyme A*. The next reaction of the TCA cycle converts this product to succinic acid. Before describing the mechanism involved, it must be stressed that some of the energy associated with the cleavage of the carbon-carbon bond of the α-ketoglutarate is carried over to succinyl coenzyme A, being preserved in the high energy C $\sim$ S bond. The structure and energetic characteristics of *succinyl* coenzyme A are very similar to those of *acetyl* coenzyme A:

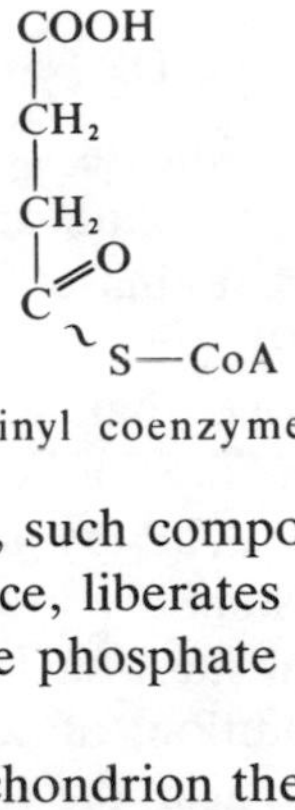

COOH
|
CH_2
|
CH_2
|
C(=O)$\sim$S—CoA

Succinyl coenzyme A

CH_3
|
C(=O)$\sim$S—CoA

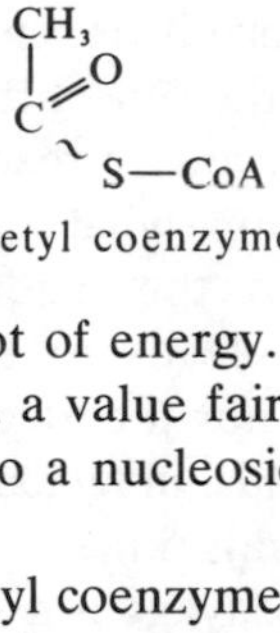

Acetyl coenzyme A

Upon hydrolysis, such compounds release a lot of energy. Succinyl coenzyme A, for instance, liberates about 42 kJ/mol, a value fairly close to that absorbed when one phosphate group is added to a nucleoside diphosphate (Section 7.9).

Within the mitochondrion the energy in succinyl coenzyme A is, however, not squandered by hydrolysis; instead, it is used to synthesize ATP. Two closely coupled reactions are responsible for the conservation of this energy. In one of them, succinyl coenzyme A is converted to free succinic acid and free coenzyme A. The energy made available by this reaction is used to drive the second reaction, in which a nucleoside diphosphate similar to ADP (it is actually *guanosine diphosphate*, GDP) unites with inorganic phosphate to give a nucleoside triphosphate (gaunosine triphosphate, GTP) (see page 184).

This coupled reaction involves only a small overall energy change and is consequently freely reversible. It is catalysed by the enzyme *succinyl coenzyme A synthetase* (working in reverse).

Effectively, the consequence of this coupling is that the high energy of a (disappearing) thioester bond enables an extra 'high energy phosphate bond' to be built on to the nucleotide. ATP is now formed by a simple transfer of $\sim$Ⓟ from GTP to ADP, making GDP available once more to accept energy

Succinyl coenzyme A: $COOH-CH_2-CH_2-C(=O)\sim S-CoA$ (High energy thioester bond) → ENERGY (about 33 kJ/mol)

Succinic acid: $COOH-CH_2-CH_2-COOH$ + H—S—CoA (Coenzyme A)

G—Ⓟ ~ Ⓟ + P_i (GDP) → G—Ⓟ ~ Ⓟ ~ Ⓟ (GTP) (Additional high energy phosphate bond)

from succinyl coenzyme A:

$$\text{GTP} + \text{ADP} \rightleftharpoons \text{ATP} + \text{GDP}$$

The enzyme responsible for this reaction is known as *nucleoside diphosphate kinase*, since it transfers a high energy phosphate from one nucleoside diphosphate to the other. It will be recalled that kinases are always involved in *high energy* transfer reactions (Section 7.9).

It follows from what has been said above that some of the energy originating from the oxidative cleavage of the carbon-carbon bond of α-ketoglutarate leads to the generation of ATP by a process in which neither oxygen nor the electron transport system are involved. This type of phosphorylation is, in fact, another example of *substrate level phosphorylation* (compare pyruvate kinase, Fig. 7.5a). In addition, of course, some of the energy of the cleavage is used to reduce lipoic acid, the reoxidation of which *does* involve oxidative phosphorylation.

9.8 Succinate to oxaloacetate

The remaining steps of the cycle serve two purposes: they ensure that succinate is converted eventually to oxaloacetate so as to complete the cycle, and at the same time they create more ATP. Three reactions are, in fact, responsible for the remaining transformations (Section 9.1f). Some readers may find it helpful to observe that a strikingly similar sequence of three reactions forms a major part of the pathway responsible for the oxidation of fatty acids (Chapter 10).

In the first reaction, succinate is oxidized to fumarate, the catalyst being succinate dehydrogenase and FAD the electron acceptor (Section 8.4).

Succinate dehydrogenase comprises two subunits. The larger of these has a molecular weight of about 70 000 and contains not only the FAD but also four atoms each of iron and sulphur. The iron is not in the form of a porphyrin (contrast, for example, the cytochromes, Chapter 8) but is probably associated with the sulphur atoms. The smaller subunit is also a non-haem iron protein, but contains no FAD.

Succinate dehydrogenase forms one of the electron transport particles (II) on the inner mitochondrial membrane. This implies that succinate formed elsewhere in the mitochondrion must diffuse to the inner membrane. The product (fumarate) presumably then diffuses back to the remaining TCA cycle enzymes. Note that succinate oxidation leads to the generation of two molecules of ATP by oxidative phosphorylation (Section 8.9). The fumarate is then converted to a hydroxy acid, L-malic acid, by the addition of water in a reaction catalysed by the enzyme fumarate hydratase. The reaction is freely reversible *in vitro*, though it tends to move in the direction of L-malic acid synthesis *in vivo* because the TCA cycle is made unidirectional by the irreversible decarboxylation of α-ketoglutarate.

Finally, oxaloacetic acid is formed in the presence of malate dehydrogenase by the oxidation of the secondary alcohol group of malic acid. NAD is simultaneously reduced to NADH, so yielding three molecules of ATP by oxidative phosphorylation.

COOH–CH_2–CH_2–COOH (Succinic acid) ⇌ [Succinate dehydrogenase; FAD → $FADH_2$] HOOC–CH=CH–COOH (Fumaric acid) ⇌ [Fumarate hydratase; H_2O] HO–CH(COOH)–CH_2–COOH (L-Malic acid) ⇌ [Malate dehydrogenase; NAD^+ → NADH $+H^+$] COOH–C(=O)–CH_2–COOH (Oxaloacetic acid)

From what has been said so far it may seem surprising that malate can reduce NAD since its standard redox potential is lower than that of NAD on the redox scale (Table 7.3). The reason why compounds like malate, lactate, and ethanol are able to reduce NAD in the presence of suitable enzymes (dehydrogenases) is that the actual redox potentials operative *in vivo* are more negative than the standard values E'_0. When there is more reduced form (e.g., malate) than oxidized form (oxalacetate), the redox potential is more negative than the standard value of E'_0. If the ratio

$$\frac{[\text{Reduced form}]}{[\text{Oxidized form}]}$$

is great enough, the effective redox potential is such that NAD^+ is readily reduced. In other words, the reducing power when a large proportion of the

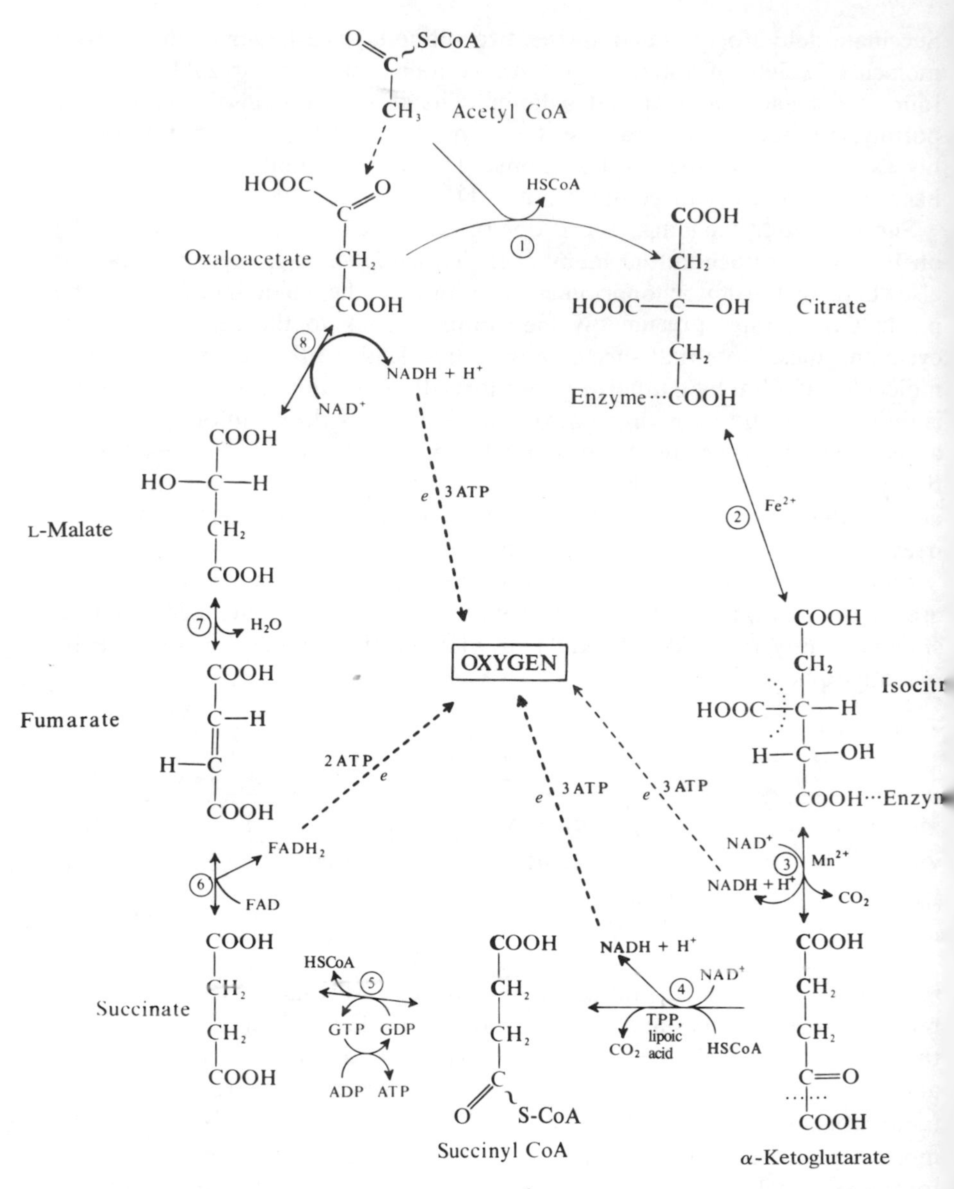

Fig. 9.2 The tricarboxylic acid cycle. [$\overset{e}{\dashrightarrow}$ represents the flow of electrons from NADH or $FADH_2$ along the electron transport chain. Enzyme attachment (top right) distinguishes two ends of the citrate molecule]

reduced form is present is greater than when a smaller proportion is present. Oxaloacetate is very readily removed by formation of citrate and this ensures that the ratio of malate to oxaloacetate is large enough to enable the oxidation of L-malate by NAD to proceed.

The standard redox potentials for the other two NAD-linked reactions of the cycle (i.e., the oxidations of isocitrate and of α-ketoglutarate) are more negative than the standard redox potential of the system $NAD^+ \rightarrow NADH$ (Table 7.3) so both reactions tend to move readily in the direction of substrate oxidation and NAD reduction.

9.9 Energy captured by the cycle

We are now in a position to look at the cycle as a whole and to calculate the net gain of ATP to the cell. In one complete turn of the cycle, one acetyl group is incorporated and two molecules of carbon dioxide are eliminated (Fig. 9.2); NAD is reduced to NADH at three stages, the enzymes catalysing the reductions being (1) isocitrate dehydrogenase, (2) α-ketoglutarate dehydrogenase complex, and (3) malate dehydrogenase. In addition, FAD is reduced to $FADH_2$ by succinate dehydrogenase. Oxidative phosphorylation of ADP yields three molecules of ATP when one molecule of NADH is oxidized (Section 8.9). Moreover, two molecules of ATP are produced by the same process when the $FADH_2$ prosthetic group of succinate dehydrogenase is reoxidized. So the number of molecules of ATP produced by oxidative phosphorylation for each acetyl group incorporated into the cycle in the form of acetyl coenzyme A is $(3 \times 3) + 2$ or 11 ATP. In addition, one more ATP is produced by substrate level phosphorylation when succinyl coenzyme A is converted to succinate, making a *total of 12 ATP produced for each complete turn of the cycle.*

The TCA cycle and the associated electron transport chains together provide the machinery which produces most of the energy available to heterotrophic cells for endergonic processes.

9.10 Irreversibility of the TCA cycle and the effect of inhibitors

Any of the acids of the cycle having six carbon atoms can be readily converted to any of the others; however, the four-carbon acids (succinate, L-malate, fumarate, and oxaloacetate) can be converted to the six-carbon or five-carbon acids only if acetyl coenzyme A is available to enter the cycle since the almost irreversible α-ketoglutarate decarboxylation makes the cycle a unidirectional process.

The cycle can be blocked by inactivating any enzyme in it. For example, fluoroacetate (FH_2CCOOH) prevents the cycle functioning because it unites with oxaloacetate in the same way as does acetyl coenzyme A. The fluorocitrate thus formed stops the cycle because aconitate hydratase cannot act upon fluorocitrate. Similarly, malonate stops the cycle at the succinate dehydrogenase step (page 114). Both malonate and fluoroacetate have been used experimentally as inhibitors for studying the reactions of the cycle. Malonyl coenzyme A is an important cell metabolite—it is a precursor of long-chain fatty acids (Chapter 10)—but it is mainly extramitochondrial in location.

It is of some interest to note that the carbon dioxide eliminated in a single turn of the cycle is *not* derived from the acetyl coenzyme A incorporated in the same turn. This can be seen from Fig. 9.2, where the carbon atoms of the acetyl group are followed through the cycle in heavy type. At first sight this may seem surprising, because citric acid is a symmetrical molecule and therefore a differentiation between carbon atoms immediately derived from acetyl coenzyme A and those derived from oxaloacetate might not be expected. The explanation is that citrate and isocitrate are bound to the enzymes in such a way that the carbon atoms just incorporated are necessarily situated at the other end of the molecule from where decarboxylation subsequently occurs. By the time succinate is reached, however, no such distinction exists, because succinic acid is not only a symmetrical molecule but it also breaks away from the enzyme surface at which it is formed. Thus acetyl coenzyme A labelled with radioactive carbon (^{14}C) does not produce radioactive carbon dioxide until one turn of the cycle has been completed—the carbon dioxide is, in fact, derived from those carbon atoms of the citrate which originate from its oxaloacetate precursor.

9.11 Entry of carbon skeletons of amino acids into the TCA cycle

In animals, the principal function of dietary proteins is to provide amino acids for body building, i.e., the synthesis of new proteins. If, however, excessive amounts are ingested, the amino acids cannot be stored. Consequently, they are oxidized, with loss of their nitrogen, to carbon skeletons which, *inter alia*, can be used as an immediate energy source by decomposition by the TCA cycle. In addition, unwanted body proteins of all types of organism are degraded to amino acids which can be deaminated and the carbon skeletons metabolized to provide energy.

Many amino acids yield carbon skeletons by a process which results in the formation of α-keto acids. The mechanism known as transamination is discussed in Chapter 15 but, briefly, it results in the removal of an amino group and a hydrogen atom from the α-carbon atom, these two monovalent groups being replaced by an oxygen atom. For example, alanine gives *pyruvic acid*:

$$CH_3—CH(NH_2)—COOH \longrightarrow CH_3—CO—COOH$$

Similarly, glutamic acid gives *α-ketoglutaric* acid and aspartic acid ($HOOCCHNH_2CH_2COOH$) gives *oxaloacetic* acid.

Serine and cysteine lose water and hydrogen sulphide respectively in one of two consecutive reactions leading to their decomposition to *pyruvic acid:*

$$HS-CH_2-CH(NH_2)-COOH \xrightarrow[\text{Cysteine desulphydratase}]{-H_2S}$$

Cysteine

$$H_2C{=}C(NH_2)COOH \rightleftharpoons H_3C-C({=}NH)-COOH \xrightarrow{H_2O \;\; NH_3} H_3C-C({=}O)-COOH$$

Imino compound (two forms) — Pyruvate

Most of the remaining amino acids break down to give nitrogen-free skeletons by longer and more complex routes. The positions of entry of amino acids into the TCA cycle are shown in Fig. 9.3.

Thus, whatever the pathway of deamination and metabolism most of the amino acids give products which are either

(a) acids of the TCA cycle, or
(b) one or other of the two common inputs to the cycle.

In the second instance, some amino acids give rise to *pyruvate*, which can then be converted to acetyl coenzyme A (just as it can be when it is produced from carbohydrates, Fig. 11.6). Others resemble fatty acids in that they give *acetyl coenzyme A*.

Whatever their method of entry into the TCA cycle, the carbon skeletons (or substances formed from them) can be broken down to yield NADH in a coupled reaction. This, of course, can then be oxidized by the electron transport system (Chapter 8) to provide energy in the form of ATP. The latter can be transported out of the mitochondrion (see Section 9.12) to provide energy to drive extramitochondrial reactions.

The enzymes responsible for the removal of nitrogen from most amino acids are present in both mitochondria and the cytosol, so the carbon skeletons may be formed in either of these locations. Those outside the mitochondrion must then be transported across the inner membrane to the inner cavity of the mitochondrion, where most of the enzymes catalysing the TCA cycle are to be found (some are actually attached to the membrane). This transportation is effected by one of several rather specific transport systems which are described in the next section.

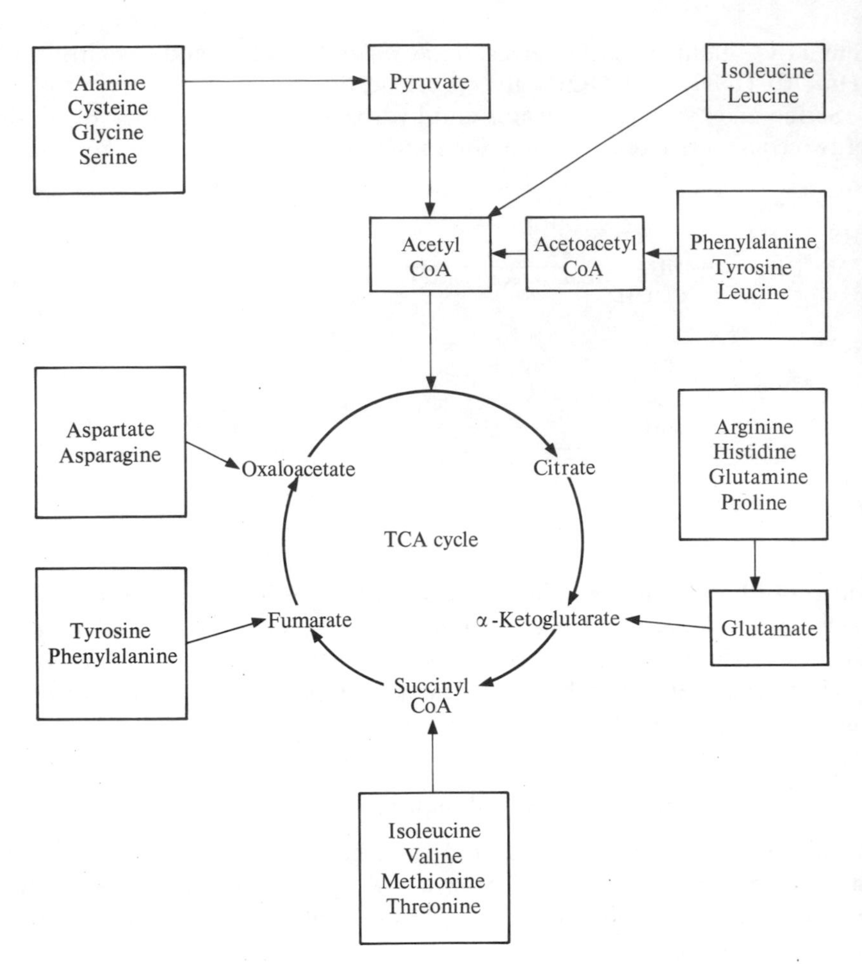

Fig. 9.3 Entry of carbon skeletons of amino acids into the TCA cycle

9.12 Movement of substances through mitochondrial membranes

Most physiologically important ions and organic solutes of low molecular weight can move freely in both directions through the *outer* membrane of the mitochondrion. The *inner* membrane, however, is a very different matter, for if ions and small molecules could diffuse readily through it, the soluble acids of the TCA cycle and many essential cofactors and activators would be lost by outward diffusion.

The inner membrane is impermeable to NAD, coenzyme A, and sugars. In addition, it is not *freely* permeable to a range of other substances—that is to say, while they *can* enter or leave the mitochondrion, they have to do so in a manner which is regulated by a number of *carrier systems*. Compound moved across the membrane by carrier proteins include several TCA cycle acids, as well as ATP, ADP, and inorganic phosphate. In addition, glutamic acid (which plays a unique role in amino acid metabolism, Chapter 15) and fatty acyl-carnitine complexes (which are important in fat metabolism) move across the inner membrane attached to specific carriers. Table 9.1 lists some of the main carrier systems; it will be noted that often the entry of one molecular species is linked to the exit of another.

In all but the last of these an equimolecular and equi-ionic exchange occurs of substances entering for substances waiting to come out. The fatty acid transport system is different in that the carnitine is part of the carrier system and is left attached to the carrier when the fatty acid is deposited (on to coenzyme A) in the mitochondrial inner cavity (Fig. 10.2).

The movement of transported substances is sometimes a passive process, the direction of movement being determined by the concentration gradient. Thus ATP will normally be transported out, while phosphate (which is used up as ATP is formed from ADP) will normally be transported in. It is, however, possible for movement to occur against a concentration gradient when such (endergonic) movement is coupled with the energy liberated when NADH is oxidized by the mitochondrial electron transport system.

The controlled movement of organic and inorganic ions through the mitochondrial membrane offers an important device for the regulation of cellular metabolism (Chapter 13). For example, if for some reason ADP is not readily available, mitochondrial ATP cannot leave by transportation on the ATP carrier. Another consequence of the regulated movement of TCA cycle acids outwards into the cytosol is that these acids are not only able to serve in an energy-producing capacity within the mitochondrion but can act as building bricks from which amino acids and other metabolically important substances can be synthesized outside the mitochondrion.

Table 9.1 Transport systems associated with the inner membrane of the mitochondrion

Type of carrier	Substances transported	
	One direction	Opposite direction
ATP carrier	ATP	ADP
Phosphate carrier	$H_2PO_4^-$	OH^-
Tricarboxylate carrier	(a) Citrate	Isocitrate
	(b) A tricarboxylic acid	A dicarboxylic acid
Dicarboxylate carrier	(a) Succinate, fumarate, or malate	Succinate, fumarate, or malate
	(b) Succinate, fumarate, or malate	Phosphate
	(c) α-Ketoglutarate	Malate
Glutamate carrier	Glutamate	Aspartate
Acyl carnitine carrier	Fatty acyl-carnitine complex	Carnitine

9.13 Some historical aspects of the TCA cycle

The elucidation of the mechanism of the TCA cycle provides an elegant example of the application of scientific reasoning. Some years before this elucidation occurred essential background work by such researchers as T. Thunberg had shown that anaerobic preparations of animal tissues in the presence of methylene blue to act as an electron acceptor could oxidize succinate, fumarate, malate, and citrate.

Much of the work of A. Szent-Györgyi and of H. A. Krebs was done on sliced or minced tissue, for it was soon discovered that the enzymes responsible for aerobic oxidation (unlike those of the earlier-investigated process of glycolysis) were associated with *water-insoluble* constituents. As we have seen, it is now recognized that these enzymes are, in fact, within the mitochondria.

Using aerobic tissue preparations, the next major advance was the observation that certain substances, including citrate, isocitrate, α-ketoglutarate, succinate, malate, oxaloacetate, and pyruvate, could 'under some conditions, especially when added in small quantities... have effects on metabolism which are not in stoichiometric proportion to the added substrate and are not necessarily accompanied by the disappearance of the added substrates' (Krebs, 1943). In other words, they had a catalytic effect on oxygen uptake. For example, F. Stare and C. Baumann found that when fumarate was added to pigeon breast muscle the oxygen uptake increased by five times the amount that would have been necessary to convert the whole of the added fumarate to carbon dioxide and water.

It remained for Krebs in the period 1932 to 1937 to show that pyruvate, or something readily formed from it, reacted with oxaloacetate to give citrate. It was then possible to integrate all the separate facts so as to formulate a cycle essentially similar to that accepted today. Later, it was recognized that 'active acetate' (acetyl coenzyme A) was the actual fuel for the cycle. This was an important new development, for it changed the emphasis of the TCA cycle from a system associated primarily with carbohydrate metabolism to one equally relevant to the catabolism of both carbohydrates and fats.

Work is still continuing today, with particular reference to the factors which control the rate at which substrates are metabolized and ATP is produced. Of special importance are the precise details of how energy-linked transportation of metabolites, ions, and phosphates through the mitochondrial inner membrane is mediated. The still little-understood mechanism by which ATP is formed by the process of oxidative phosphorylation is also being vigorously investigated.

10

Lipid metabolism

It was seen in Chapter 2 that certain lipids, especially phospholipids and steroids, contribute to the architecture of the cells since they are present in lipoprotein membranes. The neutral glycerides may also be components of some membranes, but, more important, when broken down under oxidative conditions, they provide the cell with a large quantity of energy. A molecule of the long-chain fatty acid, palmitic acid [$CH_3(CH_2)_{14}COOH$], for example, yields eight molecules of acetyl coenzyme A. Some ATP is synthesized as a result of this process; far more, however, if formed by the subsequent oxidation of the acetyl coenzyme A by the TCA cycle (Chapter 9).

In animals, adipose tissue is specially adapted for the storage of fats and oils. The insolubility of these lipids in aqueous cytoplasm keeps them isolated from the general metabolism, yet they can be mobilized readily when needed. In terms of their mass they represent a particularly efficient way to store energy; complete oxidation of 1 g of fat produces some 39 kJ of heat, while the same mass of carbohydrate or protein yields only 17 kJ. Nearly 10 per cent of the weight of a normal man is fat, much of which can be drawn upon to provide energy during periods of starvation. Somewhat similarly, hedgehogs and other hibernating animals build up a considerable deposit of fat for the autumn, a deposit which provides thermal insulation up to the time it is catabolized and both energy and water thereafter. In plant cells, starch is frequently more common than fat as an energy reserve, but this is by no means always the case and sometimes, as in the castor oil seed, it is fat which is stored to provide the energy needed during germination.

Lipids should not be regarded as a static energy deposit, for, in animals, fats are constantly being mobilized and redeposited, as well as being drawn upon for catabolic purposes. Acetyl coenzyme A, formed when fat is catabolized, not only generates ATP (Chapter 9), but may also take part in several important biosyntheses which lead to the production of other fats and lipids. In this chapter the mechanisms which result in the formation of acetyl coenzyme A from fat will be considered in detail and some consideration will also be given to the ways in which fats can provide the raw materials for the synthesis of new glycerides and certain other lipids.

10.1 Cellular location of fat metabolism

An understanding of the mechanisms of synthesis and catabolism of fats depends on a knowledge of the sites in the cell where the various events occur. The spatial separation of different processes appears to be a fundamental aspect of regulated metabolism. Before describing the details of the various pathways it is helpful to appreciate the outline of what is happening and where (Fig. 10.1).

Unesterified free fatty acids (FFA) readily enter the cell from the blood or extracellular fluid by diffusion (process 1 in Fig. 10.1). Fats themselves do not enter the cell directly but must first undergo hydrolysis catalysed by a lipase which is loosely associated with the plasmalemma. As a result, the fats

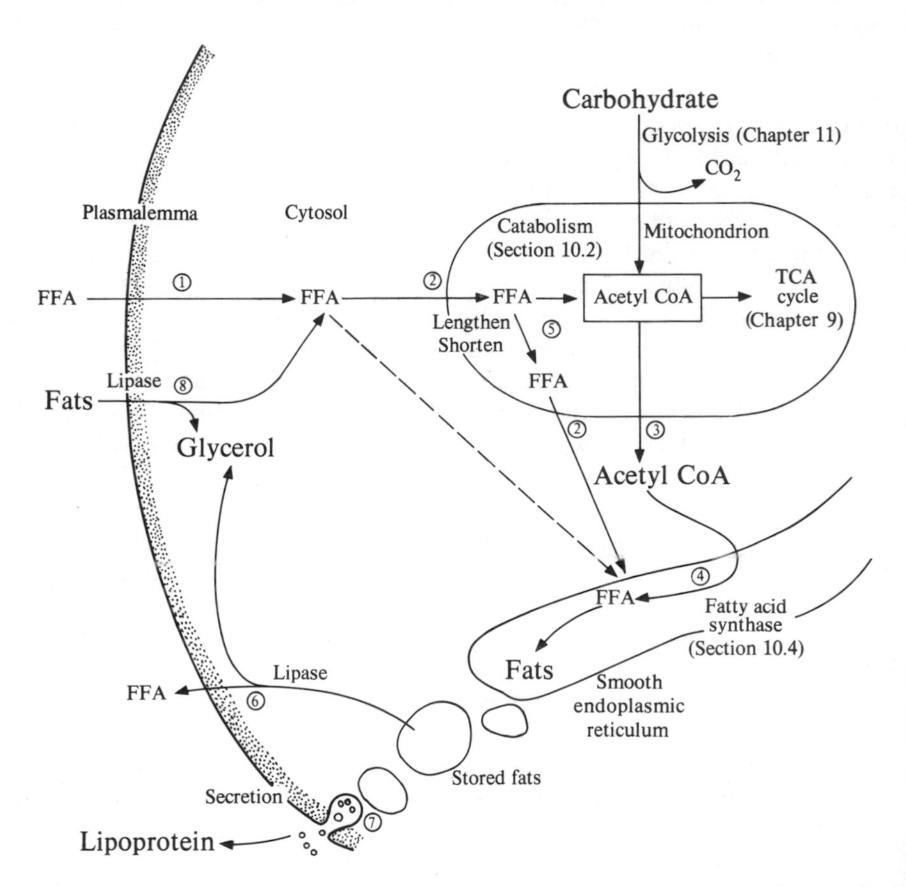

Fig. 10.1 Outline of the cellular location of fatty acid metabolism in a typical animal cell

enter the membrane and FFA and glycerol are passed into the cytosol. Fatty acids participate in several different processes in the cell. A major pathway, described in the next section, is the catabolism of the fatty acids which occurs in the *inner matrix of the mitochondrion.* This yields acetyl coenzyme A, the initial fuel of the TCA cycle (Chapter 9). But fatty acids do not readily pass through the inner mitochondrial membrane (process 2) so a specific mechanism operates which involves *carnitine.* This enables them to be carried into or out of the mitochondrion. This is discussed later (Fig. 10.2).

Many cells, and particularly those of the liver, are able to synthesize long-chain fatty acids from acetyl coenzyme A (Section 10.4). This is indicated by process 4 in Fig. 10.1. The acetyl coenzyme A usually originates in the mitochondria and the catabolism of carbohydrates is a major source (Chapter 11). Two mechanisms exist to enable intramitochondrial acetyl coenzyme A to leave the cell organelle and to pass into the cytosol. This movement (process 3 in Fig. 10.1) is necessary because fatty acid synthesis occurs in the *smooth endoplasmic reticulum.* Fats are also synthesized here (Section 10.5). The various fatty acids which undergo esterification to form fats and phospholipids may arise from the *de novo* synthesis just indicated, but, in addition, acids entering directly from outside the cell may be utilized. A third source are the fatty acids which have undergone a process of lengthening or shortening in the mitochondrion (process 5).

Fats are generally stored in the cell within lipoprotein vesicles derived from endoplasmic reticulum. They can, of course, undergo subsequent catabolism in the same cell. They may also be released into the blood to supply other cells either as free fatty acids resulting from the action of a lipase associated with the endoplasmic reticulum (process 6) or as lipoprotein complexes (process 7).

10.2 The fatty acid oxidation spiral

In animal and plant cells, the oxidation of fatty acids occurs exclusively in mitochondria. The oxidative process was known to involve β-oxidation long before the details of the individual reactions were known. Historic experiments by F. Knoop (1904) suggested that the overall process of stepwise degradation could be expressed in the following highly simplified form:

$$\text{R—(CH}_2)_n\text{—}\overset{\beta}{\text{C}}\text{H}_2 \vdots \overset{\alpha}{\text{C}}\text{H}_2\text{—COOH} \xrightarrow[\text{(now known to involve four reactions)}]{\beta\text{-Oxidation}} \text{R—(CH}_2)_n\text{—}\overset{\beta}{\text{C}}\text{OOH} + \overset{\alpha}{\text{C}}\text{H}_3\text{—COOH}$$

The principal feature of β-oxidation is the sequential removal of pairs of carbon atoms (as acetyl coenzyme A) from fatty acids to give shorter-chain fatty acids:

$$R{-}CH_2 \vdots CH_2 \overset{6}{-} CH_2 \vdots CH_2 \overset{5}{-} CH_2 \vdots CH_2 \overset{4}{-} CH_2 \vdots CH_2 \overset{3}{-} CH_2 \vdots CH_2 \overset{2}{-} CH_2 \vdots CH_2 \overset{1}{-} COOH$$

It is now known that free fatty acids are not, in fact, metabolized. Instead, a fatty acid is first changed to a high energy form by conversion to an acyl coenzyme A derivative. The fatty acyl coenzyme A then acts as the substrate for the first of four consecutive reactions which collectively account for the process of β-oxidation. All the intermediates taking part in the remaining three reactions of the sequence are likewise coenzyme A derivatives.

All these coenzyme A derivatives are high energy compounds (page 152). When the free fatty acid is converted to fatty acyl coenzyme A, the energy needed to forge the thioester linkage is derived from ATP, although the synthesis actually occurs in two steps. In the first of these, a priming reaction, the free fatty acid reacts with ATP to give a fatty acyl-AMP compound, in which the carboxyl group of the fatty acid is linked to the phosphate of the mononucleotide:

$$\underset{\text{Fatty acid}}{R{-}\overset{O}{\overset{\|}{C}}{-}OH} + \underset{\text{ATP}}{Ⓟ\sim Ⓟ\sim Ⓟ{-}\text{ribose}{-}\text{ADENINE}} \xrightarrow[\text{PP (pyrophosphate)}]{} \underset{\text{Fatty acyl-AMP}}{R{-}\overset{O}{\overset{\|}{C}}\sim Ⓟ{-}\text{ribose}{-}\text{ADENINE}}$$

This is a high energy compound, for, in effect, much of the energy derived from the cleavage of pyrophosphate from the ATP is retained in the acyl-AMP linkage. This reaction is displaced to the right by the hydrolysis of a second product, pyrophosphate (page 210), which is converted to inorganic orthophosphate by enzymes known as *pyrophosphatases*. Several similar reactions will be encountered later.

The fatty acyl-AMP then reacts with coenzyme A with the formation of fatty acyl coenzyme A and AMP:

$$R{-}\overset{O}{\overset{\|}{C}}\sim Ⓟ{-}\text{ribose}{-}\text{ADENINE} + H{-}S{-}CoA \longrightarrow R{-}\overset{O}{\overset{\|}{C}}\sim S{-}CoA + AMP$$

The formation of acyl coenzyme A has been described as a priming reaction, for although ATP is used initially, the overall process of β-oxidation has the function of generating ATP. The reaction is essential because coenzyme A has the effect of making the α- and β-carbon atoms of the fatty acid more reactive than might be expected from the saturated paraffin-like structure.

It has been emphasized that fatty acids and their coenzyme A derivatives are large molecules which are not readily transferred across the inner mitochondrial membrane. The initial priming reaction and the reaction leading to the formation of acyl coenzyme A both occur predominantly in the cytosol. The acyl group is carried into the mitochondrion linked to a carrier molecule, *carnitine*, the structure of which is

$$(H_3C)_3\overset{+}{N}—CH_2—\underset{\displaystyle OH}{\overset{\beta}{C}H}—CH_2—COOH$$

In the cytosol the acyl group is transferred from the fatty acyl coenzyme A to the β-hydroxyl group of the carnitine yielding fatty acyl carnitine and free coenzyme A (Fig. 10.2). The acyl carnitine is transferred, probably by a carrier mechanism, into the mitochondrion. There, a reversal of the previous reaction results in the formation of intramitochondrial fatty acyl coenzyme A and free carnitine. The latter can diffuse back into the cytosol. The carnitine transfer mechanism is also used for movements of other acyl groups. For example, acetyl coenzyme A may be transferred from the mitochondrial matrix to the cytosol by the mechanism shown in Fig. 10.2, but working in reverse.

Once the fatty acyl coenzyme A has reached the inner matrix of the mitochondrion the oxidation mechanism can proceed. The overall effect of the four reactions constituting each turn of the spiral of fatty acid oxidation results in the formation of one molecule of acetyl coenzyme A and the release of four protons and four electrons. Two of these electrons reduce FAD and the other two, liberated at a later stage of the spiral, reduce NAD. When these coenzymes are reoxidized by the mitochondrial electron transport particles ATP is produced by oxidative phosphorylation (Chapter 8).

The whole catabolic process involves only four reactions which are essentially those shown in Fig. 10.3 and are very similar to those by which succinic acid is converted to oxaloacetate.

The first reaction is a dehydrogenation and is catalysed by a flavoprotein enzyme, *fatty acyl coenzyme A dehydrogenase.* This catalyses the removal of one proton and one electron from each of the α- and β-carbon atoms of the fatty acyl coenzyme A; these are transferred to FAD, so reducing it to $FADH_2$ (Section 8.4). The reoxidation of the $FADH_2$ by the mitochondrial electron transport chain probably yields 2 ATP by oxidative phosphorylation. The similarity of this reaction to the succinate dehydrogenase reaction should be apparent. The product is the α, β-unsaturated fatty acyl coenzyme A.

The second reaction of the sequence is catalysed by a hydratase, *unsaturated fatty acyl coenzyme A hydratase*, which adds water to the unsaturated fatty acyl coenzyme A (Fig. 10.3). Although there are two possible stereoisomers only the L form is produced. The formation of L-β-hydroxyacyl coenzyme A recalls the stage in the TCA cycle where fumarate hydratase converts fumarate to L-malic acid. The hydratase has the high turnover number of about a million (Section 6.1).

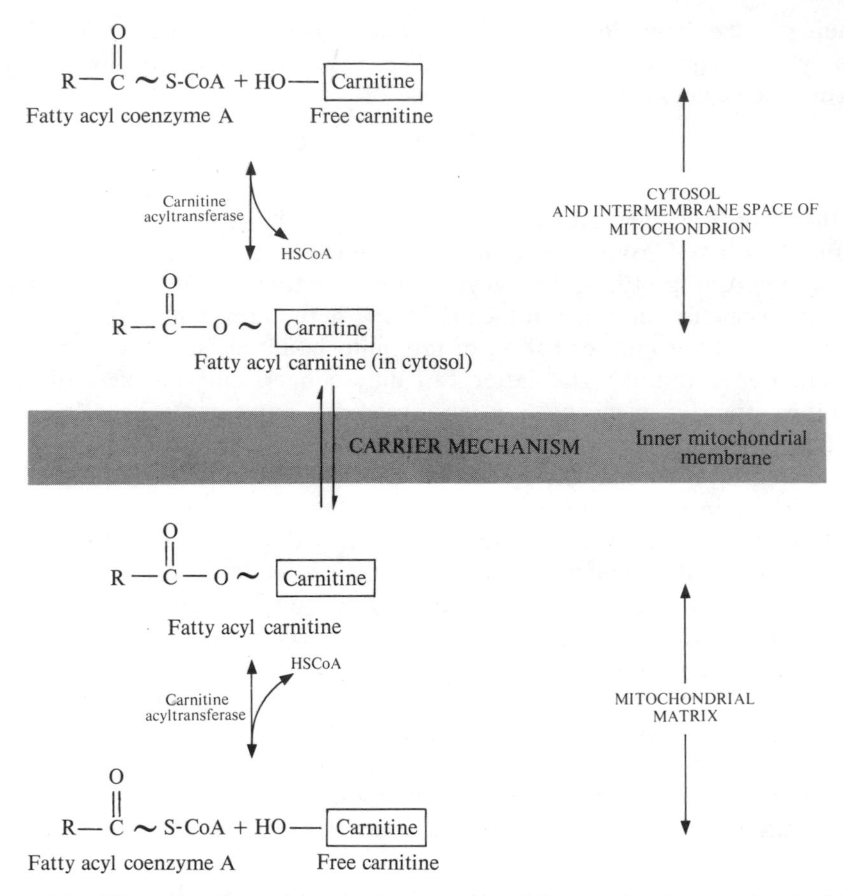

Fig. 10.2 The role of carnitine in the transfer of fatty acids into and out of the mitochondrion

The third enzyme, L-β-*hydroxyacyl coenzyme A dehydrogenase*, is specific for both NAD and the L form of the substrate. The products are the β-ketoacyl coenzyme A and reduced NAD (Fig. 10.3). The reaction is important energetically since 3 moles of ATP can be formed from each mole of NADH by oxidative phosphorylation.

The bond between the α- and β-carbon atoms is cleaved in the fourth and final reaction by the action of the enzyme β-*ketoacyl coenzyme A thiolase* (Fig. 10.3), and it will be recalled that the cleavage of a carbon–carbon bond is generally an exergonic reaction (Section 9.6). By coupling the cleavage with the acylation of coenzyme A, the energy released by the β-ketoacyl coenzyme A thiolase reaction is not lost as heat but is conserved in a new thioester linkage. This is accomplished by the intervention of a molecule of

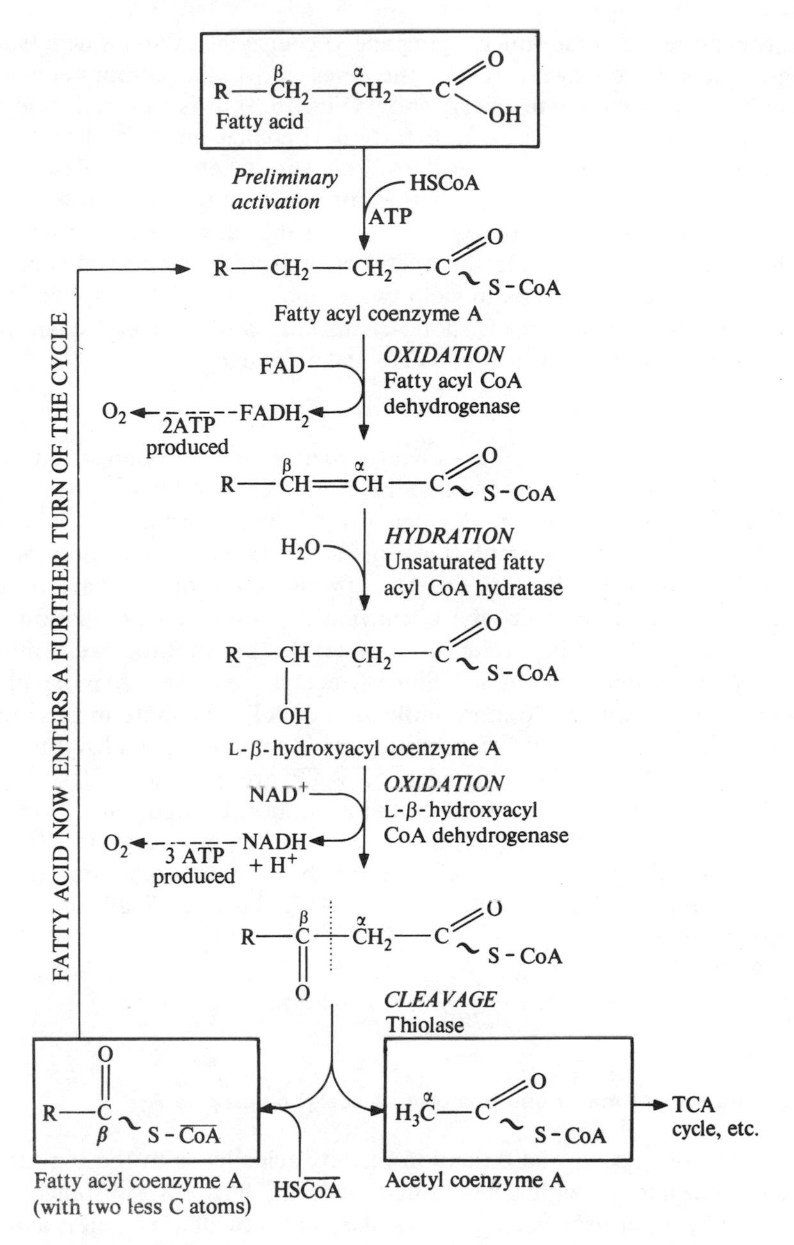

Fig. 10.3 Fatty acid oxidation spiral. (The reduced FAD and NAD pass the electrons to oxygen through the electron transport chain, producing ATP)

free coenzyme A, the products being acetyl coenzyme A and a new (shorter) long-chain acyl coenzyme A. As the lines above the components of the formula of unbound coenzyme A show (Fig. 10.3), it is this molecule which helps to form the new *long-chain* fatty acyl coenzyme A, and is therefore the activator for the *next* turn of the spiral. The β-ketoacyl thiolase reaction thus shortens the carbon chain of the fatty acid by two carbon atoms. The new long-chain fatty acyl coenzyme A acts as the substrate for a further turn of the *catabolic spiral*, while the acetyl coenzyme A undergoes subsequent oxidation by the TCA cycle, to yield twelve molecules of ATP (page 187).

The complete oxidation of a long-chain fatty acid to acetyl coenzyme A yields considerable quantities of ATP. Palmitic acid,

$$H_3C—(CH_2)_{14}—COOH$$

for instance, yields 8 molecules of acetyl coenzyme A through seven turns of the catabolic spiral. This means that 14 molecules of ATP are produced as a result of the oxidation of 7 molecules of FAD by the mitochondrial electron transport chain and 21 molecules of ATP result from oxidation of 7 molecules of NADH. So (14 + 21) or 35 molecules of ATP are produced during the breakdown to acetyl coenzyme A. Subsequent oxidation of the latter by the TCA cycle produces 12 molecules of ATP for every molecule of acetyl coenzyme A used up. Since 8 acetyl coenzyme A molecules are formed from palmitate, 96 more molecules of ATP originate in this manner. Hence, when one molecule of palmitic acid is completely oxidized to carbon dioxide and water, 131 molecules of ATP are produced, less the one molecule of ATP used up in the initial activation by thiokinase.

These 130 moles of ATP represent 33×130 kJ of energy (Section 7.4). Since the standard free energy change for the complete oxidation of palmitate to carbon dioxide and water is 9780 kJ, the overall efficiency of the energy capture is

$$\frac{33 \times 130}{9780} \times 100 = 44\%$$

10.3 The central metabolic position of acetyl coenzyme A

As has just been seen, the oxidation of fatty acids leads to the formation of acetyl coenzyme A within the mitochondrion, but this substance is also produced by other processes. In particular, and provided oxygen is available, it is produced by the decarboxylation of pyruvate (Fig. 8.1). Since pyruvate arises from the degradation of carbohydrates (Chapter 11) acetyl coenzyme A possesses a very special biochemical significance, for it is thus a common product of carbohydrate and fat metabolism, a point of convergence of the two principal routes whereby energy is made available to the cell. It is the subsequent oxidation of acetyl coenzyme A by the machinery of the TCA cycle which leads to the production of a large proportion of the ATP needed by cells.

In addition to this catabolic function, acetyl coenzyme A also has important anabolic functions, for it plays a part in the synthesis of fatty acids and the synthesis of steroids. Both of these anabolic processes require reducing power in the form of NADPH. It will be recalled (Section 5.2) that this cofactor resembles NADH except that it possesses an additional phosphate group. NADPH is predominantly *extra*mitochondrial and results from the reduction of NADP in many plant and animal cells by a reaction route known as the pentose phosphate pathway, which is located in the cytosol (Chapter 14). In plants it is also produced by the light reaction of photosynthesis (Section 14.4). In addition, there are several enzymes in the cytosol which resemble those of the mitochondrion but have NADP as cofactor instead of NAD. In the present context *isocitrate dehydrogenase* is an important example. Like the mitochondrial enzyme of the TCA cycle it oxidizes isocitrate to α-ketoglutarate but the reducing power is captured as NADPH.

Since considerable quantities of NADPH are required for fatty acid synthesis from acetyl coenzyme A it is to be expected that the site of synthesis in the cell is close to the production of NADPH. In green plants, fatty acids are synthesized in the chloroplasts during illumination. In animal cells the biosynthesis takes place in the smooth endoplasmic reticulum close to the pentose phosphate system of enzymes. This distinction in the location of fatty acid synthesis is clearly illustrated in the single-celled alga *Euglena gracilis.* When grown in the light the chloroplastic route of synthesis operates. In the dark, however, these cells can live heterotrophically and develop a high molecular weight fatty acid synthetase resembling that found in animals and situated in the endoplasmic reticulum.

In most organisms studied, fatty acid synthesis is initially a process of converting acetyl coenzyme A to palmitate ($C_{15}H_{31}COOH$). The latter may then be lengthened or shortened as a result of enzymes present in the mitochondrion, the reactions being similar to those shown in Fig. 10.3, working either in the forward or reverse direction. Initially the palmitate enters the mitochondrion by the carnitine mechanism (Fig. 10.2) and the new fatty acid, formed after lengthening or shortening, leaves the organelle in the same way.

The acetyl coenzyme A needed for the *extramitochondrial* biosynthesis of long-chain fatty acids is unlikely to be the acetyl coenzyme A formed directly from the intramitochondrial catabolic processes, for the molecule of acetyl coenzyme A is too large readily to diffuse through the mitochondrial membrane. Instead, intramitochondrial acetyl coenzyme A probably first condenses with oxaloacetate, exactly as it does in the first stage of the TCA

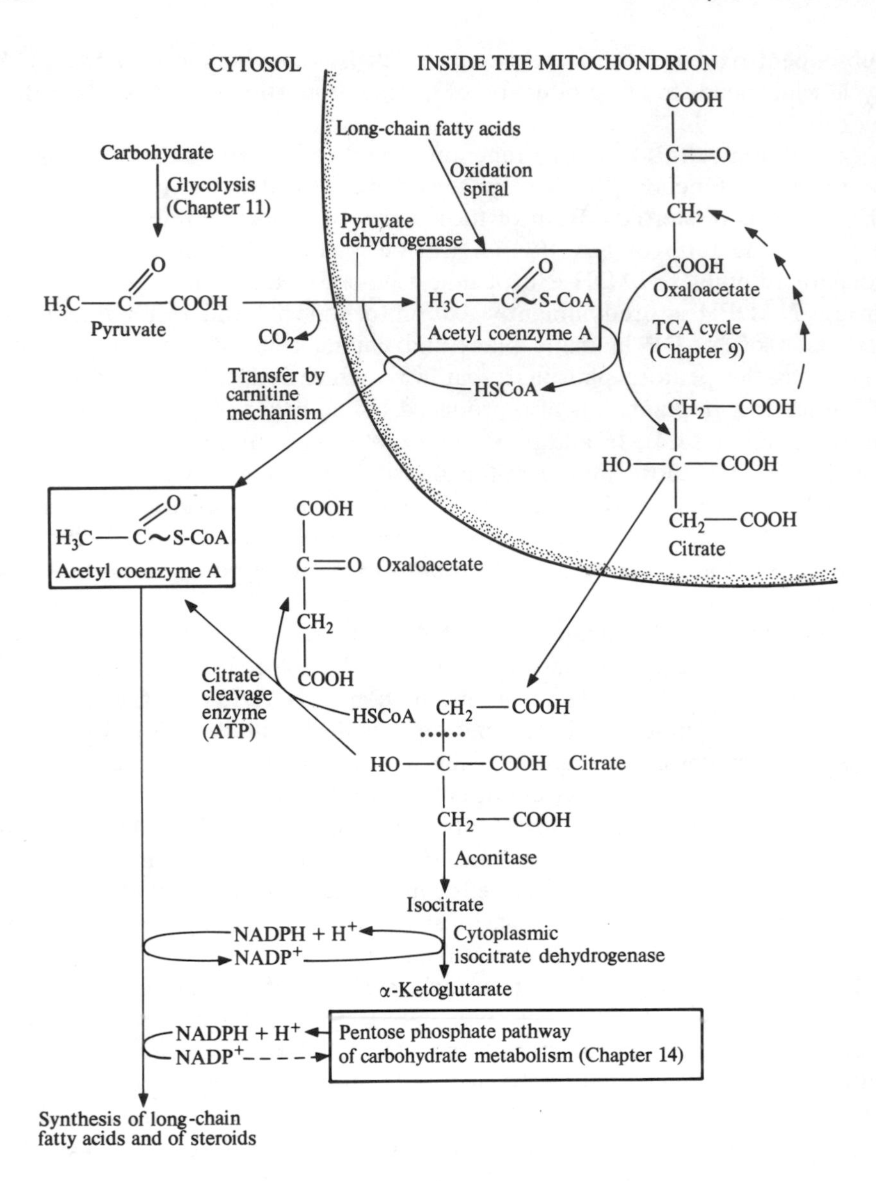

Fig. 10.4 The cellular location of some of the major processes involving acetyl coenzyme A

cycle; the citrate so formed, having shed the coenzyme A, then diffuses into the extramitochondrial cytoplasm. Once in the cytosol, the citrate probably undergoes cleavage to form acetyl coenzyme A and oxaloacetate once more. Extramitochondrial coenzyme A and also ATP are cofactors for this reaction. The carnitine mechanism has also been demonstrated as a means of transfer of acetyl coenzyme A from the mitochondrion to the cytosol. In this instance the reactions shown in Fig. 10.2 operate in the reverse direction. Figure 10.4 summarizes several of the points described here.

10.4 The biosynthesis of long-chain fatty acids

It has been found that, *in vitro*, palmitate synthesis requires a number of cofactors, including NADPH, carbon dioxide, and biotin, none of which participates in the fatty acid oxidation spiral. Thus, even if the difference in cellular location were to be disregarded, the anabolic route cannot simply be the catabolic pathway operating in reverse. It is, in fact, not uncommon to find that biosynthesis occurs by a different route from catabolism. One reason for this is that the cell has to avoid a large endergonic step which would frequently have to be overcome if anabolism were the direct reverse of catabolism (cf. pages 209 and 246).

Experiments using radioactive carbon (as $C^{14}O_2$) have shown that, while carbon dioxide is necessary for fatty acid synthesis, the carbon of this essential factor is not, in fact, incorporated into the final product. Moreover, *malonyl coenzyme A* proves to be a more potent precursor of palmitate than does acetyl coenzyme A. These facts were explained when it was realized that long-chain fatty acids are built up from shorter ones by successive addition of two-carbon units derived from malonyl coenzyme A with release of carbon dioxide.

Malonyl coenzyme A is synthesized in two steps by the carboxylation of acetyl coenzyme A. In the first, carbon dioxide is activated by forming a derivative of *biotin*. This derivative is sometimes called 'active carbon dioxide' and the energy needed for its synthesis is supplied by ATP. The structures of biotin and of 'active carbon dioxide' are shown in Fig 10.5. Second, malonyl coenzyme A is produced by a reaction in which the carbon dioxide is transferred from the biotin complex to a molecule of acetyl coenzyme A. Biotin is essential for several carboxylation reactions in addition to the one just described. Since animals cannot synthesize biotin, it is another example of a vitamin (Table 6.4). Raw egg white contains avidin, a substance which combines with biotin and inactivates it.

The enzyme, *acetyl coenzyme A carboxylase*, which catalyses the synthesis of malonyl coenzyme A for fatty acid anabolism is an example of an *allosteric* protein. The basic unit is a monomer consisting of four peptides, one of which contains a bound molecule of biotin. This monomeric form is inactive

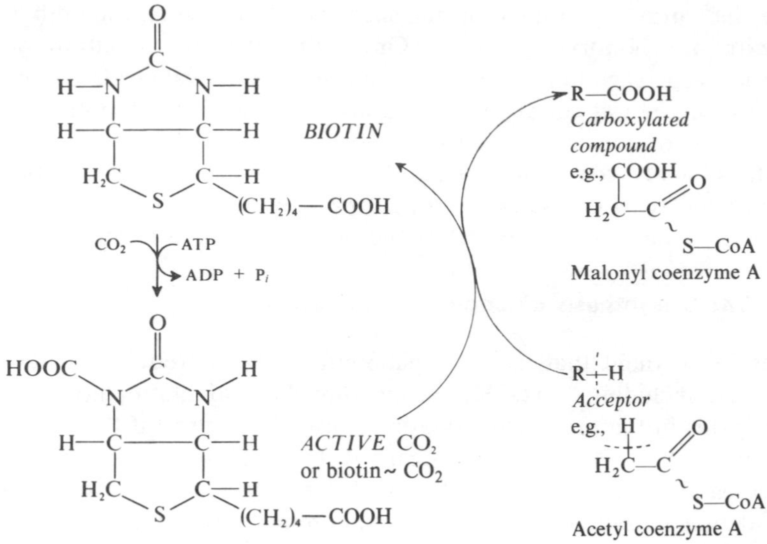

Fig. 10.5 Conversion of biotin to 'active carbon dioxide' and the synthesis of malonyl coenzyme A. Other examples of similar carboxylation reactions involving 'active carbon dioxide', considered in Sections 12.4 and 12.5, are (a) Pyruvate → oxaloacetate and (b) Propionyl coenzyme A → methylmalonyl coenzyme A. In both cases, the general reaction is R—H → R—COOH

as an enzyme. However, in the presence of citrate or isocitrate ions the monomeric units link together in chains up to twenty units long. The polymeric form of the enzyme has been demonstrated in the cytosol of cells in adipose tissue using electron microscopy and is the *active* enzyme. Activation and deactivation of the enzyme (Fig. 10.6a) appear to depend largely on the presence or absence of citric acid and are important in regulating the rate of synthesis of fatty acids (Section 13.9).

The process of fatty acid synthesis consists essentially of the addition of two-carbon units *derived from malonyl coenzyme A* to a derivative of a fatty acid. However, the synthetic spiral differs from that which oxidizes fatty acids in a second important respect, for, in each of the organisms so far investigated, the intermediates concerned in the anabolic sequence are not attached to coenzyme A, but instead are joined to a protein, *acyl carrier protein* or ACP. This protein has as its prosthetic group the sulphydryl compound, 4′-phosphopantotheine, a compound which, it will be recalled, also forms that part of the molecule of coenzyme A which possesses the sulphydryl group (Fig. 5.9). When a fatty acid forms a thioester with the

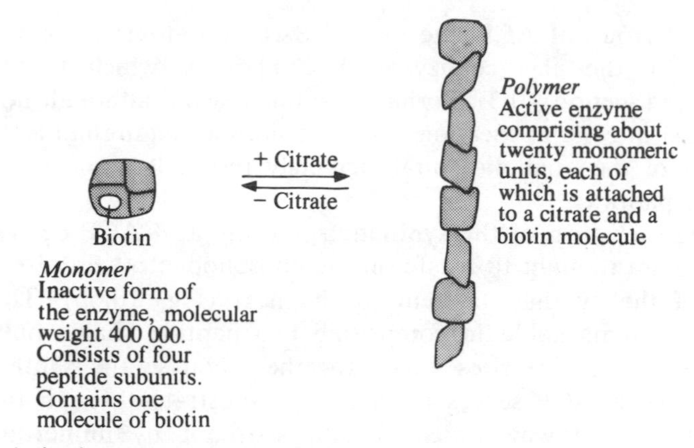

(a) Activation and deactivation of acetyl coenzyme A carboxylase

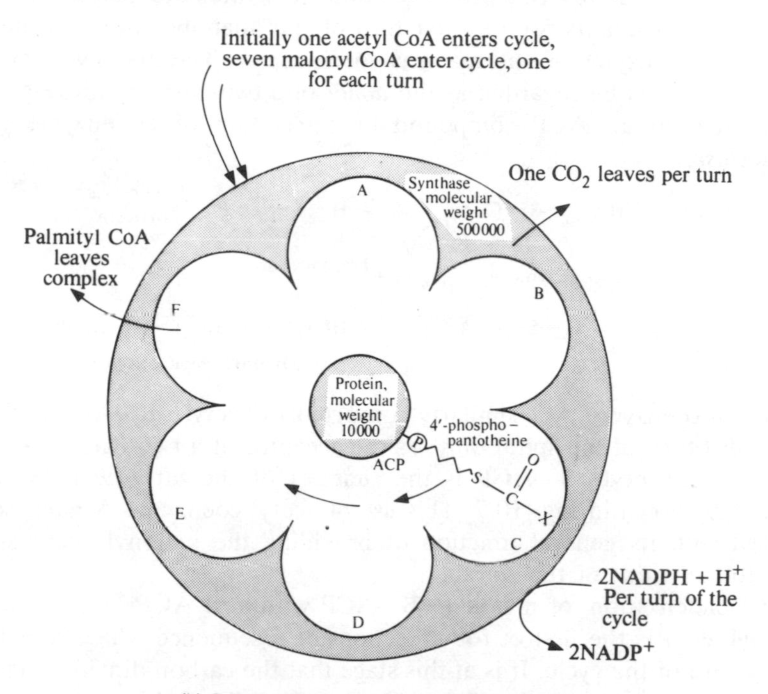

(b) Schematic model of fatty acid synthesis.
The acyl derivatives of the intermediates of the pathway (—C(=O)—X) are attached through a thioester bond to ACP. They are thus swung from one active centre to the next (A, B, C, D, etc.) of the synthase complex. (For details see Fig. 10.7)

Fig. 10.6 Structures of the enzymes which synthesize fatty acids in animals

sulphydryl group of ACP, the α- and β-carbon atoms are activated just as they are in the acyl coenzyme A compounds which take part in fat catabolism (Section 10.3). Perhaps for this reason, although notable differences exist in cellular location and in coenzyme requirements, the principal reactions of the anabolic spiral strikingly resemble some of those of the catabolic pathway.

During each turn of the synthetic spiral the acyl-ACP derivatives of the fatty acid are thought to rotate on the phosphopantotheine from one active centre of the synthase system to the next (Fig. 10.6b). The fatty acid synthase is remarkable in comprising two peptide chains only. Each has several of the active sites which together catalyse the synthetic reaction sequence. The ACP serves to direct the substrate through the successive steps of the pathway which is catalysed, not by numerous individual enzymes but mainly by two proteins whose structures are polyfunctional.

The initial reactants for the first turn of the anabolic spiral are derived from acetyl coenzyme A and malonyl coenzyme A. The malonyl coenzyme A, which can best be regarded as the *donor* of a two-carbon unit (Fig. 10.7) is converted to an ACP compound by the action of an enzyme called *transacylase*:

$$\underset{\text{Malonyl coenzyme A}}{HOOC{-}CH_2{-}\overset{}{\underset{\underset{O}{\|}}{C}}{\sim}S{-}CoA} + \underset{\text{Acyl carrier protein}}{H{-}S{-}ACP} \xrightarrow{\text{Transacylase}} H{-}S{-}CoA + \underset{\text{Malonyl–ACP complex}}{HOOC{-}CH_2{-}\overset{\overset{O}{\|}}{C}{\sim}S{-}ACP}$$

The acetyl coenzyme A is similarly converted to acetyl ~ S—ACP and acts, for the *first* turn of the spiral only, as the acceptor of a two-carbon unit—in other words, acetyl ~ S—ACP is the simplest of the fatty acyl ~ S—ACP compounds shown in Fig. 10.7. This use of acetyl coenzyme A must not be confused with its general function of providing the malonyl coenzyme A donor for *all* turns of the spiral.

The 'condensation' of malonyl ~ S—ACP with acyl-ACP (acetyl, butyryl, hexanoyl, etc.) is the first of four reactions of a sequence which constitutes any one turn of the cycle. It is at this stage that the carbon dioxide originally inserted into acetyl coenzyme A by 'active carbon dioxide' is split away from the malonyl ~ S—ACP. The process of 'condensation' would probably be endergonic, like many reactions which result in the formation of a new carbon–carbon bond, if it were not facilitated by the energy released by the simultaneous cleavage of a nearby carbon–carbon bond as decarboxylation occurs. It is this elegant 'carbon-gain by carbon-loss' mechanism which explains why *malonyl* rather than *acetyl* coenzyme A is the vital two-carbon donor; it also explains why radioactive carbon dioxide is not incorpo-

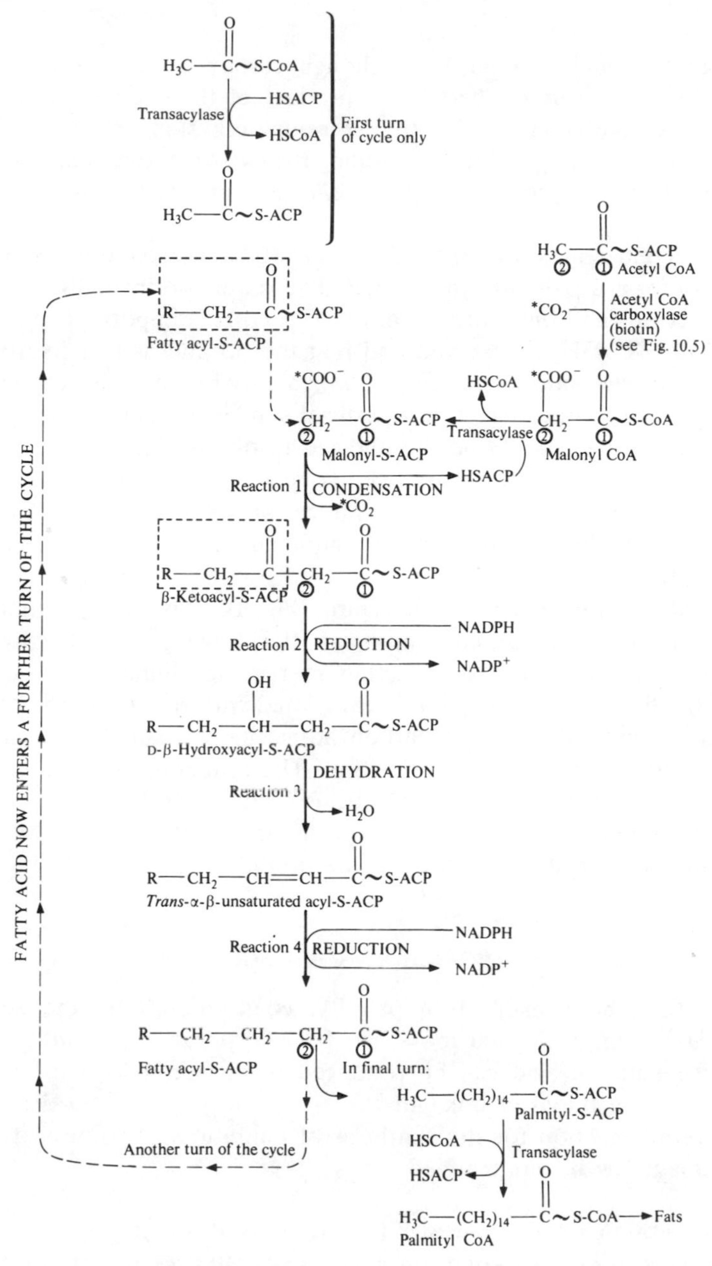

Fig. 10.7 Fatty acid synthetic spiral. In any turn of the cycle two carbon atoms, 1 and 2, derived from acetyl coenzyme A, are added to a fatty acid in the form of the acyl-S-ACP derivative, the donor being malonyl coenzyme A. For the first turn the acid is acetate; for each subsequent turn it is the fatty acid produced in the previous turn. Note the similarity of reactions 2, 3, and 4 to those shown in Fig. 10.3

rated into the final product, even though carbon dioxide is an essential cofactor for fatty acid synthesis. The product of this 'condensation' is the β-ketoacyl derivative of ACP. In the remaining steps of the spiral this compound undergoes reactions resembling the two oxidative changes and one hydrative change described on page 177, but proceeding in the reverse direction.

In the second reaction of the spiral (Fig. 10.7), the β-keto compound is reduced to form a β-hydroxyacyl-ACP. The reduction involves the transfer of two electrons and one proton from NADPH. It is important to note that this, and not NADH, is involved and that the product is D-β-hydroxyacyl-ACP. In the catabolic route (Fig. 10.3), a L-β-hydroxyacyl coenzyme A participates. Thus, once again, a superficial similarity between the catabolic and anabolic routes should not be allowed to obscure the great differences between them. Indeed, the occurrence in one route of a L reactant and in the other of a D product of the same substance (present, of course, in different complexes) provides a remarkable example of the stereospecificity of enzymes (Section 6.4).

The third reaction of the anabolic spiral (Fig. 10.7) is a dehydration which results in an α,β unsaturated compound. It is analogous to the hydratase reaction of the catabolic route acting in reverse. Finally, in the fourth reaction the first turn of the spiral is completed with the formation of a fatty acyl-ACP compound with two carbon atoms more than the original fatty acid which took part in the first reaction. The reduction is achieved by the addition of two protons and two electrons donated by NADPH.

Thus, beginning with acetyl-ACP, each turn of the spiral adds a two-carbon unit so that after seven turns the product is palmityl-ACP,

$$H_3C{-}(CH_2)_{14}{-}\overset{\overset{\large O}{/\!/}}{C}{\sim}S{-}ACP.$$

This can then be converted to palmityl coenzyme A by the action of *transacylase* (Fig. 10.7) and may then be incorporated into *fats*. Alternatively, free palmitic acid can be produced from palmityl coenzyme A by *hydrolysis* of the thioester linkage.

The overall reaction for the synthesis of palmitic acid from acetyl coenzyme A may now be summarized:

$$\begin{aligned}&8H_3CCOSCoA + 7CO_2 + 14NADPH + 14H^+ + 7ATP \rightarrow \\ &\quad H_3C(CH_2)_{14}COOH + 8HSCoA + 7CO_2 + 14NADP^+ + 7ADP + 7P_i + 6H_2O\end{aligned}$$

The stoichiometry shown in the equation indicates that 1 mole of palmitate is derived from 14 moles of reduced NADP and 7 moles of ATP. The process requires a great deal of energy since 14NADPH is equivalent to 14 × 3 high energy phosphate bonds. Moreover, 8 acetyl coenzyme A, which would lead to 96ATP on oxidation, is also used up, and becomes unavailable for immediate catabolic purposes.

10.5 The biosynthesis of triglycerides

Free fatty acids do not usually accumulate in the cell to any great extent, but are largely incorporated into glycerides and phospholipids. The enzyme lipase does not catalyse net synthesis of triglycerides from glycerol and fatty acids, unless one of these is in great excess, because the equilibrium favours hydrolysis rather than esterification. In the cell, triglyceride synthesis bypasses this unfavourable equilibrium by two devices; first, coenzyme A derivatives of the fatty acids, rather than free fatty acids, are involved, and second,

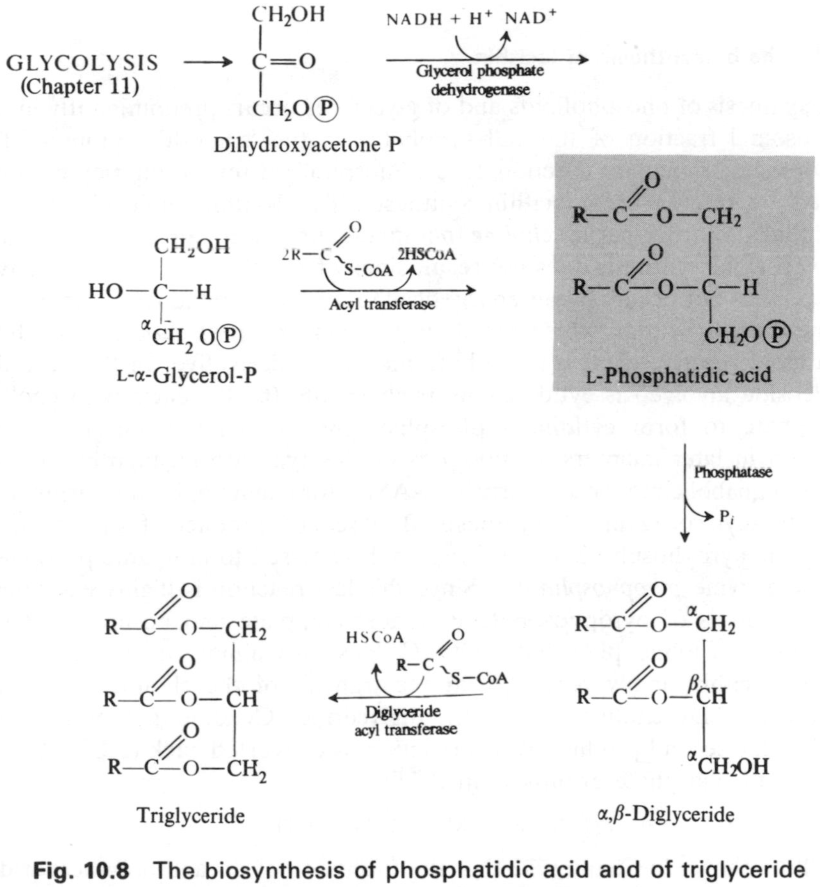

GLYCOLYSIS (Chapter 11)
Dihydroxyacetone P
NADH + H$^+$ NAD$^+$
Glycerol phosphate dehydrogenase
L-α-Glycerol-P
2R–CO–S-CoA 2HSCoA
Acyl transferase
L-Phosphatidic acid
Phosphatase
P_i
α,β-Diglyceride
HSCoA R–CO–S-CoA
Diglyceride acyl transferase
Triglyceride

Fig. 10.8 The biosynthesis of phosphatidic acid and of triglyceride

instead of glycerol, it is glycerol phosphate which is esterified. Hence both reactants are at a higher energy level than in the lipase reaction (compare page 27).

The first step, the synthesis of L-α-*glycerol phosphate*, usually occurs by a reduction of dihydroxyacetone phosphate, which is an important intermediate product of the glycolytic breakdown of glucose (Chapter 11). The reduction results from the passage of two electrons from NADH to the dihydroxyacetone phosphate (Fig. 10.8). The L-α-glycerol phosphate then reacts successively with two molecules of the acyl coenzyme A derivatives of the appropriate fatty acids. The product is a *phosphatidic acid*. The synthesis of triglyceride is completed by two reactions; first, the phosphate group of the phosphatidic acid is hydrolysed, forming an α,β-diglyceride and orthophosphate (enzymes which catalyse the hydrolysis of organic phosphates in this way are termed *phosphatases*), and finally the diglyceride is esterified by the transfer of the acyl group from a fatty acyl coenzyme A (Fig. 10.8).

10.6 The biosynthesis of lecithin

The synthesis of phospholipids and of glycerides occurs predominantly in the microsomal fraction of the cell—probably in the 'smooth' regions of the endoplasmic reticulum (Section 1.10). Phospholipid formation may be illustrated by reference to lecithin synthesis. The lecithin molecule consists essentially of two parts, *choline phosphate* linked to an α,β-*diglyceride*. However, the synthesis does not result simply from the reaction of these two compounds but occurs by an energetically more favourable route; there is a preliminary activation which results in the conversion of choline phosphate to a more reactive choline phosphate-nucleotide derivative. In this case the nucleoside involved is cytidine—its triphosphate (CTP) reacts with choline phosphate to form cytidine diphosphate-choline (CDP-choline). As will be seen in later chapters, compounds of this type are frequently encountered in anabolic reactions—fatty acyl-AMP, for example, is an intermediate in fatty acyl coenzyme A synthesis. The second product of the reaction, inorganic pyrophosphate, is itself rapidly hydrolysed to inorganic phosphate by the enzyme *pyrophosphatase*. Since this last reaction is highly exergonic, the hydrolysis of pyrophosphate is almost complete and ensures that the reaction of choline phosphate with CTP is also almost irreversible (Fig. 10.9). Lecithin finally results from the transfer of the choline phosphate moiety of CDP-choline to an α,β-diglyceride. Cytidine monophosphate (CMP), the second product of this reaction, is converted back to CTP by the transfer of phosphate groups from ATP:

$$2ATP + CMP = 2ADP + CTP$$

It will be shown in Chapter 12 that the formation of di- and polysaccharides

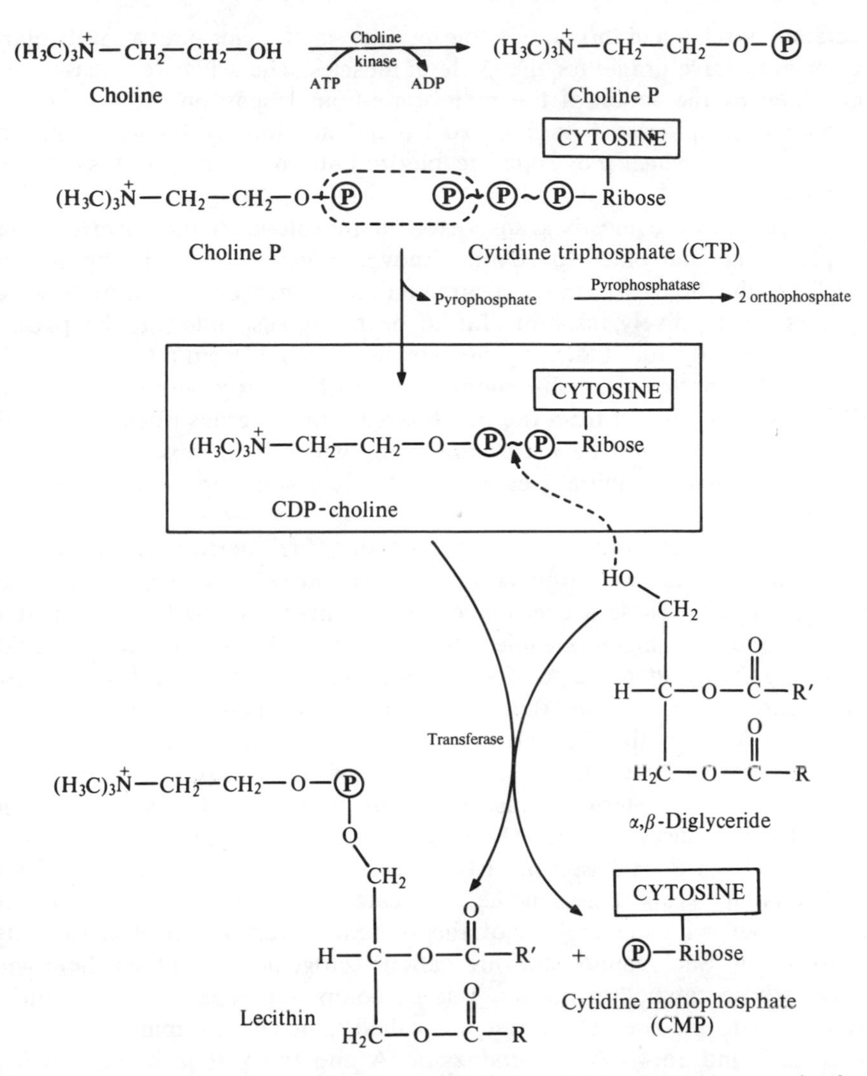

Fig. 10.9 The biosynthesis of lecithin. Choline is converted to an 'active' form, CDP-choline, a form in which it is readily transferred to an α,β-diglyceride

from monosaccharides occurs by reactions very similar to those by which lecithin is synthesized from CTP and choline.

10.7 Blood plasma lipids and the 'ketone bodies'

Mammalian blood plasma contains several lipids which are carried *from the liver and adipose tissue* to supply fuel for the other tissues such as the heart, skeletal muscle, and brain. Of these, free fatty acids are a particularly important source of fuel for the skeletal muscles. The acids are released into the blood by the action of the membrane-bound lipase in the hepatic and adipose cells (process 6 in Fig. 10.1), and are loosely bound to plasma albumen, from which they separate before entering cells by diffusion (process 1 in Fig. 10.1).

Triglycerides are mostly transported in the blood in the form of small droplets (diameter of 20 to 70 nm), known as very low density lipoprotein (VLDL), the lipid of which is surrounded by hydrophilic protein which enables the relatively insoluble fat to be freely suspended in the plasma. Process 7 in Fig. 10.1 indicates the release of VLDL from a liver cell while process 8 indicates the subsequent uptake of the triglyceride by muscle. VLDL is quite distinct from the much larger chylomicrons (diameter of 200 to 500 nm) which are the major means by which lipids absorbed *from the alimentary canal* are initially carried to the liver and adipose tissue (Section 17.10).

A third important means of transporting fuel derived from lipids is provided by the two four-carbon acids, *acetoacetic acid* and D-*β*-*hydroxybutyrate.* These are produced by the liver, especially in the fasting animal. They are highly soluble and very easily diffuse into cells, quickly providing them with a source of acetyl coenzyme A. The brain and heart are particularly dependent on these acids at times when the level of blood glucose is low, for they can provide as much as 70 per cent of the energy requirements of these organs. The acids are usually called 'ketone bodies', a rather inappropriate term because β-hydroxybutyrate is clearly not a ketone.

The liver produces the ketone bodies from acetyl coenzyme A, most of which is derived from β-oxidation of fatty acids in the fasting animal. In addition, certain amino acids (e.g., leucine) are catabolized by the removal of the amino group with metabolism of the remaining carbon skeleton to acetyl coenzyme A. Such amino acids are called 'ketogenic' to contrast them with others whose catabolism produces some compound capable of subsequent conversion to glucose. The latter are called 'glucogenic' amino acids (Sections 12.5 and 15.4). Acetyl coenzyme A and the ketone bodies are not glucogenic since animals lack the enzymes capable of converting them to glucose (Section 12.6).

Acetyl coenzyme A is readily converted to acetoacetyl coenzyme A by an enzyme catalysed aldol-type addition reaction which releases free coenzyme A. This is the initial reaction in the mitochondrion of the liver cell for forming the ketone bodies (Fig. 10.10). It might be expected that acetoacetate would result from simple hydrolysis of the thioester link. This is apparently unfavourable energetically and the cell uses a more complex mechanism instead. The acetoacetyl coenzyme A undergoes a further aldol-type addition reaction, the products being a six-carbon coenzyme A compound,

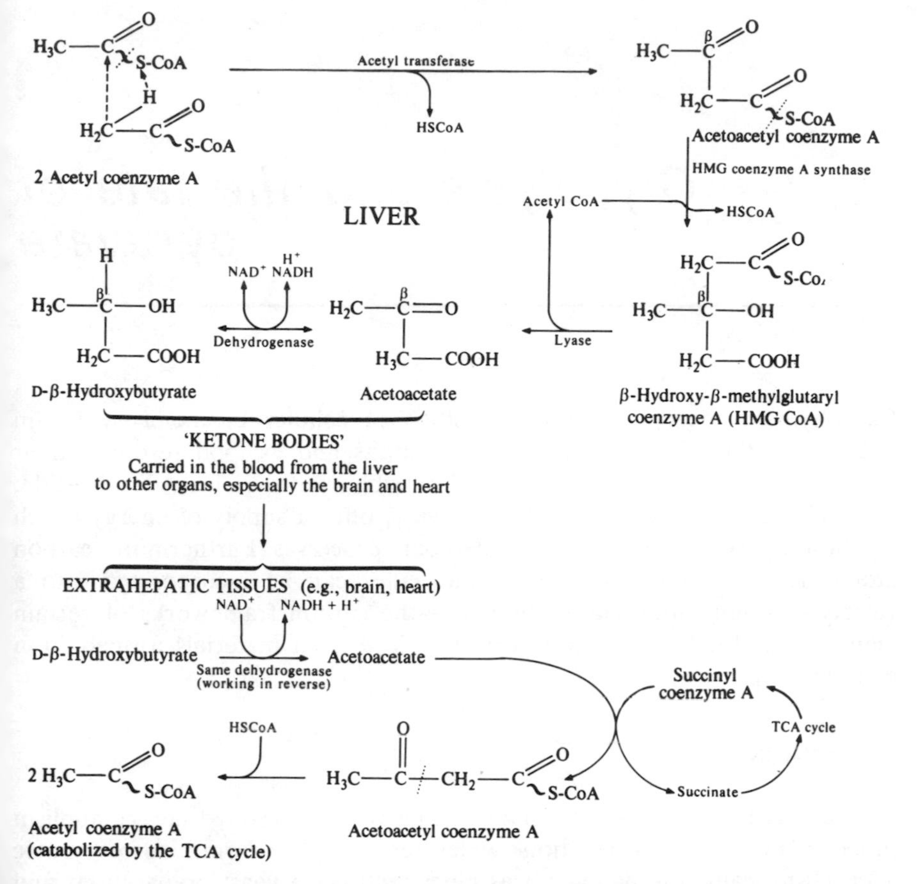

Fig. 10.10 The synthesis of 'ketone bodies' in the liver, and the initial reactions leading to their catabolism in other tissues

β-hydroxy-β-methylglutaryl coenzyme A (HMG coenzyme A), and free coenzyme A. A specific lyase now cleaves the HMG coenzyme A into acetoacetic acid and acetyl coenzyme A. Some three-quarters of the acetoacetic acid is usually converted to D-β-hydroxybutyrate by a NAD-dependent dehydrogenase located in the inner mitochondrial membrane (Fig. 10.10). The product is clearly distinct from the L-hydroxybutyrate participating in the catabolic spiral (Fig. 10.3).

The two ketone bodies easily diffuse out of the hepatic cells to be carried to other tissues. On arrival, they enter the mitochondria and are converted to acetyl coenzyme A by reactions indicated in Fig. 10.10. This is subsequently catabolized by the TCA cycle.

11

Glycolysis and the fate of pyruvate

Carbohydrates occupy a central position in cellular biochemistry for in addition to their functions as structural units and as food reserves, their breakdown by one or other of several possible routes yields ATP, NADH, or NADPH. All of these, in different ways, offer a supply of energy which can be used by the cell to drive endergonic processes. Furthermore, carbon atoms derived from carbohydrate molecules can be incorporated into a variety of other biological compounds—the carbon frameworks of certain amino acids, for instance, can be synthesized from materials formed from carbohydrates.

11.1 Glycolysis

In most types of cells, the main pathway of carbohydrate catabolism proceeds by a series of reactions which results in the formation of pyruvic acid. Historically, the pathway was elucidated using yeast preparations and muscle extracts incubated under anaerobic conditions, but it is now known that with the exception of the last step of the sequence, the reactions are essentially the same in aerobic tissues. For this same historical reason, glycolysis ('sugar-splitting') is, strictly speaking, the sequence of reactions leading all the way from carbohydrates, *via* pyruvate, to lactate, but it will, in practice, be more convenient to consider the fate of pyruvate separately.

The pathway of glycolysis illustrates a fascinating and simplifying aspect of biochemistry—namely, that processes which, in different tissues or organisms, appear superficially different, may nevertheless proceed by pathways which over most of their course are similar. The reactions of glycolysis are, in fact, essentially the same whether they occur in the context of anaerobic fermentation or form part of a mechanism by which aerobic respiration occurs. Only the fate of pyruvate distinguishes these separate activities, and it is for the moment only necessary to recognize that under aerobic conditions pyruvate enters mitochondria and is converted to acetyl coenzyme A

(Chapter 9) which is then completely oxidized to carbon dioxide and water. It is at this stage (and not during glycolysis) that the major part of the energy originally present in the carbohydrate is released and in part conserved as ATP by the mechanism of the electron transport system (Chapter 8).

11.2 Discovery of glycolysis

The glycolytic pathway was the first major metabolic route to be elucidated, and the historical setting is not only important in its own right but also because it led to the establishment of ideas and of methodology which were of crucial importance to the development of biochemistry.

In 1897, E. Büchner discovered, rather fortuitously, that cell-free extracts of yeast were able to ferment sugars, thus demonstrating for the first time that enzymes do not necessarily have to function within intact cells. Soon afterwards (1905) A. Harden and W. Young showed that fermentation of glucose by yeast extract soon slowed down unless inorganic phosphate was added. This simple observation was the first step towards the establishment of a fundamental principle, namely, that the mechanism of certain metabolic reactions involved the incorporation of inorganic phosphate into organic compounds. In the present instance, Harden and Young eventually identified fructose-1,6-diphosphate as an intermediate in glycolysis. They also discovered that yeast extract contained both a heat-labile component (a mixture of enzymes originally termed 'zymase') and a heat-stable moiety. The latter is now known to have contained NAD, ATP, ADP, and various inorganic ions. The discovery of heat-stable coenzymes was eventually to lead to the discovery of *vitamins* by F. Gowland Hopkins.

In the period from 1930 to 1940 the principal steps of the pathway were proposed by Embden and later confirmed by the independent but complementary work of many contributors, outstanding among which were W. Meyerhof, J. Parnas, C. Cori and G. Cori, and O. Warburg. Thc pathway is sometimes known as the Embden–Meyerhof pathway.

11.3 The four stages of glycolysis

Sugars almost always react biochemically in a phosphorylated form, and in *stage 1* of glycolysis the carbohydrate to be catabolized is converted to fructose diphosphate (Fig. 11.1). In fact, this substance can, more properly, be regarded as the real starting point for the pathway. Its formation is a multi-step process and differs slightly according to whether it is derived from a carbohydrate store, such as glycogen or starch, or is formed directly from monosaccharide entering the cell (as occurs when yeast is grown in glucose solution or when glucose enters a muscle cell from the blood). When

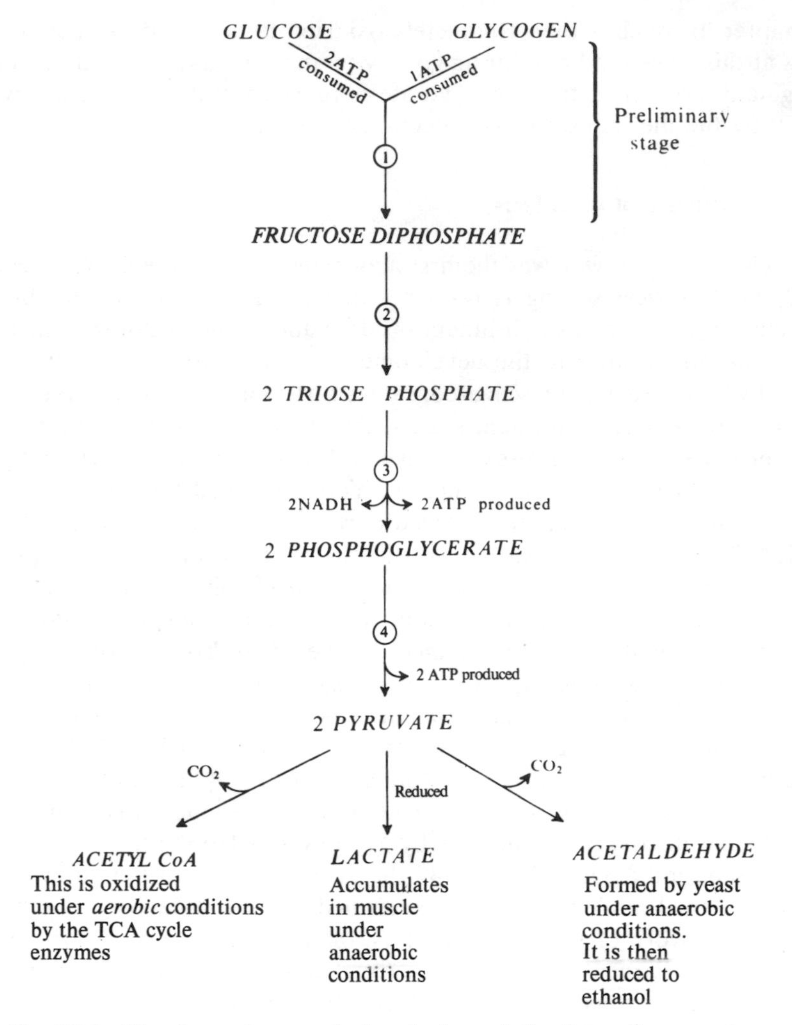

Fig. 11.1 The four stages of glycolysis and the fate of pyruvate

glycogen and starch are metabolized to fructose diphosphate, only one high energy bond is used up for each molecule of the diphosphate formed, but when glucose is the precursor, two molecules of ATP are used up per molecule of fructose diphosphate. The reason is that the polysaccharides are at a higher energy level than glucose, and ATP is used up in their synthesis (Chapter 12).

In *stage 2*, each fructose diphosphate molecule splits into two molecules of triose phosphate, thus reducing to half the size of the molecules participating in subsequent steps. Incidentally, since these later steps commence with triose phosphate, it is inevitable that fructose diphosphate, and not the monophosphate, should be the precursor.

In *stage 3*, which is another multi-step process, the aldehydic form of triose phosphate (glyceraldehyde-3-P) is oxidized to phosphoglyceric acid, much of the energy so liberated being conserved by the coupled synthesis of ATP from ADP. Lastly, in *stage 4*, which again involves a sequence of reactions, the phosphoglyceric acid is converted to pyruvic acid with a further release of energy, much of which is again conserved by synthesis of ATP. As Fig. 11.1 shows, it is only after this point that the pathways of fermentation by yeast, anaerobic respiration in muscle, and the more common aerobic respiration diverge from one another.

11.4 Stage 1: formation of fructose diphosphate

All the enzymes of the glycolytic pathway are present in the *cytosol* and all the metabolic intermediates between the initial carbohydrates and the final pyruvic acid are *phosphorylated* compounds. As stated in the previous section, the reactions of stage 1 can conveniently be regarded as preliminary reactions whereby several possible carbohydrates are converted, *via* glucose-6-phosphate, to fructose-1,6-diphosphate.

The shortest route starts from glucose; this will be described first. For most types of cells, however, this is quantitatively less important than the slightly more complex route which enables insoluble carbohydrates in storage sites to be mobilized for metabolic purposes. Glucose is first phosphorylated to glucose-6-phosphate by the action of *hexokinase*, an enzyme requiring magnesium ions as cofactor (Fig. 11.2).

There are many enzymes bearing the suffix '-kinase'. It is useful to remember that this term indicates that the enzyme catalyses the transfer of a phosphate group from the high energy compound ATP (or some similar nucleoside triphosphate) to another compound—in the present case to the *hexose*, glucose. The glucose-6-phosphate is so named because it is the hydroxyl group on carbon atom 6 (i.e., in the straight-chain form, the atom farthest from the aldehyde group) which is phosphorylated:

$^6CH_2OH$, 5, O, 4, OH, 1, HOH, HO, 3, 2, OH
Glucose

CH_2OⓅ, O, OH, HOH, HO, OH
Glucose-6-P

CH_2OH, O, H, OH, HO, O Ⓟ, OH
Glucose-1-P

Hexokinase is relatively non-specific in its action, in that it can phosphorylate not only glucose but also several other sugars. (A glucose-specific

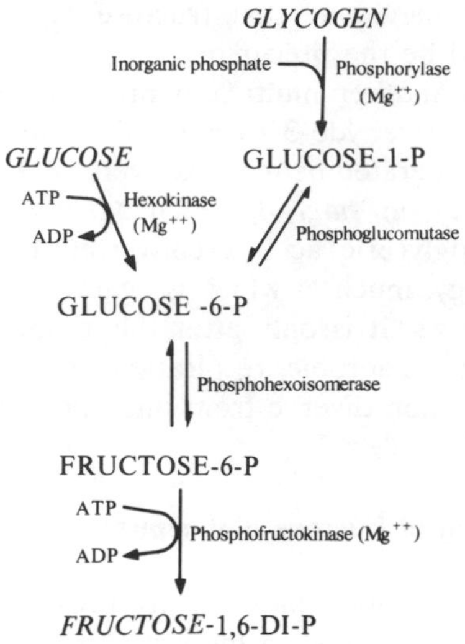

Fig. 11.2 Glycolysis—stage 1. Synthesis of fructose-1,6-diphosphate

kinase, glucokinase, is present in the mammalian liver, but it does not normally play a significant role in glycolysis.) An interesting and important property of hexokinase is that it can be inhibited by the glucose-6-phosphate it produces. Thus, if it is working so rapidly that it produces more glucose-6-phosphate than the cell can immediately utilize, the accumulating reaction product slows down the rate of the hexokinase reaction, making the process self-regulating. An enzyme which is so placed in a biological pathway that its rate of reaction determines the extent to which food reserves are mobilized is frequently under some sort of metabolic regulation. Phosphorylase, which is responsible for mobilizing insoluble carbohydrates, is another example (see below). Control mechanisms are discussed in more detail in Chapter 13.

The second reaction of the sequence from glucose to fructose-1,6-diphosphate involves the isomerization of glucose-6-phosphate with the formation of fructose-6-phosphate. The enzyme responsible is termed *phosphohexoisomerase*, and at equilibrium about one-third of the phosphohexose is in the form of fructose-6-phosphate:

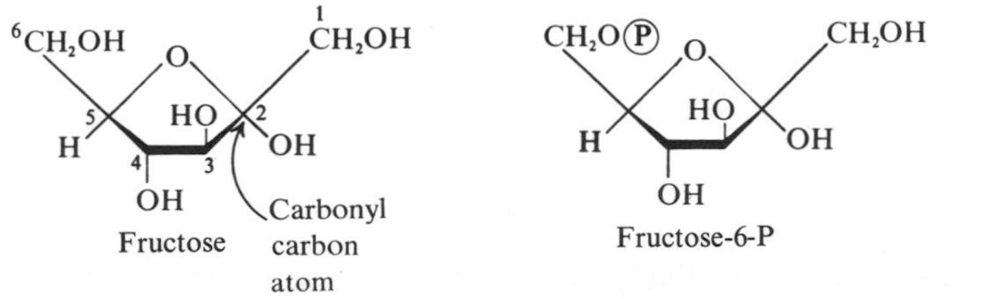

In the final reaction of stage 1 (Fig. 11.2), further phosphorylation takes place at the expense of ATP. The enzyme responsible is specific for fructose-6-phosphate and is therefore termed *phosphofructo*kinase. The product is fructose-1,6-diphosphate, a fructose molecule phosphorylated at each end of the straight-chain form:

Fructose-1,6-di-P

When the carbohydrate to be catabolized is endogenous glycogen or starch, glucose-6-phosphate is again formed but by a *two-step* process. The first step is the cleavage of the bond linking the terminal glucose unit, situated at the non-reducing end of the polysaccharide chain. The enzyme responsible is *polysaccharide phosphorylase*; phosphorolysis resembles hydrolysis except that it results in the attachment of an inorganic phosphate group to carbon atom 1 of the cleaved glucose molecule, and so leads to the formation of glucose-1-phosphate. Note that the phosphate is not derived from ATP (phosphorylase is therefore not a kinase) and that inevitably it is glucose-1-phosphate, not glucose-6-phosphate, which is produced by the reaction:

Non-reducing end of starch or glycogen molecule

$H_2PO_4^-$ H^+ Phosphorylase

Glucose-1-P

Non-reducing end of polysaccharide chain ready for further phosphorylase activity

There are two important polysaccharide phosphorylases: *starch phosphorylase* in plants and *glycogen phosphorylase* in animals. The glycogen phosphorylase present in muscle has been most studied. It exists in two forms. One of these, phosphorylase **a**, consists of four subunits and is very active, while the other, phosphorylase **b**, is a dimer of much lower activity. What is more, the interconversion of the two forms is under elaborate biochemical control. The several factors involved in that control are described in Chapter 13.

Next, the glucose-1-phosphate is readily converted by the enzyme *phosphoglucomutase* to glucose-6-phosphate. In this reaction the two isomeric glucose phosphates form an equilibrium mixture containing about 94.5 per cent glucose-6-phosphate. The name phosphogluco*mutase* implies that the enzyme transfers a phosphate group from one part of the glucose molecule to another. Glucose-1,6-diphosphate is a cofactor for the reaction.

From this point the pathway of glycolysis is exactly the same whether glucose, starch, or glycogen is the primary substrate. The glucose-6-phosphate is converted by phosphohexoisomerase to fructose-6-phosphate and this is then changed by phosphofructokinase to fructose-1,6-diphosphate in the manner already described (Fig. 11.2).

In stage 1 of glycolysis, there is a net loss of two molecules of ATP for every glucose molecule metabolized. If glycogen is the source of the fructose diphosphate only one molecule of ATP is lost, since the second phosphate group is derived from inorganic phosphate through the phosphorylase reaction. This loss of ATP during a process whose primary function is to produce ATP need cause no concern since, as will be seen in stages 3 and 4, there is nevertheless a net overall synthesis of ATP. The ATP used up in stage 1 is analogous to the energy used in priming a gun—when the cap is detonated there is a much greater release of energy than was needed initially to pull the trigger. A similar situation was encountered earlier (Section 10.2) in relation to the 'priming' of fatty acids.

11.5 Stage 2: formation of triose phosphates

Stage 2 of glycolysis is the cleavage of the fructose diphosphate produced in the previous stage to yield two molecules of triose phosphate (Fig. 11.3). The first enzyme involved is called *aldolase* because it catalyses a reaction known in organic chemistry as the aldol 'condensation'. The simplest example of this type of condensation—better thought of as an addition reaction—occurs when acetaldehyde is treated with NaOH. Two molecules of acetaldehyde unite to form one molecule of aldol in a reaction in which one of the two carbonyl groups is converted into a hydroxyl group by reaction with a 'framework' hydrogen atom of the other molecule:

$$H_3C{-}CHO + H{-}CH_2{-}CHO \longrightarrow H_3C{-}CH(OH){-}CH_2{-}CHO$$

Addition of one molecule of aldehyde to the carbonyl group of the other

Aldol

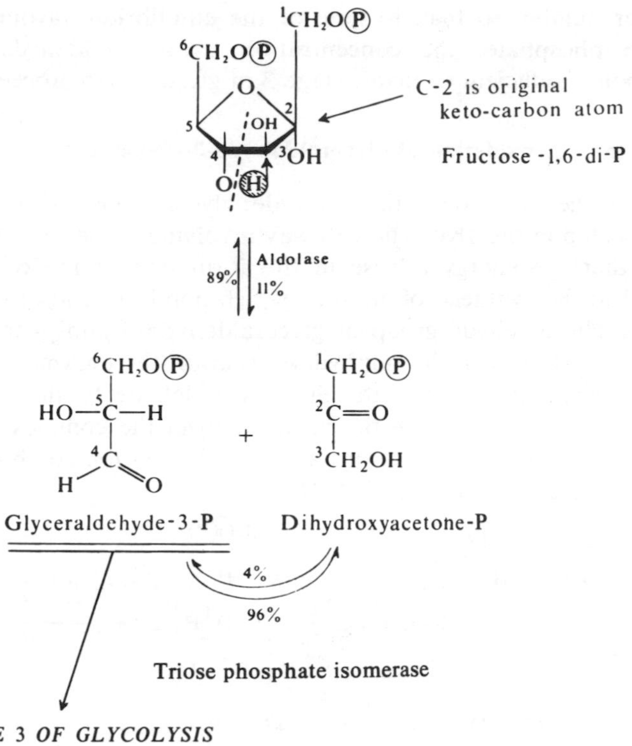

$^{1}CH_2O$Ⓟ
$^{6}CH_2O$Ⓟ
C-2 is original keto-carbon atom
Fructose-1,6-di-P

Aldolase
89%
11%

$^{6}CH_2O$Ⓟ
HO—^{5}C—H
^{4}C
H O
+
$^{1}CH_2O$Ⓟ
^{2}C=O
$^{3}CH_2OH$

Glyceraldehyde-3-P Dihydroxyacetone-P

4%
96%

Triose phosphate isomerase

TO STAGE 3 OF GLYCOLYSIS

Fig. 11.3 Glycolysis—stage 2. Cleavage of fructose-1,6-diphosphate

Aldolase catalyses a comparable reaction between a molecule of glycer-ldehyde-3-phosphate, acting as the hydrogen acceptor, and one of dihy-lroxyacetone phosphate. In other words, aldolase causes molecules of the wo isomeric forms of triose phosphate to join together, the carbonyl group of the glyceraldehyde-3-phosphate being changed to a hydroxyl group. The eaction is reversible but it tends to favour the formation of fructose liphosphate rather than triose phosphate. At equilibrium, the mixture ontains some 89 per cent fructose diphosphate but only 11 per cent triose hosphate. This reversibility is of importance, since in some circumstances ldolase catalyses the synthesis of hexose from triose phosphate (Chapter 2).

The aldolase reaction produces equimolecular quantities of glycer-ldehyde-3-phosphate and dihydroxyacetone phosphate; the final reaction f stage 2 allows interconversion of these two compounds. This is necessary ecause glyceraldehyde-3-phosphate alone is the reactant for stage 3. The wo trioses are readily interconverted through the intervention of the nzyme *triose phosphate isomerase*, the equilibrium mixture containing about 6 per cent dihydroxyacetone phosphate. The enzyme has a very high turnover number so that, in spite of the equilibrium favouring dihydroxy-acetone phosphate, the concentration of glyceraldehyde phosphate is nevertheless sufficient to enable stage 3 of glycolysis to proceed unhindered.

11.6 Stage 3: oxidation of glyceraldehyde-3-phosphate

Stage 3 is the oxidation of the aldehyde, glyceraldehyde-3-phosphate, and is the only step in the glycolytic pathway involving oxidation. It occurs in such a way that the energy release in the formation of the carboxylic acid is coupled to the synthesis of an energy-rich bond, instead of being dissipated as heat. The aldehyde group of glyceraldehyde-3-phosphate undergoes an addition reaction with the sulphydryl group of the enzyme *phosphoglyceraldehyde dehydrogenase*, so attaching the aldehyde to the enzyme surface. Oxidation then occurs with the removal from the complex of two protons and two electrons. The electrons and a hydrogen ion combine with NAD^+, which is thereby reduced to NADH.

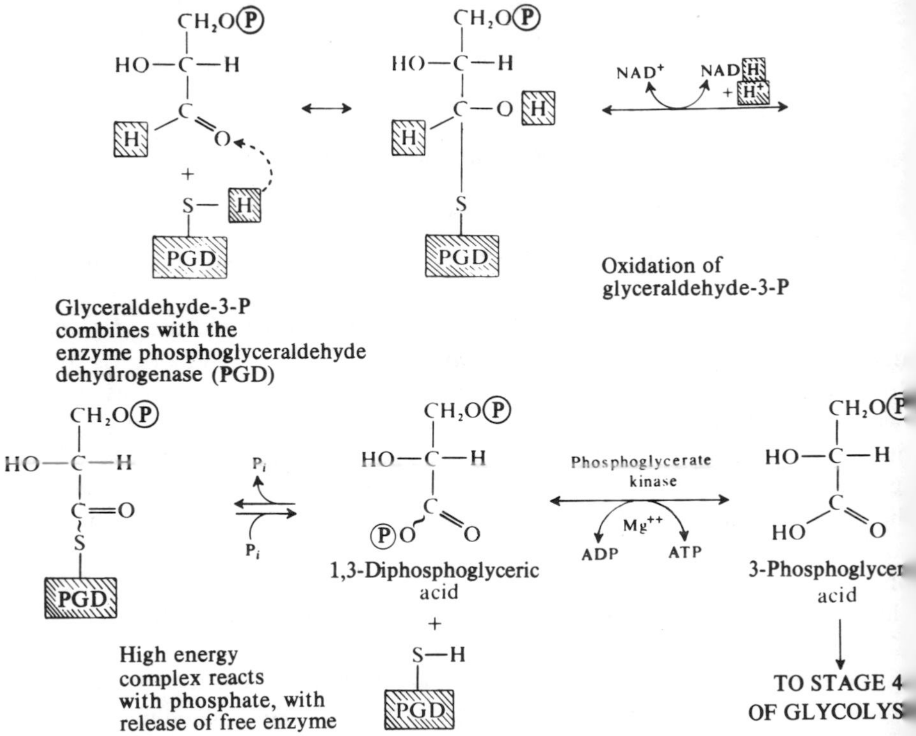

Glyceraldehyde-3-P combines with the enzyme phosphoglyceraldehyde dehydrogenase (PGD)

Oxidation of glyceraldehyde-3-P

High energy complex reacts with phosphate, with release of free enzyme

1,3-Diphosphoglyceric acid

Phosphoglycerate kinase

3-Phosphoglycer[illegible] acid

TO STAGE 4 OF GLYCOLYS[illegible]

Fig. 11.4 Glycolysis—stage 3. Oxidation of glyceraldehyde-3-phosphate an[illegible] the several steps leading to 3-phosphoglyceric acid. All the steps ar[illegible] reversible. Note that two molecules of ATP are formed per origina[illegible] molecule of hexose

Since phosphoglyceraldehyde dehydrogenase couples oxidation of glyceraldehyde-3-phosphate with reduction of NAD^+ it is better called D-glyceraldehyde-3-phosphate:NAD^+ *oxidoreductase*, but the older name is still widely used. The high energy complex of enzyme and oxidized glyceraldehyde-3-phosphate is next attacked, not by water but by inorganic orthophosphate, and is thereby split into free enzyme and a carboxyl phosphate of high energy, 1,3-diphosphoglyceric acid (Fig. 11.4).

1,3-Diphosphoglyceric acid is highly reactive, and, *in vitro*, it is found that its hydrolysis yields 3-phosphoglyceric acid, phosphoric acid, and much energy in the form of heat. It is, therefore, a high energy compound. In the living cell, a comparable reaction, catalysed by the enzyme *phosphoglycerate kinase*, couples the energy released with the transfer of phosphate to ADP. This enzyme, like almost all known kinases, requires the presence of magnesium ions. In the present reaction the enzyme is, in a sense, 'working in reverse', since ATP is being formed rather than destroyed. It provides a further example of substrate-level phosphorylation (See Sections 8.1 and 9.7).

In terms of the hexose unit from which *two* molecules of triose phosphate were formed, there is, at this stage, a gain of two molecules of ATP.

11.7 Stage 4: synthesis of pyruvate

Stage 4 of glycolysis is the conversion of 3-phosphoglyceric acid to pyruvic acid. In the first of three reactions the phosphate attached to the 3-phosphoglycerate is transferred to the central carbon atom, thus freeing the hydroxyl group on the terminal carbon in readiness for a subsequent reaction (Fig. 11.5). This transfer of a phosphate group from one part of the molecule to another is catalysed by the enzyme *phosphoglyceromutase.* It should be observed that the reaction is closely similar to that catalysed by *phosphoglucomutase* in stage 1. In just the same way as the latter enzyme only functions in the presence of glucose-1,6-diphosphate, so phosphoglyceromutase likewise needs 2,3-diphosphoglycerate as cofactor.

The 2-phosphoglyceric acid is now dehydrated by the elimination of the elements of water from positions 2 and 3. The unsaturated product, phosphoenolpyruvate (PEP) is interesting since, like ATP, it possesses a high energy phosphate bond (Fig. 7.5a). In this respect it should be contrasted with many phosphate esters of alcohols and of similar hydroxylated compounds; glucose-1-phosphate and 2-phosphoglycerate, for example, liberate only a small amount of energy when the phosphate group is hydrolysed (Section 7.9). The difference may be related to the fact that the enol form of pyruvic acid is very unstable. The dehydration of 2-phosphoglycerate converts a low energy phosphate bond to a high energy bond—the energy of the exergonic dehydration, in a certain sense, is diverted to the phosphorus bond, the

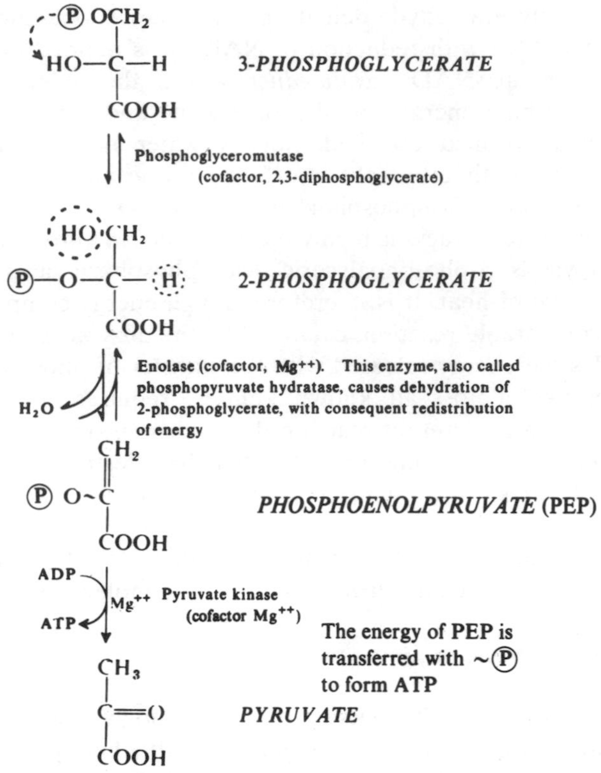

Ⓟ OCH_2
HO—C—H *3-PHOSPHOGLYCERATE*
COOH

Phosphoglyceromutase
(cofactor, 2,3-diphosphoglycerate)

$HOCH_2$
Ⓟ—O—C—H *2-PHOSPHOGLYCERATE*
COOH

H_2O ← Enolase (cofactor, Mg^{++}). This enzyme, also called phosphopyruvate hydratase, causes dehydration of 2-phosphoglycerate, with consequent redistribution of energy

CH_2
Ⓟ O~C *PHOSPHOENOLPYRUVATE* (PEP)
COOH

ADP → ATP, Mg^{++} Pyruvate kinase
(cofactor Mg^{++})

The energy of PEP is transferred with ~Ⓟ to form ATP

CH_3
C=O *PYRUVATE*
COOH

Fig. 11.5 Glycolysis—stage 4. Synthesis of pyruvate. (Note that in stage 4, two molecules of ATP are formed per original molecule of hexose)

change in molecular configuration consequent upon dehydration providing the trapping device. The reaction is catalysed by an extensively investigated enzyme usually called *enolase*, though a better name is phosphopyruvate hydratase. It requires the presence of magnesium ions and is inhibited by fluoride ions. The fluoride forms a stable complex with magnesium and phosphate ions, thus effectively diminishing the availability of magnesium ions to the active centre of the enzyme.

The next reaction involves the removal of the phosphate group from the molecule of phosphoenolpyruvate (Fig. 11.5). The enzyme *pyruvate kinase* catalyses the transfer of the phosphate group to ADP—and with it is transferred much of the energy associated with the cleavage of this group from the phosphoenolpyruvate. This reaction is therefore a further example of substrate-level phosphorylation. The two products are ATP and the *eno* form of pyruvic acid; the equilibrium lies very much in favour of the products of this reaction, so the process is virtually irreversible. The enol form of pyruvic acid is unstable and tends to change spontaneously, by internal rearrangement, to the more familiar keto form of the molecule, pyruvic acid.

During the fourth stage of glycolysis, two molecules of ATP are thus produced for each original hexose unit. Therefore, when glucose is the source of energy, the overall synthesis of ATP from ADP per mole of hexose degraded is $(-2 + 0 + 2 + 2)$ or 2 moles ATP, while reducing power in the form of 2 moles of NADH is created. This NADH is in the *cytosol* but if its reducing power can, directly or indirectly, become available within the *mitochondrion*, each molecule is, of course, potentially able to produce three molecules of ATP by oxidative phosphorylation—so long, of course, as oxygen is available.†

11.8 Fate of pyruvate under aerobic conditions

Under normal *aerobic* conditions almost all cells convert pyruvate to acetyl coenzyme A. This is energetically a very rewarding process, for not only are three molecules of ATP formed for each molecule of pyruvate which is decarboxylated, but, even more important, the subsequent oxidation of each molecule of acetyl coenzyme A by the TCA cycle yields 12 molecules of ATP.

The process is an example of *oxidative decarboxylation* and is brought about by a multienzyme complex known as *pyruvate dehydrogenase*. It should be compared with the reaction catalysed by α-ketoglutarate dehydrogenase discussed in Section 9.6:

$$\alpha\text{-Ketoglutarate} \xrightarrow{\text{Oxidative decarboxylation}} \text{succinyl coenzyme A}$$

$$\text{Pyruvate } (\alpha\text{-ketopropionate}) \xrightarrow[\text{decarboxylation}]{\text{Oxidative}} \text{acetyl coenzyme A}$$

The oxidative decarboxylation of pyruvate is of great importance, for it enables much of the potential energy within carbohydrate molecules to be conserved. (It will be recalled that the whole pathway from carbohydrate to pyruvate yields only six or seven molecules of ATP per glucose residuc under aerobic conditions, and only two or three in the absence of air.) NADH is produced by the oxidative decarboxylation of pyruvate, while both NADH and FADH are formed during the subsequent metabolism of acetyl coenzyme A (Chapter 9). The reoxidation of NADH and $FADH_2$ results in the production of ATP by the mitochondrial electron transport system (see Section 8.9).

The pyruvate dehydrogenase complex is located within the inner mitochondrial membrane. In structure, function, and cellular location, the

† Three ATP molecules are produced from one molecule of NADH, but the process of transferring the reduced coenzyme into the mitochondrion probably consumes one molecule of ATP per NADH molecule (page 233).

complex is, in most organisms, closely similar to the complex which brings about the oxidative decarboxylation of α-ketoglutarate. In consequence, the two systems will be considered together (Table 11.1). Each comprises *three enzymes.* One, *dihydrolipoyl dehydrogenase,* is probably identical in the two complexes. The other two are substrate specific (i.e., are specific for *either* pyruvate *or* α-ketoglutarate) but carry out analogous reactions and require the same cofactors.

Electron micrographs show that the three types of enzymes in each complex are symmetrically arranged, sixty dihydrolipoyl transacylase molecules lying near the centre with smaller numbers of molecules of the other two types of enzymes towards the outside. The *acyl*-specific transacylase can be imagined as a sort of 'distributor' which (a) removes from the TPP complex the acyl groups produced by the *substrate-specific* dehydrogenase, and (b) deposits them on to coenzyme A. Additionally, the transacylase collaborates with dihydrolipoyl dehydrogenase so that the lipoic acid, changed in the previous two steps, can be restored to its original form.

As the individual steps of oxidative decarboxylation are now discussed, the reader should refer constantly to the overall scheme shown in Fig. 11.6.

(a) Pyruvate dehydrogenase reaction

The reaction commences with the linking of enzyme-bound TPP to pyruvate. At this step, in which carbon dioxide is eliminated, the energy of the breaking carbon-carbon bond is retained in the TPP-aldehyde addition complex:

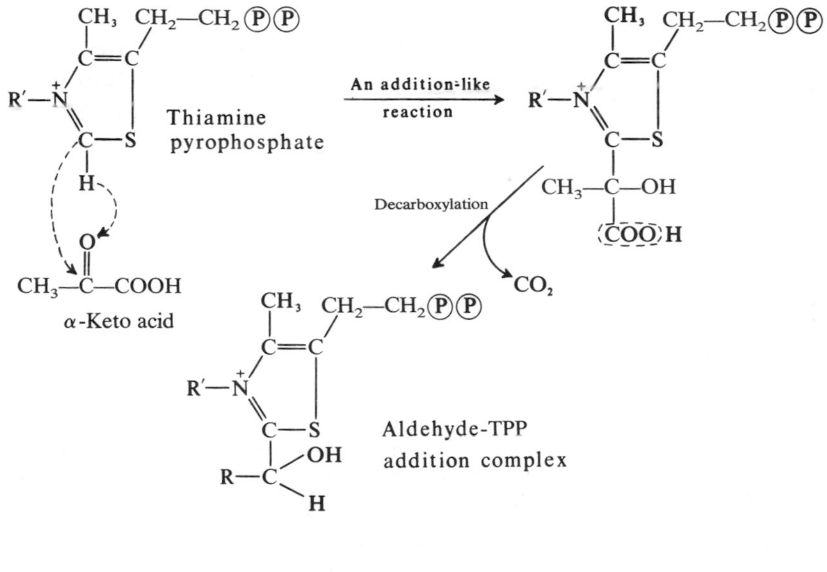

Thiamine pyrophosphate

α-Keto acid

An addition-like reaction

Decarboxylation

CO_2

Aldehyde-TPP addition complex

Table 11.1 Comparison of enzymes and cofactors concerned with the oxidative decarboxylation of pyruvate and of α-ketoglutarate

Pyruvate dehydrogenase complex	α-Ketoglutarate dehydrogenase complex	Prosthetic groups and cofactors
Pyruvate dehydrogenase	α-Ketoglutarate dehydrogenase	TPP
Dihydrolipoyl transacetylase	Dihydrolipoyl transsuccinylase	Lipoic acid and coenzyme A
Dihydrolipoyl dehydrogenase	Dihydrolipoyl dehydrogenase	FAD, NAD

(b) Dihydrolipoyl transacetylase reaction

Lipoic acid has the following structure:

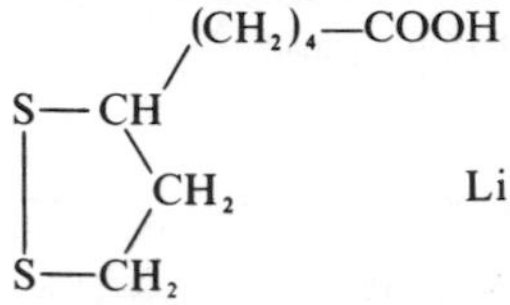

$$\text{S—CH—(CH}_2)_4\text{—COOH},\ \text{S—CH}_2\text{—CH}_2\text{—CH (ring, S—S)}$$

Lipoic acid (oxidized form)

By formation of an amide linkage, the acid group of the lipoic acid is combined with a lysine side chain of transacetylase. The *lipoamide* so formed provides the mechanism whereby acetyl groups are transferred. It has been postulated that the long chain of lipoamide:

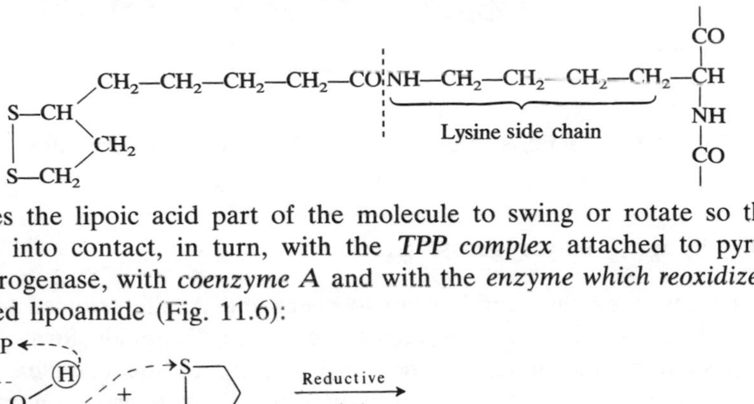

$$\text{(S—CH—CH}_2\text{—CH}_2\text{—S ring)—CH}_2\text{—CH}_2\text{—CH}_2\text{—CH}_2\text{—CO} \vdots \underbrace{\text{NH—CH}_2\text{—CH}_2\text{—CH}_2\text{—CH}_2}_{\text{Lysine side chain}}\text{—CH(—CO—)(—NH—CO—)}$$

enables the lipoic acid part of the molecule to swing or rotate so that it comes into contact, in turn, with the *TPP complex* attached to pyruvate dehydrogenase, with *coenzyme A* and with the *enzyme which reoxidizes* the reduced lipoamide (Fig. 11.6):

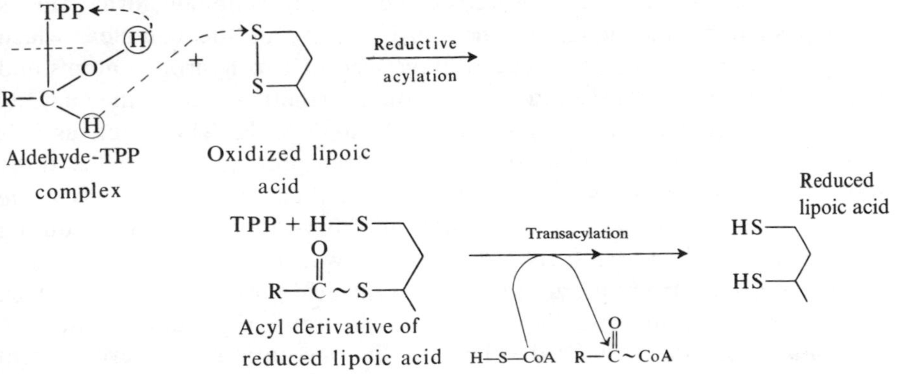

Aldehyde-TPP complex + Oxidized lipoic acid —(Reductive acylation)→ Acyl derivative of reduced lipoic acid (TPP + H—S—, $\text{R—C(=O)}\sim\text{S—}$) —(Transacylation; H—S—CoA → $\text{R—C(=O)}\sim\text{CoA}$)→ Reduced lipoic acid (HS—, HS—)

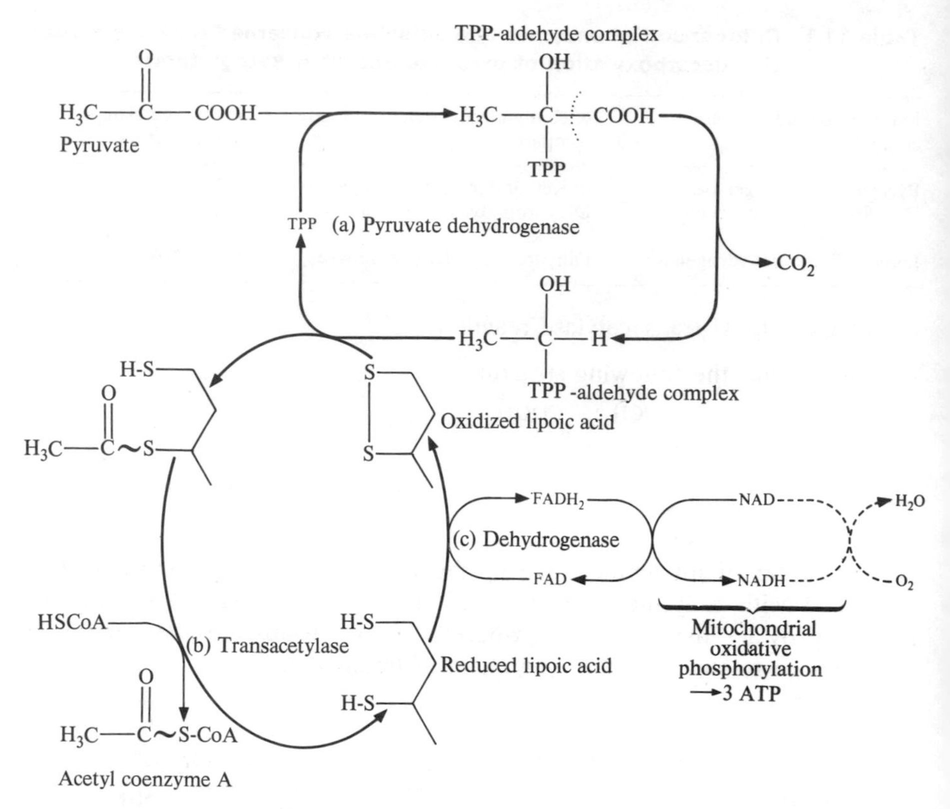

Fig. 11.6 Reactions of the pyruvate dehydrogenase complex

(c) Dihydrolipoyl dehydrogenase reaction

Having transferred the acetyl group to coenzyme A, the reduced form of lipoic acid, the dithiol, must be reoxidized to the disulphide form. This is accomplished by the action of the third enzyme of the complex, which catalyses the removal from reduced lipoic acid of two hydrogen atoms and enables them to be transferred to FAD, the prosthetic group of the enzyme. The $FADH_2$ is then able to reduce NAD^+ to NADH. The latter, as it is reoxidized *via* the mitochondrial electron transport system, yields three molecules of ATP. Thus, when pyruvate is linked to the TCA cycle *via* oxidative decarboxylation (12 + 3) molecules of ATP are produced during its complete oxidation to carbon dioxide and water.

Since it is the carbon atom of the carboxylic acid group of pyruvate which is lost, and this group arises from the aldehyde group of glyceraldehyde-3-P, the carbon dioxide produced originates, in one turn of the cycle, from carbon atoms 3 and 4 of the original hexose unit undergoing catabolism. This has been shown by experiments using glucose, the carbon atoms of which were specifically labelled at positions 3 and 4. In this respect the catabolism of glucose by the glycolytic pathway differs from its metabolism by a second pathway, the pentose phosphate pathway, which is described in Chapter 14.

11.9 Fate of pyruvate under anaerobic conditions

Even a temporary lack of oxygen can be a serious hazard for many types of cells; vertebrate brain cells, for example, are particularly susceptible to damage caused by *anoxia.* In the absence of oxygen, the mitochondrial electron transport system is unable to function, and so the TCA cycle, which generates ATP, comes to a halt. Furthermore, the fuel for this cycle is acetyl coenzyme A, itself often formed by the oxygen-dependent oxidative decarboxylation reaction.

Since fats also give rise to acetyl coenzyme A *via* the oxygen-dependent process of β-oxidation, fat catabolism also stops under anaerobic conditions and, if the cell is to survive, it must turn to glycolysis as the means for producing the bulk of its ATP. This, in effect, means that an anaerobic mechanism must somehow bring about the reoxidation of the NADH produced by the activity of phosphoglyceraldehyde dehydrogenase in stage 3 of glycolysis; there is only a limited amount of NAD^+ in the cell, and glycolysis would also stop under anaerobic conditions if all this coenzyme were to be converted to the reduced form. Various types of cells, and especially different microorganisms, can achieve the anaerobic oxidation of NADH in a variety of ways, and the mechanism will now be illustrated by considering the case of muscle undergoing strenuous exercise and that of yeast growing under anaerobic conditions.

In the *presence* of air, the electrons removed from an oxidizable substrate eventually reduce oxygen, the ultimate electron acceptor, to water. This reduced form of the electron acceptor is, paradoxically, inconspicuous owing to the abundance of water in all living cells. In the *absence* of air, however, some electron acceptor other than gaseous oxygen must be present and consequently some other reduction product than water must be formed. In muscle suffering oxygen lack, that product is lactate, while in yeast fermentation it is alcohol. They are both formed from pyruvate, the end-product of glycolysis, and their formation provides an elegant device which allows muscle and yeast to solve their redox problems when anaerobic conditions prevail.

When a muscle is heavily exercised, the blood is unable to supply oxygen rapidly enough for its requirements. *Under these conditions of temporary anoxia,* the muscle uses pyruvate as an electron acceptor, thus enabling

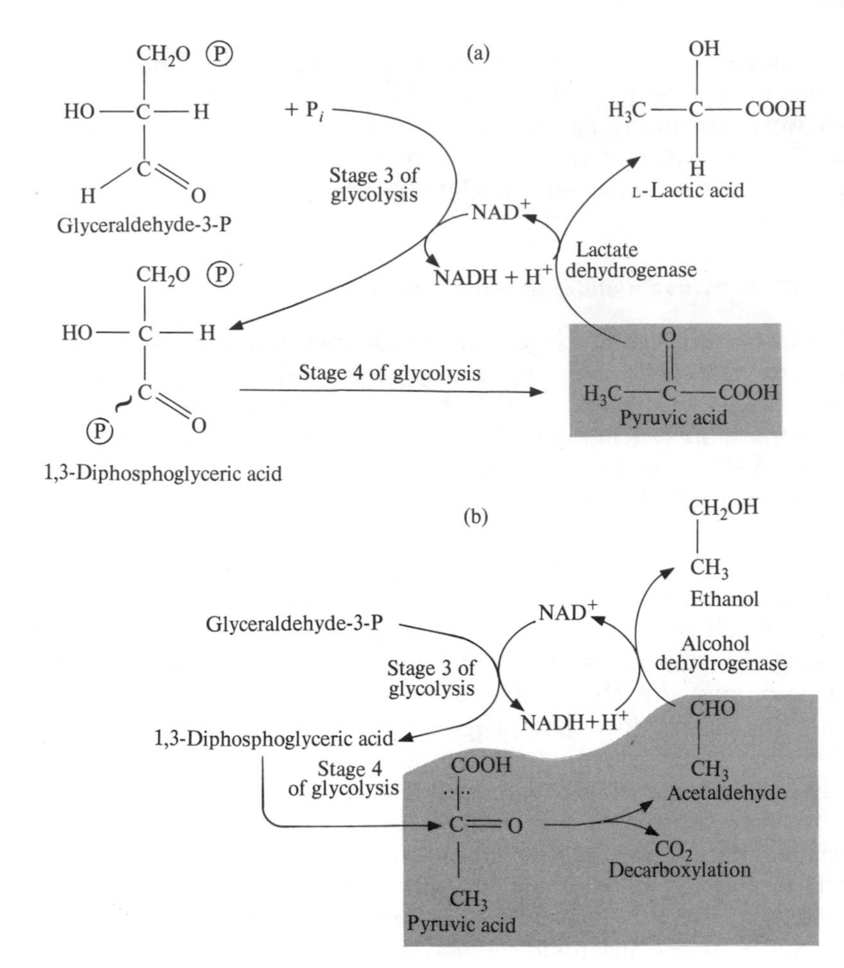

Fig. 11.7 The fate of pyruvate.
(a) Muscle under anaerobic conditions
(b) Fermentation of yeast (anaerobic)

NADH, formed in stage 3 of glycolysis, to revert back to NAD^+. In effect, the *oxidation* of aldehyde by the phosphoglyceraldehyde dehydrogenase reaction is coupled, *via* the NAD^+/NADH system, to the *reduction* of the ketone group of pyruvate (Fig. 11.7a). The enzyme responsible for the reduction of pyruvate is lactate dehydrogenase. Thus the formation of lactic acid gives the muscle a temporary boost in the form of ATP derived from glycolysis, at a time when it is most needed.

Lactate, therefore, is the final product of anaerobic muscular activity, not because muscle has any need of this substance biochemically, but because it has an imperative need to *utilize pyruvate as an electron acceptor.* The system nevertheless has its limitations, as every athlete familiar with muscle fatigue will know; the muscle is said to be 'fatigued' when its activity is so prolonged that lactate accumulates at the site of its formation.

The process whereby alcohol is formed when yeast ferments sugar is essentially similar to the process occurring in muscle under anaerobic conditions. Yeast normally respires in the ordinary aerobic manner utilizing the electron transport system to oxidize NADH; fermentation begins when the carbon dioxide evolved during the aerobic process is unable to escape from the vessel and there is inadequate access of oxygen. Whereas animals in such circumstances would suffocate, yeast switches to the anaerobic process of fermentation.

In industrial fermentation processes, fructose diphosphate, the primary reactant for glycolysis and hence for alcohol production, is formed from exogenous glucose (which is often made by man from starch before the yeast is added). Some cells faced with anaerobic conditions in nature may use exogenous glucose, or, of course, they can call upon internally stored carbohydrate. The pyruvate formed by the pathway of glycolysis is converted to ethanol for the same reason that it is converted to lactate by anoxic muscle. The difference rests in the fact that, whereas pyruvate in muscle is *directly* reduced by NADH, in yeast it is first *decarboxylated,* and the product of the reaction, acetaldehyde, acts as the electron acceptor.

The (non-oxidative) decarboxylation of pyruvate by yeast growing under *anaerobic* conditions should be both compared to, and constrasted with, the oxidative decarboxylation which occurs under *aerobic* conditions and which leads to the formation of acetyl coenzyme A (Section 11.8). The anaerobic reaction is catalysed by *pyruvate decarboxylase*; this enzyme, like pyruvate dehydrogenase, requires thiamine pyrophosphate (TPP) and manganous ions as cofactors. The reaction, in fact, resembles the dehydrogenase reaction in its initial stages, for an 'aldehyde-TPP complex' is formed, and this complex is then decarboxylated. However, unlike the pyruvate dehydrogenase reaction, the complex does not pass the aldehyde to lipoic acid, but releases it as acetaldehyde:

$$\underset{\text{Pyruvate}}{H_3C{-}\overset{\overset{\displaystyle O}{\|}}{C}{-}COOH} + TPP \longrightarrow CH_3{-}\overset{\overset{\displaystyle OH}{|}}{\underset{\underset{\displaystyle TPP}{|}}{C}}{-}COOH \xrightarrow{-CO_2} CH_3{-}\overset{\overset{\displaystyle OH}{|}}{\underset{\underset{\displaystyle TPP}{|}}{C}}{-}H \longrightarrow \underset{\text{Acetaldehyde}}{H_3C{-}C(=O)H} + TPP$$

Since the reaction does not involve the reduction of lipoic acid, no ATP is formed at this stage (in contrast to the pyruvate dehydrogenase reaction).

However, in the presence of alcohol dehydrogenase, the acetaldehyde which is produced acts as electron acceptor for the NADH produced by the phosphoglyceraldehyde dehydrogenase reaction. By maintaining the balance between NAD^+ and NADH, the reduction of acetaldehyde thus fulfils the same function as the reduction of pyruvate in anaerobic muscle—it keeps the glycolysis pathway functioning and so allows continued anaerobic production of ATP.

Needless to say, the yeast has no use for the alcohol formed by fermentation. As in the case of lactate in muscle, it is a *waste product* and can actually be disadvantageous; as its concentration increases it acts as a poison, eventually preventing further growth.

11.10 Energy capture resulting from glucose catabolism

It is now possible to assess the energy captured by the system $ADP + P_i \rightarrow ATP$ as a result of glucose catabolism. The *anaerobic* catabolism of glucose to ethanol or to lactate will be considered first. In both cases, 4 moles of ATP result from the conversion of 1 mole of fructose diphosphate to pyruvate. Two triose phosphate molecules arise from 1 molecule of fructose diphosphate but 2 molecules of ATP are required for the synthesis of fructose diphosphate from glucose. Thus when 1 mole of glucose is metabolized either to lactate in muscle or to ethanol in yeast, 4 minus 2 moles of ATP are synthesized.

It can be shown that the standard free energy change for the reaction

$$\text{Glucose} \longrightarrow 2\ \text{lactate} + 2\text{H}^+$$

is about -196 kJ/mol, and for the reaction

$$\text{Glucose} \longrightarrow 2\ \text{ethanol} + 2\text{CO}_2$$

is about -234 kJ/mol. If it is assumed that ΔG^0 for ATP formation from ADP is of the order of 33 kJ/mol, it is possible to estimate the approximate efficiency with which glycolysis captures the available energy. When glucose is converted to lactate, $\Delta G^0 = -196$ kJ/mol and 2 ATP molecules are synthesized, representing 2×33 kJ/mol. Thus, the approximate efficiency of energy capture is $66/196 \times 100$, or 34 per cent. For conversion of glucose to ethanol, the corresponding figure is 28 per cent.

We have said 'approximate efficiency', for standard free energy, ΔG^0, refers to idealized reactions occurring in molar solutions under circumstances where the conditions are so maintained that the reaction never departs significantly from true equilibrium ($\Delta G = 0$, page 144). These circumstances are, to say the least, unbiological—glucose and lactate are present at different levels in different cells, never there in molar concentrations, and three of the reactions of the glycolytic steady-state sequence operate a long way from equilibrium. Allowing for these deviations from standard conditions, some authors have put the true efficiency of energy capture in both reactions at about 50 per cent.

Nevertheless, these figures should not disguise the *inefficiency* with which the total potential energy of the glucose molecule is utilized in the absence of air. When 1 mole of glucose is oxidized aerobically to carbon dioxide and water under standard conditions, 2870 kJ/mol of free energy are liberated, whereas when it is converted to lactate or ethanol, only about 209 kJ/mol are made available. In other words, irrespective of whether biological efficiency of these anaerobic reactions is 30, 50, or even 100 per cent, something like $[(2870 - 209)/2870] \times 100$, or 93 per cent, of the energy of the glucose is uselessly locked up within the molecules of lactate or ethanol for so long as the anaerobic circumstances persist.

When 1 molecule of glucose is metabolized under *aerobic* conditions, the number of ATP molecules produced is much larger than under anaerobic conditions. As each pyruvate molecule is converted to acetyl coenzyme A, 3 molecules of ATP are produced at the pyruvate dehydrogenase step. Moreover, when each acetyl coenzyme A molecule is subsequently metabolized, 12 molecules of ATP are generated. But two pyruvate molecules arise from one of glucose and so $(2 \times 3) + (2 \times 12)$, or 30 molecules, of ATP are produced as a result of the pyruvate oxidation. Since, in addition, 2 molecules of ATP are formed during the glycolytic production of pyruvate from 1 molecule of glucose, a total of 32 molecules of ATP result from each glucose molecule catabolized. In addition, 2 molecules of NADH are produced in the cytosol, and mechanisms exist whereby this reducing power can indirectly be transferred to the mitochondrion. Each of these two molecules indirectly yields another 2 or 3 ATP, making the total yield of ATP, under aerobic conditions, 36 or 38 molecules. If the ΔG^0 for glucose is taken as 2870 kJ/mol, if 38 ATP are formed, and ~P in ATP is equivalent to 33 kJ/mol, the approximate efficiency of energy capture under aerobic conditions is $(38 \times 33/2870) \times 100$, or 43 per cent. It will also be noticed that 38/2 or 19 times as much energy is available as a result of aerobic glucose metabolism as becomes available during anaerobic degradation of glucose.

11.11 The Pasteur effect and the ADP/ATP ratio

It was shown in the previous section that the yeast cell is able to obtain a very much larger amount of ATP from a given amount of glucose if air is present than is possible anaerobically. It therefore does not need to metabolize so much glucose to obtain a *given amount* of energy, and it is found that the quantity of glucose disappearing in the presence of air is only a small fraction of that metabolized under anaerobic conditions. Clearly, some sort of compensatory mechanism restricts glycolysis when oxygen is available. This phenomenon was first observed by Pasteur during a study of the metabolism of yeast cells. It is therefore often called the *Pasteur effect.*

When the TCA cycle is operating, the amount of ATP in mitochondria goes up and the amount of ADP consequently falls. Since a specific transport system allows ADP and ATP to be transferred from mitochondrion to cytosol (Section 9.12), the ratio of the concentration of ATP to that of ADP in the cytosol also increases. This has at least two effects. First, ATP is an allosteric inhibitor of *phosphofructokinase*, the enzyme responsible for the formation of fructose-1,6-diphosphate. An excess of ATP in the cytosol thus decreases the activity of this important enzyme and slows down the entry of carbohydrate into the glycolytic (and therefore the TCA) pathway. Second, the low concentration of ADP puts a brake upon the two reactions of glycolysis for which this substance is an essential cofactor. These and other regulatory devices are considered in more detail in Chapter 13.

12

Gluconeogenesis and the synthesis of carbohydrates

In this chapter the biosynthesis of carbohydrates will be considered. There are two possibilities: they may either be formed from other carbohydrates or they may be formed from non-carbohydrate precursors. We propose first to deal with the interconversion of carbohydrates, but it must be remembered that carbohydrates undergoing conversion may arise not only from pre-existing stored carbohydrates (or, in animals, from dietary carbohydrates) but also from non-carbohydrate sources by the pathways described later in the chapter.

12.1 Biosynthesis of disaccharides from monosaccharides

Disaccharides are broken down to monosaccharides by enzyme-catalysed hydrolytic reactions which are essentially irreversible under biological conditions. The reverse pathway involves an important new type of high energy compound termed a *nucleoside diphosphate sugar.*

These compounds are formed by the reaction of a nucleoside triphosphate (ATP, UTP, GTP, CTP) with a sugar-1-phosphate (often glucose-1-phosphate). According to which nucleotide base is present, the fate of the sugar phosphate will differ, as the following examples will illustrate. For the synthesis of sucrose in the sugar-cane, for example, the first step is the formation of uridine diphosphate glucose:

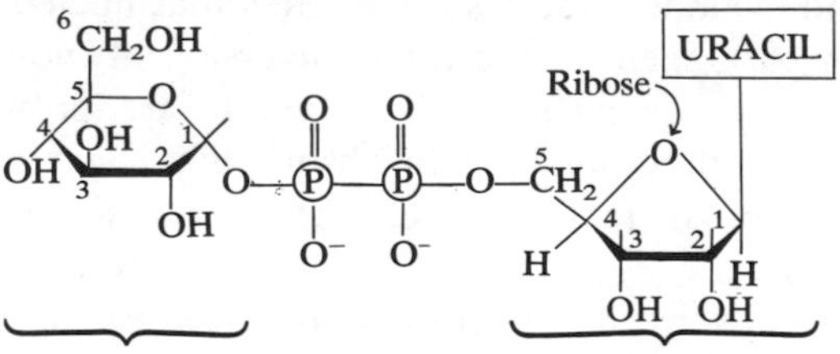

Glucose Uridine

Uridine diphosphate glucose (UDPG)

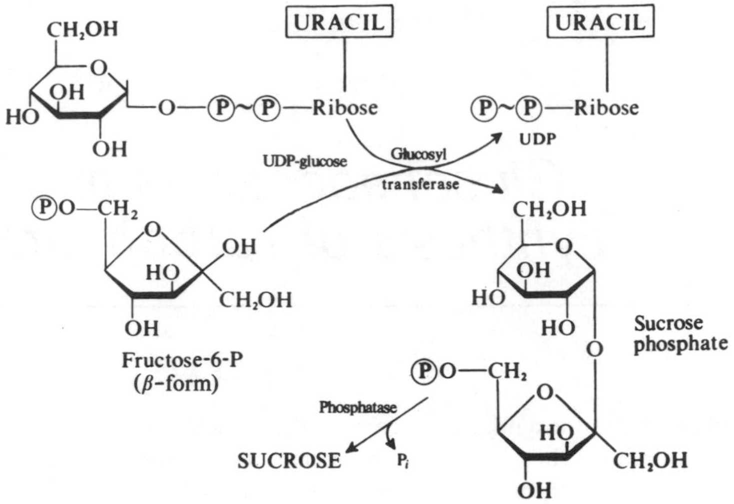

Fig. 12.1 The synthesis of sucrose

The production of UDP-glucose is represented by the equation:

$$\text{UTP} + \text{glucose-1-P} \rightleftharpoons \text{UDP-glucose} + \text{PP}$$

As written, the reaction is freely reversible, but, in practice, it favours the formation of UDP-glucose because the *second product*, inorganic pyrophosphate, is hydrolysed to orthophosphate by the action of a specific *pyrophosphatase.*

The second step in sucrose synthesis is the transfer of the glucose unit to fructose-6-phosphate, to give sucrose phosphate, which is then hydrolysed to sucrose by the enzyme phosphatase (Fig. 12.1). The glucosyl transferase, which catalyses the transfer of the glucose, is specific for *uridine* disphosphate glucose. In most other plants, where sucrose concentrations are lower than in the sugar-cane, a similar system operates but with fructose, not fructose-6-phosphate, as the glucose acceptor.

Other disaccharides and also polysaccharides are synthesized by a series of reactions essentially similar to those above. Lactose (milk sugar) synthesis provides a second example, and also illustrates that nucleoside diphosphate sugars also often participate in the interconversion of monosaccharides.

When certain microorganisms such as yeasts are grown in a medium containing galactose, they are able to absorb this sugar and convert it to glucose and glycogen. In mammals, too, galactose is readily converted to glucose or to lactose.

In both yeast and liver cells galactose metabolism usually commences with phosphorylation, the reaction being catalysed by *galactokinase.* This enzyme transfers a phosphate group from ATP to the 'carbonyl' hydroxyl group of galactose, with the result that galactose-1-phosphate is formed:

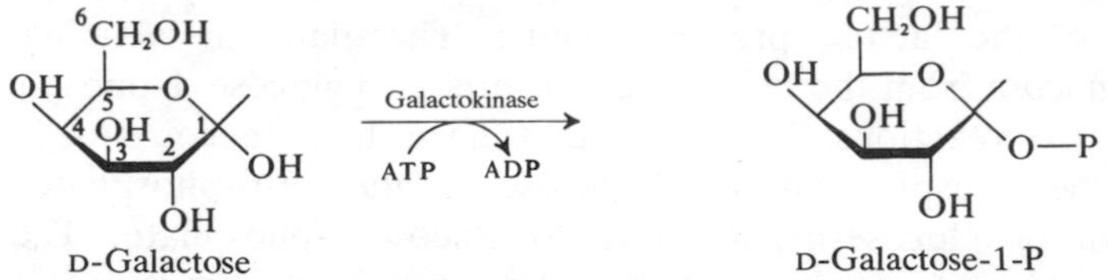

$$\text{D-Galactose} \xrightarrow[\text{ATP} \rightarrow \text{ADP}]{\text{Galactokinase}} \text{D-Galactose-1-P}$$

It is noteworthy that this reaction differs from the phosphorylation of glucose in the reaction catalysed by hexokinase for, as was seen earlier (Fig. 11.2), this leads to the production of glucose-6-phosphate.

Whether the original galactose is to be built up into lactose or changed into glucose, the galactose-1-phosphate is next converted to UDP-galactose:

$$\text{UTP} + \text{galactose-1-P} \rightleftharpoons \text{UDP-galactose} + \text{PP}$$

This is a necessary preliminary for the stereochemistry of galactose to be altered so that it is eventually converted to *glucose.* Once in this 'active' form, an enzyme called *UDP-galactose-4-epimerase* catalyses a rearrangement of the groups around carbon atom 4 of the galactose ring, producing UDP-glucose (Fig. 12.2). Such a rearrangement is called *epimerization*; it will be seen from the figure that a hydroxyl group written 'up' before epimerization occurs is written 'down' afterwards—i.e., the mirror image arrangement of the groups around this *one particular* carbon atom is somehow adopted.

The epimerase has NAD^+ bound to it, present in equimolecular amount to the protein, and one possible mechanism of epimerization is that it involves oxidation at carbon atom 4 (with the formation of a carbonyl group) followed by reduction of this *unsaturated, and therefore optically inactive,* group.

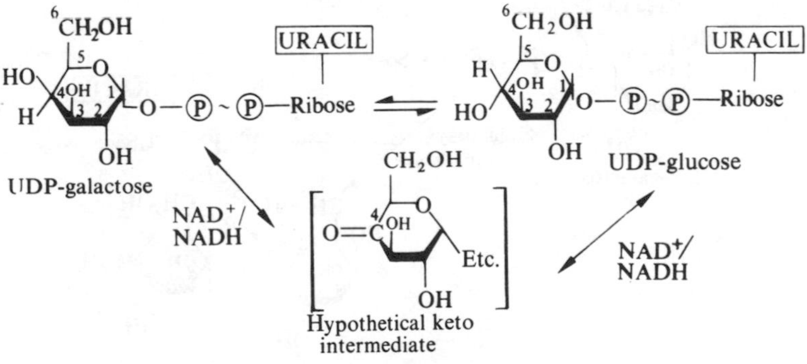

Fig. 12.2 The interconversion of UDP-galactose and UDP-glucose, catalysed by UDP-galactose-4-epimerase

Operating in the opposite direction to that just described, UDP-galactose-4-epimerase is important in the mammary gland since it participates in the synthesis of the lactose present in milk. The gland in lactating animals absorbs *glucose* from the blood and converts it to glucose-1-phosphate. This involves two reactions, both of which have been encountered already, namely, the phosphorylation of glucose to glucose-6-phosphate and the conversion of glucose-6-phosphate to glucose-1-phosphate. The first is catalysed by hexokinase and the second by phosphoglucomutase. Glucose-1-phosphate is then converted to UDP-glucose, the latter of which undergoes epimerization to UDP-galactose.

Lactose is finally formed by a reaction similar to that described for the synthesis of sucrose phosphate in plants (Fig. 12.1), for the UDP-galactose reacts with glucose to form the β-galactoside, lactose, the enzyme responsible being galactosyl transferase (Fig. 12.3).

Lactose plays an important role in infant nutrition for it is the major carbohydrate of milk. Dietary lactose is readily hydrolysed to galactose and glucose by the enzyme *lactase* present in the intestinal epithelium (Section 17.9). The galactose is potentially toxic, and in the normal infant is converted to galactose-1-phosphate by *galactokinase* present in the liver. The galactose-1-phosphate then reacts with UDP-glucose, in the presence of a specific transferase:

$$\text{UDP-glucose} + \text{galactose-1-P} \rightleftharpoons \text{UDP-galactose} + \text{glucose-1-P}$$

The UDP-galactose, by reversal of the epimerization reaction mentioned

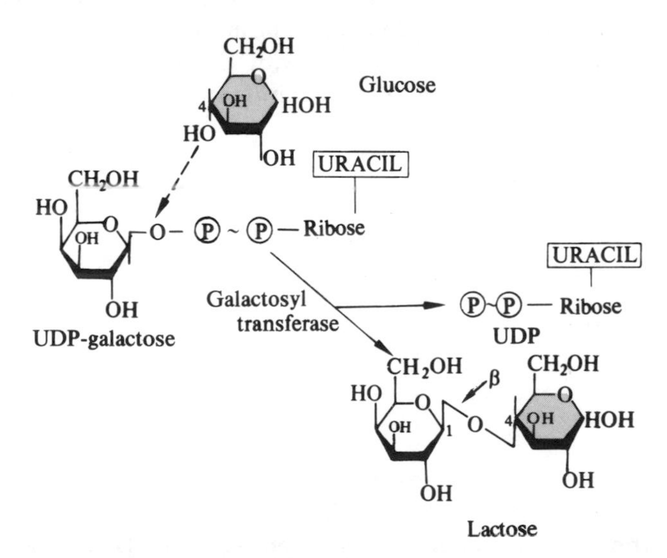

Fig. 12.3 Synthesis of lactose from glucose and UDP-galactose

above, is converted to UDP-glucose, which is then able to give rise to glycogen in the way described in the next section.

There are, however, two genetic disorders, each known as *galactosaemia.* Infants afflicted by these conditions lack enzymes essential for the conversion of galactose to glucose derivatives, with the result that the level of galactose in their blood rises when they are fed milk. This can cause mental disorders and other serious symptoms. Such infants escape these difficulties if fed on diets other than milk.

12.2 Biosynthesis of polysaccharides from monosaccharides

When starch or glycogen undergoes glycolysis, the first reaction, catalysed by phosphorylase, involves the removal of glucose residues situated at the ends of the branches of polysaccharide molecules. In this reaction, inorganic phosphate is incorporated in such a way that the glucose residue is released as glucose-1-phosphate (page 219):

$$\underset{\substack{\text{Starch or}\\ \text{glycogen}}}{[\text{Glucose}]_n} + \text{P}_i \xrightarrow{\text{Phosphorylase}} \text{glucose-1-P} + \underset{\substack{\text{Starch or glycogen}\\ \text{with one less unit}\\ \text{of glucose}}}{[\text{glucose}]_{n-1}}$$

It is unlikely that significant quantities of polysaccharide are synthesized from glucose-1-phosphate by the reverse of the phosphorylase reaction. The position of the equilibrium, for concentrations of the reacting substances present under *biological* conditions, lies very much in favour of glucose-1-phosphate formation. Net overall *synthesis* occurs, in fact, by a quite different mechanism, involving the participation of the appropriate nucleoside diphosphate glucose.

Glucose, in preliminary reactions, can be converted by hexokinase to glucose-6-phosphate (Fig. 11.2), which, by the phosphoglucomutase reaction, gives glucose-1-phosphate. Equally important in most organisms, however, is glucose-1-phosphate originating from fructose-6-phosphate. The latter may arise in animals by the pathway of gluconeogenesis or, in plants, from the photosynthetic carbon cycle (Fig. 14.8). Whatever its origin, the glucose-1-phosphate then reacts with a nucleoside triphosphate. The reaction is shown in general form in Fig. 12.4.

Figure 12.5 shows the final reaction whereby a glucose residue is *transferred to a starch* 'primer' molecule in starch synthesis. In this case the enzyme involved, ADP-glucose: starch *glucosyl transferase,* is specific for *adenosine* diphosphate glucose (ADP-glucose), which is formed from ATP and glucose-1-phosphate via the reactions shown in Fig. 12.4. *Glycogen* synthesis in animals and fungi is similar except that the nucleoside involved is

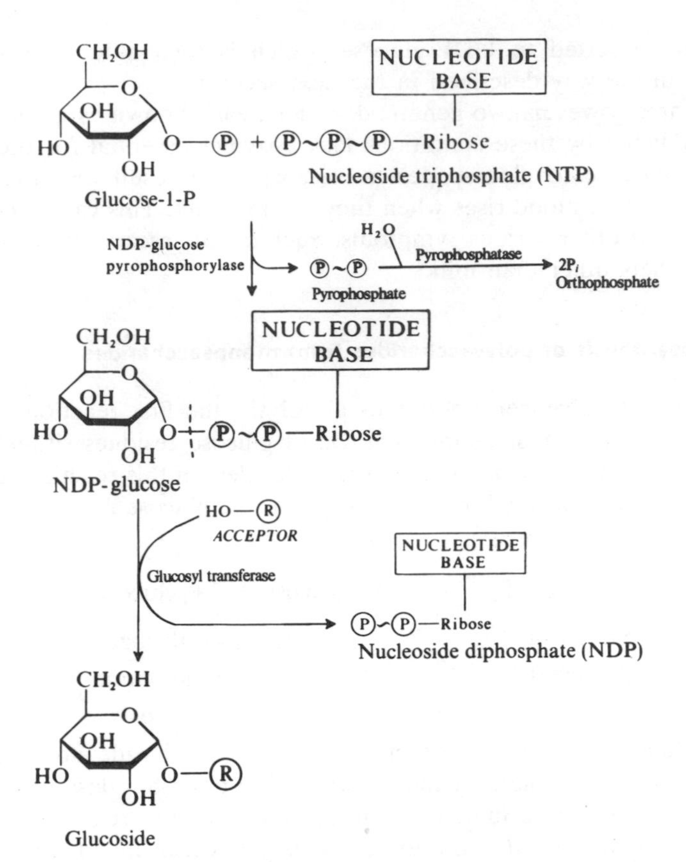

Fig. 12.4 The role of nucleoside diphosphate glucose in the synthesis of a glucoside, e.g., synthesis of starch or glycogen

uridine; uridine triphosphate (UTP) reacts with glucose-1-phosphate, the UDP-glucose so formed eventually transferring the glucose to a glycogen 'primer' molecule. Attachment occurs, as in the case of starch, to carbon atom 4 of the terminal glucose unit of a side chain; the other product is UDP:

$$\text{UTP} + \text{glucose-1-P} \rightleftharpoons \text{UDP-glucose} + \text{pyrophosphate}$$

$$\text{Pyrophosphate} + H_2O \longrightarrow 2\ \text{orthophosphate}$$

$$\text{UDP-glucose} + \underset{\substack{\text{Glycogen 'primer'}\\ \text{(a partly built}\\ \text{glycogen molecule)}}}{[\text{glucose}]_n} \longrightarrow \text{UDP} + \underset{\substack{\text{Glycogen with}\\ \text{additional}\\ \text{glucose unit}}}{[\text{glucose}]_{n+1}}$$

Note that, for both starch and glycogen, attachment occurs to a smaller preexisting molecule or 'primer'; as the molecule becomes larger it starts to branch (Section 12.3), and then parts of it break away to form new 'primer

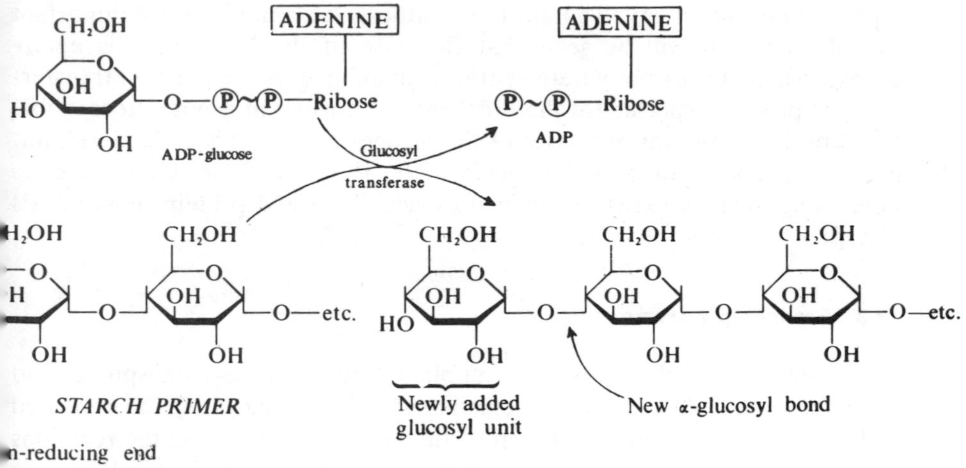

Fig. 12.5 The addition of a glucose unit to a starch molecule

molecules. The glucosyl transferases are often called *starch synthase* and *glycogen synthase* respectively.

Cellulose is present in plant cell walls and its synthesis resembles those of starch and glycogen except that the nucleoside of the NDP-glucose is in this case *guanosine.* Moreover, the final glucosyl transferase reaction results in the formation of β-glucoside linkages, for unlike starch and glycogen, cellulose comprises β-glucoside residues.

In leaves, the synthesis of starch, sucrose, and cellulose may at certain times be going on simultaneously, but the NDP-glucose compound is different in each case (ADP-glucose for starch, UDP-glucose for sucrose, and GDP-glucose for cellulose). Table 12.1 shows some of the nucleoside

Table 12.1 Acceptor molecules and NDP-sugars involved in the synthesis of carbohydrates

Carbohydrate	Acceptor molecule	Nucleoside diphosphate sugar donor
Animals		
Glycogen	Glycogen primer	UDP-glucose
Lactose	Glucose	UDP-galactose
Chondroitin	Chondroitin primer	UDP derivatives
Glycoproteins	Glycoprotein primer	UDP and CDP derivatives
Plants		
Cellulose	Cellulose primer	GDP-glucose or UDP-glucose
Starch	Starch primer	ADP-glucose
Sucrose	Fructose-6-P	UDP-glucose
Xylan	Xylan primer	UDP-xylose

diphosphate sugars involved in the synthesis of a number of important carbohydrates. It will be seen that the base of the NDP-sugar is nearly always uridine for carbohydrate synthesis in animals. Since glucosyl transferases are probably specific for the different nucleosides, it is not unreasonable to suppose that the various bases of the nucleosides act as a 'code' determining which carbohydrate is to be formed. If this is indeed the case an interesting analogy exists between polysaccharide and protein biosynthesis (Chapter 16).

12.3 Branching enzyme

If the leaf cells of plants are supplied with glucose-1-phosphate and ATP they are able to elongate existing starch molecules by the repeated addition of glucose units, each step of the elongation involving the reactions shown in Fig. 12.5. This is probably the way in which amylose, the unbranched form of starch, is formed. Most plant starches, however, contain a branched-chain form, amylopectin (Section 4.6), as well as amylose.

The branching is produced by an enzyme which acts upon a branch of amylopectin which has previously elongated, without further branching, until it contains some forty-two glucose units. The enzyme hydrolyses an α-1,4-glucoside bond and transfers the shorter section of the glucose chain to another part of the amylopectin molecule, attaching it to a glucose residue so as to form a 1,6-glucoside bond (Fig. 12.6). The enzyme, sometimes known as Q-enzyme, is a glucosyl transferase, since it transfers a section comprising glucoside units from one part of the amylopectin molecule to another. The extent of branching and the number of glucose residues between branches are both determined by the enzyme.

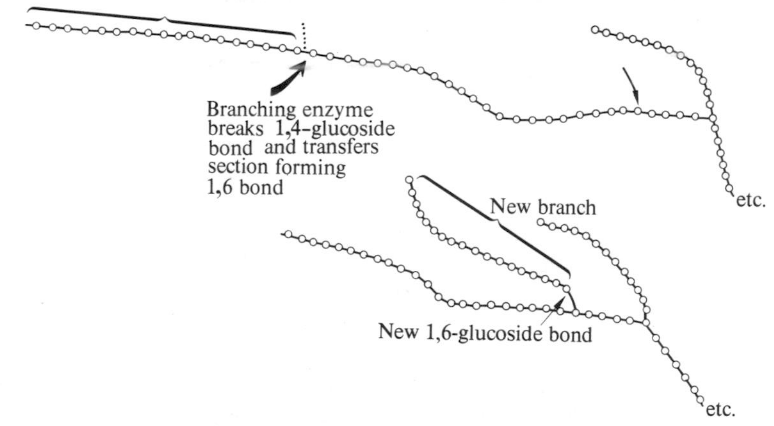

Fig. 12.6 The branching of amylopectin by 'branching enzyme'. (The circles represent glucose units)

Glycogen is a more highly branched molecule than amylopectin but here again the branching is achieved by a glucosyl transferase which removes a straight-chain section of glucose units by breaking a 1,4-α-glucoside linkage, and then attaches the sections so removed to another part of the glycogen molecule by a 1,6-α-glucoside linkage. Thus, both when glycogen is synthesized in animal tissues and amylopectin is made in plants, the glucosyl transferases which catalyse the formation of the initial polysaccharide chain and the branching enzymes responsible for the characteristic branching of the chains of the molecules cooperate together in the task of forming these important polysaccharides.

12.4 Formation of glucose and other carbohydrates from non-carbohydrate sources

It was seen in Chapter 11 that a number of carbohydrates can be converted to fructose-6-phosphate. This then enters a major route for carbohydrate breakdown, the glycolytic pathway. The terminal product of this pathway is pyruvic acid, the fate of which varies according to the conditions (Sections 11.8 and 11.9).

A correspondingly valuable and versatile trunk route allows the synthesis of glucose and other carbohydrates from pyruvate and oxaloacetate or from any metabolite which gives rise to these acids. This trunk route is superficially similar to glycolysis and is called the pathway of *gluconeogenesis.* The enzymes responsible are located in the cytosol; in animals they are principally in liver and kidney cells. Hardly any gluconeogenesis occurs in cells of the brain.

The main pathway of gluconeogenesis is shown in Fig. 12.7. It will be observed that in the part of the pathway lying between pyruvate and fructose-6-phosphate there are two major differences between the reactions of gluconeogenesis and those of glycolysis. In addition, there are further differences, the nature of which depends on the subsequent fate of fructose-6-phosphate. All of these differences have one thing in common—they overcome what would be an insurmountable energy barrier if the process were simply glycolysis in reverse. Glycolysis is, in fact, a strongly exergonic process, and consequently is virtually irreversible.

Gluconeogenesis, as the name implies, was historically studied in the context of synthesis of glucose from non-carbohydrate precursors. It is, in fact, of major importance in animals as a means of maintaining the level of glucose in the blood. It can, however, lead to other carbohydrates, since fructose-6-phosphate can be converted to glucose-1-phosphate, from which several carbohydrates can be made *via* nucleoside diphosphate glucose (Section 12.1). The feeder pathways into the main trunk route are also of varying quantitative importance. In plants, the photosynthetic carbon cycle

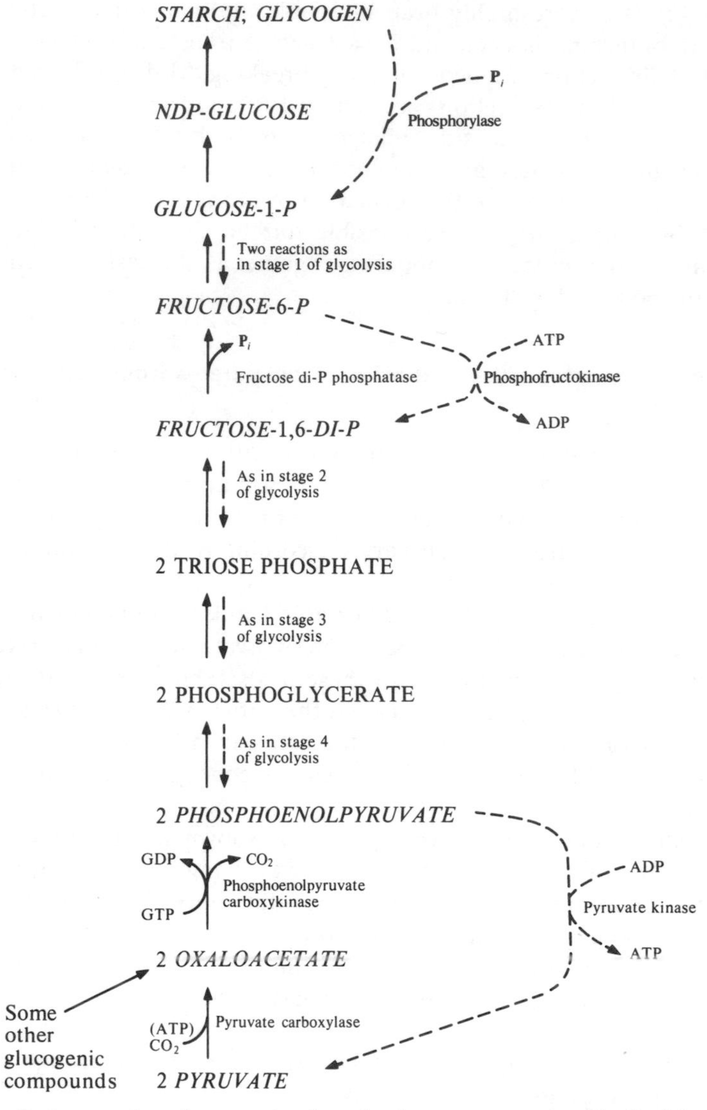

Fig. 12.7 Pathway for the synthesis of glycogen or starch (bold arrows) contrasted with the glycolytic pathway (broken arrows)

(Chapter 14) reduces carbon dioxide to phosphoglyceric acid, an intermediate of gluconeogenesis. In animals, lactate, formed during muscular activity, is released into the blood and converted to pyruvate in the liver. The gluconeogenic pathway therefore allows this product of muscular exercise to be converted back to glycogen. In most, if not all organisms, amino acids and TCA cycle intermediates can be the precursors of carbohydrates.

TCA cycle intermediates yield oxaloacetate, a member of the gluconeogenesis pathway. Similarly, the nitrogen-free carbon skeletons left after deamination of at least some of the amino acids arising from protein degradation are, or give rise to, intermediates of gluconeogenesis (in animals, only *glucogenic* amino acids can do this, but in plants and many microorganisms, probably all amino acids can by one process or another be converted to carbohydrates). Some of these feeder pathways will be described in Section 12.5.

Plants and microorganisms are more versatile than higher animals in that they can not only build up carbohydrates from pyruvate and from TCA cycle intermediates but from acetyl coenzyme A as well. Since acetyl coenzyme A is the normal product of fatty acid oxidation, this means that they can synthesize carbohydrates from fatty acids. This is made possible by a cycle, the *glyoxylate cycle*, which *does not operate in higher animals* since they lack two enzymes essential for the process. In consequence, plants and many microorganisms can, by the mechanism described in Section 12.6, convert acetyl coenzyme A into succinate, from which, *via* oxaloacetate, gluconeogenesis leads to carbohydrate synthesis. In higher animals, *no net synthesis of carbohydrate from fatty acids is possible.*

Commencing with pyruvate, we will now look at the central route of gluconeogenesis in more detail. Pyruvate is converted to oxaloacetate by a reaction which involves carbon dioxide 'fixation'. The process closely resembles that by which malonyl coenzyme A is formed from acetyl coenzyme A in the course of fatty acid biosynthesis (Fig. 10.5). As in that reaction, carbon dioxide is first activated by forming a derivative of *biotin*, ATP being also essential. The carbon dioxide is then transferred from the 'active carbon dioxide' to pyruvate, so producing oxaloacetate:

$\overset{*}{C}O_2$ + ATP

Biotin

$\overset{*}{C}OOH-CH_2-C(=O)-COOH$ Oxaloacetate

P_i + ADP

Biotin ~ $\overset{*}{C}O_2$

'Active carbon dioxide'

$H-CH_2-C(=O)-COOH$ Pyruvate

The whole reaction is catalysed by the enzyme *pyruvate carboxylase.* This enzyme should not be confused with pyruvate decarboxylase which, in yeast, converts pyruvate to acetaldehyde (page 231), nor with the complex of enzymes called pyruvate dehydrogenase, which is responsible for the oxidative decarboxylation of pyruvate (page 226).

At this point, it may well be asked why pyruvate, arising, perhaps, from the lactate formed during vigorous exercise, should pass by this circuitous route (pyruvate → carboxylation to oxaloacetate → decarboxylation of oxaloacetate) to phosphoenolpyruvate, rather than undergoing direct phosphorylation. As is so often the case, energetic aspects of the reaction must be taken into consideration, for it will be recalled that when *pyruvate kinase* catalyses the conversion of phosphoenolpyruvate to pyruvic acid, the reaction is highly exergonic (Fig. 7.5a):

$$\text{Phosphoenolpyruvate} + \text{ADP} \rightarrow \text{pyruvate} + \text{ATP}$$
$$(-\Delta G^\circ = 24\ \text{kJ/mol})$$

The reverse reaction—the phosphorylation of pyruvate by ATP—is therefore correspondingly endergonic. The cell bypasses this unfavourable energy barrier by the two exergonic reactions described above:

$$\text{Pyruvate} + CO_2 + \text{ATP} \underset{\text{(biotin)}}{\overset{\text{Pyruvate carboxylase}}{\rightleftharpoons}} \text{oxaloacetate} + \text{ADP} + P_i$$
$$(-\Delta G^\circ = 4.5\ \text{kJ/mol})$$

$$\text{Oxaloacetate} + \text{GTP} \underset{\text{carboxykinase}}{\overset{\text{Phosphoenol-pyruvate}}{\rightleftharpoons}} \text{phosphoenolpyruvate} + \text{GDP} + CO_2$$
$$(-\Delta G^\circ = 5.6\ \text{kJ/mol})$$
$$[\text{ATP} + \text{GDP} \rightleftharpoons \text{ADP} + \text{GTP}]$$

Note that in the single-reaction 'direct' route from phosphoenolpyruvate to pyruvate only one ATP is produced, whereas in the 'indirect' reverse direction the energy of two high energy phosphate groups is involved. The energetic principles we have just considered are applicable in more or less modified form to other instances where anabolic and catabolic routes differ, and for this reason they are very important.

Phosphoenolpyruvate is produced from oxaloacetate in a reaction catalysed by *phosphoenolpyruvate carboxykinase.* As the word 'kinase' suggests, the enzyme transfers a high energy phosphate group—in the present case from guanosine triphosphate (GTP)—to the oxaloacetate; at the same time it catalyses decarboxylation with the release of carbon dioxide:

$$\underset{\text{Oxaloacetate}}{\text{HOOC}-CH_2-C(=O)-COOH} \xrightarrow[\text{Phosphoenolpyruvate carboxykinase}]{\text{GTP} \rightarrow \text{GDP}} CO_2 + \underset{\text{Phosphoenol-pyruvate}}{H_2C=C(O\text{Ⓟ})-COOH}$$

The enzyme is, in fact, named after the reverse reaction in which phosphoenolpyruvate is carboxylated and dephosphorylated to form oxaloacetate, GTP being simultaneously formed from GDP.

The phosphoenolpyruvate produced from oxaloacetate is next converted to glyceraldehyde-3-phosphate by a series of reactions identical to those occurring in stages 2 and 3 of glycolysis (but operating in the reverse direction). Indeed, the enzymes involved are the same as those catalysing the glycolytic reactions (Chapter 11). Triose phosphate isomerase ensures that sufficient dihydroxyacetone phosphate is produced from glyceraldehyde-3-phosphate to enable aldolase to convert the two triose phosphate molecules to fructose-1,6-diphosphate. As was pointed out previously (Fig. 11.3), the position of equilibrium of the aldolase reaction favours synthesis of hexose diphosphate.

In the next step, another major difference exists between the pathways of glycolysis and gluconeogenesis, and here, as before, the difference has an energetic explanation. If glycolysis were simply reversed in the conversion of fructose-1,6-diphosphate to fructose-6-phosphate, the following reaction would be somewhat endergonic from left to right and hence the equilibrium mixture would contain little fructose-6-phosphate:

$$\text{Fructose-1,6-di-P} + \text{ADP} \underset{\text{Phosphofructokinase}}{\rightleftharpoons} \text{fructose-6-P} + \text{ATP}$$

$$(+\Delta G^\circ = 12.6\ \text{kJ/mol})$$

This reaction is, however, bypassed by an exergonic and, therefore, an energetically more favourable process. A phosphatase catalyses the hydrolytic removal of a phosphate group from the fructose diphosphate, producing inorganic orthophosphate and fructose-6-phosphate:

$$\text{Fructose-1,6-di-P} \overset{\text{Fructose di-P phosphatase}}{\rightleftharpoons} \text{fructose-6-P} + P_i$$

$$(-\Delta G^\circ = 16.7\ \text{kJ/mol})$$

Fructose-6-phosphate is then converted to glucose-6-phosphate by phosphohexoisomerase. The reaction is identical to that occurring in stage 1 of glycolysis, but acting in reverse (Fig. 11.2). The fate of the glucose-6-phosphate then depends upon which carbohydrate is being synthesized. In gluconeogenesis in the strictest sense, glucose is the end-product. Again, for energetic reasons, the reaction is not the reverse of glycolysis. Instead of the hexokinase reaction, direct hydrolysis occurs, the enzyme responsible being glucose-6-phosphatase:

$$\text{Glucose-6-P} + H_2O \rightarrow \text{glucose} + P_i$$

$$(-\Delta G^\circ = 13.8\ \text{kJ/mol})$$

More usually glycogen (or in plants, starch or cellulose) are to be the final products, and in these cases it is not necessary for glucose to be formed *en route*. Instead, the glucose-6-phosphate is converted to glucose-1-phosphate by phosphoglucomutase (Fig. 11.2). The glucose-1-phosphate is then converted to a nucleoside diphosphate glucose. The NDP-glucose then reacts with appropriate acceptor or primer molecules to form di- or polysaccharides as described in Sections 12.1 to 12.3.

12.5 Origin of oxaloacetate

In the previous section some of the sources of the pyruvate and oxaloacetate fed into the main pathway of gluconeogenesis were briefly mentioned. We now come to a more detailed discussion of these feeder pathways which link together carbohydrate and protein metabolism. In some organisms, fat metabolism is also linked with gluconeogenesis, and this will be treated separately in the next section.

Figure 12.8 summarizes some of the main feeder routes. That leading from pyruvate to oxaloacetate has already been discussed as it is a part of the main pathway. TCA cycle intermediates all give rise to oxaloacetate, since this is the last member of the cycle. Thus any amino acid or other substance that breaks down to give α-ketoglutarate, succinate, or fumarate will give rise to oxaloacetate. For example, oxidative deamination of

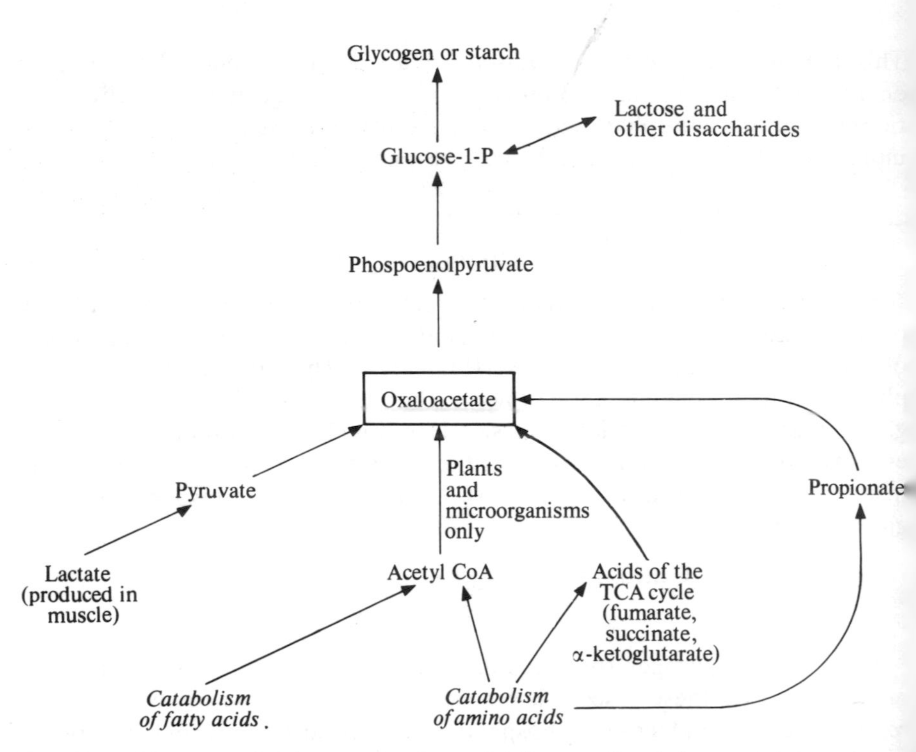

Fig. 12.8 The overall pattern of polysaccharide synthesis from various precursors

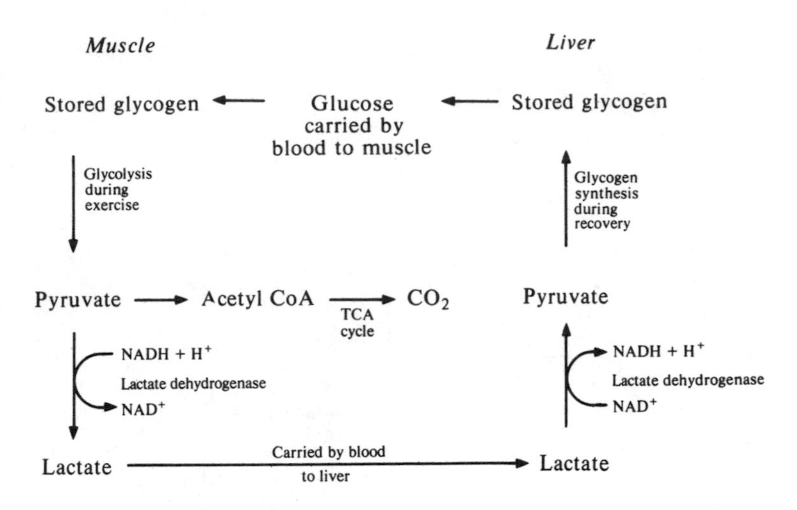

Fig. 12.9 The relationship between glycogen breakdown and its resynthesis during muscular exercise and the recovery phase

glutamic acid will give α-ketoglutarate:

$$\text{HOOC—CH}_2\text{—CH}_2\text{—CHNH}_2\text{—COOH} \xrightarrow{-\text{NH}_3+\text{O}} \text{HOOC—CH}_2\text{—CH}_2\text{—CO—COOH}$$

By a different mechanism, namely, that of transamination (Section 15.2), deamination of aspartic acid gives oxaloacetate:

$$\text{HOOC—CH}_2\text{—CHNH}_2\text{—COOH} \rightarrow \text{HOOC—CH}_2\text{—CO—COOH}$$

A major source of pyruvate in animals is *lactic acid* produced by skeletal muscle (Section 11.9). The more energetic the exercise, the greater the amount of lactate which accumulates, leading in extreme cases to muscular fatigue. During and immediately after exercise, lactate diffuses out of muscle and is carried by the blood to the liver, where it is oxidized to pyruvate. The pathway of gluconeogenesis, which, it will be recalled, is very active in the liver, converts pyruvate to glucose. The latter can then reenter the blood and pass back to skeletal muscle to start the process all over again. The whole process, sometimes known as the Cori cycle, is summarized in Fig. 12.9.

Finally, *propionate* is glucogenic because it can be converted, in animals, plants, and microorganisms, to succinyl coenzyme A. It can arise from the catabolism of several amino acids, including methionine, threonine, and valine. In addition, propionate plays a particularly important role in ruminants such as cattle and sheep. The gastric microorganisms are able to digest cellulose and other components of fodder, a major product being propionic

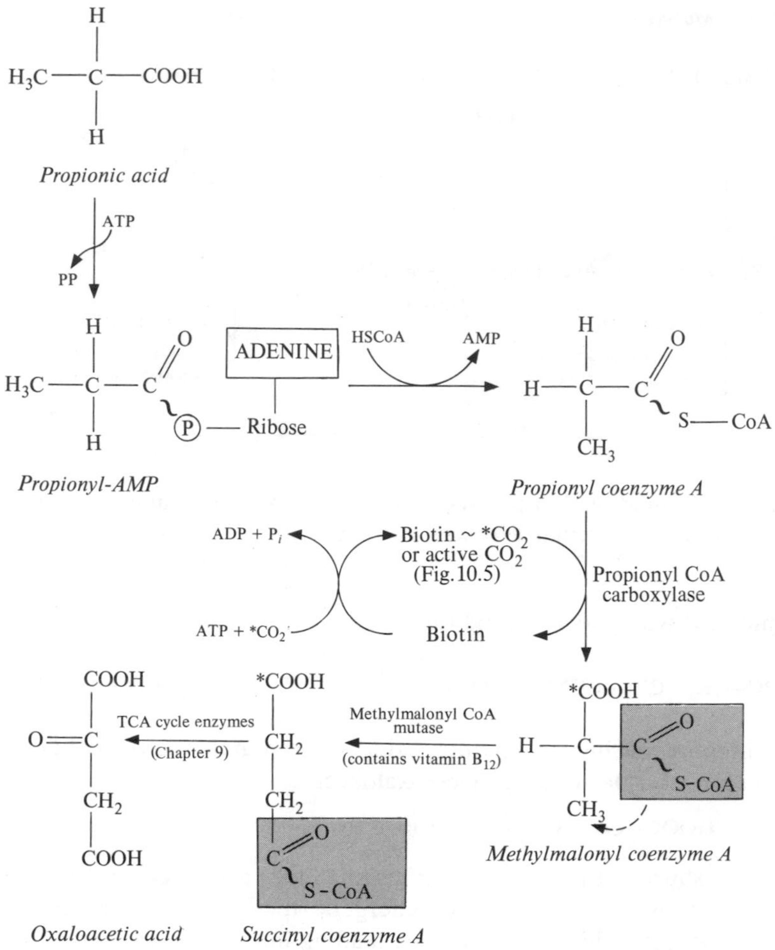

$$H_3C-CH_2-COOH$$

Propionic acid

↓ ATP → PP

Propionyl-AMP

HSCoA → AMP

Propionyl coenzyme A

Propionyl CoA carboxylase

ADP + P_i ← ATP + $^*CO_2$; Biotin ~ $^*CO_2$ or active CO_2 (Fig. 10.5); Biotin

Methylmalonyl coenzyme A

Methylmalonyl CoA mutase (contains vitamin B_{12})

Succinyl coenzyme A

TCA cycle enzymes (Chapter 9)

Oxaloacetic acid

Fig. 12.10 Conversion of propionate to oxaloacetate

acid. This is absorbed by the alimentary canal and is eventually glucogenic in the liver.

For propionic acid to be converted to oxaloacetate, free propionic acid must first be converted to the active form, propionyl coenzyme A. The reactions concerned are similar to those occurring when a free fatty acid is 'primed' prior to its catabolism (page 196). The next step is similar to the reaction in which pyruvate is carboxylated. It involves the catalytic action of *propionyl coenzyme A carboxylase*, working in conjunction with biotin (which acts as a carrier of 'active' carbon dioxide). Carbon dioxide is thus transferred to the propionyl coenzyme A, so producing methylmalonyl coenzyme A (Fig. 12.10). The last-named substance then undergoes a remarkable isomerization to succinyl coenzyme A in a reaction catalysed by an enzyme whose cofactor is vitamin B_{12}; the enzyme is called *methylmalonyl coenzyme A mutase.* The succinyl coenzyme A so formed is then converted to oxaloacetate by the machinery of the TCA cycle in the manner described in Chapter 9.

12.6 The synthesis of oxaloacetate from acetyl coenzyme A

In *plants and many microorganisms* fatty acids derived from glycerides or phospholipids can act as precursors of oxaloacetate and are therefore potentially glucogenic. The fatty acids are first broken down to acetyl coenzyme A by the pathway described in Chapter 10. The acetyl coenzyme A is then fed into a cycle known as *the glyoxylate cycle,* the enzymes of which occur in the mitochondria of plants and in the equivalent structures of bacteria. The cycle resembles the TCA cycle (Chapter 9) in many ways except that, instead of the two carbon atoms of acetyl coenzyme A being oxidized to carbon dioxide, reactions leading to a *net synthesis of oxaloacetate* take place. Net synthesis of oxaloacetate from acetyl coenzyme A is impossible in the TCA cycle, since for each molecule of acetyl coenzyme A fed into the cycle, two carbon atoms are always lost as carbon dioxide.

The first reaction of the glyoxylate cycle, as in the TCA cycle, is the 'condensation' of acetyl coenzyme A and oxaloacetate to form citric acid. This is catalysed by citrate synthase. The citric acid is now converted to isocitric acid by aconitate hydratase. The next step is quite different from the TCA cycle for, instead of being oxidized, the isocitrate undergoes cleavage. The enzyme responsible is called *isocitrate lyase* and the products are succinate and *glyoxylic acid* (Fig. 12.11). The last-named substance gives the cycle its name, for it does not occur in the TCA cycle. An enzyme called *malate synthase* now catalyses an aldol-type 'condensation' (page 178) between a second molecule of acetyl coenzyme A and the glyoxylate, the product of the reaction being L-malate. Thus two molecules of acetyl coenzyme A and the original molecule of oxaloacetate are converted to one molecule each of succinate and L-malate. These are now both converted to oxaloacetate, as in the TCA cycle, by reactions catalysed by succinate dehydrogenase, fumarate hydratase, and malate dehydrogenase (only the latter being necessary, of course, for the L-malate). One of the oxaloacetate molecules effectively replaces the one used initially to form the citrate, but the *overall effect* of the glyoxylate cycle is to convert two acetyl coenzyme A molecules to one molecule of oxaloacetate.

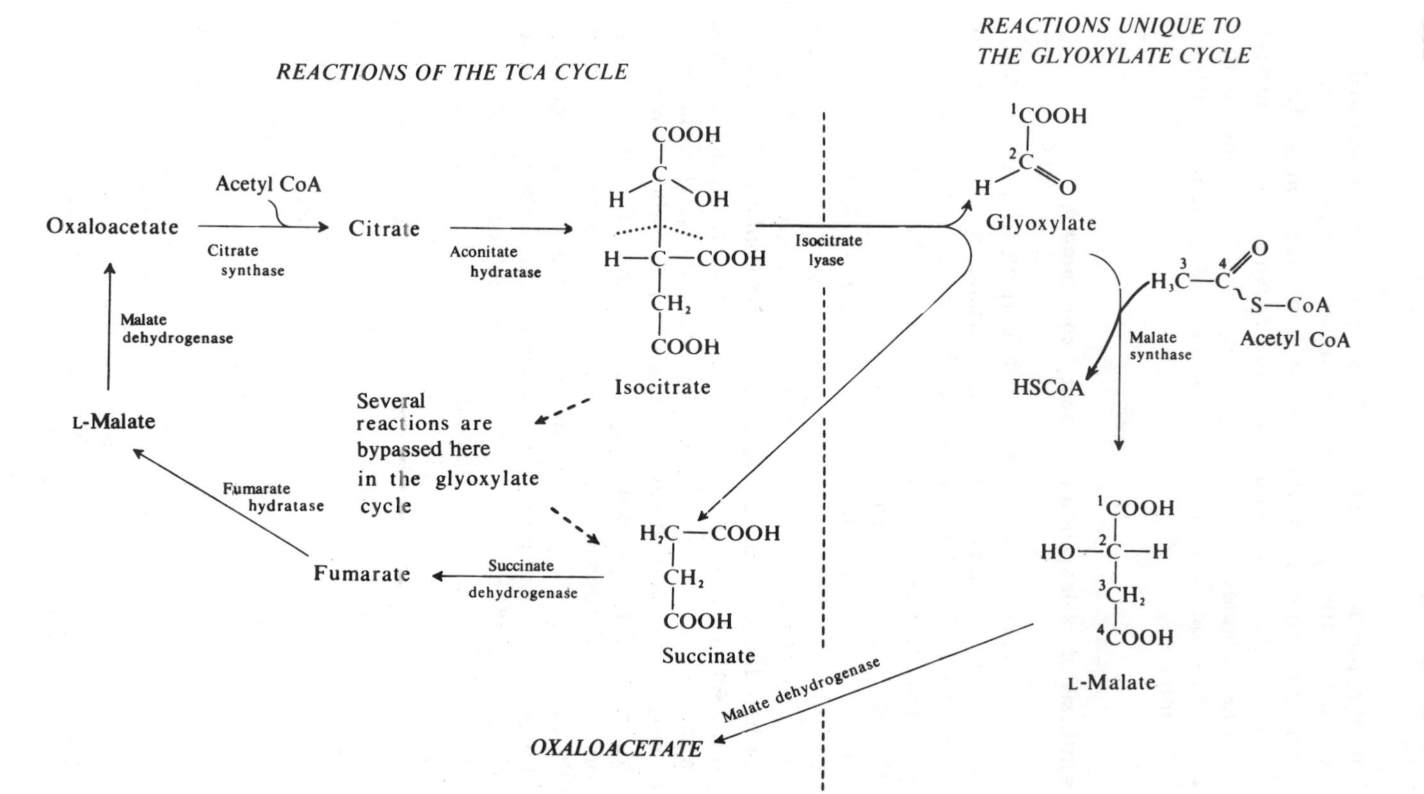

Fig. 12.11 The glyoxylate cycle. [In this cycle, the overall reaction is 2 acetyl coenzyme A→oxaloacetate. Most of the reactions occur also in the TCA cycle (Chapter 9). Two enzymes, isocitrate lyase and malate synthase, occur in the glyoxylate cycle but not in the TCA cycle]

Animal tissues lack the two enzymes which are peculiar to the glyoxylate cycle (isocitrate lyase and malate synthase) and so they are *unable to catalyse a net conversion of acetyl coenzyme A to oxaloacetate.* This has an important consequence, for it means that substances such as fatty acids and certain amino acids (especially *leucine*), whose catabolism yields acetyl coenzyme A, cannot be converted to oxaloacetate. Consequently, such substances are *not glucogenic.* As discussed in Chapter 10, the acetyl group of acetyl coenzyme A is usually oxidized by the TCA cycle or converted to ketone bodies.

12.7 A general outline of carbohydrate metabolism in green plants

In the present and previous chapters several aspects of carbohydrate biochemistry in plants have been described and at this point it is convenient to indicate some of the connections between them as well as some other metabolic processes to be considered later.

In many plants the fructose-6-phosphate produced by photosynthesis (Chapter 14) is first converted to the highly soluble sugar, sucrose, which acts as a *temporary* carbohydrate store and is also frequently the major form in which carbohydrate is translocated from one part of the plant to another. Fructose-6-phosphate is converted to sucrose by the reactions indicated in Fig. 12.1 but, for this to be possible, a proportion of it must first be made into glucose-1-phosphate, the immediate precursor of UDP-glucose. Two enzymes, previously mentioned in Chapter 11, phosphohexoisomerase and phosphoglucomutase, are responsible for the interconversion of fructose-6-phosphate and glucose-1-phosphate (Fig. 12.12).

Once sucrose has been formed in the leaf it is often transported (in the phloem) to other parts of the plant. Upon arrival, the sucrose is hydrolysed by sucrase to glucose and fructose, each of which must be phosphorylated by appropriate kinases before undergoing further metabolism. The resulting fructose-6-phosphate and glucose-6-phosphate are easily interconverted, and, once formed, numerous other substances can be synthesized from them which can subsequently produce not only ATP but a number of other important compounds as well. These include acetyl coenzyme A which, in turn, is a precursor of fatty acids, and acids of the TCA cycle from which many amino acids for protein synthesis arise (Chapter 15). Glucose-6-phosphate is the initial reactant for a cycle known as the *pentose phosphate cycle*, a metabolic pathway which has numerous anabolic functions (Chapter 14). Alternatively, the glucose-6-phosphate may be converted to glucose-1-phosphate from which cellulose or starch can be formed (Section 12.1).

Clearly, the formation of fructose-6-phosphate by photosynthesis initiates a large number of biochemical reactions throughout the plant. This is true also of the breakdown of storage compounds such as the food materials in

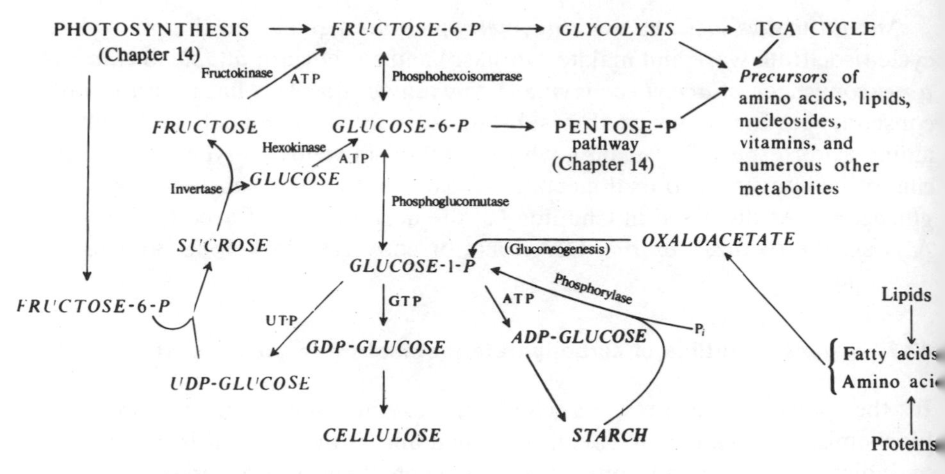

Fig. 12.12 Outline of carbohydrate metabolism in the green plant

germinating seeds, potato tubers, tap roots, corms, and bulbs. Stored carbohydrates are usually hydrolysed to their component hexose units which, after phosphorylation, are converted to sucrose or some other soluble sugar for transportation to the growing parts of the plant. The fatty acids of stored lipids are converted by the fatty acid oxidation spiral (Chapter 10) to acetyl coenzyme A and this, like the breakdown products of amino acids, can be converted to oxaloacetate and hence to glucose-1-phosphate. This is then able to initiate any of the numerous processes indicated in Fig. 12.12.

13

The control of metabolism

Several biochemical control mechanisms have been encountered in earlier chapters (see Section 6.9 and *control mechanisms* in the index). In the present chapter a general review of some of the mechanisms is presented, and the control of glycolysis, gluconeogenesis, the TCA cycle, and fat metabolism are considered in somewhat more detail.

13.1 Types of control mechanism

The most important developments in our understanding of metabolic control were made after 1960 yet already a vast literature testifies to the great interest aroused by this subject. Here we can only indicate some of the main features.

Figure 13.1 shows a hypothetical pathway which demonstrates the main types of control mechanism. The rate at which a pathway provides products is dependent, *inter alia*, on the supply of some of the *reactants.* In the pathway, reactions 1 and 3 depend on the supply of A and J. A deficit of either will slow or stop the whole sequence. Similarly, *cofactor availability* is important. For example, reaction 6 can only proceed if the [NAD]/[NADH] ratio is appropriate for the redox reaction concerned. In many pathways control may be exerted by the supply of ATP or ADP or coenzyme A.

A very important means of regulation, especially in bacteria, is effected by controlling the *quantity of enzyme* present. Reaction 2 in Fig. 13.1 is regulated by the synthesis and/or catabolism of the enzyme. Increased levels of the enzyme may result from stimulation of protein synthesis. Alternatively, reaction 2 may be slowed down by the proteolytic destruction of the enzyme by intracellular proteases or by a reduction in the rate at which it is synthesized in the ribosomes (Section 16.6).

Cellular compartmentation (reaction 7 in Fig. 13.1) is another important regulatory device. Reactions are usually localized in particular subcellular regions (Table 1.1). Pathways are often regulated by the rate of transfer of metabolites from one part of the cell to another, e.g., from the cytosol to

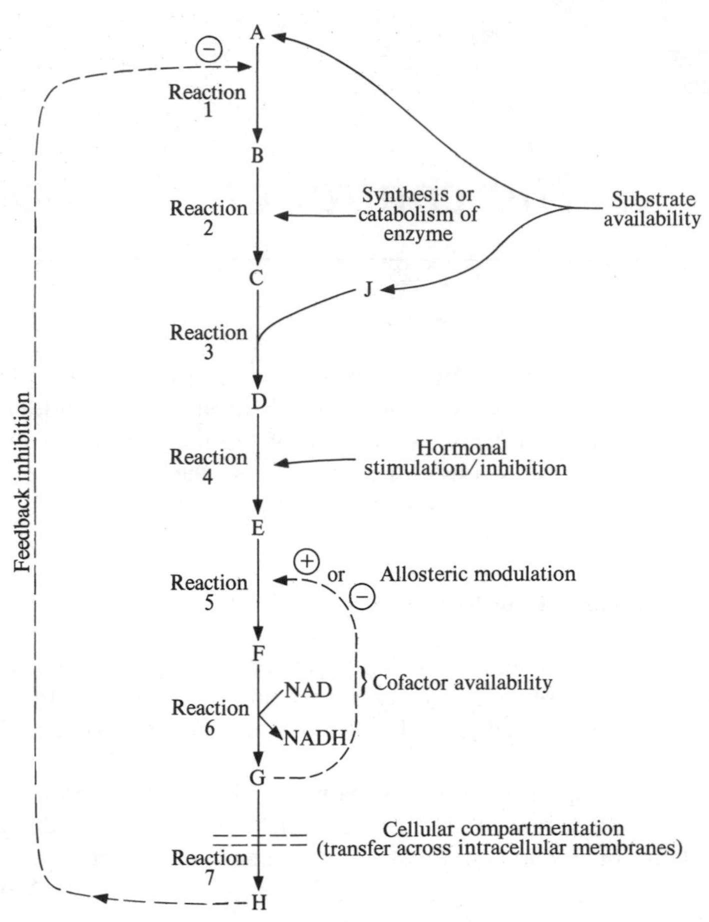

Fig. 13.1 A hypothetical biochemical pathway showing various control mechanisms

the mitochondrion, or by the entry of substances into the cell from the surrounding environment (Sections 9.12, 10.1, 10.3).

Finally, *activation or inactivation of enzymes* is a major means of control and is an important theme of this chapter. Enzymic reactions may often be inhibited or stimulated by small molecules which are not necessarily the reactants directly involved. For example, G in Fig. 13.1 may combine with enzyme 5 to alter its configuration so as to change its rate of catalysis. Such modification is called 'allosteric' alteration (Section 3.9). The name means 'another shape'—an allosteric change normally occurs as a result of modulation by a substance *other than the substrate* at a site *other than that* which the substrate can occupy. If G stimulates a greater rate of reaction this is said to be *positive allosteric modulation* (often indicated by ⊕). Alternatively, G may be a *negative allosteric modulator* (indicated as ⊖) in which case it would slow reaction 5. Feedback inhibition (Section 6.9) is a common example of this. Thus an accumulation of H in Fig. 13.1 slows reaction 1 so as to prevent excessive and wasteful conversion of A to H. Some important examples of feedback inhibition occur in the synthesis of amino acids and the synthesis of purines and pyrimidines (Sections 15.13 and 15.14).

Related to allosteric modulation of enzyme activity is *covalent modification* of enzyme structure. This involves either activation or inactivation of enzymes by forming or breaking covalent bonds. A familiar example is the activation of trypsin. This proteolytic enzyme is stored in the pancreas in an inactive form, trypsinogen. It is activated in the intestine by the breaking of a covalent bond, thus removing part of its peptide chain and producing the free enzyme. In the present chapter several important examples of covalent modification through phosphorylation or dephosphorylation of enzymes are described (Section 13.7). A number of hormones regulate biochemical pathways through this mechanism in the higher animals.

13.2 Allosteric enzymes

For most enzymes a plot of the initial velocity of reaction v against the substrate concentration, $[S]$ is hyperbolic (Fig. 6.5). Allosteric enzymes usually show a different relationship, the shape of the curve often being *sigmoid* (Fig. 13.2a). The general explanation for this is that the enzymes possess more than one catalytic site. In many instances the enzyme has several subunits, most of which have an active centre for catalysing the reaction. As one molecule attaches to a catalytic site the shape of the subunit changes slightly, and this induces an alteration in the other sites so that they become more reactive. This phenomenon is known as *cooperativity*.

Most allosteric enzymes probably possess *regulatory sites or regulatory subunits* which are distinct from the *catalytic sites or subunits*. These regulatory sites specifically bind to the *modulators*, which are small molecules not themselves directly involved in the enzyme's major reaction. Regulatory sites or subunits may be inhibitory or they may have an activating effect. Some enzymes contain several examples of both types of regulation so that there are positive modulators and negative modulators each with their own regulatory sites.

Inhibitory modulators decrease the activity of the enzyme for any given substrate concentration, the degree of inhibition being related to the concentration of the inhibitor. Conversely, positive allosteric modulators raise the enzyme's activity (Fig. 13.2). These effects are probably brought about by changes in the shape of the enzyme as a whole in such a way that the

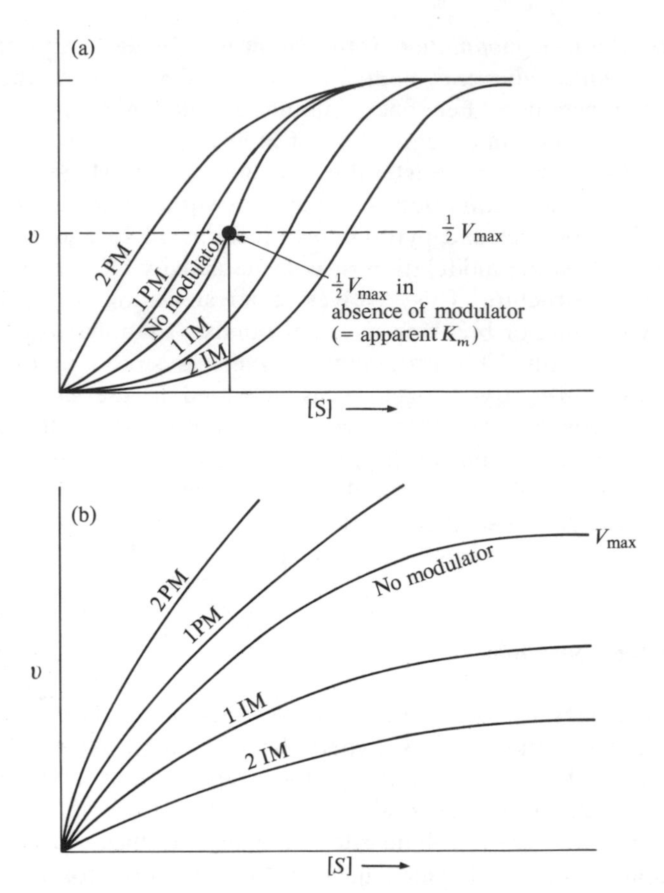

Fig. 13.2 The relationships between the initial velocity of enzyme reactions and the substrate concentration for allosteric enzymes. PM is positive modulator; 2PM is double the concentration of PM; IM is negative modulator; 2IM is double the concentration (a) for enzymes showing a change in apparent K_m and (b) for enzymes showing a change in V_{max}

catalytic sites are inactivated or activated to varying degrees. Most modulators, whether positive or negative, are neither the reactants nor products of the enzyme's major reaction and are called *heterotropic* modulators. In some cases, however, the modulator, though specifically combining with a regulatory site, is also a potential substrate for the enzyme. Such modulators are called *homotropic* and they may be either stimulatory or inhibitory.

This general description of allosteric enzymes applies to most examples. Effectively, the modulators change the apparent K_m of the enzyme; positive modulators reduce the K_m, while negative ones raise it (Fig. 13.2a). There are also a few enzymes for which modulators appear to change the V_{max} value (Fig. 13.2b).

An important characteristic of allosteric enzymes is that they are sensitive to changes in the concentrations of the modulators. These changes serve to signal the need to increase or decrease the rate of the catalysed reaction in order to regulate the whole biochemical pathway in which the enzyme takes part. The following sections illustrate how allosteric modulation controls several important pathways. Readers are advised to revise the subject matter of Chapters 8 to 12 before tackling these sections in detail.

13.3 Control of glycolysis

The major purpose of the conversion of glucose to acetyl coenzyme A by the glycolytic pathway and the subsequent oxidation of the acetyl groups to carbon dioxide by the TCA cycle is the generation of ATP for driving the many endergonic processes in the cell (Chapters 8 and 10). A fine control is present in most cells to ensure that these pathways function at just sufficient rates to maintain the required level of ATP. This control is largely through small molecules acting as allosteric modulators. Among these are certain common metabolites such as citrate, glucose-6-phosphate, fructose diphosphate, acetyl coenzyme A, and the adenosine phosphates. Cells generally attempt to keep the ratio

$$\frac{[\mathrm{ATP}]}{[\mathrm{ADP}] + [\mathrm{AMP}]}$$

above about 50. If ATP is being used up and converted to ADP, the cell's metabolism is stimulated to produce more ATP. This occurs, for example, in muscle cells during the work of contraction and in cells synthesizing protein, for both processes require a large supply of ATP. On the other hand, in 'resting' cells with a low rate of utilization of ATP, glycolysis is slowed down or even reversed (gluconeogenesis) to prevent uneconomic use of fuel.

In the glycolytic conversion of glucose to pyruvate there is growing evidence that most of the reactions are close to equilibrium, with three notable exceptions. These reactions are the *pacemakers* of the pathway and can be speeded up or slowed down as required. The reactions are those catalysed by *hexokinase*, *phosphofructokinase*, and *pyruvate kinase*. Since the other reactions are more or less freely reversible with their rates and direction depending only on the concentrations of their substrates, it is these pacemakers which determine the overall rate of the metabolism of glucose by the pathway.

There are three main indications to suggest that the three enzymes are the pacemakers of glycolysis. The first is based on the fact that each shows a relatively lower activity *in vitro* than the other glycolytic enzymes. Clearly, the overall rate of the pathway is limited by the rate of its slowest reaction. Second, *the mass action ratios* for these enzymes are far from equilibrium *in vivo* or in perfused tissues. This can be explained by taking *phosphofructokinase* as an example. It catalyses the reaction:

$$\text{Fructose-6-P} + \text{ATP} \rightleftharpoons \text{fructose-di-P} + \text{ADP}$$

(Equilibrium constant $K = 380$)

By rapidly freezing perfused animal tissues to −80°C with liquid nitrogen and analysing the reactants and products (the so-called freeze clamp technique), an indication of the concentrations as they were in the functioning cells is obtained. The mass action ratio for the reaction is calculated from the relation:

$$\frac{[\text{Fructose-di-P}][\text{ADP}]}{[\text{Fructose-6-P}][\text{ATP}]}$$

In a number of tissues, values ranging from 0.03 to 0.7 are obtained, showing that the reaction in the cell is far from a state of equilibrium (where the ratio would be 380). For the enzymes not thought to be the pacemakers of glycolysis the mass action ratios are fairly close to the equilibrium constants.

The third criterion is that the enzymes are allosteric and show varied activity depending on the metabolic circumstances. Taking again phosphofructokinase as an example, it has been shown that increasing the rate of glycolysis of the heart or diaphragm muscle of the rat by lack of oxygen (anoxia) leads to a fall in the concentration of the substrate fructose-6-phosphate. This indicates that the enzyme must have been working faster under anoxia since the rate of glycolysis was greater in spite of the drop in the concentration of the enzyme's substrate. For enzymes whose rate of catalysis depends only on the concentration of the reactants a decrease in the substrate concentration would normally lead to slower reaction.

We will now consider in more detail how the activities of the three pacemakers of glycolysis are modulated to control the rate of glycolysis. Probably the most important point of regulation is the step catalysed by *phosphofructokinase.* An interesting feature of the enzyme is that it is inhibited by concentrations of ATP above a certain optimal level. The enzyme has binding sites for ATP other than those participating in the catalysed reaction. Attachment of ATP to these sites slows the enzyme's activity, an example of ATP acting as a negative allosteric modulator. This effect is reduced, however, by high concentrations of AMP (Fig. 13.3a).

Stimulation of the enzyme's activity by AMP and its inhibition by ATP links the rate of reaction directly to the energy needs of the cell. In many

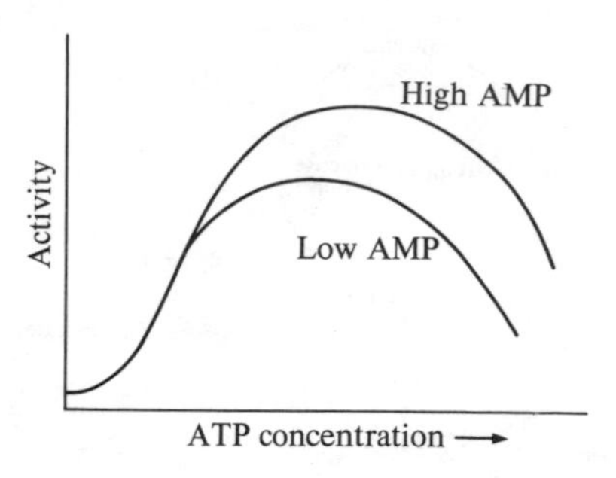

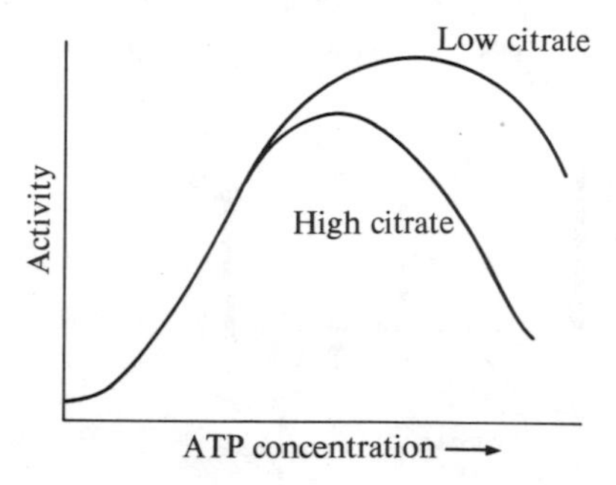

Fig. 13.3 The effect of allosteric modulators on the activity of phosphofructokinase (PFK).
(a) The effect of high and low AMP levels
(b) The effect of high and low citrate concentrations

cells it is probably a *rise in the concentration of AMP* rather than a fall in that of ATP which is most important in stimulating the synthesis of fructose diphosphate. Actively contracting skeletal muscle, for example, shows very little or no reduction in its ATP content. This is because there are at least two reactions which restore the level of ATP as it is depleted:

(a) Creatine ~ Ⓟ + ADP = creatine + ATP (see Section 20.4)

(b) 2ADP = AMP + ATP

The second reaction is catalysed by an enzyme called *adenylate kinase* present not only in muscle cells but also in cells of nearly all eukaryotes. It ensures that when ADP is formed, some is restored to ATP while the rest, as AMP, can stimulate phosphofructokinase and the rate of glycolysis. This is the basis of the 'Pasteur effect' whereby fermentation in yeast is increased under anaerobic conditions (page 233).

Besides AMP, *citrate* also acts as an important regulator of phosphofructokinase, but it has the opposite effect to AMP for it is a negative allosteric modulator (Fig. 13.3b). High concentrations of citrate in the cell decrease

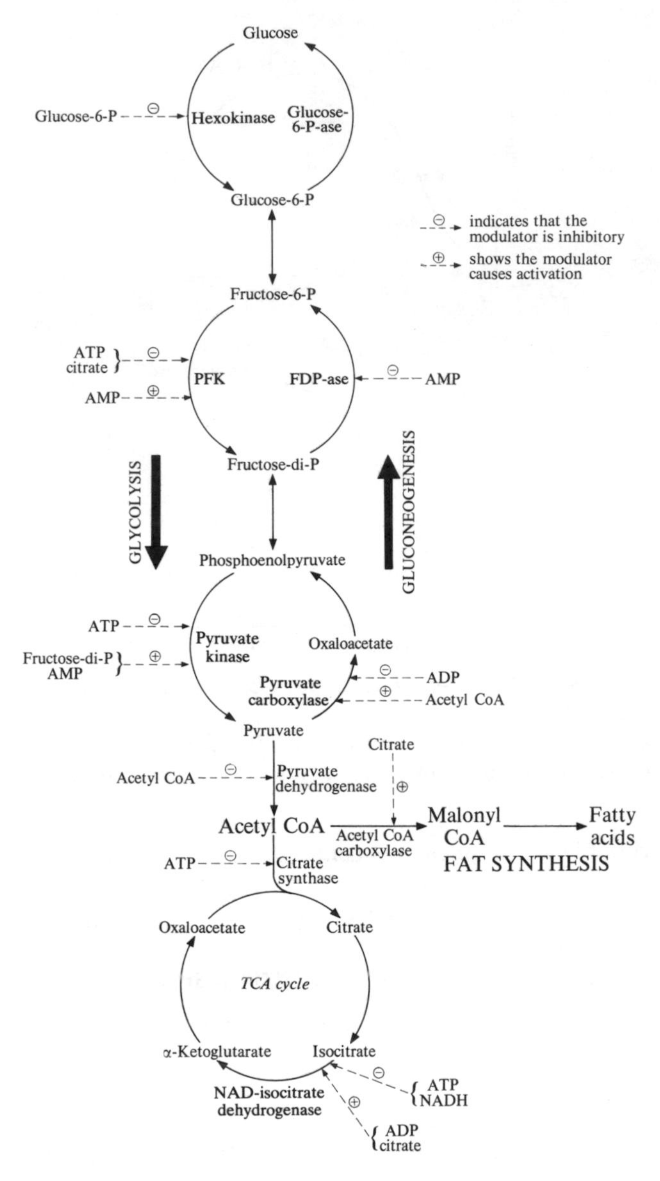

Fig. 13.4 The role of allosteric modulators in the control of glycolysis, gluconeogenesis, and the TCA cycle

the enzyme's rate of reaction, so slowing the whole glycolytic pathway. When glycolysis (or the catabolism of fatty acids) is functioning faster than required to fulfil the energy needs of the cell, citrate tends to accumulate in the mitochondria as a result of acetyl coenzyme A condensing with oxaloacetate (Section 9.2). Some of the citrate leaves the mitochondria, as indicated in Fig. 10.4, and once in the cytosol it slows down the reaction catalysed by phosphofructokinase (Fig. 13.4).

Turning now to the control of the *hexokinase* reaction of glycolysis we find that at this point regulation is simpler than for the phosphofructokinase reaction. Hexokinase is inhibited by high concentrations of its product, i.e., glucose-6-phosphate. The importance of hexokinase as a regulator of glycolysis appears to be secondary to that of phosphofructokinase. When the activity of the latter is reduced, perhaps by a rise in the citrate concentration, an increase in the level of fructose-6-phosphate occurs. This is readily converted to glucose-6-phosphate by the phosphohexoisomerase reaction (Fig. 11.2). As the concentration of glucose-6-phosphate increases it slows down the hexokinase reaction so preventing any further accumulation of hexose phosphates in the cell:

Glucose —(HK; ATP → ADP)→ glucose-6-P ⇌(PHI) fructose-6-P —(PFK; ATP → ADP)→ fructose-di-P

glucose-6-P ⊖ → HK; Citrate ⊖ → PFK

where HK is hexokinase, PHI is phosphohexoisomerase, and PFK is phosphofructokinase. Together, the control of the activity of these two kinases regulates the rate of formation of fructose diphosphate in many cells and so determines the rate of the subsequent reactions of glycolysis.

Mammalian liver cells present an interesting exception to this generalization. They need to phosphorylate glucose *even when glycolysis is inhibited.* As glucose reaches the liver cells from the intestine it must be phosphorylated prior to conversion to glycogen (Section 12.2). Hepatic cells contain an enzyme, *glucokinase,* whose reaction is superficially identical to that catalysed by hexokinase:

$$\text{Glucose} + \text{ATP} \rightarrow \text{glucose-6-P} + \text{ADP}$$

However, its reaction is not slowed down by an accumulation of glucose-6-phosphate and it is therefore able to function even when phosphofructokinase is strongly inhibited.

The third and least understood pacemaker of glycolysis is *pyruvate kinase.* This enzyme is activated in liver cells by fructose diphosphate and so a rise in the cellular concentration of the product of stage 1 of glycolysis increases the rate of the fourth stage (Fig. 11.5). The enzyme is also stimulated by a rise in the [AMP]/[ATP] ratio with both ATP and AMP acting separately as allosteric modulators—ATP inhibiting and AMP stimulating the reaction. In plant cells, but probably not in animal cells, citrate inhibits the enzyme. This exemplifies a common feature of control mechanisms. The principal pacemakers in biochemical pathways are usually the same enzymes in quite unrelated organisms, yet the allosteric modulators and their mode of action often vary.

13.4 Control of reactions in the mitochondria

The fate of pyruvate is under metabolic control which depends on the type of organism and the availability of oxygen (Section 11.8). Under anaerobic conditions, pyruvate is reduced to lactate in the cytosol of animal cells to enable the NADH produced in the phosphoglyceraldehyde dehydrogenase reaction to be reoxidized. It appears that it is mainly the $[NAD^+]/[NADH]$ ratio in the cytosol that determines whether lactate is produced or whether the pyruvic acid is decarboxylated in the mitochondrion to form acetyl coenzyme A. When oxygen is plentiful the cytosolic ratio of the two forms of the coenzyme is of the order of 1000. This enables the NAD^+-dependent reaction catalysed by phosphoglyceraldehyde dehydrogenase to move in the direction of the net synthesis of 1,3-diphosphoglycerate. A fall in the ratio (i.e., an increase in the relative amount of NADH) would slow the reaction and so decrease the rate of the whole glycolytic pathway. This is, as we have seen, avoided by intervention of the lactate dehydrogenase reaction (Section 11.9).

Pyruvate dehydrogenase is the key enzyme complex linking glycolysis with the TCA cycle. Being located in the mitochondria and functioning as a bridge between the cytosolic glycolytic system and mitochondrial metabolism (Section 11.8), it is just the sort of enzyme one would expect to be subject to metabolic control. It is strongly inhibited allosterically by high concentrations of acetyl coenzyme A, an inhibition which is prevented by the presence of free coenzyme A. Thus the free coenzyme stimulates the formation of acetyl coenzyme A, but an accumulation of acetyl coenzyme A resulting from glycolysis or fatty acid catabolism inhibits the enzyme and so *prevents excessive decarboxylation* of pyruvate. This leaves any excess pyruvate free to be converted to glucose by the process of gluconeogenesis (Section 13.5).

The details of the control of the individual reactions of the *TCA cycle* are still poorly understood. Although any of the TCA cycle acids can be fed into the system (Section 9.10), acetyl coenzyme A is normally the main initial substrate. Since the supply of acetyl coenzyme A is determined mostly by the rate of (a) glycolysis and (b) fatty acid oxidation, the TCA cycle is largely regulated by the activity of these pathways. In addition the oxidation/reduction state of NAD is important. Normally under aerobic conditions the mitochondrial [NAD^+]/[NADH] ratio is about 8. This provides adequate *oxidized* coenzyme for the NAD-linked oxidoreductases taking part in the TCA cycle and in fatty acid oxidation, but provides sufficient NADH to maintain a steady rate of oxidative phosphorylation. Under anaerobic conditions, however, the ratio decreases as the NADH is no longer oxidized by the electron transport particles. This, in turn, slows down the rate of many of the mitochondrial dehydrogenase reactions. Finally, the *concentration of ADP* is important. Whenever the concentration of ADP in the mitochondrion increases under aerobic conditions, oxidative phosphorylation is stimulated (Section 8.9) and more NADH is converted to NAD^+. This, in turn, stimulates an increase in the rate of the various reactions catalysed by dehydrogenases.

Only two of the enzymes of the TCA cycle have been widely proposed as possible regulators (Fig. 13.4). The most important of these is the first member of the cycle, *citrate synthase.* In animals and higher plants this enzyme is inhibited by high concentrations of ATP, whereas in several species of prokaryotes (e.g., *Escherichia coli* and *Rhodopseudomonas*) it is inhibited by a high [NADH]/[NAD^+] ratio. Thus the same enzyme is modulated allosterically by different substances in different organisms, but the overall result is similar. High concentrations of ATP (or NADH) signal that the cycle should be slowed down. Conversely, a rise in the levels of ADP (or NAD^+) at the expense of ATP (or NADH) increases the rate of formation of citrate and so stimulates the cycle.

NAD-linked *isocitrate dehydrogenase* may be a second pacemaker in the TCA cycle. It is stimulated *in vitro* by ADP and by citrate, but inhibited by high levels of ATP or NADH. Thus a high [ATP]/[ADP] ratio may slow this step in the cycle, an appropriate response in view of the fact that a major function of the cycle is to produce NADH to enable oxidative phosphorylation to occur. Two difficulties arise here, however, for it is not always clear how far observations made on enzymes *in vitro* can be applied to the reactions as they occur in the living cell. Moreover, there are *two* isocitrate dehydrogenases in mitochondria (Section 9.5). Unlike the NAD-linked enzyme, NADP-dependent isocitrate dehydrogenase is not considered to be controlled allosterically. Thus, even if the NAD-linked enzyme is inhibited, it is conceivable that the reaction could still proceed catalysed by the other enzyme.

13.5 Control of gluconeogenesis

The control of gluconeogenesis is closely interrelated with the regulation of glycolysis; at times when it is required to have a net synthesis of glucose from pyruvate it is evidently appropriate to slow down or stop the opposing sequence, glycolysis. This is particularly apparent when one recalls that many of the enzymes are involved in both processes. In the simplest case, to synthesize glucose the cell 'turns off' those reactions unique to glycolysis and 'turns on' those unique to the pathway of gluconeogenesis. In practice, this may not really be so clear-cut as it sounds. In the only mammalian tissues capable of performing both pathways, the liver and kidney, it has been observed that it is sometimes possible for glycolysis and gluconeogenesis to occur simultaneously.

The main pacemakers of gluconeogenesis are just those reactions where gluconeogenesis and glycolysis are distinctly different. These are (a) the conversion of pyruvate to phosphoenolpyruvate, and (b) the dephosphorylation of fructose diphosphate to form fructose-6-phosphate (Fig. 12.7).

It will be recalled that to convert pyruvic acid to phosphoenolpyruvate requires two enzymes, the first of which, *pyruvate carboxylase,* converts pyruvate to oxaloacetate (Section 12.4). This is an allosteric enzyme whose activity is modulated positively by acetyl coenzyme A and negatively by ADP. For high levels of acetyl coenzyme A to stimulate pyruvate carboxylase represents an elegant device to ensure that when the mitochondrial TCA cycle is receiving a good supply of this metabolite, probably as a result of the catabolism of fatty acids, then gluconeogenesis is increased. Thus *fat catabolism indirectly promotes glucose synthesis.* Similarly, when the [ATP]/[ADP] ratio in the cell is high there is less need of glucose catabolism, and the enzyme is activated to initiate glucose synthesis. Rather neatly, the opposing glycolytic enzyme, pyruvate kinase, is simultaneously inhibited by a rise in the concentration of ATP (Fig. 13.4).

The other main pacemaker of gluconeogenesis, *fructose diphosphate phosphatase* (FDP-ase), is strongly inhibited by AMP. Thus when the cell's AMP content is high and there is a need to promote glucose catabolism, the enzyme involved in catalysing the synthesis of glucose is inhibited. Again rather neatly, the corresponding glycolytic enzyme, *phosphofructokinase* (PFK), is at the same time stimulated by this same modulator.

E. A. Newsholm has suggested that the reactions catalysed by these two enzymes form a *substrate cycle* which represents a very sensitive means of control of glycolysis and gluconeogenesis. A small increase in the activity of one enzyme and a small degree of inhibition of the other can bring about surprisingly large changes in the overall rate of reaction (Fig. 13.5). Substrate cycles of this type were at one time considered wasteful in the sense that if fructose-6-phosphate and the diphosphate were continuously undergoing interconversion the only net result would be the breakdown of ATP to ADP. They were sometimes called 'futile cycles'. The trend now is to consider that the cycles, far from being useless, actually provide an important means of regulation. Small changes in activity of the enzymes brought about by allosteric modulators could have profound effects. Other possible

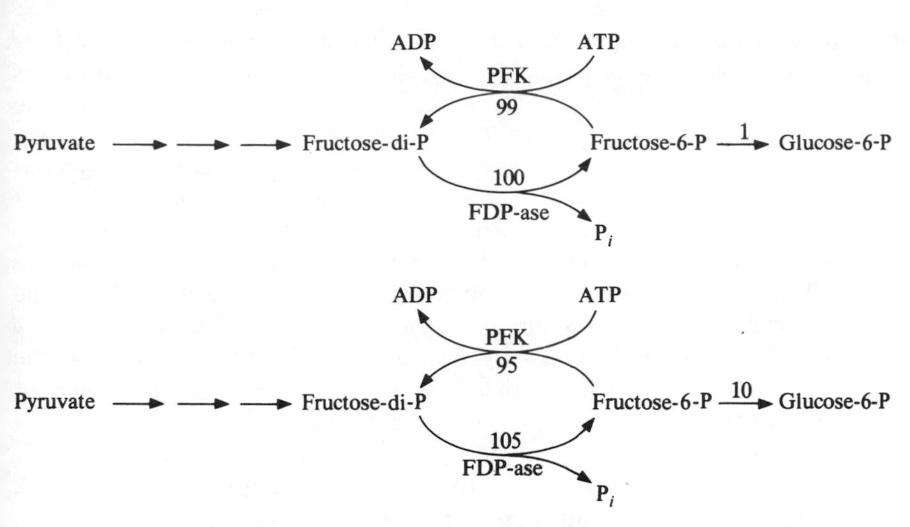

Fig. 13.5 How small changes in the activities of enzymes catalysing a substrate cycle can greatly change the amount of product formed. Arbitrary initial values are ascribed to the two enzyme activities. A 5 per cent increase in the activity of fructose diphosphate phosphatase (FDP-ase) and a similar fall in the activity of phosphofructokinase (PFK) can lead to a tenfold increase in the rate of synthesis of fructose-6-phosphate

cycles of this type include the reactions whereby pyruvate and phosphoenolpyruvate are interconverted, and the reactions catalysed by hexokinase and glucose-6-phosphatase (Fig. 13.4).

13.6 Hormonal control of metabolism

In previous sections of this chapter we considered examples of metabolic control based mainly on allosteric effects of *small molecules* on enzyme activity. The general principles involved, but not necessarily the particular details, are thought to apply to all organisms.

In higher animals, an additional form of metabolic regulation, *hormonal control*, is superimposed upon allosteric control. The evolution of highly complex multicellular animals has required that the metabolism of cells throughout the body must be coordinated so that the actions of the biochemically specialized tissues are regulated to serve the needs of the organism as a whole.

Mammals possess several dozen hormones, probably all of which produce specific effects on their receptor or 'target' organs through changes in biochemical functions. Attention will be focused here principally upon the hormones *insulin*, *glucagon*, and *adrenalin*, since the actions of these hormones relate closely with the control mechanisms already described in this chapter. Before considering their actions in detail, a brief account of the role they play in general metabolic control may be helpful.

Historically, insulin and glucagon were considered to be primarily concerned with the control of the *concentration of glucose in the blood.* After digesting and absorbing a carbohydrate meal, the blood glucose concentration rises from the fasting level of about 5 mM to double or even treble this value. This *hyperglycaemic* state normally stimulates special cells in the pancreas, called β cells, to secrete *insulin*, a small protein which functions as a hormone (Fig. 3.6). This passes by means of the blood circulation to the liver and muscles, stimulating these organs to increase their uptake of glucose and their conversion of glucose to glycogen. Subsequently, the body is able to draw on these stores of glycogen to provide energy between meals and during fasting. A number of organs—and especially the brain, heart, and testis—require a constant supply of glucose, and when the blood glucose concentration decreases below the 'normal level' the hormone *glucagon* (together when relevant with *adrenalin*) stimulates glycogen breakdown. This enables glucose to pass into the blood, so restoring its concentration there.

Glucagon is secreted by another sort of pancreatic cell, called α cells, in response to a low concentration of glucose in the blood (hypoglycaemia). It is a small protein containing twenty-nine amino acid residues. The structure of *adrenalin* is shown in Fig. 15.9. Clearly, it is not a peptide. It is secreted by cells of the *adrenal medulla*, particularly in response to stress—being, in fact, part of the 'fright and flight' response. Thus, when an animal needs to escape from a predator, adrenalin is secreted into the blood. Among its many effects is the raising of the blood glucose concentration and the stimulation of glycolysis in muscle. These changes enable the extra ATP to be produced which is needed to drive the muscles as the animal attempts to escape from danger.

13.7 Hormonal control of glycogen synthesis and breakdown

Insulin stimulates the uptake of glucose by muscle cells and promotes the synthesis of glycogen in the cells of both the liver and muscles. It also stimulates storage of fats in adipose tissue and induces protein synthesis while decreasing the catabolism of amino acids.

Conversely, *glucagon* stimulates the breakdown of glycogen in the *liver cells* and the release of glucose from these cells into the blood. *Adrenalin* more specifically affects the *muscle cells*, inducing the breakdown of glycogen and stimulating glycolysis there. Many of these effects are now recognized to be the result of a highly complex process known as the *covalent modification* of enzymes.

A number of enzymes exist in two main forms which we will call **a** and **b**. The **a** form is the active enzyme while the **b** form is either inactive or less active than the **a** form. Hormonal control of a reaction catalysed by an enzyme of this sort is exerted by regulating the relative proportions of the two forms of enzyme. In most cases studied so far, the difference between the **a** and **b** forms is due to the presence or absence of *covalently bound phosphate*. One form of the enzyme has a free hydroxyl group on a serine residue while in the other form the group is esterified to phosphate. Physiologically, the phosphate originates from ATP and is added by a *kinase*. On the other hand, a specific *phosphatase* removes the phosphate group when the phosphoserine form of the enzyme is returned to the unphosphorylated form:

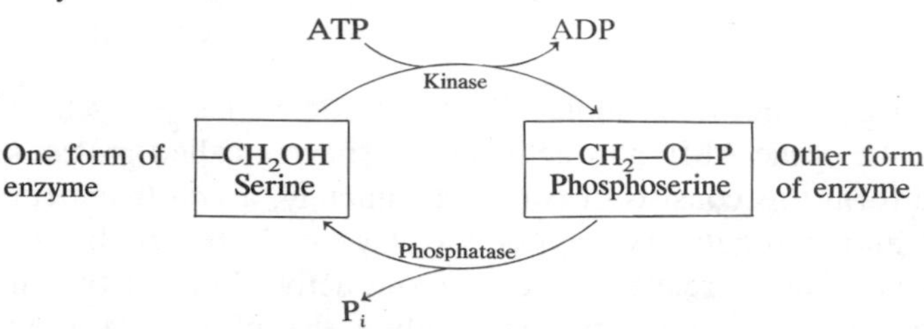

In the case of *glycogen phosphorylase* (page 219) and also the hormone-sensitive *lipase* which catalyses reaction 6 in Fig. 10.1, it is the **a** form of the enzyme which is phosphorylated. In these cases, the less active **b** forms contain no phosphate. Conversely, in *glycogen synthase* and the adipose tissue *pyruvate dehydrogenase* the active or **a** forms are not phosphorylated while the less active **b** forms are.

The rate-limiting step in glycogen synthesis from glucose-6-phosphate is the one catalysed by *glycogen synthase*. However, *glycogen phosphorylase* is the rate-limiting enzyme in the conversion of glycogen to glucose-6-phosphate. It is therefore not surprising to find that these two enzymes are under hormonal control. The mechanism of regulation is very sensitive—one molecule of glucagon can stimulate the release of as many as 300 000 glucose molecules from liver glycogen. In effect, there exists an elaborate system of amplification of the stimulus. This is sometimes called a 'cascade' for one molecule elicits changes in hundreds of others and these in turn elicit changes in many others, so magnifying the original signal in stages. This important principle was discovered by E. W. Sutherland.

Let us consider first the action of glucagon and adrenalin on the liver and muscle cells respectively. Their main effects are to *activate the phosphorylase* and to *inhibit the synthase*, thus 'switching on' glycogen breakdown and 'switching off' its synthesis.

Initially, both glucagon and adrenalin appear to bind to specific receptor sites on the *outside* of the cell membrane (Fig. 13.6). Somehow, perhaps through allosteric changes, this binding activates the enzyme *adenyl cyclase* which is located on the inside of the *plasmalemma.* This enzyme catalyses the conversion of ATP to the cyclic form of AMP, 3′,5′-cyclic AMP with the elimination of pyrophosphate:

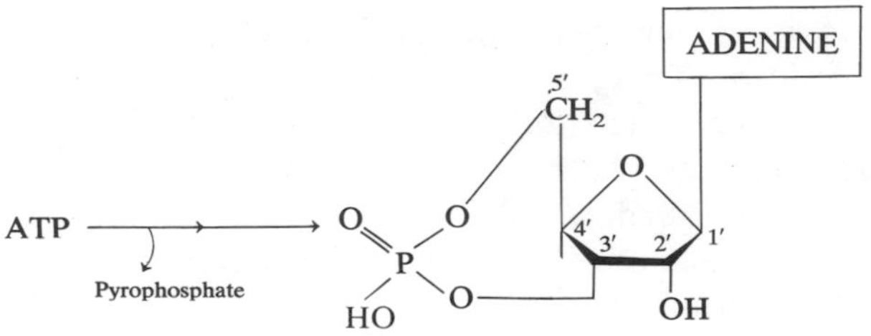

ATP → (Pyrophosphate) → 3′,5′-cyclic AMP

For each original hormone molecule, many molecules of cyclic AMP are produced. The cyclic AMP now activates an enzyme called *protein kinase.* In its inactive form this consists of two parts, namely, a *catalytic* peptide (C in Fig. 13.6) and a regulatory peptide, R. Cyclic AMP combines with the regulatory peptide, so releasing the free and active form of the enzyme, C. Like other kinases, protein kinases catalyse the phosphorylation of substrates using ATP as the source of phosphate. In this instance the *substrates are themselves enzymes,* the activities of which are modified by the phosphorylation.

At least a dozen hormones are now known to operate through cyclic AMP. In some instances they raise the intracellular cyclic AMP concentration (by stimulating adenylcyclase). In other cases, they reduce it, using *phosphodiesterase,* which converts cyclic AMP to 5′-AMP. Thus the cyclic AMP in the cell acts as a 'secondary messenger', carrying the signal *from the cell membrane* to activate or inactivate *intracellular enzymes.*

In the present case, the active protein kinase phosphorylates *two* enzymes. The first is *glycogen synthase* **a** which is (directly) *inactivated* by the phosphorylation. This is the mechanism which 'switches off' the glycogen synthesis. At the same time the protein kinase (indirectly) *activates glycogen phosphorylase.* This happens 'by proxy'—it first phosphorylates an enzyme called *phosphorylase kinase* **b**. When phosphorylated by protein kinase this enzyme is activated—it is the **a** form which contains the phosphate.

The function of phosphorylase kinase **a** is to activate *glycogen phosphorylase.* This is again a phosphorylation. Two molecules of glycogen phosphorylase **b** (each of which is a dimer) become phosphorylated at the expense of ATP to form the active enzyme phosphorylase **a**, which is a phosphorylated tetramer (Fig. 13.6). The monomers of this enzyme are not themselves

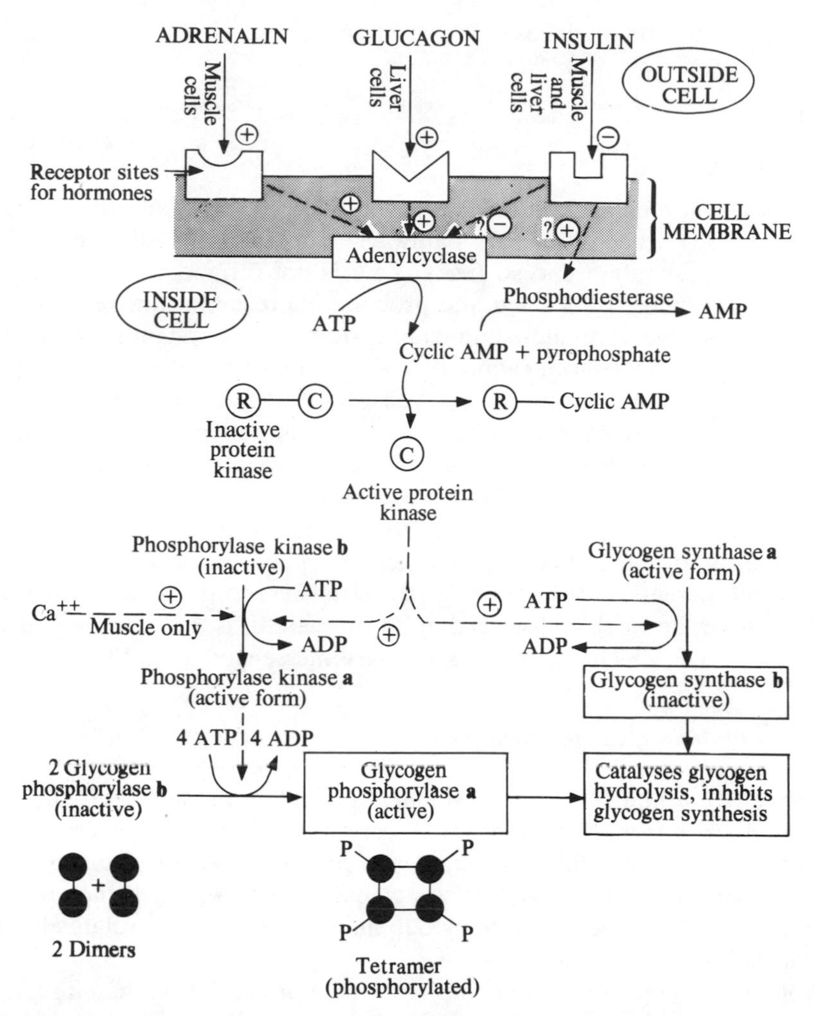

Fig. 13.6 The activation of glycogen phosphorylase and the inhibition of glycogen synthase by adrenalin and glucagon

catalytic; their molecular weight is about 90 000 so the active tetrameric phosphorylase has a molecular weight of about 360 000. Although the details may be complicated it should be appreciated that the end result is that each adrenalin or glucagon molecule reaching the cell surface elicits the activation of numerous phosphorylase molecules, so stimulating the conversion of glycogen to glucose-1-phosphate.

In the case of *liver cells* the glucose-1-phosphate produced in response to glucagon is converted to glucose and passed into the blood, so raising the blood concentration of glucose. The reactions can be summarized as follows:

$$\text{Glycogen (stored in liver cell)} \xrightarrow[\text{P}_i]{\text{Glucagon} \downarrow \text{ Phosphorylase}} \text{glucose-1-P} \xrightarrow{\text{Mutase}} \text{glucose-6-P} \xrightarrow[\text{P}_i]{\text{Glucose-6-P phosphatase}} \text{glucose (passed into blood)}$$

In *muscle*, adrenalin stimulates glycogen phosphorylase and the reactions proceed along the glycolytic pathway. Muscle lacks the glucose-6-phosphate-phosphatase and so *free glucose* is not formed.

In muscles there is a second and probably more important regulation of glycolysis than the adrenalin-stimulated system. It is a system stimulated by an increase in the concentration of *calcium ions*. At rest, nearly all the glycogen phosphorylase in muscle cells is in the inactive **b** form. The contraction of muscles is stimulated by nerve impulses and a complicated series of processes leads to a rise in the *cytosolic calcium concentration* (Chapter 20). Phosphorylase kinase **b** is directly converted to the **a** form when the calcium concentration rises to 10^{-4} mol/litre. This means that, as the muscular contraction begins, glycogen phosphorylase is activated, initiating a great increase in the rate of glycolysis. This results in the synthesis of ATP to drive the work of contraction. In this case, it is the *calcium ions* and not cyclic AMP which act as the secondary messenger (Fig. 13.6).

13.8 Insulin and glycogen synthesis

Historically the interest in the hormones controlling the levels of glucose in the blood arose from a study of the disease *diabetes mellitus*. There are two main types of this condition: patients who produce insufficient or no insulin due to abnormal β cells, and others whose target organs do not respond normally to the hormone although it may be present in relatively high concentrations in the blood.

Probably no hormonal system has received more attention than insulin yet its control of glycogen synthesis is still obscure. There is no doubt that insulin causes *activation of glycogen synthase* by dephosphorylation (**b** → **a**) and that *glycogen phosphorylase is inactivated* (**a** → **b**), thus leading to the synthesis of glycogen and a fall in the blood glucose concentration. There is, however, no clear overall explanation.

An attractive suggestion is that insulin *lowers* the cellular *cyclic AMP* concentration by inactivating the adenyl cyclase or activating phosphodiesterase (an enzyme which converts cyclic AMP to the ordinary form of AMP). Diabetic rats have high levels of cyclic AMP in the livers which would be consistent with the suggestion. There are, however, a number of reports indicating that insulin stimulates glycogen synthase without any change in the level of cyclic AMP.

Another suggestion is that insulin reacts with specific binding sites on the outside of the liver and muscle cells and that this causes the formation of a secondary messenger other than cyclic AMP. This could in turn stimulate a protein phosphatase which would reverse the various kinase reactions shown in Fig. 13.6. Unfortunately no suitable secondary messenger has yet been unequivocally demonstrated. The reader of some elementary text—maybe even ours—will undoubtedly one day win fame by solving this mystery!

13.9 Control of fat metabolism

Although most studies of metabolic regulation in mammals have been directed towards glycolysis and glycogenesis, fat metabolism is probably more important quantitatively. A normal man has enough fat stored in his adipose tissues to supply his energy needs for six weeks, but his glycogen stores are scarcely sufficient for one day of fasting. There must clearly be a system of regulation to ensure that both the storage and utilization of fat and glycogen are balanced to meet the needs of varying dietary conditions and changing energy requirements. For example, during fasting, fatty acids must be used as the main source of energy for the muscles in preference to glucose derived from glycogen; this preference conserves the glucose, which is the principal substrate for metabolism in the brain.

In the well-fed individual, the muscle, liver, and adipose cells take up fats from the blood and hydrolyse them to free fatty acids (FFA) which are reconverted to new triglycerides. FFA are also absorbed and incorporated into the fats of muscle cells (Fig. 10.1). In addition, some FFA are synthesized *de novo* in the liver and adipose tissue from acetyl coenzyme A originating from glycolysis.

Each of these processes is indirectly stimulated by *insulin.* The hormone increases the rate of glucose metabolism by glycolysis, so producing dihydroxyacetone phosphate. This is the precursor of L-α-glycerol phosphate, which must be available for the esterification of the fatty acids (page 209). Furthermore, insulin increases the activity of the *pyruvate dehydrogenase* complex in adipose tissue by a **b** to **a** change involving dephosphorylation of the first component of the complex (Section 11.8). The importance of this is that by stimulating the enzyme, more acetyl coenzyme A is produced which is, of course, the precursor of long-chain fatty acids (Chapter 10).

The major control in the synthesis of *long-chain fatty acids* from acetyl coenzyme A is its conversion to malonyl coenzyme A (Section 10.4). This reaction appears to be the pacemaker since the activity of the fatty acid synthase complex is thought to be limited only by the supply of malonyl coenzyme A and of NADPH. Acetyl coenzyme A carboxylase, which catalyses the rate-limiting step, is specifically activated by *citrate* which acts as a positive allosteric modulator (Fig. 10.6). It seems likely that when the energy requirements of the adipose or liver cell are being met adequately the diffusion of citrate from the mitochondria leads in this way to the stimulation of fatty acid synthesis and to the storage of fats.

The *stored fat* is used to provide energy when the dietary intake of 'fuel' is reduced. Thus it supplies energy during the normal overnight fast. Several hormones are involved in controlling the use of the fat. *Insulin* levels in the blood decrease as the blood concentration of glucose falls. The lower insulin level inhibits fat synthesis, probably by reducing the rate of formation of L-α-glycerol-phosphate needed for the esterification of fatty acids (Section 10.5). Meanwhile the *membrane-bound lipase* in adipose cells is activated by a rise in the blood concentration of *glucagon.* This occurs because glucagon causes an increase in the intracellular concentration of cyclic AMP which then activates protein kinase (as described in Section 13.7). The protein kinase in adipose cells activates the intracellular lipase by inducing a **b** to **a** change similar to that of glycogen phosphorylase in muscles. The activated lipase catalyses the hydrolysis of stored fats releasing FFA into the blood (process 6 in Fig. 10.1). These then provide most of the energy requirements of the muscles during the period of fasting or low dietary intake of energy, so conserving glucose for use by the brain.

A similar activation of the adipose tissue lipase by *adrenalin* also occurs and is probably most important during exercise or the 'fright and flight' response. The lipolysis leads to the release of FFA into the blood and these are taken up by the active muscles to be catabolized immediately to provide ATP. It is interesting that *nor-adrenalin* secreted by various nerves in the adipose tissue may have a similar effect. Stimulation of the nerves leads to a release of FFA from the adipose cells. This is an example of nervous activity stimulating the enzymes involved in 'ordinary' metabolism.

14

Photosynthesis, the pentose phosphate cycle, and the functions of NADPH

The major pathway for the catabolism of carbohydrates in almost all organisms is glycolysis (Chapter 11). There is, however, a second pathway present in most, if not all, organisms; it is known as the pentose phosphate pathway, or sometimes as the hexose monophosphate 'shunt'. A considerable part of this pathway is similar to many of the reactions of the carbon cycle of photosynthesis and it is tempting to imagine this alternative system of metabolism as being evolutionarily primitive. Like the photosynthetic machinery, its functions appear to be closely associated with anabolic processes. Perhaps for this reason it tends to be most highly developed in certain specialized cells which have particular anabolic functions.

14.1 The overall process of photosynthesis

Many of the earlier theories about the remarkable process which ultimately supports all life, the reduction of carbon dioxide to sugar by green plants and by certain photosynthetic bacteria, postulated various types of reactions which had little or no connection with any other facet of biochemistry. Later advances, made possible by the introduction of radio isotopes and chromatography, have however shown that photosynthesis fits into the same framework as other known biochemical processes. In particular, the reduction of carbon dioxide is now known to be brought about by the same two driving forces encountered in other anabolic processes—ATP and reducing power. It is the *origin* of these energy-yielding substances, and not their chemical nature, which is different in photosynthesis in that they are formed at the expense of the energy of sunlight.

The reactions of photosynthesis can conveniently be divided into two groups:

1. The conversion of light energy into chemical energy
2. The use of that chemical energy to reduce carbon dioxide

The sum total of all the reactions of photosynthesis is often represented by the following equation:

$$6CO_2 + 6H_2O + \text{light} \xrightarrow[\text{Enzymes}]{\text{Chlorophyll}} C_6H_{12}O_6 + 6O_2$$

This overall equation gives no clue as to how photosynthesis actually works. But the advent of radioactive carbon (^{14}C) made it possible to label the carbon of carbon dioxide and so to investigate which cellular constituents became radioactive after plants had been exposed to labelled carbon dioxide for very short periods of time. Much of the work was carried out by M. Calvin and his team at Berkeley, California—the carbon cycle of photosynthesis is consequently sometimes known as the Calvin cycle. The principal test organisms were the algae *Chlorella* and *Scenedesmus*, although leaves of higher plants were also employed. When either genus of unicellular algae was employed, the suspension of $^{14}CO_2$-treated cells was subjected to exposure to light for a few seconds, and then plunged into boiling absolute alcohol to terminate the reaction. After concentrating by low-temperature evaporation and after clean-up procedures which included centrifugation and fat extraction, the remaining aqueous layer was investigated by paper chromatography in order to identify the labelled compounds produced.

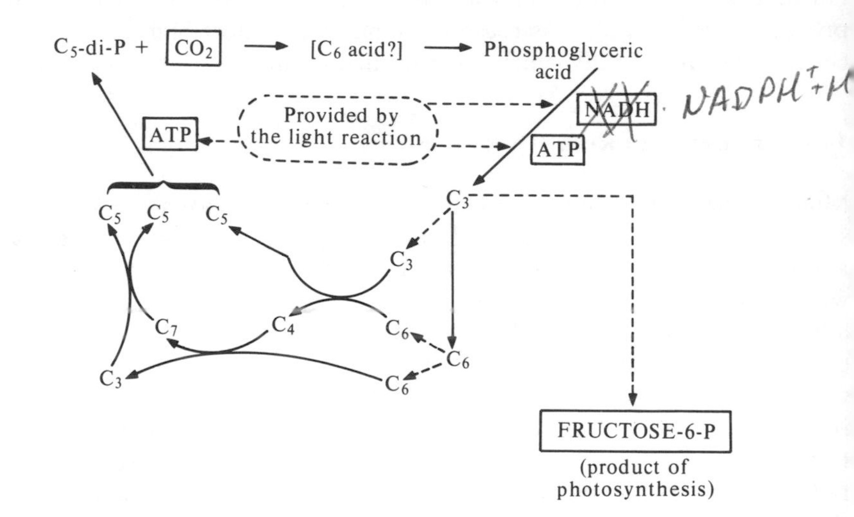

Fig. 14.1 The carbon cycle of photosynthesis in outline. [C_3, C_4, C_5, C_6, C_7 are monophosphorylated sugars; C_5-di-P, ribulose diphosphate; square brackets indicate compounds entering the carbon cycle and (lower right) leaving it. The cycle works out statistically, so stoichiometric relationships are not shown: the dotted arrows indicate that some part of the material C_3 or C_6 takes the pathway shown]

Experiments of this type revealed much of what we know about the main carbon cycle of photosynthesis (at least in some plants, alternative pathways exist). It consists of some twelve separate enzyme-catalysed reactions, each of which can occur in darkness. Clearly, therefore, the contribution of light to these syntheses must be vicarious. In fact, as will be shown in Section 14.5, the reactions of the cycle can be carried out individually in the laboratory, provided the appropriate enzyme and, where applicable, suitable cofactors and an energy supply are present. In its simplest form, the cycle can be summarized in the way shown in Fig. 14.1 where C_3, C_4, C_5, etc., represent sugar phosphates containing three, four, five, etc., carbon atoms. The contribution of the light reaction is to provide energy to drive the cycle; ATP raises the energy of reactants by phosphorylation at two places in the cycle, and reducing power, in the form of NADPH, is involved in one step, namely, the reduction of phosphoglyceric acid to triose phosphate (C_3). The light reaction is the only part of the total photosynthetic process which may have any claim to mechanistic uniqueness.

14.2 Chloroplasts

Photosynthesis takes place in cell organelles called *chloroplasts* which are present in cells of green, red, and brown algae, and in many of the cells in the leaves of higher plants. Prokaryotic cells which photosynthesize do not have differentiated chloroplasts. It has been estimated that, worldwide, the fixation of carbon dioxide by photosynthesis leads to the formation of about 10^{11} tons of organic products a year.

Chloroplasts are usually larger than mitochondria and there may be one or many of them in eukaryotic cells which photosynthesize. They contain DNA and RNA, can synthesize some proteins, and are capable of self-replication. Their structure is in some ways reminiscent of that of mitochondria, in that they possess an outer lipoprotein membrane enclosing an inner membrane system and an aqueous chloroplastic matrix (Fig. 14.2). In higher plants, but not universally throughout the plant kingdom, the double lipoprotein membranes of the inner system come together in some areas of the chloroplast and separate again in others. Since chlorophyll is associated with the lipoprotein of the inner membranes, this has the effect of producing regions of greater concentration of chlorophyll, where the double membranes, or *lamellae*, are arranged on top of one another like a pile of hollow coins. These regions are called *grana*, the hollow areas between the lamellae are called *thylakoid vesicles*, and the lamellae forming the wall of the vesicles (i.e., the outside of the hollow coin) are the thylakoid membranes. The enzymes responsible for capturing and transducing light energy are associated with the thylakoids. The areas between the grana are crossed by less densely arranged membranes and filled with a fluid containing the

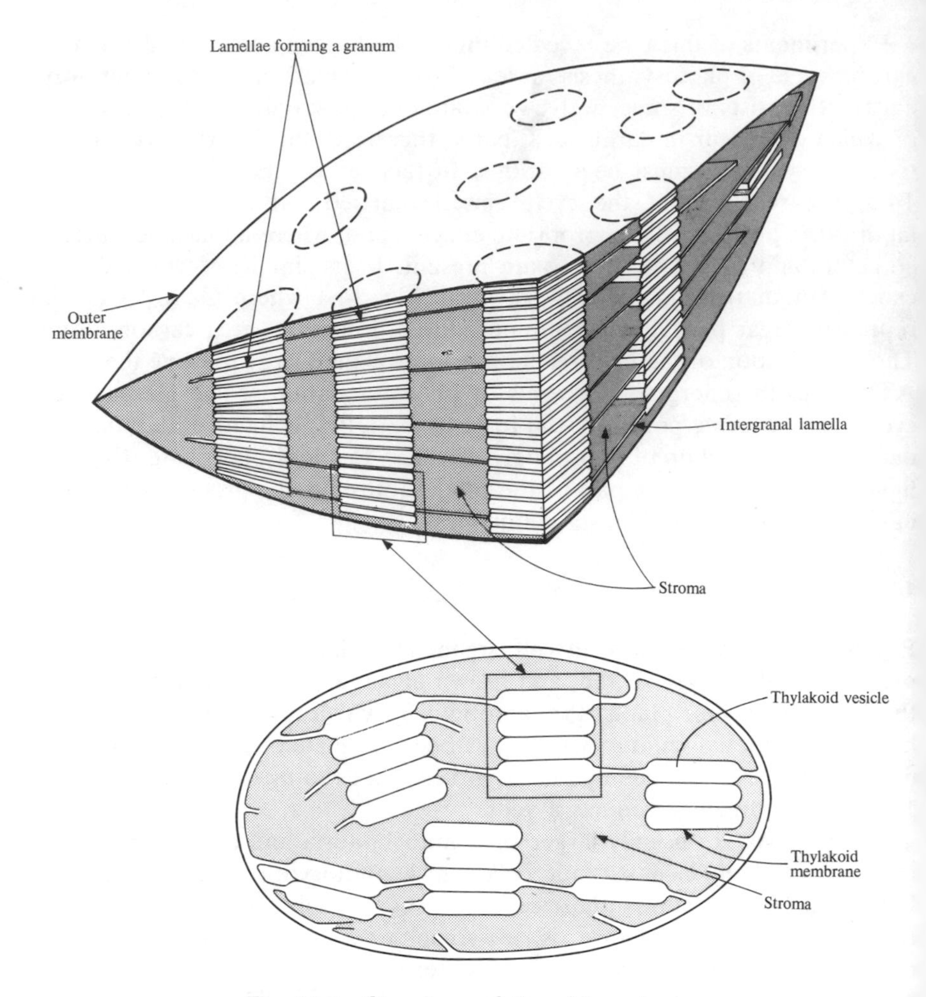

Fig. 14.2 Structure of the chloroplast

water-soluble enzymes responsible for the 'dark' reactions of the Calvin cycle. This aqueous region is known as the *stroma*.

Chloroplasts can be isolated from photosynthetic eukaryotic cells by homogenizing them in 0.3 *M* sucrose solution, removing cell wall debris by filtering through muslin, and sedimenting the chloroplasts by centrifuging at 1200*g* for 10 min (Chapter 1). It is also possible, by further disruption, to separate grana from stroma. It is found that, under appropriate experimental conditions, the isolated grana can carry out the functions of the 'light' reaction whereas the stroma, in the absence of light, are able to carry out all the reactions of the photosynthetic carbon cycle.

Chlorophyll **a** (Fig. 14.3) is almost universally present in chloroplasts of plants in all phyla of the plant kingdom, but the accessory pigments which accompany it vary considerably. Slightly different chlorophylls, designated **b**, **c**, and **d**, have been recognized in various green plants, as have several carotenoids and hydroxylated carotenoids. All the chlorophylls have a porphyrin-like structure with a central magnesium ion. When chloroplasts are ashed, 1 per cent of the residue is iron; magnesium, copper, and manganese are also present. In intact chloroplasts, the iron occurs, in part, as non-haem iron protein, but some is also complexed in haem groups of cytochromes $\mathbf{b}_6$ and **f**. These cytochromes, which are different from those concerned with electron transport in mitochondria, take part in the transport of photoactivated electrons.

In photosynthetic bacteria, the situation is somewhat different, for they lack discrete chloroplasts just as they lack discrete mitochondria (Chapter 1). Furthermore, chlorophyll **a** is absent; instead, they contain a slightly different pigment, bacteriochlorophyll (of which at least two variants exist). The difference in pigment is probably related to the fact that most photosynthetic bacteria produce no oxygen. The reason for this is that, whereas in higher plants *water* is the ultimate electron donor (Section 14.4), in various species of bacteria water is replaced by a variety of electron donors (e.g.,

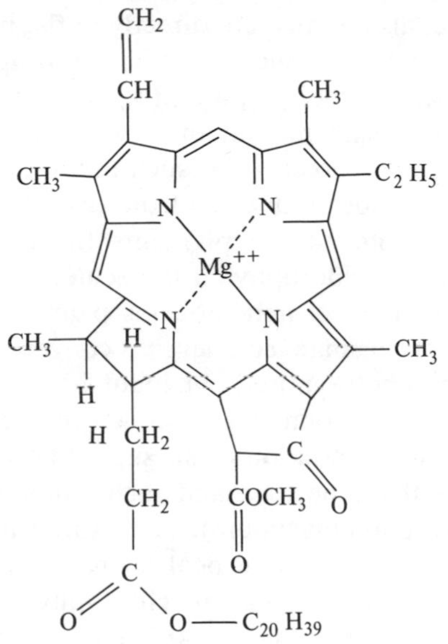

Chlorophyll **a**
(compare with structure
of protohaem, Fig 8.5)

Fig. 14.3 Structure of chlorophyll **a**

hydrogen sulphide, organic acids, isopropanol). When hydrogen sulphide is the electron donor, the overall equation for photosynthesis is

$$6CO_2 + 12H_2S \xrightarrow{\text{Light}} C_6H_{12}O_6 + 6H_2O + 12S$$

This can be compared with the reaction for photosynthesis in green plants:

$$6CO_2 + 12H_2O \xrightarrow{\text{Light}} C_6H_{12}O_6 + 6H_2O + 6O_2$$

14.3 Role of the photosynthetic pigments

In photosynthetic eukariotic cells, chlorophyll **a** is the principal light-trapping pigment. It is, however, always accompanied by other pigments, including other chlorophylls, carotenoids, and phycobilins. In common with most dyestuffs, all of these compounds contain systems of alternating double bonds, a characteristic which accounts for their colour and probably also for their photoreceptor capability. The carotenoids and phycobilins probably shield the photochemical machinery from damage caused by strong oxidizing agents arising directly or indirectly from exposure to high light intensities. However, some of these accessory pigments, and especially the widely distributed β-carotene, also participate directly in the light-trapping process, for their absorption spectra are such that they trap light at wavelengths at which the two principal chlorophylls, **a** and **b**, are not particularly effective. They are still 'accessory' pigments, for once they have trapped light energy the arrangement of pigment molecules is such that the energy is passed on to chlorophyll **a** molecules. Indeed, in one of the two photosystems soon to be described, the energy from all the pigments (including chlorophyll **a**) is channelled to a specialized chlorophyll **a** molecule, with properties different from those of some hundreds of neighbouring pigment molecules—a sort of reaction centre where the channelled energy is converted to chemical energy

When isolated, each photosynthetic pigment has a characteristic absorption spectrum. That for chlorophyll **a** is shown in Fig. 14.4. (In the intact chloroplast a more complex situation exists since chlorophyll **a** can be bound or aggregated in more than one way, and such binding alters somewhat the wavelength of the absorption maximum). The figure shows that the spectrum of chlorophyll **a** comprises two main peaks, one at each end of the visible spectrum. The absorption maxima of chlorophyll **b** overlap those of chlorophyll **a** but are closer together ($\sim$480 and 645 nm), while β-carotene absorbs in the region 420 to 520 nm.

If chlorophyll **a**, chlorophyll **b**, and β-carotene collectively trap most of the light energy used for photosynthesis, the sums of the absorptions of these pigments at each wavelength should give a composite absorption curve

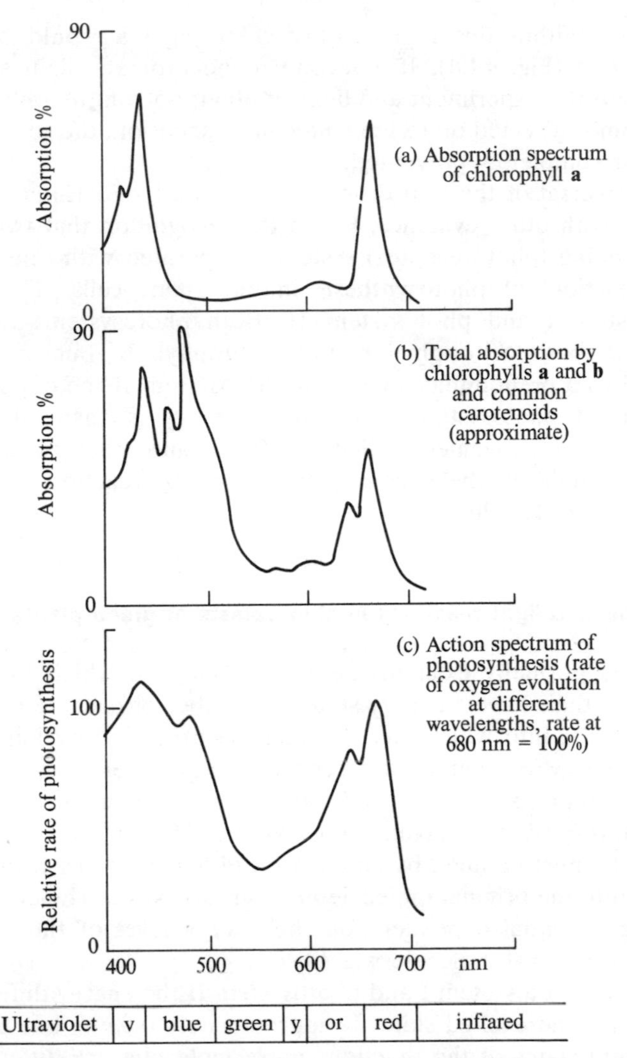

Fig. 14.4 **Absorption spectra of leaf pigments and the relative rate of photosynthesis at different wavelengths**

which roughly parallels the efficiency of photosynthesis at each wavelength (using monochromatic light). In fact, the action spectrum of photosynthesis, which measures the relative rate of photosynthesis at different wavelengths, does indeed rise at both ends of the spectrum and fall in the middle.

When unicellular algae such as *Chlorella* are exposed to monochromatic light of successively increasing wavelength within the narrow far-red spectral region of 670 to 700 nm, a sudden drop of photosynthetic efficiency is observed. Within this region, only chlorophyll **a** would be expected to absorb light (Fig. 14.4). If a second monochromatic light source is introduced into the experiment and light of about 650 nm *as well as* light of 680 to 700 nm is directed on to the *Chlorella* suspension, the so-called 'red drop' is almost completely prevented.

This reversal of the 'red drop', first noticed by R. Emerson about 1958, together with other evidence, led to the recognition that two quite distinct light-trapping (photoreceptor) systems cooperated with one another in the 'light reaction' of photosynthesis in eukaryotic cells. These are termed photosystem I and photosystem II. Both photosystems usually contain, *inter alia*, both chlorophyll **a** and chlorophyll **b**, but in photosystem I chlorophyll **a** predominates while in photosystem II the opposite applies. In the case of photosystem I, the light energy is channelled into special chlorophyll **a** molecules, called P700, because the pigment absorbs at 700 nm. Similarly, the special pigment in the reaction centre of photosystem II is called P680.

14.4 The two light reactions in chloroplasts of green plants

Molecules normally exist in the *ground state*, in which the electrons are present in their lowest energy state. A number of discrete orbital states are also possible, into which electrons can be propelled by impact of units of radiant energy, or quanta, of precisely the right energy level. The molecules are then said to be in an *excited state*, from which an electron with reducing potential may be lost, leaving a positive 'hole' in the original molecule which eventually must be filled by an electron of lower energy from elsewhere, to reconstitute the original molecule in its ground state. The electron lost has a reducing potential dependent on the energy level of the excited state to which the original molecule was raised.

For both photosystem I and photosystem II the energy difference between the ground and excited states is approximately 0.9 eV, but the positions of the ground states of the reaction centre molecules are different in the two systems. In photosystem I, the ground state of P700 is approximately $E_0' = 0.4$ V and the electron emitted on excitation by light of 700 nm has an energy of approximately -0.45 V. This is more than enough energy to allow the electron to reduce $NADP^+$ to NADPH ($E_0' \simeq -0.3$ V). In other words, photosystem I, upon excitation, splits to give a *strong reducing agent but a weak oxidizing agent.* This is illustrated by the positions of P700 and its emitted electron in Fig. 14.5.

The ground state of the molecule in the reaction centre of photosystem II is much more positive than that of photosystem I. Upon excitation by light of 680 nm an electron with a potential of about -0.05 V is emitted, leaving

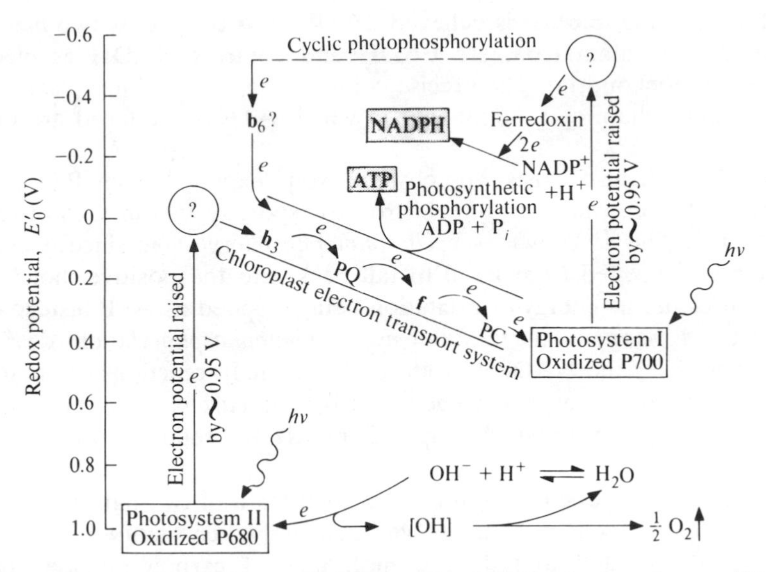

Fig. 14.5 The light reactions of photosynthesis (PQ, plastoquinone; PC, plastocyanin; hv, quantum of light).

a positive 'hole' with an electrode potential of $E_0' \doteqdot 1.0$ V. In other words, photosystem II, upon excitation, splits to give a rather weak reducing agent but a strong oxidizing agent (compare the redox scale, Table 7.3).

Figure 14.5 illustrates the way it is believed the two light reactions in green plants are interrelated. The electron with high reducing power produced by photosystem I directly or indirectly reduces a non-haem iron protein, ferredoxin. This electron carrier has a very highly negative standard redox potential and is able to reduce $NADP^+$ to NADPH. Light of shorter wavelength causes the emission of an electron from P680 which reduces an as yet unknown electron acceptor. The latter passes its electron to a series of carriers participating in a chloroplastic electron transport chain, from which it eventually emerges to reduce the positive 'hole' of activated P700 so as to return P700 to its ground state. Finally, the positive 'hole' of pigment P680 of photosystem II is filled when this highly oxidized ion, oxidizes water to oxygen, electrons, and hydrogen ions. The mechanism of this last reaction is unknown.

The chloroplastic electron transport chain comprises a series of carriers probably including cytochromes **f**, $\mathbf{b}_3$, and $\mathbf{b}_6$, which are only found in plants. Other identified carriers are plastoquinone (compare ubiquinone in mitochondria, Section 8.5) and plastocyanin (which is a protein containing copper). During the passage of electrons down this redox gradient some ATP is produced, a situation reminiscent of oxidative phosphorylation in mitochondria (Fig. 8.6). It is believed 2ATP are produced as two electrons pass down the chain (compare FADH and contrast NADH as electron donors in mitochondria). The precise composition of the chloroplastic electron transport chain and the manner in which ATP is produced are uncertain.

A mechanism also exists whereby electrons expelled from P700, upon photon excitation, can enter the electron transport system instead of reducing $NADP^+$. This is known as *cyclic photophosphorylation*, since it enables the electron expelled from P700 to fall back into the positive 'hole' from whence it came, its energy of excitation being trapped as ATP instead of as NADPH, as happens in the *non-cyclic photophosphorylation* described above. The ATP and NADPH produced by the light reactions of photosynthesis provide all the energy needed to drive the *carbon cycle*. The latter is catalysed by enzymes in the chloroplastic matrix and reduces carbon dioxide to carbohydrates.

There is an energetic reason for the participation of two light reactions in the total 'light reaction' of photosynthesis in higher plants. Twenty-four electrons are needed to reduce 6 molecules of carbon dioxide, as the following equation shows:

$$6CO_2 + 24H^+ + 24e \rightarrow C_6H_{12}O_6 + 6H_2O$$

Consequently, the minimal quantity of energy needed per equivalent of reducing power to reduce 6 mol of carbon dioxide to 1 mol of sugar is

$$\frac{\Delta G^\circ \text{ for glucose formation}}{24} = \frac{2870}{24} = 120 \text{ kJ/mol}$$

Light with a wavelength of 700 nm is associated with 169 kJ per 6×10^{23} photons, and so, if only one photon energized one electron, the conversion of light energy into reducing energy would need to operate at an efficiency of 120/169, or 71 per cent. This high level of efficiency is not achieved in cellular processes (compare, for example, the efficiency of energy capture during oxidative phosphorylation, Section 8.9), and, in somewhat different ways, two photons contribute energy to drive the carbon dioxide reduction process.

14.5 The photosynthetic carbon cycle

Unlike the light reaction, the mechanism whereby carbon dioxide is converted to sugars, and eventually to starch, fats, proteins, and nucleic acids, is for the most part well established. Only one reaction in step 1 of the sequence below is rather dubious. The carbon cycle can be divided into three steps in the following way:

Step 1. The fixation of carbon dioxide, a process which converts a five-carbon sugar, *ribulose-5-phosphate*, to two molecules of the three-carbon compound, 3-phosphoglyceric acid.

Step 2. A series of reactions, essentially the same as some of those of glycolysis operating in the reverse direction, which converts the 3-phosphoglyceric acid to fructose-6-phosphate.

Step 3. A series of sugar phosphate transformations, similar to those of the pentose phosphate pathway, regenerating ribulose-5-phosphate from the fructose-6-phosphate.

Step 1. Carbon dioxide fixation

The most convenient point at which to start considering the carbon cycle is with the reaction in which the ketopentose phosphate, ribulose-5-phosphate, reacts with ATP. The reaction is comparable to other kinase reactions, including that which leads from fructose-6-phosphate to fructose-1,6-diphosphate (Section 11.4). The enzyme *phosphoribokinase* transfers a phosphate group from the high energy donor, ATP, and places it upon carbon atom 1 of ribulose-5-phosphate (Fig. 14.6). The reaction is entirely 'conventional', only the *origin* of the ATP possibly being different from usual, for, under good lighting conditions, the ATP in plant chloroplasts largely arises from photosynthetic phosphorylation and not from oxidative phosphorylation. High concentrations of AMP inhibit the enzyme. This appears to be important in the regulation of carbon dioxide fixation since it ensures that fixation only occurs maximally when the [ATP]/[AMP] ratio is high.

We come now to the reaction in the carbon cycle which results in the 'fixation' of carbon dioxide. 'Fixation', or the union of carbon dioxide with a preexisting organic molecule, is, in itself, not unusual; carboxylation reactions dependent upon biotin as cofactor have already been referred to (Fig. 10.5), and many different carbon dioxide acceptors are known in both plants and animals. But in the particular case of photosynthesis it is the ribulose-1,5-diphosphate formed by the kinase reaction which acts as *the acceptor of the carbon dioxide.* The first identifiable product of this 'fixation' reaction is *3-phosphoglyceric acid*, but it is possible that the acid indicated in brackets in Fig. 14.6 is a transient intermediate. The reaction is not fully understood and is catalysed by an enzyme complex, known simply as 'carboxylation enzyme', or more precisely as ribulose diphosphate carboxylase.

Step 2. Conversion of 3-phosphoglyceric acid to fructose-6-phosphate

Each of the reactions in this step has been considered before, a fact which again illustrates a general simplifying feature of biochemistry. The cell shows

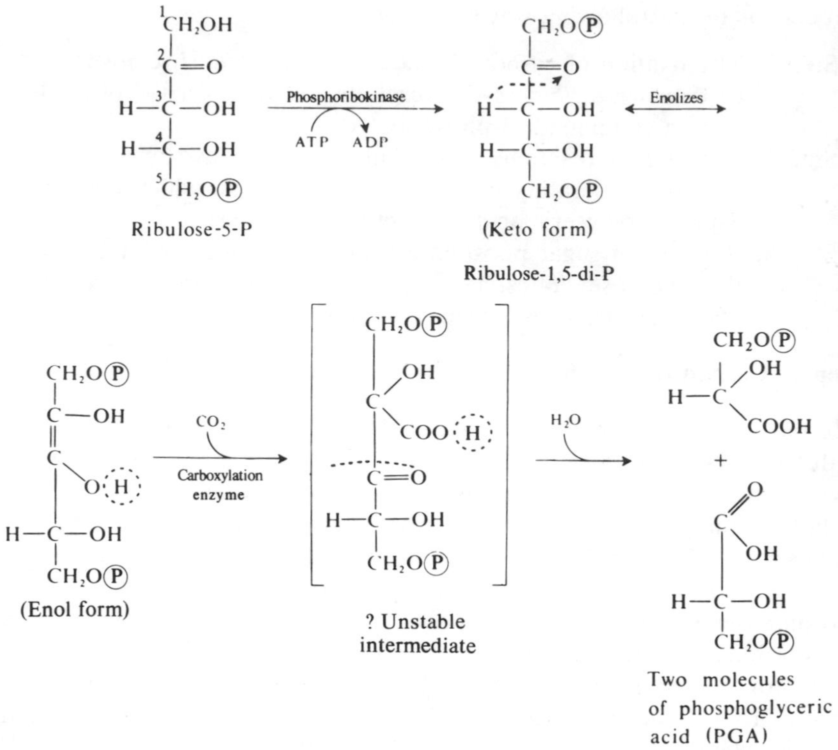

Ribulose-5-P

Phosphoribokinase

ATP ADP

(Keto form)

Ribulose-1,5-di-P

Enolizes

(Enol form)

CO_2

Carboxylation enzyme

? Unstable intermediate

H_2O

Two molecules of phosphoglyceric acid (PGA)

Fig. 14.6 Step 1 of the photosynthetic carbon cycle: the fixation of carbon dioxide

great economy in that it often uses the same set of reactions for more than one purpose. Thus the 3-phosphoglyceric acid is converted to *fructose-1,6-diphosphate* by the reactions of stages 2 and 3 of glycolysis (Sections 11.5 and 11.6). There are only two differences: the reaction proceeds in the reverse *direction* from that of glycolysis in this chloroplast-located reaction, and the chloroplast enzyme, phosphoglyceraldehyde dehydrogenase, uses NADP as cofactor instead of NAD. Since, in this instance, 1,3-diphosphoglyceric acid is reduced to glyceraldehyde-3-phosphate, the reaction requires a supply of *reduced* NADP. NADPH is, however, a product of the light reaction and so the reduction is able to proceed very readily. The involvement of the $NADP^+$/NADPH system rather than NAD is quite a common feature of *anabolic* as opposed to catabolic reaction sequences.

The fructose-1,6-diphosphate produced by the reactions mentioned above is dephosphorylated to form fructose-6-phosphate, the initial reactant of step 3 of the carbon cycle. The dephosphorylation is catalysed by *fructose diphosphate phosphatase.* This enzyme, and not phosphofructokinase, takes part in the reaction for the energetic reasons discussed in Section 12.4.

The fructose-6-phosphate has many possible fates including conversion to sucrose and other carbohydrates (Fig. 12.12), but much of it must be fed back into the cycle to regenerate ribulose-5-phosphate so that step 1 of the cycle can take place again. This regeneration involves a series of interconversions of sugar phosphates and these together form step 3 of the carbon cycle.

Step 3. The interconversion of fructose-6-phosphate and ribulose-5-phosphate

All the reactions about to be considered are freely reversible, allowing fructose-6-phosphate to be converted into ribulose-5-phosphate, as occurs in photosynthesis, but equally allowing ribulose-5-phosphate to be converted to fructose-6-phosphate. The enzymes catalysing the reactions are water soluble; they are present in the soluble fraction of the cell and in the aqueous phase of the chloroplasts.

It is best not to consider these reactions from a stoichiometric viewpoint; the direction of change and the quantity of sugar phosphate produced by any particular reaction is determined by environmental factors such as the nature and quantity of the sugar phosphate fed into the cycle and by the cellular demand for individual sugar phosphates. Such factors result in more or less of the products of any one reaction being 'siphoned off' at any time. In other words, the best mental picture of this process is of a group of reactions feeding one another with reactants or removing products, and therefore proceeding at rates determined by mass action considerations. The result is that the system settles down to a steady state (page 126). The *statistical* outcome, in *photosynthesis*, is that the removal of ribulose-5-phosphate by step 1 of photosynthesis draws the sugar transformation reactions in the direction of ribulose-5-phosphate synthesis, whereas, in the pentose phosphate oxidative pathway, the reverse conditions drive the cycle in the direction of fructose-6-phosphate synthesis.

In following the reactions about to be discussed, the reader is advised to refer to the general schemes shown in Fig. 14.1 and (in more detail) in Fig. 14.7. It should be observed that all the reactions involve sugar phosphates in which the phosphate group is attached to the terminal, highest-numbered, carbon atom. The main enzymes concerned, *transketolase* and *transaldolase*, dislodge respectively a two-carbon unit and a three-carbon unit from a *ketone* sugar and place it on the carbonyl group of an *aldehyde* sugar.

Transketolase first removes a two-carbon fragment from fructose-6-phosphate and attaches it to glyceraldehyde-3-phosphate (Fig. 14.7). The reaction, like that already encountered for aldolase (Fig. 11.3), can be

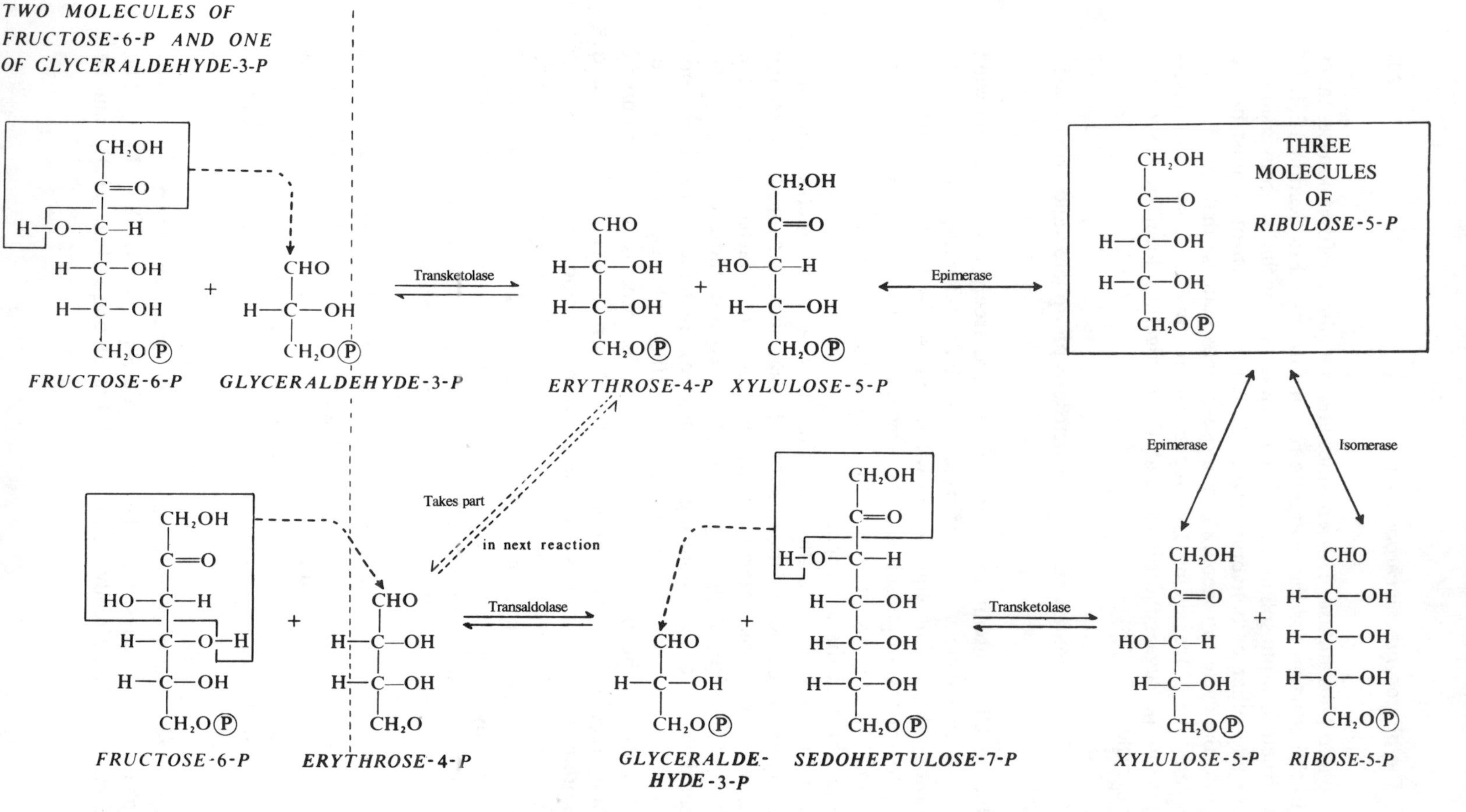

Fig. 14.7 The interconversion of fructose-6-phosphate and ribulose-5-phosphate. (The sequence of reactions from left to right represents step 3 of the carbon cycle of photosynthesis. In the reverse direction, it forms step 3 of the pentose

regarded as one in which addition occurs on to the carbonyl double bond of the glyceraldehyde-3-phosphate. Figure 14.7 demonstrates how the reaction may be envisaged, although it does not imply any particular mechanism for the reaction *in vivo*. In the present case, the products of the reaction are erythrose-4-phosphate and xylulose-5-phosphate:

$$\underset{\text{Ketose}}{\overset{\text{Minus 2C}}{C_6}} + \underset{\text{Aldose}}{C_3} \xrightleftharpoons{\text{Transketolase}} \underset{\substack{\text{Residue}\\\text{(an aldose)}}}{C_4} + \underset{\substack{\text{Addition}\\\text{product}\\\text{(a ketose)}}}{C_5}$$

The active centre of transketolase contains thiamine pyrophosphate (TPP) and it is to this that the two-carbon fragment is temporarily attached.

In the presence of an *epimerase*, the xylulose-5-phosphate is converted to another ketopentose, ribulose-5-phosphate; this involves interchanging the positions of the hydrogen atom and the hydroxyl group around one of the carbon atoms.

Next, the erythrose residue from the transketolase reaction accepts a three-carbon fraction from fructose-6-phosphate. This results in the formation of a triose phosphate and a sugar phosphate with a molecule containing seven carbon atoms. It is called sedoheptulose-7-phosphate, and the enzyme is a *transaldolase*. It is easy to remember that the enzyme catalysing the transfer of a three-carbon unit is transaldolase, by comparison with the aldolase reaction (page 220). The transaldolase reaction is represented in Fig. 14.7 and can be summarized by the equation:

$$\underset{\text{Ketose}}{\overset{\text{Minus 3C}}{C_6}} + \underset{\text{Aldose}}{C_4} \xrightleftharpoons{\text{Transketolase}} \underset{\substack{\text{Residue}\\\text{(an aldose)}}}{C_3} + \underset{\substack{\text{Addition}\\\text{product}\\\text{(a ketose)}}}{C_7}$$

It should be observed that in this and all other reactions involving sugar phosphates for which transaldolases or transketolases act as catalysts, the aldehyde acceptor of the transferred group becomes a ketone, while the ketone donor of that group becomes an aldehyde—and thus ready for the next step in the reaction sequence. On this occasion and in the presence of transketolase, sedoheptulose-7-phosphate reacts with glyceraldehyde-3-phosphate. By the transfer of a two-carbon unit, the sedoheptulose-7-phosphate is converted to ribose-5-phosphate, while glyceraldehyde-3-phosphate is changed to xylulose-5-phosphate. Specific enzymes then change both pentose phosphates to ribulose-5-phosphate (Fig. 14.7). All these reactions, working together, have therefore produced 3

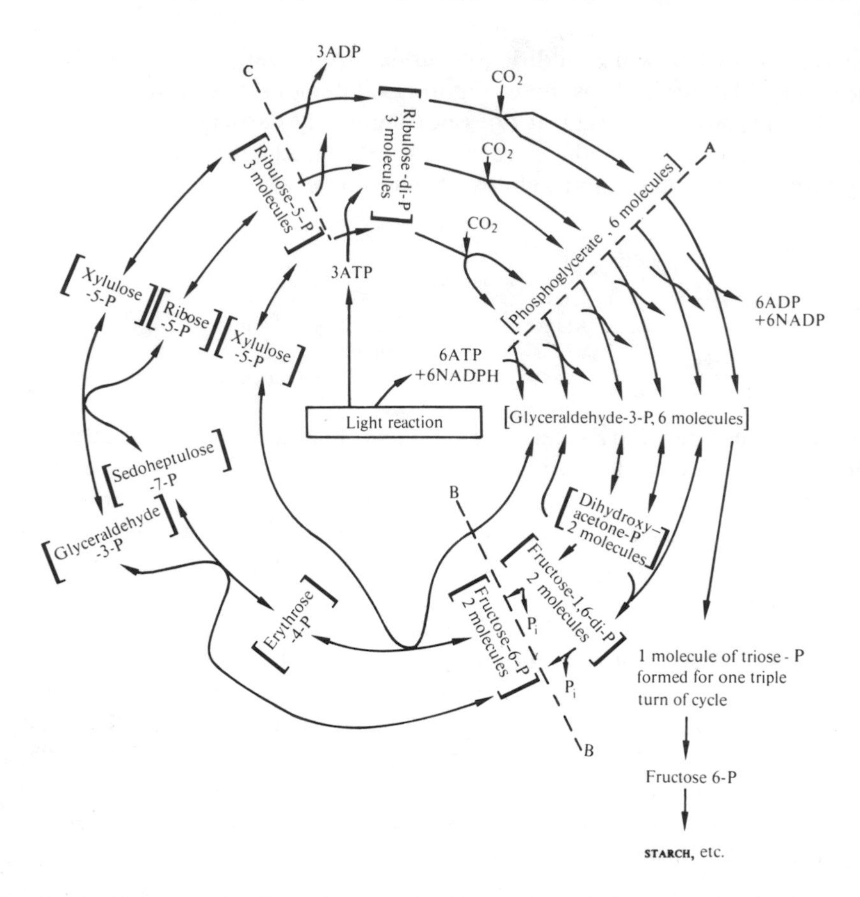

Fig. 14.8 Photosynthetic carbon cycle. From C to A is stage 1 of photosynthesis. From A to B resembles the glycolytic pathway working in reverse. The part of the cycle B to C represents the third step of the carboncycle and from C to B (anticlockwise) the third step of the pentose phosphate pathway.

molecules of ribulose-5-phosphate; these are converted to ribulose-1,5-diphosphate in the first step of the next turn of the cycle.

The statistical result of these equilibrated transformations, shown in Fig. 14.8, is that, for 3 molecules of carbon dioxide fed into the cycle by reaction with 3 molecules of ribulose-1,5-diphosphate, *3 molecules of ribulose diphosphate and 1 molecule of triose phosphate* are produced:

$$3C_5 + 3C_1 \rightarrow 3C_5 + 1C_3$$

Since 2 molecules of triose phosphate can unite to give fructose-1,6-

diphosphate (the aldolase reaction), this equation can also be written in the following way:

$$6C_5 + 6C_1 \rightarrow 6C_5 + 1C_6$$

It must, however, be remembered that *any* of the intermediate sugar phosphates can equally well be 'siphoned off' if a cellular demand for any one of them exists. For the production of nucleic acids, for example, pentoses are needed, and in this case the equation would read:

$$5C_5 + 5C_1 \rightarrow 5C_5 + 1C_5$$

For each molecule of carbon dioxide fixed in step 1 of the carbon cycle *one ATP molecule* is required for the phosphoribokinase reaction. In addition, *2 molecules of ATP and 2 of NADPH* are required in step 2 to convert the 2 molecules of phosphoglyceric acid, formed in step 1, to glyceraldehyde-3-phosphate. The ATP and NADPH are produced by the light reaction in the living plant, although non-photosynthetic sources of these compounds can be used in *in vitro* studies, showing that, given these sources of energy, the whole carbon cycle can be carried out in darkness. Some of the ATP could arise from oxidative phosphorylation *in vivo*, but since photosynthesis can take place under good lighting conditions at twenty times the rate of respiration, the respiratory contribution is normally small.

14.6 The pentose phosphate, or hexose monophosphate, pathway

The principal catabolic route for carbohydrates is the glycolysis pathway, followed, under aerobic conditions, by the TCA pathway. The glycolytic enzymes are present in the soluble fraction of the cell but the route from pyruvate is located in the mitochondria.

There is, however, an alternative route by which the oxidation of carbohydrates can take place. Quantitatively, it usually accounts for a somewhat small percentage of the oxygen taken up by the cell but, as will be seen later, it serves a qualitatively different purpose from the main pathway, and at least one of the products, NADPH, is of vital importance in cell anabolism. The enzymes concerned are present in the cytosol; it should be noted that the mitochondrion is not directly involved in any part of the cycle. The pentose phosphate pathway consists of two oxidative steps, followed by the same series of interconversions of sugar phosphates encountered as step 3 of photosynthesis.

The pathway is also known as the hexose monophosphate pathway (or 'shunt') because the initial reactant is *glucose-6-phosphate.* As in glycolysis this substance can be formed from glucose by the hexokinase reaction, or from starch or glycogen by the action of phosphorylase and phosphoglucomutase (Chapter 11).

Step 1. Oxidation of glucose-6-phosphate to 6-phosphogluconic acid

In this process, the secondary alcohol grouping associated with carbon atom 1 of glucopyranose-6-phosphate is oxidized to a carbonyl group (Fig. 14.9). Because of the proximity of the pyrane ring's oxygen atom, this oxidation has an interesting consequence—the sugar phosphate is oxidized to the oxidation level of a carboxylic acid, yet the six-membered ring structure is nevertheless retained. In other words, the oxidation produces an internal ester, formed between a carboxyl group and a hydroxyl group in the *same* molecule; such an internal ester is called a *lactone.* The enzyme catalysing the production of the lactone form of 6-phosphogluconic acid is *glucose-6-phosphate dehydrogenase,* and a very important feature of this enzyme is that its cofactor is NADP. In this respect it differs from the enzymes of glycolysis and the TCA cycle, for almost all the dehydrogenases of these pathways are dependent on NAD. Consequently, the glucose-6-phosphate dehydrogenase reaction, together with the reaction of step 2 below, provides the cell with *most of its NADPH.*

The dehydrogenase reaction just described is reversible *in vitro,* but *in vivo* the oxidation is followed at once by a hydrolytic reaction catalysed by *lactonase.* By opening the ring system to give the straight-chain form of 6-phosphogluconic acid (Fig. 14.9), the position of the equilibrium is pulled towards the right.

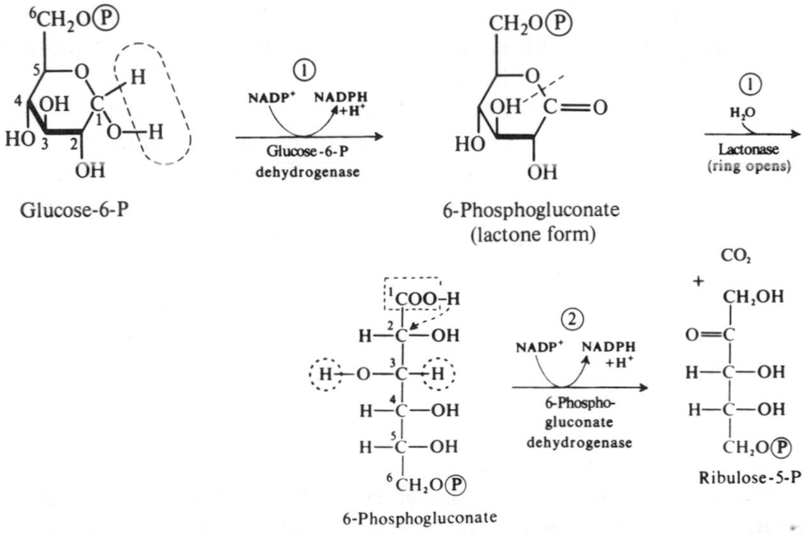

Fig. 14.9 Steps 1 and 2 of the pentose phosphate pathway. (Note that in both of the steps NADP is reduced to NADPH)

Step 2. Oxidative decarboxylation of 6-phosphogluconate

In this reaction the enzyme *6-phosphogluconate dehydrogenase* brings about oxidation of the secondary alcohol group on carbon atom 3 of the 6-phosphogluconic acid molecule at the same time as the elements of carbon dioxide are removed from the unphosphorylated end of the molecule. Consequently, as the original carbon atom 1 is released as carbon dioxide, the third carbon atom of the original molecule becomes the ketone carbon atom of a ketopentose (Fig. 14.9). This ketopentose is *ribulose-5-phosphate*, and, since 6-phosphogluconate dehydrogenase is NADP-linked, its formation is coupled with the reduction of $NADP^+$ to NADPH.

Thus, in these two steps of the pentose phosphate pathway, glucose-6-phosphate is converted to ribulose-5-phosphate with the simultaneous reduction, outside the mitochondrion, of two molecules of $NADP^+$ to NADPH. It is of interest to note that it is carbon atom 1 of the original glucose molecule which appears as (respiratory) carbon dioxide when its oxidation follows this pathway, whereas, in the glycolytic pathway, the carbon dioxide released when pyruvate is decarboxylated arises from carbon atoms 3 and 4 of the original hexose (page 228). Glucose labelled either at carbon atom 1 or at carbon atom 3 has been used to distinguish between the two routes and so to determine the contribution of each route to the metabolism in any particular cell or tissue. Although this technique has provided useful information, its overall reliability has been questioned.

Step 3. Interconversion of ribulose-5-phosphate and glucose-6-phosphate

The ribulose-5-phosphate produced by the first two steps of the pathway may be the precursor of ribose and deoxyribose. However, for the pentose pathway to be a complete cycle, much of the ribulose-5-phosphate must be converted back to glucose-6-phosphate.

Apart from its localization in the cytosol, the mechanism whereby ribulose-5-phosphate is converted back to glucose-6-phosphate differs in only very minor respects from step 3 of the carbon cycle of photosynthesis, operating in the opposite direction. This is best followed by reference to Fig. 14.7 and to Fig. 14.8, commencing at C and proceeding anticlockwise to B. An epimerase first converts two molecules of ribulose-5-phosphate to xylulose-5-phosphate, while an isomerase converts a third molecule to ribose-5-phosphate. *Transketolase* then enables the ribose-5-phosphate to accept a two-carbon fragment from xylulose-5-phosphate, the products being sedoheptulose-7-phosphate and glyceraldehyde-3-phosphate:

$$\underset{\text{Ketose}}{C_5} + \underset{\text{Aldose}}{C_5} \xrightleftharpoons[]{\text{Transketolase}} \underset{\substack{\text{Residue}\\\text{(an aldose)}}}{C_3} + \underset{\substack{\text{Addition}\\\text{product}\\\text{(a ketose)}}}{C_7}$$

(Minus 2C: arrow from the ketose C_5 to the aldose C_5)

The products of the last reaction next react, in the presence of *transaldolase*, to give erythrose-4-phosphate and fructose-6-phosphate:

$$\underset{\text{Ketose}}{C_7} + \underset{\text{Aldose}}{C_3} \xrightleftharpoons[]{\text{Transaldolase}} \underset{\substack{\text{Residue}\\\text{(an aldose)}}}{C_4} + \underset{\substack{\text{Addition}\\\text{product}\\\text{(a ketose)}}}{C_6}$$

(Minus 3C from C_7 to C_3)

The sugar with four carbon atoms in its molecule, erythrose-4-phosphate, next accepts two carbon atoms from the remaining molecule of xylulose-5-phosphate in a reaction catalysed by *transketolase.* The products are another molecule of fructose-6-phosphate and a molecule of glyceraldehyde-3-phosphate. The fructose-6-phosphate is readily converted to glucose-6-phosphate by the enzyme *phosphohexoisomerase* (page 218). Similarly, the glyceraldehyde phosphate can readily give glucose-6-phosphate via fructose-1,6-diphosphate, the enzymes responsible including aldolase and fructose diphosphate phosphatase (Fig. 12.7).

Not only can ribulose-5-phosphate be formed from hexose by steps 1 and 2 of the pentose phosphate cycle, but any excess ribulose-5-phosphate can be converted back to hexose phosphate without loss of the reducing power of NADPH formed during the oxidative steps. The outcome of six turns of the pentose phosphate cycle is that, statistically, 1 molecule of hexose is completely metabolized to 6 molecules of carbon dioxide in such a way that 12 molecules of NADPH are formed from 12 molecules of $NADP^+$ at the same time. Naturally, $NADP^+$ must be regenerated from NADPH or the process would soon stop (compare NADH, Section 11.9) and it is therefore necessary to consider now a matter of considerable importance—the ways in which NADPH can function within the cell as a reducing agent.

14.7 Differences between the pentose phosphate pathway and the main pathway of sugar degradation

The pentose phosphate pathway differs in several important respects from the glycolytic pathway (followed by the TCA cycle), for not only are the products different but so are the cofactor requirements and the functional characteristics of the two systems. Glycolysis, like the pentose phosphate pathway, is located in the cytosol, but whereas the major functions of glycolysis are to produce ATP and pyruvate (the oxidation of which, *via* acetyl coenzyme A, yields additional ATP by mitochondrial oxidative phosphorylation) the pentose phosphate pathway produces no ATP directly but instead results in the reduction of NADP. It is unlikely that NADPH, produced in the cytosol, acts as a direct source of electrons for mitochondrial ATP production, for, apart from the difficulty relating to location, a prior transhydrogenation reaction of the type

$$\text{NADPH} + \text{NAD}^+ \rightleftharpoons \text{NADP}^+ + \text{NADH}$$

would need to occur. While such systems are known, it appears likely that the principal functions of the pentose phosphate pathway are (a) to produce various non-hexose sugars and (b) to produce NADPH, which provides reducing power needed for a number of anabolic processes.

The pentose phosphate pathway has few cofactor requirements. NADP is necessary, of course, and thiamine pyrophosphate is a prosthetic group for transketolase. In contrast, as was seen in earlier chapters, the main catabolic route requires numerous cofactors including inorganic phosphate, NAD, lipoic acid, thiamine pyrophosphate, and coenzyme A.

Differences in enzyme and coenzyme requirements of the main and pentose phosphate oxidative systems are reflected in their different vulnerability to several poisons. The glycolytic pathway is highly sensitive to fluoride ions, because of the effect of this poison on enolase (Section 11.7). Similarly, the TCA cycle is inhibited by fluoroacetate and malonate (Section 9.10). Again, the principal energetic function of the main pathway—the production of ATP—is prevented by dinitrophenols (which uncouple oxidation from phosphorylation) and by such substances as cyanide ions and hydrogen sulphide (which inactivate cytochrome oxidase). The pentose phosphate pathway, in contrast, operates in the presence of $10^{-1}\,M$ fluoride and is also rather insensitive to cyanide ions. On the other hand, the pentose phosphate system appears to be sensitive to copper ions, and both of the dehydrogenases of the cycle are inhibited by thioperazine.

The pentose phosphate pathway is often well developed in lower organisms and plants, but most vertebrate tissues also possess some activity. Up to 40 per cent of the carbohydrate oxidation which takes place in the lactating mammary gland and 20 per cent in the liver occurs *via* the pentose phosphate pathway. However, its activity in brain and muscle is often very small, although these tissues contain some of the enzymes of the cycle.

14.8 Utilization of sugars produced by the pentose phosphate pathway

One of the main functions of the pathway is to produce pentoses which are components of nucleotides and are therefore essential for the synthesis of nucleic acids, NAD, ATP, GTP, and other similar electron and energy carriers.

The pentose phosphates occurring in the cycle can be converted readily to *ribose-5-phosphate.* This does not directly combine with the nucleotide bases but undergoes a prior activation. In this process a pyrophosphate group is added from ATP to carbon atom 1 of the furanose ring, so producing AMP and 5′-phosphoribose-1′-pyrophosphate (PRPP). The latter reacts with nucleotide bases by a reaction catalysed by specific pyrophosphorylases, pyrophosphate being released. This synthesis of the nucleoside monophosphate is drawn towards completion by the hydrolysis of the pyrophosphate to orthophosphate (Fig. 14.10).

Deoxyribose mononucleotides for DNA synthesis are largely formed from pre-formed *ribo*nucleotides. In other words, deoxyribose is not synthesized from free ribose but the reduction occurs at the nucleotide level. The process of reduction takes place as indicated below, the parent ribonucleoside being in the form of the diphosphate. The electrons and protons involved in the reduction are derived from NADPH—a further instance where this cofactor provides reducing power for an anabolic process:

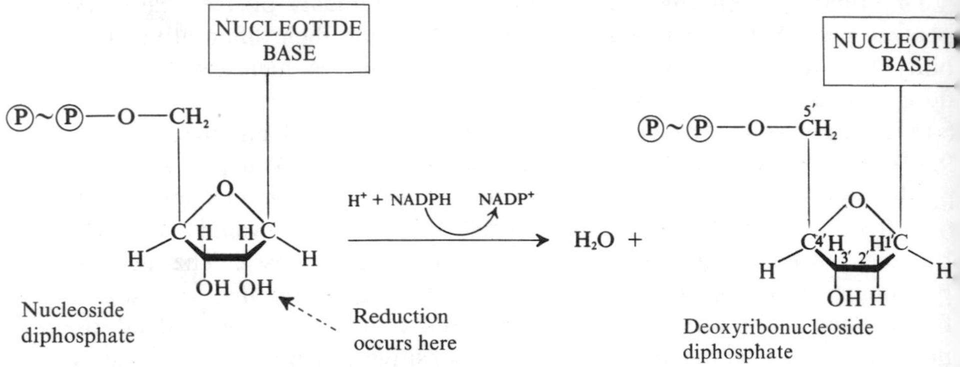

NUCLEOTIDE BASE

Ⓟ~Ⓟ—O—CH_2

C H H C

H H

OH OH

Nucleoside diphosphate

H^+ + NADPH → $NADP^+$

Reduction occurs here

H_2O +

NUCLEOTIDE BASE

Ⓟ~Ⓟ—O—5′CH_2

C4′H H1′C

3′ 2′

H H

OH H

Deoxyribonucleoside diphosphate

Erythrose-4-phosphate, another sugar produced by the pentose phosphate pathway, is of considerable importance as a precursor of part of the carbon skeleton of three amino acids, phenylalanine, tyrosine, and tryptophan. However, these can only be synthesized by plants and certain microorganisms, for animals lack the necessary enzymes. In plants, 6-phosphogluconic acid is a precursor of L-ascorbic acid (vitamin C).

14.9 Some processes requiring a supply of reduced NADP

Since both the pentose phosphate pathway and the light reaction of photosynthesis are principal routes for producing NADPH, it is appropriate that an indication should now be given of some of the major reactions, other than the reduction of carbon dioxide by plants, which require this form of reducing power.

NADPH, rather than NADH, is frequently used as the source of reducing power for anabolic processes. This is undoubtedly associated with the relative abundance of the *reduced* form of the two cofactors, for in the cytosol of most cells NAD is predominantly in the oxidized form whereas NADP is re

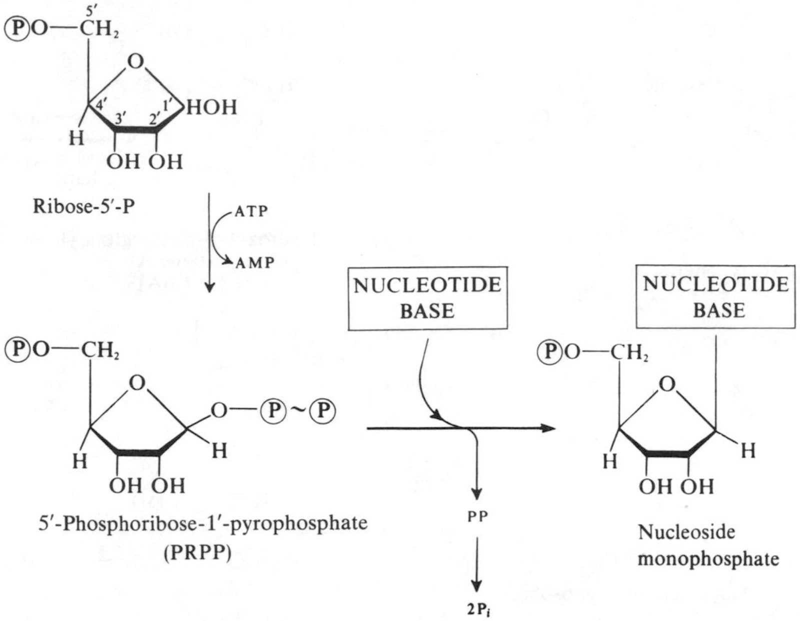

Fig. 14.10 Synthesis of mononucleotides from ribose-5-phosphate

duced. For example, in the cytosol of liver, the ratio [NAD]/[NADH] is about 1000 but the ratio [NADP]/[NADPH] is about 0.01. When synthetic reactions requiring NADPH are occurring, the pentose phosphate pathway is stimulated, for it appears that the main control of the NADP-linked oxidoreductases is the [NADP]/[NADPH] ratio. A rise in the NADP concentration leads to greater enzyme activity which tends to restore the ratio to normal.

NADPH is of particular importance in the anabolism of lipids for, in addition to its use in the formation of *long-chain fatty acids* (Chapter 10), considerable quantities are needed for the synthesis of *steroids* and *carotenoids*. The steroids (and, in plants, the carotenoids and rubber latex) are synthesized from *isopentenylpyrophosphate*. This five-carbon compound is formed by a series of reactions which are outlined in Fig. 14.11. Three molecules of acetyl coenzyme A give rise to 1 molecule of β-hydroxyl-β-methylglutaryl coenzyme A (HMG—CoA) with the elimination of 2 molecules of coenzyme A. The acyl group of HMG—CoA undergoes *reduction* reactions which require four equivalents of reducing power derived from 2 molecules of NADPH. The acyl linkage of acyl coenzyme A is thereby converted to an alcohol group with the release of free coenzyme A. The dihydroxy acid so formed, *mevalonic* acid, then undergoes reactions which involve the loss of the elements of water and carbon dioxide. Three

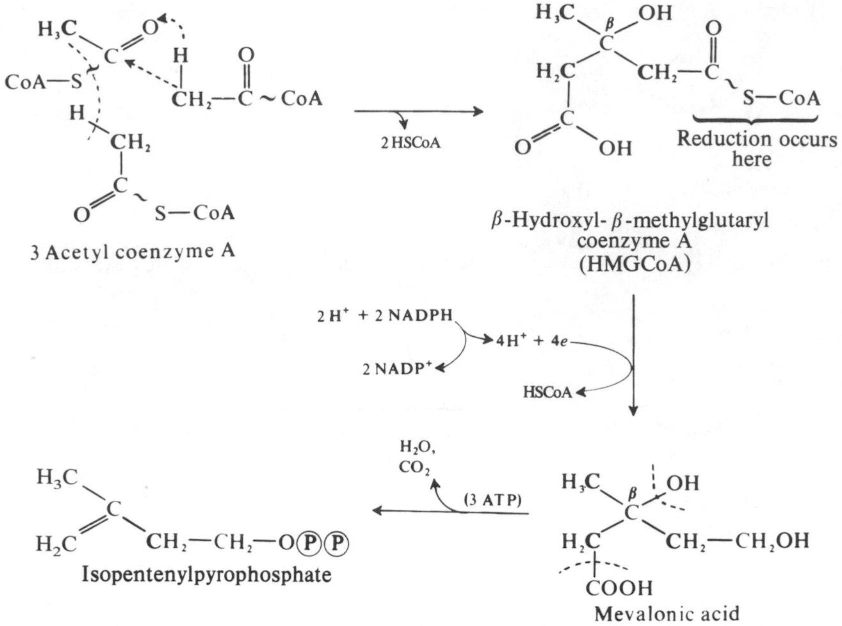

Fig. 14.11 Synthesis of isopentenylpyrophosphate from acetyl coenzyme A

molecules of ATP provide the necessary energy and also provide the pyrophosphate contained in the product, isopentenylpyrophosphate. Three of the carbon atoms of this compound have been shown by the use of radioactive carbon to arise from the methyl group of acetate (shown in heavy type in Fig. 14.11) while the other two are derived from the carboxyl group.

Isopentenylpyrophosphate is itself the precursor of numerous lipids. Rubber latex is a polymer consisting of thousands of isopentenyl groups linked together (Fig. 14.12). The side chain of ubiquinone (Fig. 8.4) is also derived from it.

Cholesterol contains twenty-seven carbon atoms and each of these has been shown to arise from acetyl coenzyme A; labelling techniques have shown that mevalonic acid and isopentenylpyrophosphate are again intermediates in the reaction sequence. Fifteen of the carbon atoms are derived from the methyl group of acetate (shown in heavy type in Fig. 14.12), the other twelve originating from the carboxyl group. The cholesterol molecule (and those of other steroids) is formed by the combination of 6 molecules of isopentenylpyrophosphate, and the formation of 6 molecules of this precursor requires 12 molecules of NADPH. In addition, other reactions in the sequence by which cholesterol is synthesized appear to use the reducing power of NADPH.

The carotenoids are common plant pigments; β-carotene (Fig. 14.12) is a typical example and contains forty carbon atoms which arise by the combination of eight isopentenyl units. These, of course, require 16 molecules of NADPH for their synthesis. Higher animals cannot synthesize carotenoids and for this reason β-carotene is an important factor they must acquire in their diet. Animals are able to convert β-carotene to vitamin A by splitting the molecule at the point of symmetry. Vitamin A is a precursor of the photosensitive pigment of the eye (Chapter 20, page 42).

As well as the reactions described above, NADPH is an essential cofactor in a series of metabolic changes whereby lipid-soluble organic compounds of

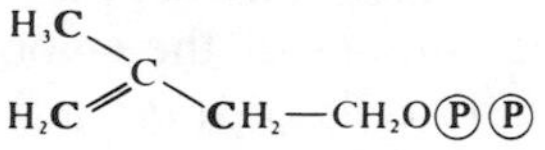

$$H_3C-C(=CH_2)-CH_2-CH_2O\text{Ⓟ}\text{Ⓟ}$$

Isopentenylpyrophosphate
(carbon atoms in heavy type derived from CH_3 group of acetyl CoA)

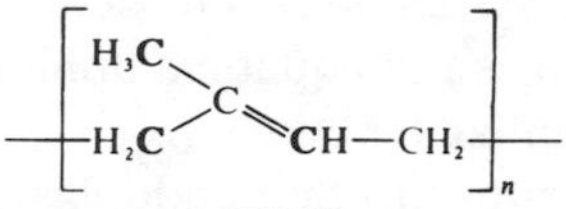

$$\left[-H_2C-C(CH_3)=CH-CH_2-\right]_n$$

RUBBER

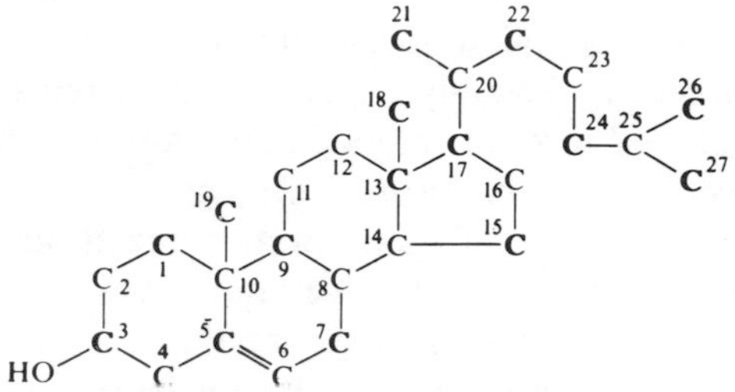

CHOLESTEROL
(hydrogen atoms omitted)

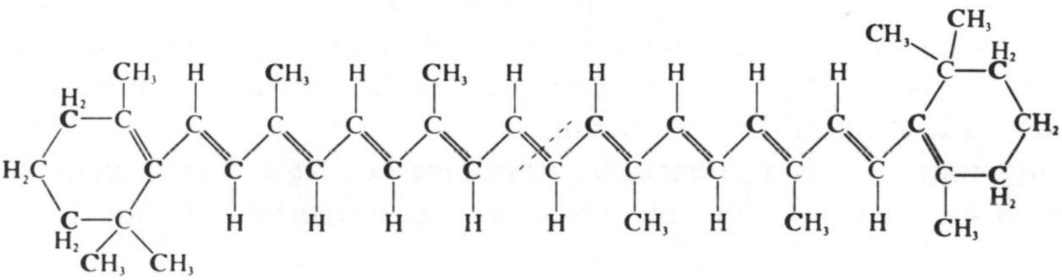

β-CAROTENE

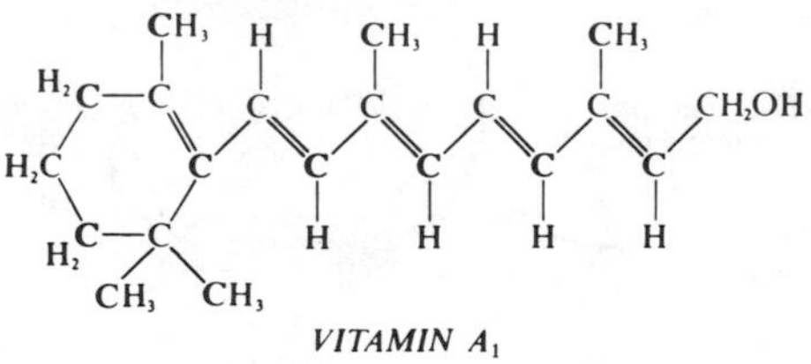

VITAMIN A_1

Fig. 14.12 Four derivatives of isopentenylpyrophosphate

both endogenous (internal) and exogenous origin are oxidized or hydroxylated. The versatile metabolic system responsible for these changes and some of the functions it fulfils are described in the next section.

14.10 The microsomal electron transport system

We have so far encountered two electron transport systems which serve different functions, namely, the mitochondrial electron transport system and the electron transport system located in the chloroplasts (Sections 8.9 and 14.4). Associated with the endoplasmic reticulum of many vertebrate cells is a third type of electron transport system. It is particularly well developed as an integral part of the membranes of the smooth endoplasmic reticulum of liver cells, but also occurs in cells of the adrenal bodies and sex organs. It participates in the biosynthesis of specific steroids and also helps to render more water soluble, and therefore more readily excretable, a range of potentially toxic foreign or *xenobiotic* substances which would otherwise tend to accumulate in the animal body in consequence of their solubility in fat deposits. Possibly for evolutionary reasons, the system seems to be more developed in warm-blooded land animals than in fish and aquatic amphibia. Insects and plants possess systems which serve similar functions.

When liver cells are homogenized and centrifuged, the endoplasmic reticulum breaks up to give minute vesicles which sediment at approximately 100 000 g for 60 min. This fraction formed on centrifugation is termed the *microsomal fraction.* Hepatic microsomes, supplemented by NADPH, are able to bring about a wide range of oxidations and hydroxylations. The hepatic microsomal system probably includes a flavoprotein and a non-haem iron protein, both of which can undergo reversible oxidation and reduction, as well as a unique cytochrome termed *cytochrome* P_{450}. The latter gets its name from the fact that, on poisoning with carbon monoxide, it develops a characteristic spectral line at 450 nm.

Cytochrome P_{450} is a remarkable substance in that it shares some of the properties of haemoglobin and some of the properties of mitochondrial

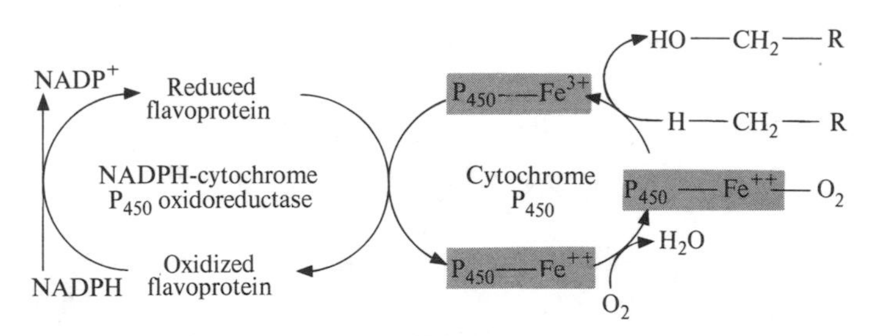

Fig. 14.13 Suggested role of NADPH in hepatic microsomal oxidations

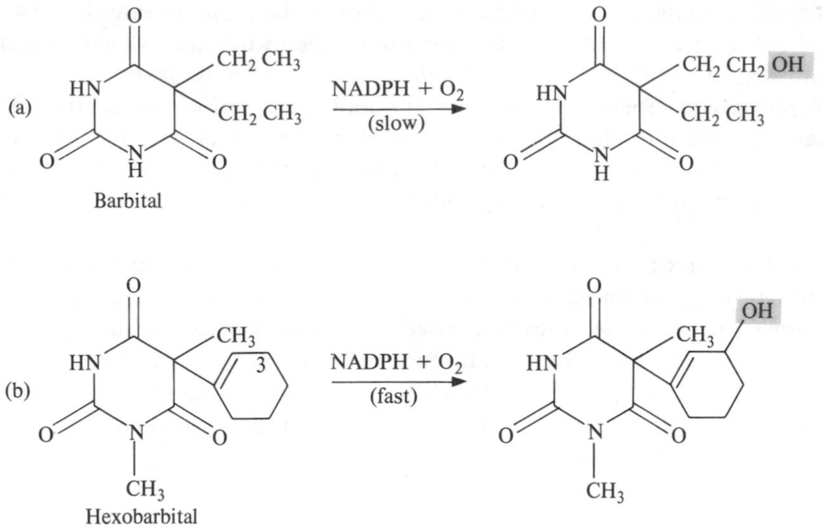

Fig. 14.14 Microsomal hydroxylation of two barbiturate drugs. (a) Metabolism of barbital (side-chain hydroxylation) and (b) metabolism of hexobarbital (ring hydroxylation)

cytochromes. In common with haemoglobin, it exists in a ferrous iron form which produces a complex with molecular oxygen (compare oxyhaemoglobin, Chapter 18). However, like typical cytochromes, it can exist in ferrous and ferric forms. The result is that P_{450} forms a cycle comprising three components instead of the redox pair involved in other cellular cytochromes. Several variants of the scheme shown in Fig. 14.13 have been proposed, differing principally with regard to the position of entry of the lipophilic substrate, RCH_3. All such schemes agree, however, that the paradox of the dependence of oxidation on a powerful *reducing agent*, NADPH, is that the latter converts ferric iron in P_{450} to a ferrous form. They all agree also that atmospheric oxygen is the ultimate oxidant and that this forms an oxyhaemoglobin-like complex with reduced P_{450}.

The 'mixed function' oxidase system plays a part in medical science, for it determines the period over which many drugs (e.g., the barbiturates) are effective. It is also important in agriculture and in environmental chemistry, for microsomal oxidation is one of several factors determining the persistence, rate of degradation, and selective toxicity of poisons used to control pests.

The hydroxylation of two barbiturate drugs will serve to illustrate the way in which the NADPH-dependent microsome system works. Barbital is not greatly metabolized in most mammals; it is excreted largely unchanged and consequently tends to be a long-acting anaesthetic. In contrast, hexobarbital is rapidly deactivated and is therefore a short-acting anaesthetic (Fig. 14.14).

The P_{450} system also plays an important role in the later stages of steroid synthesis. The earlier steps of *cholesterol* synthesis from isopentenyl-pyrophosphate (Section 14.9) lead eventually to a long-chain, partly unsaturated compound called *squalene*. This is converted to an epoxide by the microsomal system in liver cells. Rearrangement of the double bonds of squalene epoxide leads to ring closure and cholesterol is eventually produced.

Cholesterol can be regarded as the precursor of numerous important steroids in higher animals. These steroids include the androgens and oestrogens, progesterone, the glucocorticoids, and corticoids such as aldosterone. Some of these are mentioned in Chapter 2. The last two kinds of steroids are made principally by cells of the adrenal cortex. Many of the steps in the synthesis of each of these groups of steroids involve the microsomal oxidase system.

15

Nitrogen metabolism and the biochemistry of amino acids

15.1 Nitrogen fixation

Atmospheric nitrogen is the ultimate source of the nitrogen present in the proteins, nucleic acids, and other vital nitrogenous components of all living cells. However, neither animals nor higher plants are able directly to utilize atmospheric nitrogen, but depend instead upon the activities of a relatively few types of bacteria and blue-green algae which are capable of converting atmospheric nitrogen to ammonia. This process, termed nitrogen fixation, is a fundamental step in the *nitrogen cycle*, for the ammonium compounds are further transformed to nitrites and nitrates in the soil and into organic nitrogenous compounds in plants. When plants and animals die and decay, much of this nitrogen returns to the soil or the sea. Animal excreta also contain nitrogenous materials which are decomposed to simple inorganic nitrogen compounds. On the other hand, a proportion of the nitrogen fixed by nitrifying bacteria finds its way back into the atmosphere as a result of the activities of denitrifying bacteria.

In recent times, man has increasingly supplemented the natural process of nitrogen fixation by such industrial nitrogen fixation reactions as the Haber process. The ammonium compounds and nitrates obtained by chemical processes are used as fertilizers to replace nitrogenous compounds removed from arable land when crops are harvested. Nevertheless, microbial nitrogen fixation is very largely responsible for maintaining the level of combined nitrogen in the biosphere. In view of man's dependence on plant protein, the importance of natural nitrogen fixation cannot be overestimated. For this reason, biologists are now trying to identify the genetic system in legumes responsible for accepting nitrogen-fixing bacteria. Genetic engineering is often regarded with some apprehension, but perhaps no other genetic discovery could match in importance the construction of a wider range of plants with the capacity to incorporate nitrogen-fixing bacteria. The saving in energy and in fertilizers would be immense, and even the poorest nations might then become self-sufficient in proteins. The mechanisms of microbial nitrogen fixation pose some intriguing problems for the biologist and biochemist, for molecular nitrogen is an extremely unreactive substance at ordinary temperatures and pressures, the activation energy (Section 6.6) of the reaction by which nitrogen is reduced to ammonia being very high. Despite this barrier to reaction, which is still considerable at high temperature unless pressures of several hundred atmospheres are applied, it is a remarkable fact that several types of microorganisms, quite unrelated to one another, are able to reduce nitrogen to ammonia at a temperature, pressure, and pH compatible with the existence of life.

Some nitrogen-fixing organisms are present in soil, others in root nodules of plants, and yet others live in water. Some of the soil bacteria capable of fixing nitrogen are aerobic (e.g., *Azotobacter*) but it may well be that anaerobic species make a quantitatively greater contribution to nitrogen fixation. The latter include species of *Clostridium*, a genus almost universally distributed in soils, and it is upon these that most of the recent investigations into the biochemistry of nitrogen fixation have been performed. Other microorganisms, belonging to the genus *Rhizobium*, enter into a symbiotic relationship with many leguminous plants (clovers, peas, beans, etc.). Various species of *Rhizobium* are associated with the nodules on the roots of these plants and are able to convert nitrogen gas into the amino groups of amino acids, ammonium ions probably being formed as an intermediate. Root nodules are remarkable in that they contain a form of haemoglobin, leghaemoglobin, the presence of which often causes them to have a pink colour. Curiously, the pigment lies outside the bacteria associated with the root nodules, though it is never present in bacteria-free *Leguminosae.* A third important example of nitrogen fixation is provided by certain blue-green algae, for the production of rice in Asiatic countries relies upon their presence in the water of paddy fields. Eventually the algae die and decompose, liberating ammonia and nitrate ions which are subsequently absorbed by the roots of the growing rice. Several groups of photosynthetic bacteria also contribute to the nitrogen store in water.

Despite its importance, the biochemical mechanism of nitrogen fixation is still very imperfectly understood, and the account which follows must therefore be regarded as incomplete and, in places, speculative. It is, however, established that three interrelated processes are involved. The first of these produces the energy for the nitrogenase (nitrogen-reducing) reaction; this process, which is ATP-dependent, appears to lead to the formation of the reduced form of *ferredoxin*, a protein containing non-haem iron as well as cysteine and a more labile form of sulphur. This type of electron carrier is of interest in that it structurally resembles, but is certainly not identical with, the non-haem iron proteins which participate in the light reaction of photosynthesis, in mitochondrial electron transport, and proba

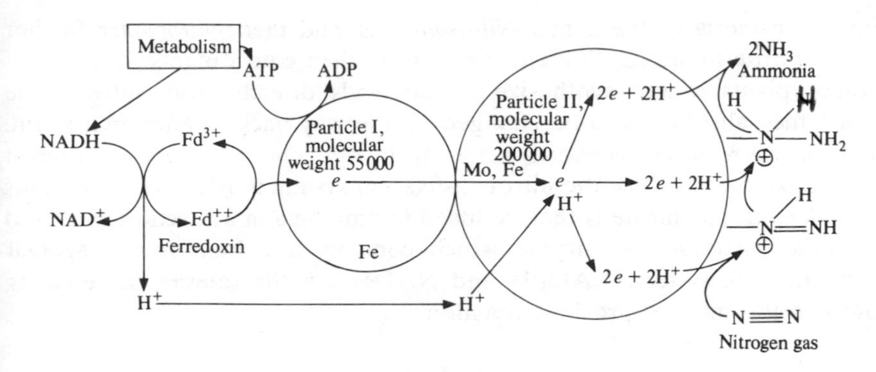

Fig. 15.1 A model of the nitrogen-fixation system

bly in microsomal electron transport (Sections 8.8, 14.4 and 14.10).

In addition to ferredoxin, two highly complex protein particles are essential components of the nitrogen-fixation system. Particle I is the smaller and contains *non-haem iron.* Particle II contains two *molybdenum* ions and about thirty atoms of both iron and labile sulphur. There is as yet no complete explanation as to how the particles function. Figure 15.1 illustrates a generalized hypothesis. Metabolism produces reducing power in the form of NADH. The electrons are then transferred to ferredoxin, the iron of which undergoes ferrous/ferric valency changes. They then pass *via* particle I to particle II, whose function is to transfer them to intermediates leading from nitrogen gas to ammonia. As in many other processes of biological reduction, for every pair of electrons transferred, two protons are also required. Three reductive steps can be identified:

$$N_2 \xrightarrow[+2H]{+2e} HN{=}NH \xrightarrow[+2H]{+2e} H_2N{-}NH_2 \xrightarrow[+2H]{+2e} 2NH_3$$

The intermediates (di-imide and hydrazine) may never be formed as such in the cell but are present as compounds linked to particle II (Fig. 15.1).

Besides electrons and protons the system of nitrogen fixation requires a large supply of ATP:

$$N_2 + 6e + 6H^+ + 12ATP \rightarrow 2NH_3 + 12ADP + 12P_i$$

The exact function of the ATP is uncertain although it has been suggested that it raises the energy levels of the protein particles by inducing conformational changes and so activating the nitrogen intermediates.

It may seem strange that, in practice, almost all the nitrogen entering plant roots from the soil is actually in the form of *nitrate.* The explanation is that, once atmospheric nitrogen has been 'fixed' in the form of ammonia, other soil organisms oxidize ammonia, *via* nitrite, to nitrate. They do this in order to utilize the energy of the exergonic oxidative reaction. Oxidation to *nitrite* is done by bacteria of the genus *Nitrosomonas*, and then *Nitrobacter* further oxidize nitrite to nitrate. Some of this nitrate then enters plants roots.

Since plants cannot synthesize amino acids directly from nitrate, the nitrate they absorb has to be changed all the way back to ammonia again. First, *nitrate reductase* converts nitrate to nitrite. This enzyme is of interest in that, like particle II of the nitrogen-fixation system in bacteria, it contains *molybdenum*. The nitrite is then reduced to ammonia in a reaction catalysed by *nitrite reductase*, an enzyme which contains iron built into a special porphyrin ring system. NADPH and NADH are the sources of reducing power in this and the previous reaction.

15.2 Synthesis of amino groups and their transfer to keto acids

Cells of most organisms are able to convert ammonia to the α-amino group of amino acids. This activity is perhaps greatest in plants and micro-organisms since animals rely in large measure on pre-formed dietary amino acids for their protein synthesis. Plant roots readily absorb ammonium or nitrate ions from the soil. Absorbed nitrate is reduced to ammonia by enzymes in plant tissue prior to the formation of amino groups (Section 15.1).

When labelled ammonium salts (e.g., $^{15}NH_4Cl$) are absorbed by the roots of plants, or when microorganisms such as yeasts are incubated in a medium containing similarly labelled ammonium compounds, amino acids and other nitrogen compounds are soon found to contain ^{15}N. Quantitatively, the most important initial reaction is the conversion of ammonia to *glutamic acid*. In plant and animal cells the enzyme concerned, *glutamate dehydrogenase*, is situated mainly in the mitochondria, where the α-ketoglutarate, which is the acceptor of the ammonia, is synthesized by the enzymes of the TCA cycle. Glutamate dehydrogenase requires either NADH or NADPH as cofactor for the synthesis of glutamate, the cofactor requirement depending on the sources of the enzyme. The reaction, which if *fully reversible*, probably involves the formation of an imino acid from the α-ketoglutarate and ammonia. This is then reduced to glutamate by electrons donated by NADH or NADPH:

$$\underset{\alpha\text{-Ketoglutaric acid}}{HOOC{-}CH_2{-}CH_2{-}C(=O){-}COOH} + NH_3 \underset{}{\overset{-H_2O}{\rightleftharpoons}} \underset{\text{Imino acid}}{HOOC{-}CH_2{-}CH_2{-}C(=NH){-}COOH} \underset{\text{Glutamate dehydrogenase}}{\overset{NAD(P)H + H^+ \;\; NAD(P)^+}{\rightleftharpoons}} \underset{\text{L-Glutamic acid}}{HOOC{-}CH_2{-}CH_2{-}CH(NH_2){-}COOH}$$

Very soon after labelled nitrogen derived from $^{15}NH_3$ appears in glutamic acid, it is found also in the other α-amino acids of proteins. The amino group of glutamic acid is, in fact, *readily transferred* to a variety of α-keto acids which are thereby converted to α-amino acids. The process is known appropriately as *transamination* and is catalysed by enzymes called *transaminases* or *amino transferases.* The typical reaction, which is freely reversible, may be represented in the following way:

$$\underset{\text{L-Glutamic acid}}{HOOC{-}CH(NH_2){-}CH_2{-}CH_2{-}COOH} + \underset{\alpha\text{-Keto acid}}{\boxed{R{-}CO{-}COOH}} \underset{\text{transaminase}}{\overset{\text{Specific}}{\rightleftharpoons}} \underset{\text{L-}\alpha\text{-Amino acid}}{\boxed{R{-}CH(NH_2){-}COOH}} + \underset{\alpha\text{-Ketoglutarate}}{HOOC{-}CO{-}CH_2{-}CH_2{-}COOH}$$

If considered in the *reverse direction* to that shown above, it is clear that transaminases remove the elements of ammonia from L-α-amino acids and pass them to α-ketoglutarate. Transaminases only act on the L forms of amino acids and a specific transaminase exists for each of these. Pyridoxal phosphate, a derivative of vitamin B_6, is an essential cofactor (Section 15.8).

It should be emphasized that the α-ketoglutarate/glutamate interconversion (followed by the glutamate dehydrogenase reaction) *provides a mechanism which is unique in amino acid metabolism.* It enables the majority of amino acids to be deaminated by the donation of their amino groups to α-ketoglutarate, the action being catalysed by specific transaminases. The glutamic acid so formed is then broken down to produce α-ketoglutarate and ammonia by the reaction catalysed by glutamate dehydrogenase. This ammonia, as will be seen later, is very toxic to cells and if it is not to be excreted at once, it is largely transformed to other useful or excretable nitrogenous compounds (Sections 15.6 and 15.7).

Conversely, transamination also occurs during the *synthesis* of most (if not all) amino acids present in *plant* proteins. The process is more restricted in *animals,* for not all the corresponding α-keto acids occur in the cell—although, if provided artificially in the diet, the keto acids which are *normally absent* can, in most cases, readily accept amino groups by transamination to form 'essential' amino acids.

L-*Aspartate,* like L-glutamate, can act as the donor of amino groups for amino acid synthesis—i.e., specific transaminases enable L-aspartate as well as L-glutamate to be the donor in transamination reactions (Fig. 15.2). Much of the aspartate may itself originate by the transamination of oxaloacetate, L-glutamate being the donor. Oxaloacetate is, of course, readily synthesized by the reactions of the TCA cycle (Section 9.8).

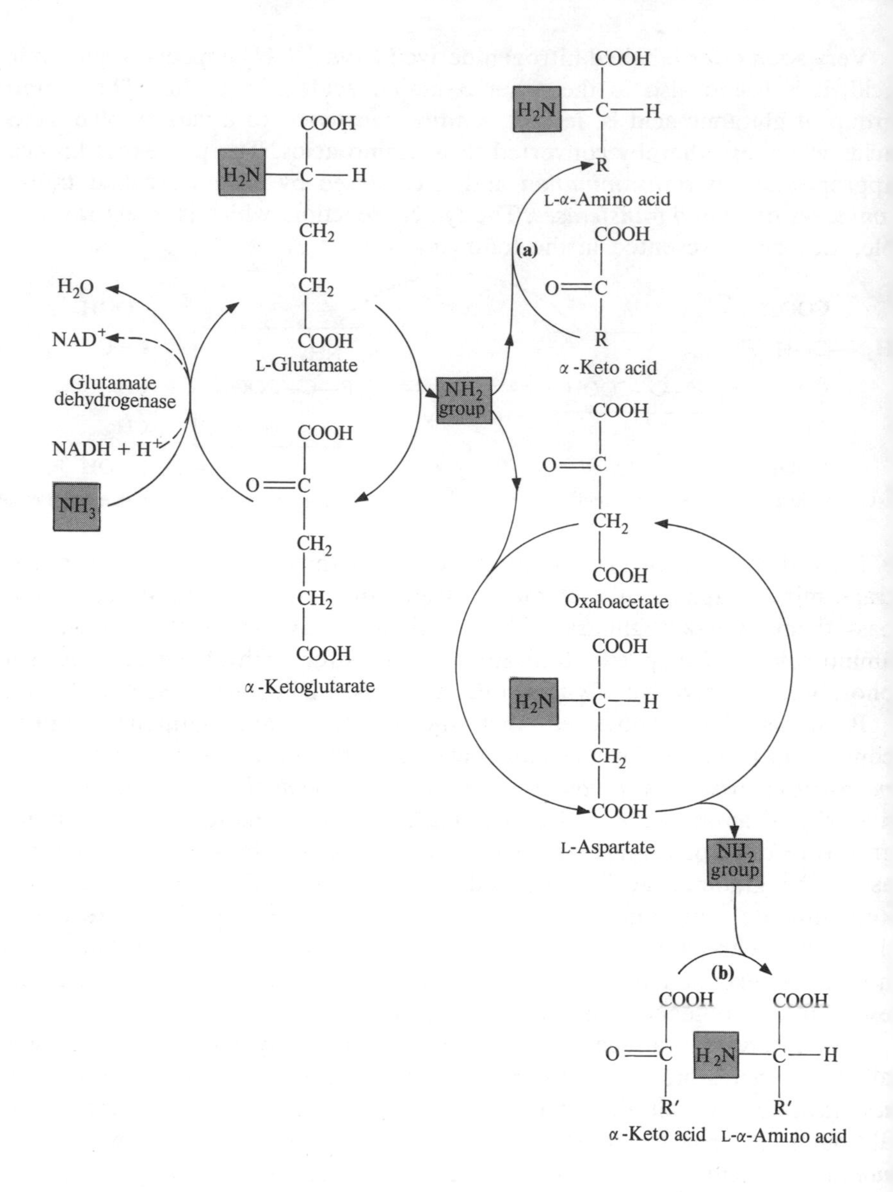

Fig. 15.2 The synthesis of glutamate from α-ketoglutarate and the formation of an α-amino acid from the corresponding α-keto acid (a) by direct transamination from glutamate and (b) by transamination via the oxaloacetate/aspartate system. Each of the reactions is reversible

15.3 General aspects of amino acid metabolism

The L-amino acids occupy a central position in cellular metabolism since almost all biochemical reactions are catalysed by enzymes comprising L-amino acid residues. Moreover, the routes of synthesis and catabolism of amino acids link up with the biochemical pathways followed by carbohydrates and fatty acids, so amino acid biochemistry cannot be isolated from that of these other two groups of biologically important compounds.

Green plants and many bacteria can synthesize each of the L-amino acids rom relatively simple carbon compounds such as triose phosphate or acids of the TCA cycle and these, in turn, are the products of lipid or of carbohydrate metabolism (Chapters 9, 10, and 11). Oxaloacetic acid, for example, can act as the direct precursor of the carbon skeletons of some five amino acids. Animals, however, have lost the ability to synthesize about nine amino acids and they must, therefore, obtain these 'essential' amino acids from their diet (Section 15.4).

The catabolism of amino acids leads to the formation of ammonia and other nitrogen compounds. The carbon skeletons which remain are mostly converted to acids of the TCA cycle, to pyruvate, propionate, or to acetyl coenzyme A. As discussed in Chapter 12, several of these nitrogen-free acids can act as the precursors of polysaccharides although, in animals, acetyl coenzyme A can only be oxidized by the TCA cycle or used for the ynthesis of lipids. In addition, several amino acids provide the raw material or the synthesis of other types of important compounds such as the purines nd porphyrins. Some of these metabolic changes are summarized in Fig. 5.3. Protein synthesis is a somewhat special topic and is considered eparately in Chapter 16.

Animals depend directly or indirectly on plants for all their food materials nd thus for the amino acids to build proteins. An adult non-pregnant, on-lactating animal requires a certain level of protein in its diet in order to naintain and repair its tissues and to allow for growth of hair and nails. In he adult human, the daily protein requirement is about 1 g per kilogram of ody weight. If the dietary protein level is increased above the minimum mount needed to maintain health, the animal shows no great increase in the verall nitrogen content of its body, but instead, the excess amino acids bsorbed from the intestine are catabolized, the nitrogen of the amino roups being excreted—in different types of animals—as ammonia, urea, or ric acid. Thus the nitrogen excretion greatly increases when excess protein consumed. At the same time the carbon skeletons of amino acids are onverted, directly or indirectly, either to oxaloacetate (and hence to lycogen) or to acetyl coenzyme A (which may be oxidized by the TCA ycle or used for the synthesis of lipids). The pathway by which any articular amino acid is degraded will determine whether it is glucogenic or etogenic (Section 15.4).

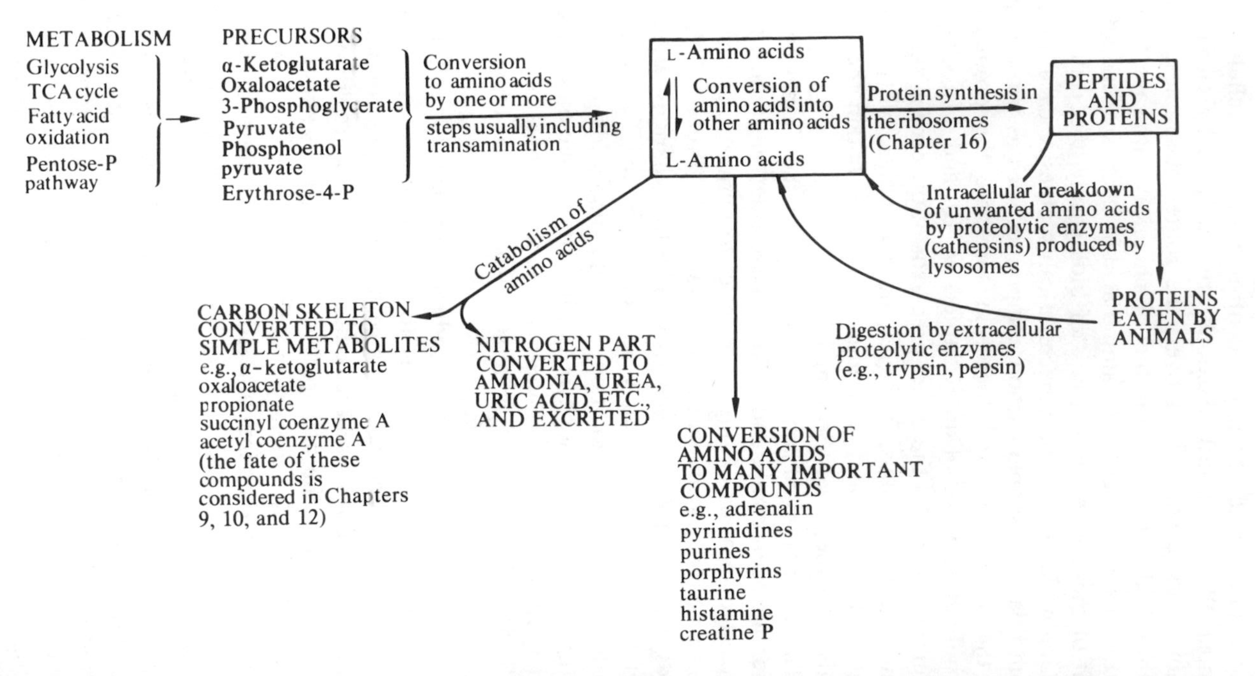

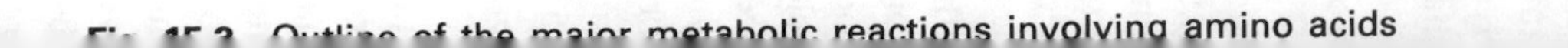
Fig. 15.3 Outline of the major metabolic reactions involving amino acids

The proteins of an animal's body are constantly being degraded and new ones built up. In other words, there is a state of metabolic flux; a perpetual turnover is occurring in which tissue proteins are constantly being broken down and replaced. This turnover may be quite rapid in the case of liver or plasma proteins where the half-life is about 7 to 10 days, but the proteins of muscle and skin often have a half-life of several months. The amino acids which are released as the tissue proteins are broken down are largely reincorporated into new proteins, although only after some have been converted to other amino acids. This process is not completely efficient and therefore some of the amino acids are catabolized in the process, their nitrogen being excreted. Thus each animal requires dietary protein to make good this loss. If the animal is given a diet inadequate in protein the problem of inefficiency is aggravated, the animal being obliged to break down relatively dispensable proteins—especially those in the liver—to provide amino acids required for various indispensable enzymes. Under these conditions, the nitrogen excretion decreases to a very low level while the typical symptoms of emaciation appear.

During complete starvation, the animal at first utilizes its reserves of glycogen and lipid to provide energy. Eventually, however, as these reserves become depleted, it has to *obtain energy* by resorting to the catabolism of its proteins. The more dispensable proteins are utilized first, but eventually even essential proteins are consumed. At this stage of starvation, nitrogen excretion rises rapidly as the amino groups of amino acids are discarded. Death soon results once the essential proteins are appreciably depleted.

Young growing animals and pregnant females require a relatively greater dietary protein intake than normal adults. The results of feeding a low protein diet, or of complete starvation, are consequently much more pronounced and appear at an earlier stage.

15.4 Essential and non-essential amino acids

It will be recalled (Chapter 3) that each protein has a definite primary structure, usually comprising residues of most of the twenty or so amino acids. If for some reason a particular amino acid is not made available to the ribosomes, the protein cannot be formed there. If a deficit of certain amino acids temporarily exists, *plants* and many microorganisms can synthesize them from other amino acids which happen to be present in temporary excess.

In higher *animals*, however, some part of this biosynthetic versatility has been lost, for most of them are unable to make the carbon skeletons of about nine amino acids, so these or their keto acid precursors must be supplied in the diet. The animal is unable to survive unless it is provided

Table 15.1 Essential and non-essential amino acids

Non-essential	Essential
Alanine	Methionine
Arginine	Threonine
Aspartic acid	Tryptophan
Cysteine	Valine
Glutamic acid	Leucine†
Glycine	Isoleucine‡
Histidine	Lysine‡
Hydroxyproline	Phenylalanine‡
Proline	Tyrosine‡
Serine	

† Amino acid is ketogenic (Section 10.7).
‡ Amino acid is glucogenic and ketogenic.
Amino acids not marked are glucogenic only (Section 12.4).

with them, and hence they are known as *essential* amino acids. Those amino acids the animal is able to synthesize for itself are termed *non-essential* amino acids.

The amino acids are classified as essential or non-essential in Table 15.1. The list of essential acids could be expanded, for it is found that in young rats, for example, arginine and histidine are *partially essential* since the growth of the animal is so great that although it can synthesize them it is unable to do so at a sufficiently high rate. Similarly, young chicks require glycine yet the adult is able to synthesize sufficient for its needs. Moreover, cysteine is not essential provided sufficient sulphur in the form of methionine is provided. Animals are able to synthesize tyrosine from phenylalanine, itself an essential animo acid. Therefore, provided the animal's diet contains sufficient phenylalanine to supply the necessary quantity of these two amino acids, tyrosine is not itself essential. It is noteworthy that all the non-essential amino acids are glucogenic; the converse, however, is not true.

In order to test whether a particular amino acid is essential the procedure normally adopted is to feed the animal a special diet which is complete in every way except that the amino acid under consideration is absent. This can be done by feeding no protein and using pure amino acid mixtures to replace it. Under these conditions, if the amino acid is indeed essential, the animal develops a negative nitrogen balance which means that its excretion of nitrogen exceeds the nitrogen intake. This results from the breakdown of the more dispensable proteins in order to release the deficient amino acid for the synthesis of more urgently needed proteins. The excess nitrogen excretion comes from the catabolism of the more common of the non-essential amino acids released from dispensable protein. Thus a diet deficient in an essential amino acid produces symptoms similar to those obtained by feeding a very low protein diet. Long-term lack of an essential amino acid prevents growth and produces deficiency diseases and even death—especially in growing animals which need to synthesize a large quantity of protein.

Many plant proteins contain relatively small quantities of some amino acids which are essential to animals. Methionine and lysine, in particular, are often only present in small quantities. This implies that an animal could develop deficiency diseases if fed exclusively on a poorly balanced vegetarian diet.

Many animal proteins, including those in muscle, eggs, and milk, contain all the essential amino acids in suitable quantities for adequate animal nutrition.

Ruminant animals such as cattle and sheep present a special case. They do not suffer from deficiency of essential amino acids under normal conditions even though their fodder may not contain adequate amounts of them. The reason is that the microorganisms in their specially adapted stomachs are able to synthesize all the essential amino acids and vitamins, and these are eventually made available to the animal when it subsequently digests the bacteria.

15.5 The general pattern of amino acid synthesis and catabolism

In *plants*, the carbon skeleton of each of the twenty or so amino acids is derived from simple precursors. Thus two acids of the TCA cycle, namely, *oxaloacetate* and *α-ketoglutarate*, each give rise to some five amino acids. An intermediate compound of glycolysis, *3-phosphoglyceric acid*, is the precursor of four amino acids, while three more arise from *pyruvate*. D-*Erythrose-4-phosphate*, synthesized in the pentose phosphate pathway (Chapter 14), and *phosphoenolpyruvate* together provide carbon atoms for phenylalanine, tyrosine, and tryptophan. The origin of the carbon skeletons of amino acids is summarized in Table 15.2. Two amino acids, lysine and histidine, do not readily fit into the scheme. The synthesis of lysine differs in various groups of organisms—it is an essential amino acid in animals. The purine ring of ATP provides most of the atoms for histidine synthesis.

Considerable variability in the pattern of amino acid metabolism is to be found in nature, and only the major pathways in the majority of animals and some microorganisms are indicated here. Pathways represented by dashed lines in Table 15.2 only occur in green plants or in certain microorganisms; they are probably absent in animals—with the result that the amino acids are 'essential' and must be supplied in the diet (Section 15.4). It is often convenient to consider amino acids in groups. For example, the α-ketoglutarate family of amino acids share a common precursor, have similar

Table 15.2 Origin and major catabolic fate of carbon skeletons of amino acids

Precursor	Amino acid	Major product of catabolism
α-Ketoglutarate	glutamic acid, glutamine, arginine, proline, hydroxyproline	α-ketoglutarate
α-Ketoglutarate (dashed); Oxaloacetate (dashed)	lysine	various products
Oxaloacetate	aspartic acid, asparagine	oxaloacetate
Oxaloacetate (dashed)	methionine, threonine	propionate
	isoleucine	propionyl coenzyme A + acetoacetyl coenzyme A†
3-Phosphoglycerate	glycine, serine, cysteine, cystine	pyruvate
Pyruvate	alanine	pyruvate
Pyruvate (dashed)	valine	succinyl coenzyme A
	leucine	acetoacetate† + acetyl coenzyme A†
Erythrose-4-P + Phosphoenolpyruvate (dashed)	phenylalanine ↓ tyrosine	acetoacetic acid† + fumarate
	tryptophan	various products
Purine ring of ATP	histidine	α-ketoglutarate

† Indicates the product is potentially ketogenic (Section 10.7).
(Dashed arrows indicate that the pathways are largely absent in animals.)

fates, and considerable interconversion occurs between amino acids within the group.

Table 15.2 also shows the major products of catabolism of the carbon skeletons of the amino acids. Most of these can be converted to carbohydrate, as described in Chapter 12, or undergo complete oxidation in the TCA cycle (Section 9.11).

15.6 Nitrogen excretion in animals

In the catabolism of amino acids the first reaction is in most cases the removal of the α-amino group by transamination, the product being the corresponding keto acid. The amino group is thereby transferred to either oxaloacetate or α-ketoglutarate, so producing aspartate or glutamate respectively:

$$\text{RCHNH}_2\text{COOH} + \text{oxaloacetate} \underset{}{\overset{\text{Specific transaminases}}{\rightleftharpoons}} \text{RCOCOOH} + \text{aspartate}$$

$$\text{RCHNH}_2\text{COOH} + \alpha\text{-ketoglutarate} \rightleftharpoons \text{RCOCOOH} + \text{glutamate}$$

Aspartate can, in turn, pass on the amino group to α-ketoglutarate. Much of the glutamic acid so formed is then oxidized by the action of the enzyme glutamate dehydrogenase, which, as was mentioned in Section 15.2, is located in the mitochondria. The result is that *ammonia* and NADH are produced, and α-ketoglutarate is regenerated (Fig. 15.4). The NADH, upon reoxidation by the mitochondrial electron transport system, yields ATP by oxidative phosphorylation (Chapter 8).

Many aquatic animals excrete the ammonium ion. It is, in particular, the principal nitrogenous substance excreted by aquatic invertebrates, although it is also eliminated by bony fish through their gills. In other animals, several mechanisms have evolved for excreting nitrogen in less toxic forms. Thus cartilaginous fish (e.g., sharks), adult amphibians, and mammals excrete nitrogen, not as ammonia but as *urea* (Table 15.3), the amide of carbonic acid. Urea is a relatively harmless, soluble compound, so in animals where it represents the major form of nitrogen excretion it is possible for it to accumulate to quite high concentrations without deleterious effects. Thus, while an ammonium ion concentration of 0.2 mg per 100 ml in vertebrate blood is abnormally high, the normal level of urea in mammalian blood is about 40 mg per 100 ml. This has an important consequence in amphibia and mammals for it enables them to excrete nitrogen in a concentrated form

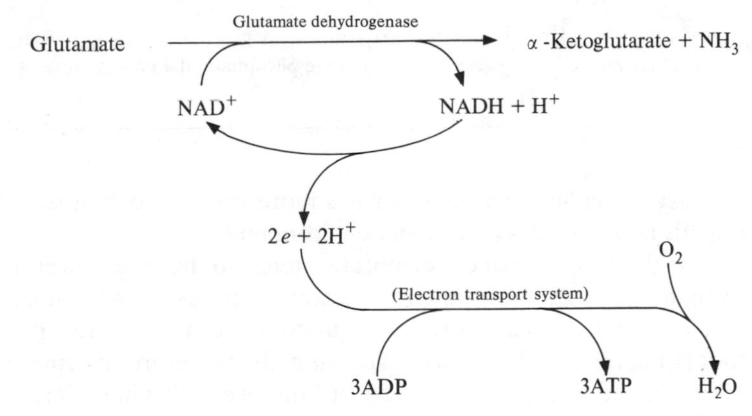

Fig. 15.4 Role of glutamate dehydrogenase in releasing ammonia from glutamate

Table 15.3 Nitrogenous excretory products of animals

Compound		Animals excreting the compound
Ammonium ion	NH_4^+	Most aquatic invertebrates; also excreted by the gills of bony fish
Urea	$H_2N-C(=O)-NH_2$	Cartilaginous fish, amphibians, and mammals
Trimethylamine oxide	$(H_3C)_3N \rightarrow O$	Excreted by the kidneys of marine bony fish
Uric acid	(H–N1, C2(=O), N3–H, C4, C5, C6(=O), N7–H, C8(=O), N9–H)	Insects, terrestrial molluscs, terrestrial reptiles, and birds. In many mammals, uric acid is the end-product of purine but not of amino acid metabolism
Guanine	(H_2N–C, N, C–OH, C–N, C–H, N–H)	Spiders
Allantoin	(H_2N–C(=O)–NH–CH; O=C–NH–C(=O)–NH)	Product of the catabolism of *purines* in many mammals (e.g., dog). An enzyme, *uricase*, oxidizes uric acid removing carbon atom number 6 to produce allantoin
Creatinine	(HN=C–NH–C(=O)–CH_2–N–CH_3)	Vertebrates. It is a product of the breakdown of creatine phosphate, the energy store of muscle

and so to survive on land where water is a more precious commodity than it is for organisms living in an aqueous environment.

The body fluids of marine vertebrates tend to have a lower osmotic pressure than sea water. Consequently, water would be drawn out of fish by osmosis if they did not have special regulatory mechanisms to oppose the water loss (Chapter 17). Like terrestrial animals, therefore, marine fish live in an environment where conservation of body water is vital. Cartilaginous fish have a high blood urea concentration which raises their osmotic pressure to that of sea water, so preventing loss of excess water by osmosis. The urine

Table 15.4 Relative amounts of nitrogenous compounds excreted by mammals on high and low protein diets

Nitrogenous compounds	High nitrogen diet	Low nitrogen diet	Difference
Urea	24	6	18
Creatinine	1	1	0
Ammonia	1	1	0
Uric acid	2	1	1
Others	~2	~1	~1
Total nitrogen excreted	30	10	20

of many marine bony fish contains trimethylamine oxide which, like urea, is a relatively non-toxic yet highly soluble compound.

From the point of view of conservation of water, birds and insects are the most successful of terrestrial animals, for the bulk of their unwanted nitrogen is excreted in the form of the triketopurine, *uric acid* (Table 15.3). This sparingly soluble substance is excreted as a solid with the result that little water is wasted in eliminating it. Such animals are described as *uricotelic* to contrast them with those which excrete nitrogen predominantly in the form of ammonia or urea—and which are known as *ammoniotelic* and *ureotelic* respectively. Spiders excrete nitrogen mainly as the purine, guanine, which, like the uric acid excreted by insects, is eliminated as a relatively dry mass.

Several nitrogenous compounds are present in mammalian urine. Of these, urea is quantitatively the most important, but small quantities of uric acid, allantoin, ammonia, and creatinine (Table 15.3) are also present. The uric acid, unlike the bulk of the urate produced by uricotelic animals, arises from the catabolism of nucleotide bases of nucleic acids. Some mammals further degrade the uric acid to allantoin. Creatinine arises from creatine, the compound which, as creatine phosphate, acts as a store of high energy phosphate in the muscles (Section 20.4). The urea, on the other hand, is the main end-product of protein catabolism. Thus when mammals are fed a very high protein diet the excretion of urea rather than of other nitrogenous compounds increases (Table 15.4).

15.7 Urea synthesis in mammals

The major nitrogenous excretory product when amino acids are catabolized in mammals is urea, $(H_2N)_2C{=}O$. Being highly soluble in water, it is easily excreted by the kidneys.

Urea is synthesized by a cycle of reactions known as the *ornithine cycle.* Ornithine is an α-amino acid which is not found in animal proteins. Its formula is $H_2N(CH_2)_3CHNH_2COOH$. Krebs first identified the cycle in

1932 when he observed that the addition of small quantities of ornithine or arginine significantly increased the synthesis of urea by liver slices. More recently, citrulline has been shown to have a similar stimulatory effect.

The synthesis of urea is an endergonic process. Three molecules of ATP are required for each molecule of urea formed. Clearly, mammals expend considerable energy simply to render the catabolic product of nitrogen metabolism innocuous. All the enzymes of the ornithine cycle are present in the liver, although some are present in other organs as well (e.g., the kidney and mammary gland). The cycle can be conveniently divided into a preliminary step followed by three others. Both the preliminary and first step occur in liver mitochondria, but the final two steps are located in the cytosol, an interesting example of compartmentation of reactions.

The preliminary step

Before entering the ornithine cycle, ammonium ions are linked to carbon dioxide and phosphate to form a high energy compound, *carbamyl phosphate.* Carbamic acid is another name for aminoformic acid, H_2NCOOH. The phosphate of carbamyl phosphate is derived from ATP, two molecules being required for each carbamyl phosphate produced:

$$NH_3 + CO_2 + 2ATP \xrightarrow[\text{synthase}]{\text{Carbamyl-P}} H_2N\text{—}C(=O)\text{—}O{\sim}P + 2ADP + P_i$$

Carbamyl ~P

Carbamyl phosphate provides the carbon atom and *one* of the nitrogen atoms of the urea molecule. The other amino group in urea is derived from aspartic acid.

Step 1

In this reaction, the carbamyl phosphate reacts with a diamino acid, *ornithine.* Ornithine, which gives the cycle of reactions its name, is an acid containing five carbon atoms in its molecule, and is thus related to *glutaric acid* and *glutamic acid.* Since ornithine is re-formed in the third and final step of the sequence, it can be regarded as a cocatalyst (Fig. 15.5).

The carbamyl group is transferred from carbamyl phosphate to ornithine by an enzyme known as *ornithine carbamyl transferase.* The amino acid so formed is called citrulline, the other product being inorganic orthophosphate. Citrulline consists of a carbamyl group attached to the terminal or δ-amino group of ornithine (Fig. 15.5).

Step 2

An amino group is next transferred from aspartic acid to the carbamyl keto group of citrulline to form the amino acid *arginine.* It should be remembered

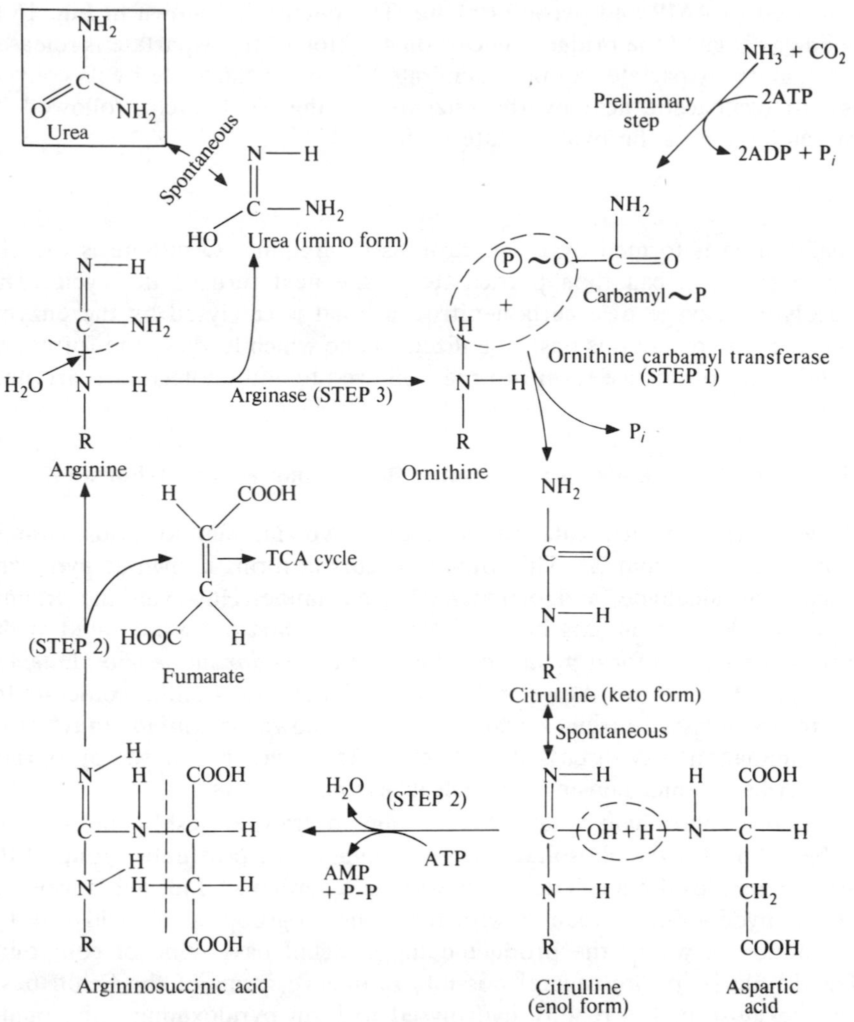

Fig. 15.5 **The ornithine cycle for the synthesis of urea. Throughout the cycle R is —CH_2—CH_2—CH_2—$CHNH_2$—COOH**

that the aspartic acid can itself originate from oxaloacetic acid by transamination from glutamic acid and so, via the glutamate dehydrogenase reaction, its nitrogen atom can originate from unwanted ammonium ions (Figs. 15.2 and 15.5). This nitrogen atom eventually becomes the second nitrogen atom of the final urea molecule. The reaction concerned is an example of one in which aspartic acid forms a 'nitrogen bridge' which is later cleaved on the opposite side from that on which it is initially formed. The intermediate double molecule is called argininosuccinic acid and the energy for the formation of the 'nitrogen bridge' originates from ATP, the latter being converted to AMP and pyrophosphate. The details are shown in Fig. 15.5. Upon cleavage of the bridge, the carbon skeleton of the aspartate is released as fumarate. Aspartate can be regenerated from the fumarate by its conversion to oxaloacetic acid by the enzymes of the TCA cycle, followed by transamination of the oxaloacetate so formed.

Step 3

Finally, urea is formed by the hydrolysis of arginine. Ornithine is thereby regenerated and can then participate in the next turn of the cycle. This hydrolytic cleavage of a carbon–nitrogen bond is catalysed by the enzyme *arginase*. The reaction is best visualized as one which leads to the formation of the imino resonance form of urea, followed by intramolecular rearrangement.

15.8 The role of derivatives of vitamin B_6 in amino acid metabolism

Many enzymes which catalyse reactions involving amino acids contain derivatives of vitamin B_6. This vitamin occurs in forms known as *pyridoxal*, which is an aldehyde, and *pyridoxamine*, an amine. However, the vitamin does not take part directly in metabolism, but is first phosphorylated at the expense of ATP to form *pyridoxal phosphate* or *pyridoxamine phosphate*, the formulae of which are shown in Fig. 15.6. These are essential cofactors for the transaminases, enzymes which are also known as *amino transferases*. Transamination was discussed in Section 15.2 and it is now appropriate to describe the mechanism by which it occurs.

Pyridoxal phosphate is attached to amino transferases by links which probably involve the phosphate group. Acting as the prosthetic group of the enzyme, pyridoxal phosphate reacts with an α-amino acid in such a way that the aldehyde group condenses with the α-amino group of the acid with the elimination of water, the product being a 'Schiff base' type of compound (Fig. 15.6). There are several possible resonance forms of the Schiff base; the one illustrated is readily hydrolysed to form pyridoxamine phosphate, still attached to the enzyme, and a free α-keto acid corresponding to the original α-amino acid. The reactions illustrated are reversible. Thus with the enzyme alanine:α-ketoglutarate amino transferase, the α-amino group of alanine may be transferred to enzyme-bound pyridoxal phosphate to form pyruvate and pyridoxamine phosphate. The amino group is transferred from the latter to *α-ketoglutarate* to form glutamic acid and thus the prosthetic group is converted back to the pyridoxal phosphate form:

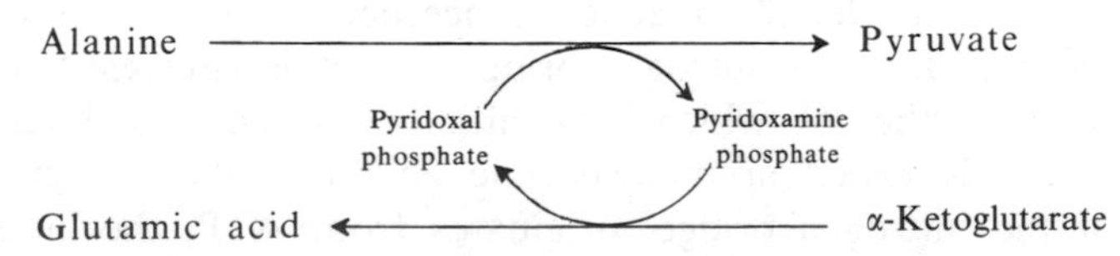

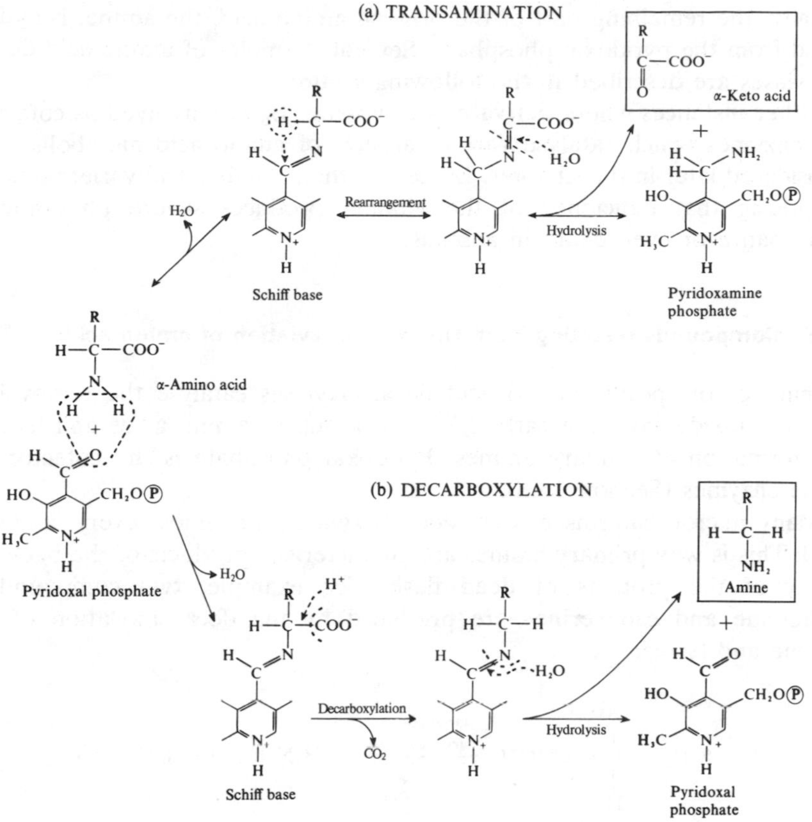

Fig. 15.6 The role of derivatives of vitamin B_6 in the mechanism of (a) aminotransferases (transaminases) and (b) amino acid decarboxylases

Thus, during transamination, the aldehyde and amino forms of the cofactor are repeatedly undergoing interconversion. The importance of the cofactor is shown by the fact that the tissues of animals which are deficient in vitamin B_6 have reduced transaminase activity, the activity being restored to normal when the vitamin is supplied in the diet.

The *amino acid decarboxylases* represent a second group of enzymes for which pyridoxal phosphate is the cofactor. These enzymes catalyse the general reaction where the α-carboxyl group of an amino acid is removed as carbon dioxide, leaving the corresponding primary amine:

$$R\text{—}CHNH_2 \vdots COOH \longrightarrow CO_2 + RCH_2NH_2$$

The reaction is brought about by the amino acid forming a Schiff base with enzyme-attached pyridoxal phosphate (Fig. 15.6). The carboxyl group of the amino acid is thereby made labile so that it breaks away as carbon dioxide.

Finally, the remaining part of the original amino acid, the amine, is hydrolysed from the pyridoxal phosphate. Several examples of amino acid decarboxylases are described in the following section.

Other instances where derivatives of vitamin B_6 are involved as cofactors for enzymes which catalyse various aspects of amino acid metabolism are considered later in this chapter. In view of their number and variety it is not surprising that deficiency of the vitamin produces severe physiological disturbance or even death in animals.

15.9 Compounds resulting from the decarboxylation of amino acids

A number of specific amino acid decarboxylases catalyse the removal of carbon dioxide from the carboxylic acid group of amino acids and lead to the formation of primary amines. Pyridoxal phosphate is the cofactor for these enzymes (Section 15.8).

Many microorganisms possess decarboxylases for almost every α-amino acid. This is why primary amines are characteristic products of the bacterial decay of the proteins of dead flesh. For example, two compounds—putrescine and cadaverine—are produced by the decarboxylation of ornithine and lysine:

$$\underset{\text{Ornithine}}{H_2N{-}(CH_2)_3{-}C(NH_2)(H){-}COOH} \xrightarrow[-CO_2]{\text{Ornithine decarboxylase}} \underset{\text{Putrescine}}{H_2N.(CH_2)_3{-}CH_2{-}NH_2}$$

$$\underset{\text{Lysine}}{H_2N{-}(CH_2)_4{-}C(NH_2)(H){-}COOH} \xrightarrow[-CO_2]{\text{Lysine decarboxylase}} \underset{\text{Cadaverine}}{H_2N{-}(CH_2)_4{-}CH_2{-}NH_2}$$

Some amino acid decarboxylases occur in almost all organisms so far investigated. Particularly common are those which catalyse the decarboxylation of serine, aspartic acid, and cysteine. The products are of great importance—ethanolamine is a precursor of choline (page 28):

$$\underset{\text{Serine}}{HO{-}CH_2{-}C(NH_2)(H){-}COOH} \xrightarrow[-CO_2]{\text{Serine decarboxylase}} \underset{\text{Ethanolamine}}{HO{-}CH_2{-}CH_2{-}NH_2} \rightarrow \text{Choline}$$

and both β-thioethylamine and β-alanine contribute to the structure of coenzyme A:

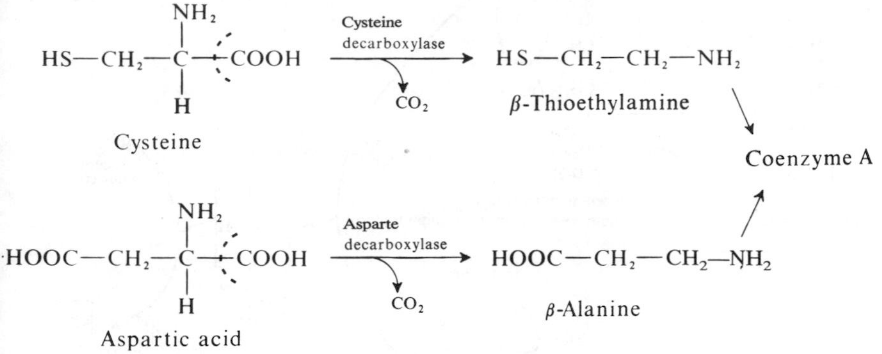

$$HS{-}CH_2{-}\underset{H}{\overset{NH_2}{C}}{-}COOH \xrightarrow[-CO_2]{\text{Cysteine decarboxylase}} HS{-}CH_2{-}CH_2{-}NH_2 \rightarrow \text{Coenzyme A}$$

Cysteine → β-Thioethylamine

$$HOOC{-}CH_2{-}\underset{H}{\overset{NH_2}{C}}{-}COOH \xrightarrow[-CO_2]{\text{Asparte decarboxylase}} HOOC{-}CH_2{-}CH_2{-}NH_2 \rightarrow \text{Coenzyme A}$$

Aspartic acid → β-Alanine

Many vertebrate tissues contain histidine decarboxylase, an enzyme which catalyses the production of the local hormone *histamine*:

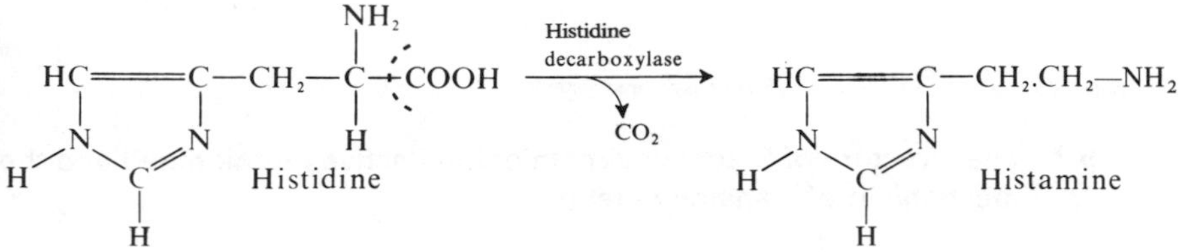

$$\text{Histidine} \xrightarrow[-CO_2]{\text{Histidine decarboxylase}} \text{Histamine}$$

Histidine: HC═C—CH₂—C(NH₂)(H)—COOH (imidazole ring: HC, N–H, C–H, N)

Histamine: HC═C—CH₂.CH₂—NH₂ (imidazole ring: HC, N–H, C–H, N)

It will be shown later that decarboxylases are also important in the synthesis of adrenalin.

15.10 The transfer of units containing only one carbon atom

In the following sections of this chapter we shall describe the synthesis of several important but complex compounds derived from amino acids. But first attention must be drawn to two biochemical systems which transfer one-carbon fractions—methyl, formyl, formimino, and formaldehyde—from one molecule to another, for such systems are frequently involved in the metabolism of amino acids.

It will be recalled (Chapter 5) that two-carbon acetate units seldom participate directly in cellular biochemistry but in the form of 'active acetate'. As has been seen, 'active acetate' owes its activity to the fact that the acetate group is actually attached to a carrier or coenzyme by a high energy bond, being, in fact, in the form of acetyl coenzyme A. The activity of one-carbon fragments is somewhat similar, though the carriers are different. One carrier functions as a carrier of active methyl groups and a second as a carrier of 'active formate' and of the two similar fragments containing one carbon atom mentioned above.

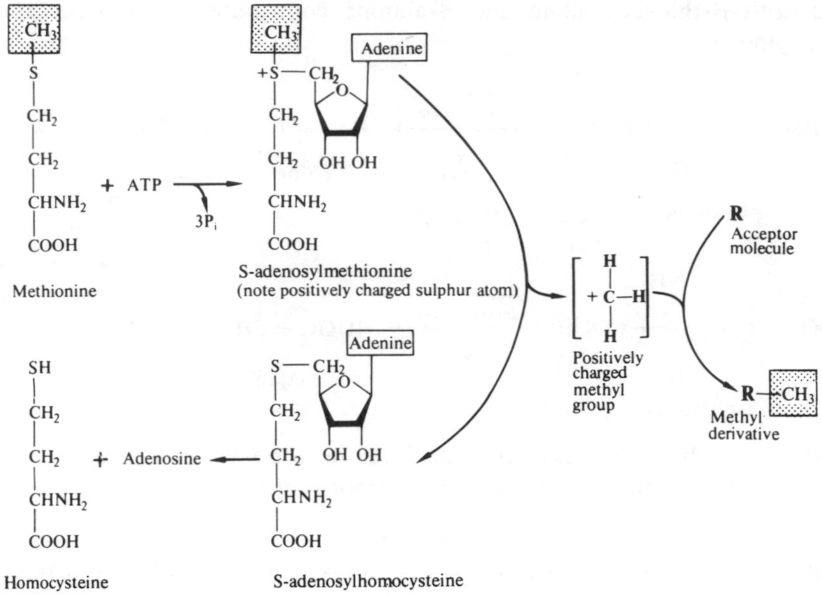

Fig. 15.7 The structure of S-adenosylmethionine ('active methionine') and the mechanism of transmethylation

Methionine is not only an essential amino acid but acts as a donor of methyl groups. In order that it can fulfil this function, its sulphur atom has to assume the charged *sulphonium* form and this is achieved by reaction with ATP. 'Active methyl' is thus usually 'active methionine' and the latter is *S-adenosylmethionine* (the S- indicating that the adenosine is attached to the sulphur atom). The structure of this compound is shown in Fig. 15.7. Following heterolytic cleavage, the positively charged methyl group can attach itself to several types of atoms possessing a suitably placed lone pair of electrons. After donating its methyl group, the methionine is liberated as a sulphydryl amino acid; this is known as homocysteine because it contains one methylene group more than cysteine:

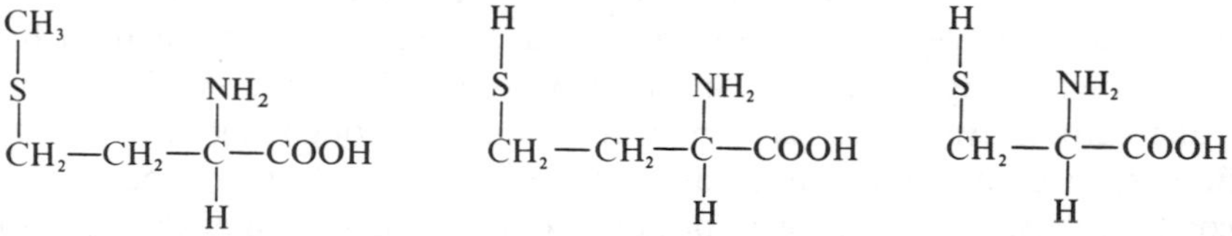

$$CH_3{-}S{-}CH_2{-}CH_2{-}CH(NH_2){-}COOH$$

Methionine

$$H{-}S{-}CH_2{-}CH_2{-}CH(NH_2){-}COOH$$

Homocysteine

$$H{-}S{-}CH_2{-}CH(NH_2){-}COOH$$

Cysteine

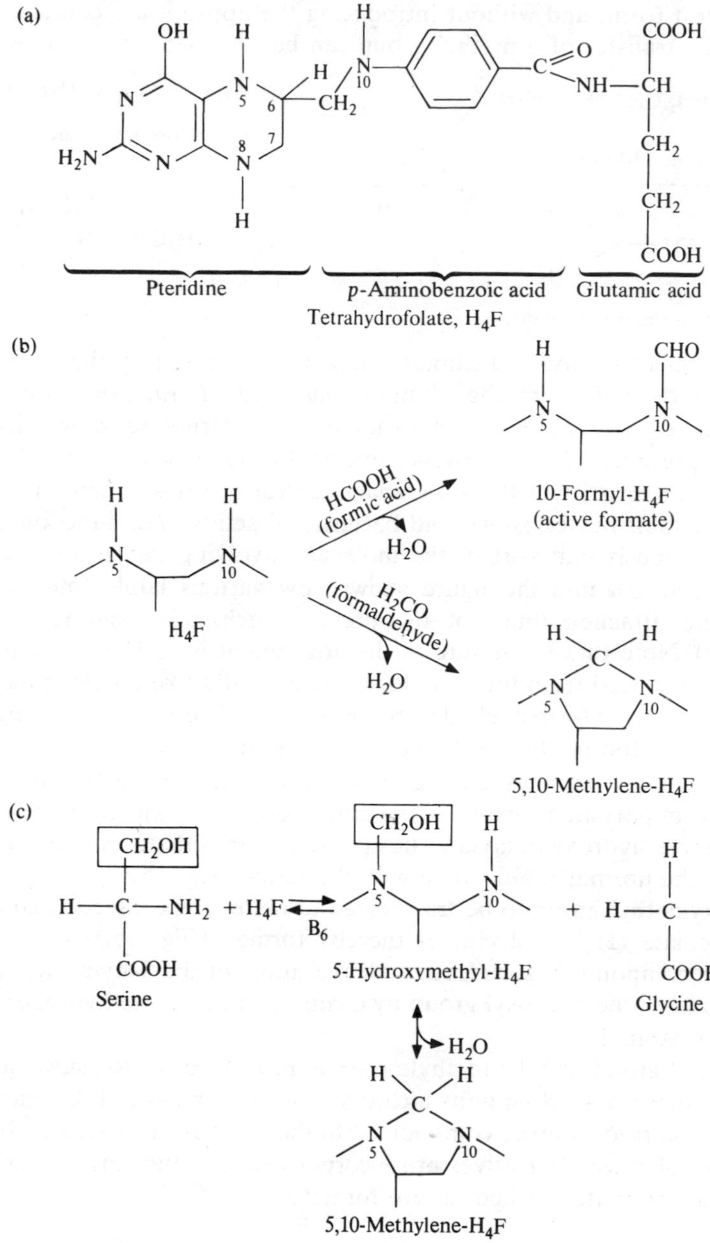

(a)

Pteridine | *p*-Aminobenzoic acid | Glutamic acid

Tetrahydrofolate, H_4F

(b)

H_4F

HCOOH (formic acid) → H_2O

10-Formyl-H_4F (active formate)

H_2CO (formaldehyde) → H_2O

5,10-Methylene-H_4F

(c)

Serine + H_4F $\rightleftharpoons$ (B_6) 5-Hydroxymethyl-H_4F + Glycine

5-Hydroxymethyl-H_4F ⇅ H_2O

5,10-Methylene-H_4F

Fig. 15.8 (a) The structure of tetrahydrofolic acid (H_4F),
(b) its conversion to two forms of a carrier of one-carbon units, and
(c) the reaction catalysed by serine hydroxymethylase

In its simplest form, and without introducing the appropriate cofactors and enzymes, the transfer of a methyl group can be represented as follows:

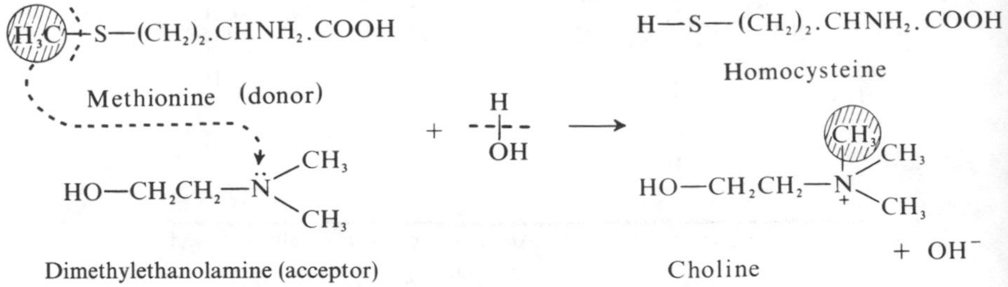

H_3C—S—$(CH_2)_2$.CHNH$_2$.COOH
Methionine (donor)

+ H—OH ⟶

H—S—$(CH_2)_2$.CHNH$_2$.COOH
Homocysteine

HO—CH_2CH_2—N(CH_3)$_2$
Dimethylethanolamine (acceptor)

HO—CH_2CH_2—N^+(CH_3)$_3$ + OH^-
Choline

The transfer of formyl and similar fractions is achieved by the mediation of a carrier derived from the vitamin *folic acid. Tetrahydrofolate*, often abbreviated to H_4F, has the somewhat complex structure shown in Fig. 15.8a. The presence of a *p*-aminobenzoic acid residue is of interest for it is the structural similarity of the sulphonamide drugs to this compound which accounts for their bacteriostatic and bactericidal action. The functional part of the coenzyme is that part of the molecule involving the nitrogen atoms marked 5 and 10, and the figure shows how various labile 'one-carbon' fractions are attached (but not the precise mechanism leading to their attachment). Note that the nature of the attachment is such that the formyl derivative, produced from formate, looks superficially like an aldehyde, and that the addition of formaldehyde produces a new five-membered ring, the carbon atom of the methylene being one of the members.

There are many reactions leading to one-carbon derivatives of tetrahydrofolate. One important example is the conversion of serine to glycine. The enzyme, serine hydroxymethylase, has pyridoxal phosphate as one cofactor. This forms the normal Schiff's base with the serine (Fig. 15.6) which enables the hydroxymethyl group to be transferred to the enzyme's second cofactor, tetrahydrofolate (H_4F). Glycine is thereby formed (Fig. 15.8c). The final one-carbon compound results from a 'dehydration' of the 5-hydroxylmethyl-H_4F which loses the hydroxyl group by combination with the hydrogen atom on nitrogen atom 10.

The methyl group of 5,10-methylene-H_4F may be converted subsequently to several other one-carbon units including =CH— and —CHO, which may then be transferred to other compounds. In the synthesis of purines (Section 15.12) both of these derivatives are incorporated into the ring systems and are often appropriately called 'active formate'.

15.11 Formation of adrenalin and nor-adrenalin

The hormone adrenalin (epinephrin) is secreted by the medulla of the adrenal gland. In addition to increasing the level of glucose in the blood

(Section 13.6), it produces a remarkable set of physiological reactions. It raises the rate of heart beat, increases the blood pressure, and causes dilatation of blood vessels in the skeletal muscles. These effects allow more oxygen and glucose to reach the muscle cells and so prolong their aerobic metabolism and enhance their ability to contract. Adrenalin is secreted in response to fear and its effect is to enable the animal to make a rapid escape from danger. Nor-adrenalin, though chemically very similar, has several physiological effects different from those of adrenalin—it has, for example, only a small effect on the blood glucose level. Its main function is as a local chemical transmitter which mediates the passage of signals across neuromotor junctions (Section 19.5).

In the presence of NADPH, phenylalanine is oxidized to tyrosine by a hydroxylase system in the endoplasmic reticulum. Tyrosinase inserts a second hydroxyl group; this enzyme is a flavoprotein which requires copper ions as cofactor. Decarboxylation then occurs to give hydroxytyramine (Fig. 15.9). This is next hydroxylated in the side chain on the carbon atom nearest the ring. The product is nor-adrenalin; two optical isomers of this substance can be synthesized in the laboratory, but as so often occurs in nature, asymmetric synthesis facilitated by enzymes leads only to the production of the active (laevo) isomer.

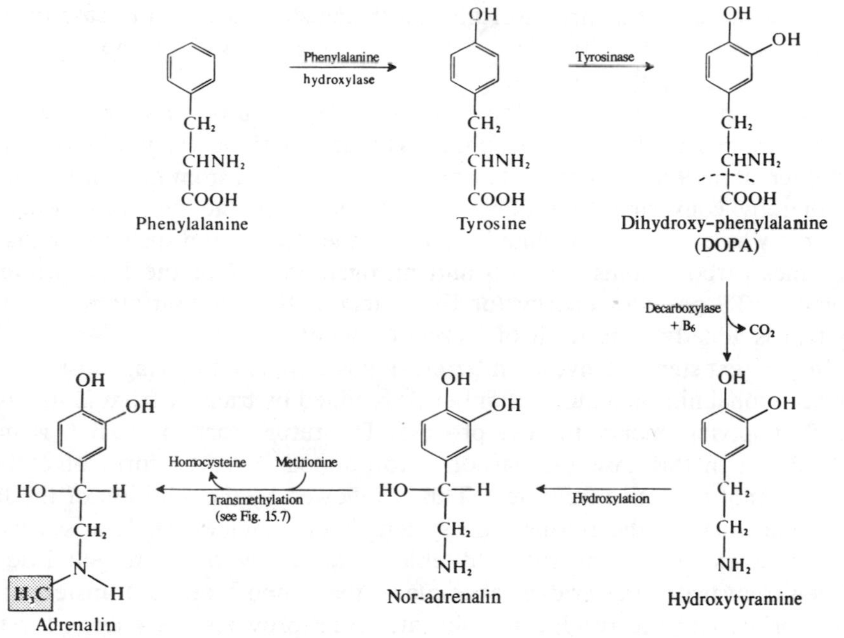

Phenylalanine → (Phenylalanine hydroxylase) → Tyrosine → (Tyrosinase) → Dihydroxy-phenylalanine (DOPA) → (Decarboxylase + B_6; CO_2) → Hydroxytyramine → (Hydroxylation) → Nor-adrenalin → (Transmethylation (see Fig. 15.7); Methionine → Homocysteine) → Adrenalin

Fig. 15.9 Biosynthesis of adrenalin and nor-adrenalin from phenylalanine

L-Nor-adrenalin is converted to L-adrenalin by the addition of a methyl group. The latter is usually derived from methionine by the mechanism shown in Fig. 15.7, the acceptor molecule in this case being nor-adrenalin.

15.12 The synthesis of purine rings

The purine structure is an excellent example of a complicated molecule constructed from numerous simpler units. The origin of the atoms in the rings may be indicated thus:

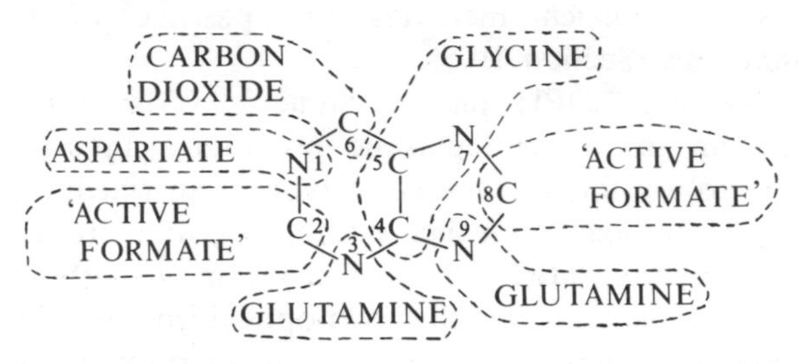

The process leading to purine synthesis is complex and only a simplified account is presented here. It can be followed by constant reference to Fig. 15.10, in which each atom of the ring is numbered in the reaction where it is inserted. The route of synthesis is such that the ribose and phosphate of the prospective nucleotide are inserted first (being shed later in the case of uric acid formation). The *smaller* of the two rings of the purine system is completed before the larger ring.

The purine begins its development by the addition of an amino group to carbon atom 1 of ribose-5-phosphate. This amino group eventually becomes nitrogen atom 9 of the completed purine. It is derived from glutamine which is thereby converted to glutamate. ATP supplies the necessary energy. Next, a whole glycine molecule is added to the structure in such a way that it provides carbon atoms 4 and 5 and nitrogen atom 7 of the final product. Again, ATP provides energy for the reaction. Now carbon atom 8 of the system is added in the form of 'active formate'.

In the next step, the five-membered ring is completed by ring closure, and an additional nitrogen atom (number 3) is added by transfer from glutamine, ATP supplying energy for the process. The future carbon atom 6 is now added but in this case the carbon incorporated is in the form of carbon dioxide (not as 'active formate'). This is followed by the addition of the last nitrogen atom of the purine ring system (atom number 1). It arises from aspartate *via* an intermediate 'double molecule' with a nitrogen bridge. When cleavage of the bridge takes place, the amino group is transferred to the purine, and the residue is fumarate. ATP provides the energy for the reaction. This process resembles the conversion of citrulline to arginine (step

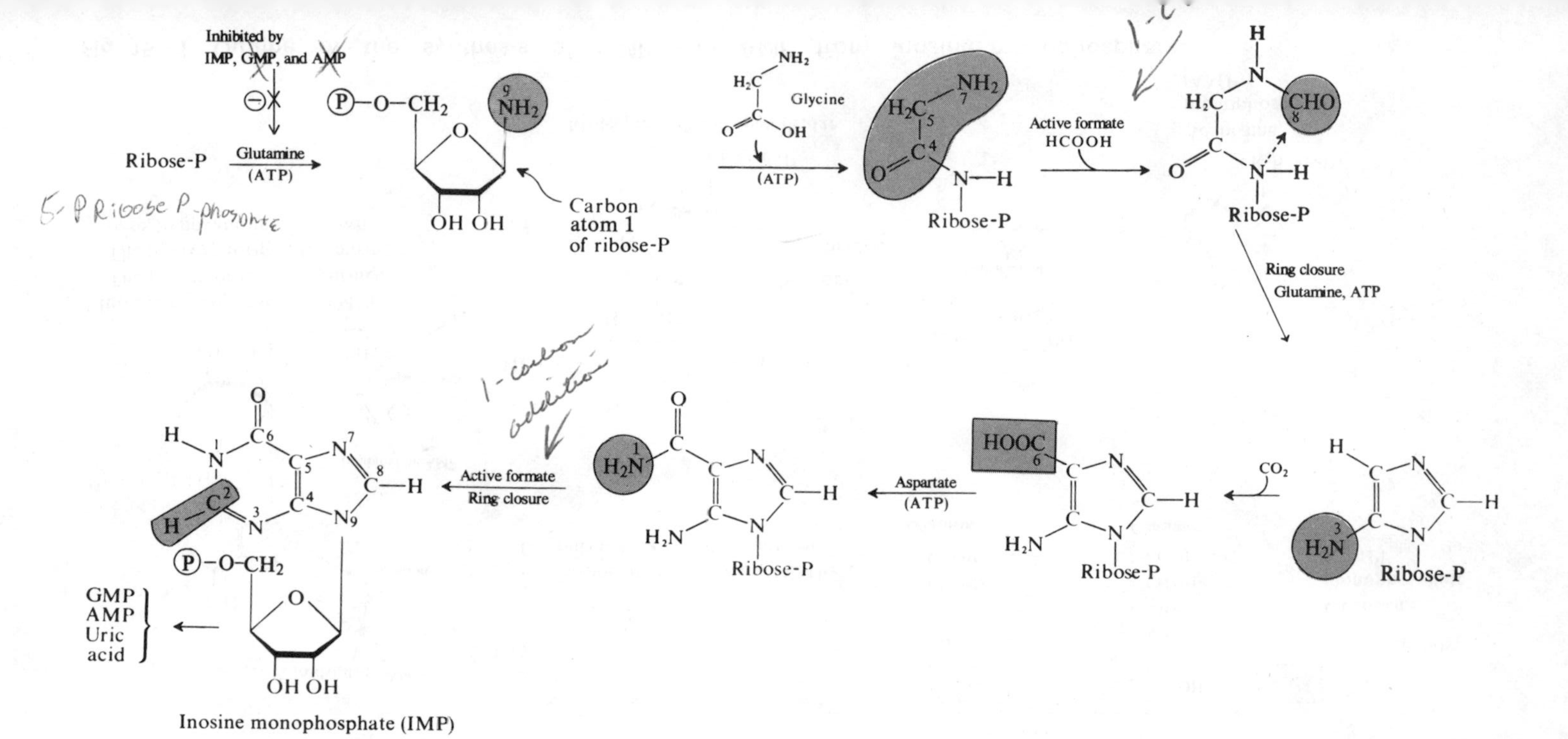

Fig. 15.10 Simplified outline of purine synthesis. (Numbers refer to atoms in the final purine molecule)

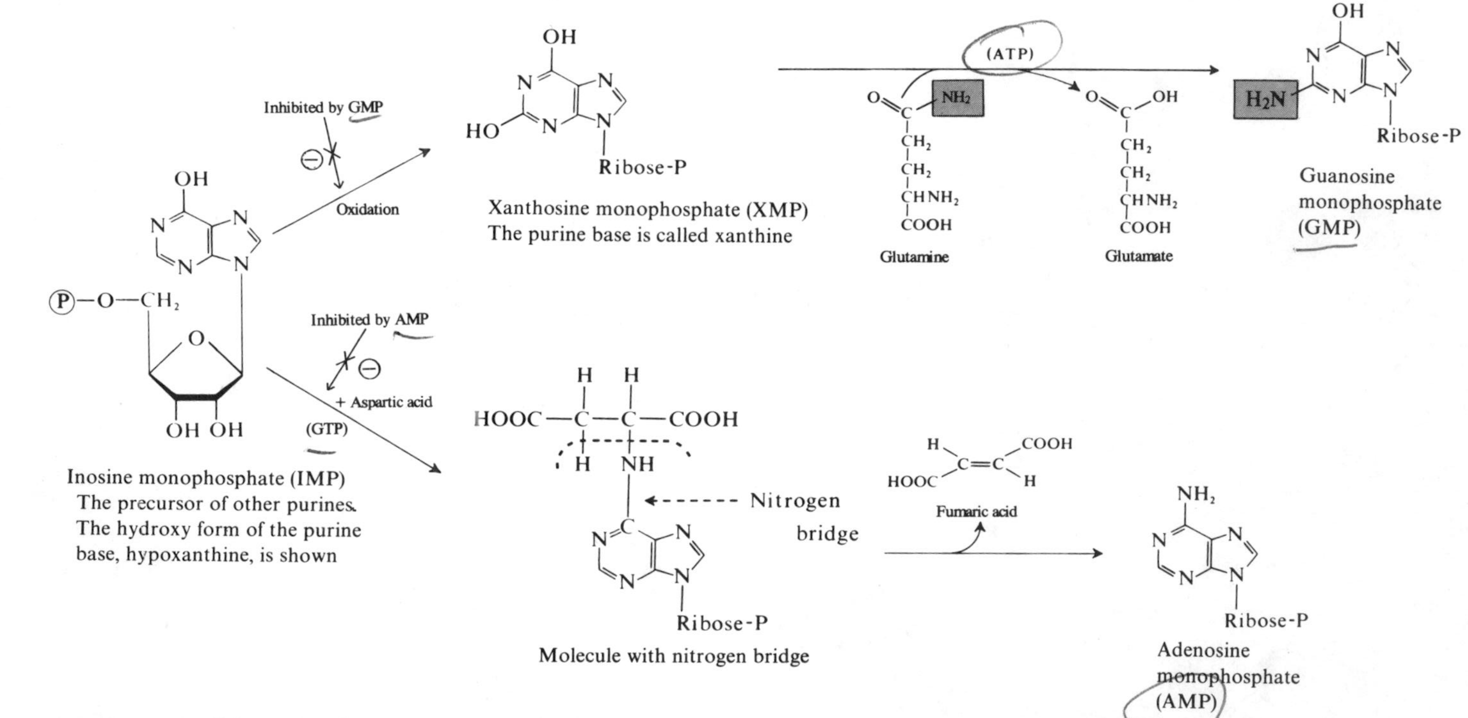

Fig. 15.11 Outline of the synthesis of GMP and AMP from inosine monophosphate

2 in Fig. 15.5) and the synthesis of AMP from IMP (Fig. 15.11). Finally, carbon atom 2 of the purine base is added—as 'active formate'—and a complete purine nucleotide, inosine monophosphate (IMP), results as closure takes place to form the larger ring.

In uricotelic animals, uric acid is synthesized by the pathway leading to IMP. The process is completed by the removal of the ribose phosphate from the purine, the latter (hypoxanthine) then being oxidized by flavoprotein enzymes to uric acid.

15.13 The biosynthesis of the nucleotides AMP and GMP

As we have just seen, the nucleotide inosine monophosphate (IMP) is synthesized from a number of relatively simple compounds. It is, in turn, the precursor of AMP and GMP. IMP is actually *hypoxanthine ribose phosphate,* hypoxanthine being a monohydroxypurine (or monoketopurine since tautomeric change is possible). It is converted to AMP or GMP by the steps shown in the Fig. 15.11; the roles, as nitrogen donors, of aspartate (which forms an intermediate compound with a nitrogen bridge) and of glutamine are noteworthy.

The pathways leading to the synthesis of IMP, AMP, and GMP have been the subject of many investigations of metabolic control. The cell needs an adequate supply of the nucleotides in view of their numerous functions (Chapter 5). Moreover, the ratio of adenine- to guanine-based compounds is critical in the composition of the nucleic acids. One very important point of control is, as is so often the case, the initial reaction of the series and involves feedback control (Section 13.1). The first reaction shown in Fig. 15.10 is strongly inhibited by excessive levels of GMP, AMP, or IMP which act as allosteric inhibitors.

A similar situation exists at the branching point where IMP is destined to undergo conversion to GMP or to AMP. For each branch the final product when in excess inhibits the initial reaction (Fig. 15.11). It is interesting to note that ATP is needed in the formation of GMP while GTP provides the energy for AMP synthesis.

15.14 The synthesis of pyrimidines and porphyrins

The pyrimidine ring system is derived from the amino acid, aspartate, and carbamyl phosphate. As in the synthesis of urea, the carbamyl phosphate provides a carbon and a nitrogen atom for the final product. In the present instance the carbamyl phosphate synthase is not located in the mitochondria as occurs in urea synthesis (Section 15.7) but in the cytosol. In mammals the cytosolic enzyme is specifically inhibited by UTP. Since uridine is a

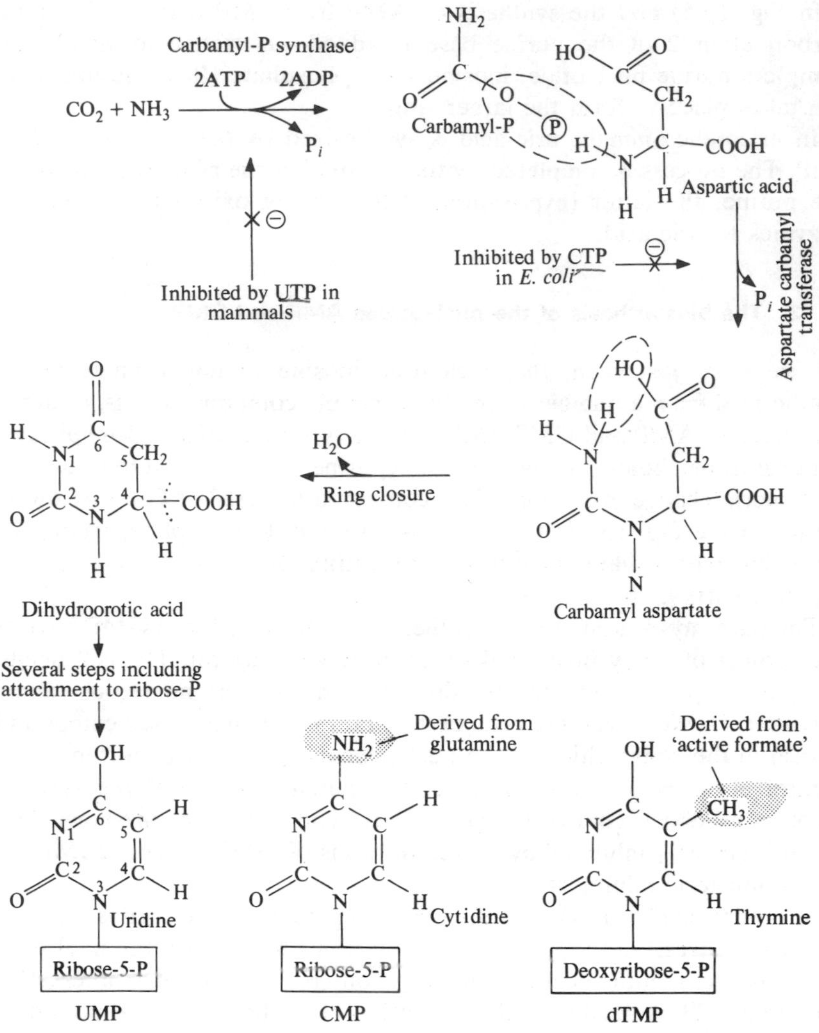

Carbamyl-P synthase
2ATP 2ADP
$CO_2 + NH_3$
P_i
Inhibited by UTP in mammals
NH_2
Carbamyl-P
Aspartic acid
Inhibited by CTP in *E. coli*
Aspartate carbamyl transferase
P_i
Carbamyl aspartate
H_2O
Ring closure
Dihydroorotic acid
Several steps including attachment to ribose-P
Uridine
Ribose-5-P
UMP
Derived from glutamine
Cytidine
Ribose-5-P
CMP
Derived from 'active formate'
Thymine
Deoxyribose-5-P
dTMP

Fig. 15.12 Simplified outline of pyrimidine synthesis

pyrimidine this is an example of the first reaction of a series being controlled by one of the final products. Some microorganisms have a different means of control. In *Escherichia coli*, for example, it is the next reaction of the series (Fig. 15.12) which is the point of control, for the enzyme, aspartate carbamyl transferase, is strongly inhibited allosterically by CTP (Section 3.9). Thus the site of control differs but the final result is equally effective.

Figure 15.12 shows some of the main reactions leading from carbamyl phosphate to UMP. In the second reaction the carbamyl phosphate unites with aspartate with the elimination of inorganic phosphate, the reaction being catalysed by aspartate carbamyltransferase. In the following reaction the carbamyl aspartate loses a molecule of water, so forming the six-membered ring. The initial pyrimidine, dihydroorotic acid, then undergoes several reactions to form UMP. These include decarboxylation and attachment to ribose-5-phosphate via PRPP (Fig. 14.10).

The other two important pyrimidine bases are derived from UMP. CMP is produced by addition of the amino group from the amido group of glutamine, a reaction requiring ATP. Thymine is produced from the deoxyribose compound, dUMP, which accepts a methyl group from 'active formate', so forming dTMP (Fig. 15.12).

The porphyrin ring system occurs in the cytochromes (Section 8.6) and in the haem part of haemoglobin (Section 18.1).

The synthesis begins with a reaction in which the TCA cycle intermediate, succinyl coenzyme A, reacts with glycine. Coenzyme A and the carboxyl group of the glycine molecule are eliminated, producing δ-aminolaevulinic acid. This acid contains five carbon atoms and could also be called γ-keto-δ-aminovaleric acid:

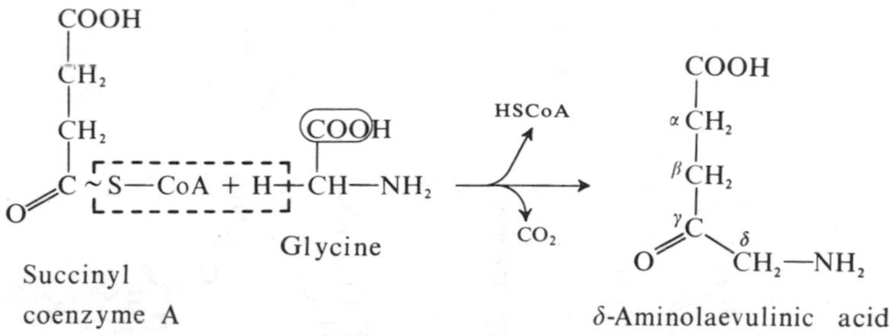

$$HOOC-CH_2-CH_2-C(=O)\sim S-CoA + H-CH(COOH)-NH_2 \xrightarrow[CO_2]{HSCoA} HOOC-{}^{\alpha}CH_2-{}^{\beta}CH_2-{}^{\gamma}C(=O)-{}^{\delta}CH_2-NH_2$$

Succinyl coenzyme A + Glycine → δ-Aminolaevulinic acid

This is the first of the series of reactions leading to the synthesis of haem. The aminolaevulinic acid synthase is inhibited by excess haem and thus demonstrates another example of an initial reaction being controlled by a final product.

The five-membered heterocyclic pyrrole ring is then formed by the condensation of 2 molecules of δ-aminolaevulinic acid, 2 molecules of water being eliminated, the oxygen atom in each case being provided by the γ-carbonyl groups. The product is known as *porphobilinogen*; two of its carbon atoms and both nitrogen atoms are derived from the two original glycine molecules (Fig. 15.13).

The porphyrin structure is derived from four porphobilinogen molecules which link together with the elimination of the side-chain amino groups. The

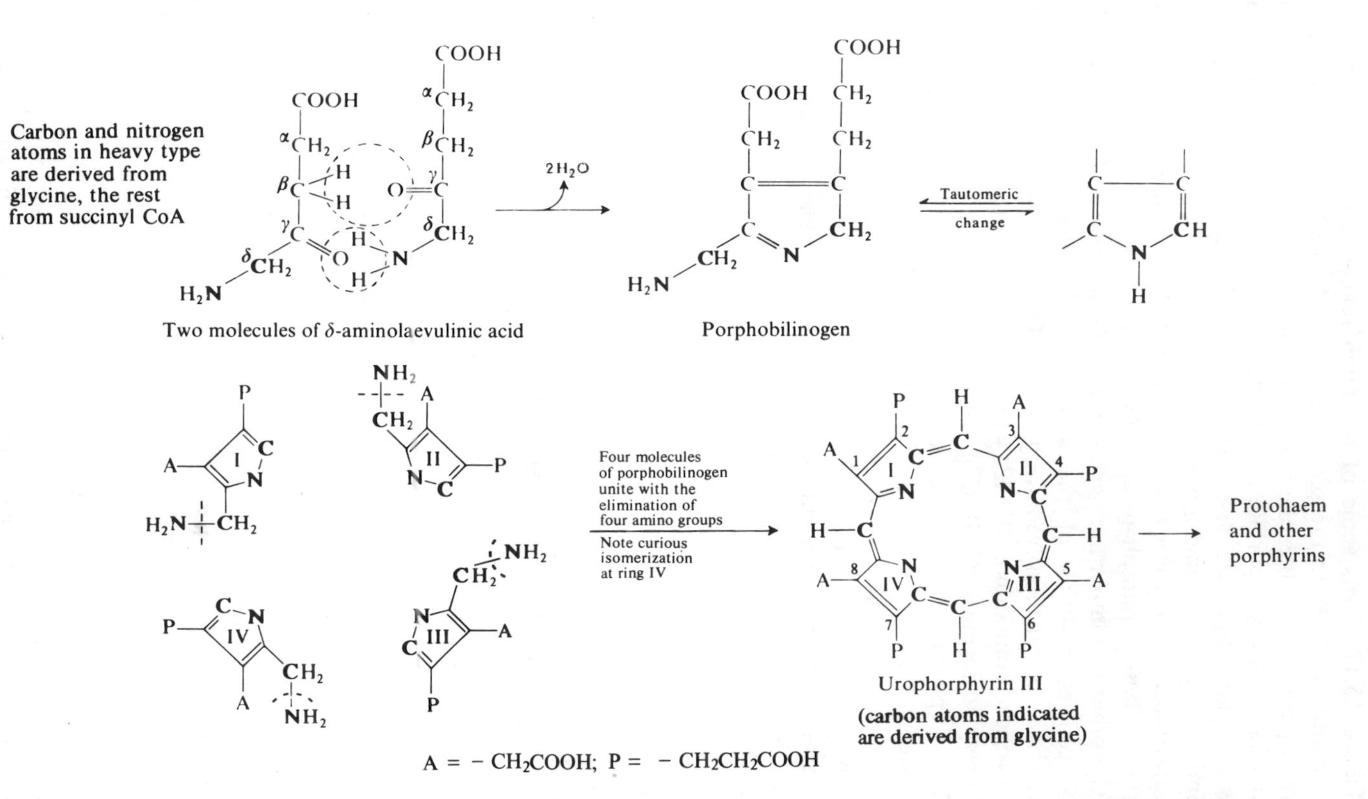

Fig. 15.13 Outline of the synthesis of the porphyrin ring system

mechanism is complicated and only partially understood. One curious feature is that an isomerization occurs in the pyrrole ring number IV (Fig. 15.13). Thus side groups 1, 3, and 5 of the first identifiable porphyrin, uroporphyrin III, are —CH_2COOH groups, but it is side group 8, and not 7, which has this structure. The figure shows that each of the methine groups linking the pyrrole rings is derived from the original glycine precursor, as, of course, are the nitrogen atoms of the pyrrole rings. The only other compound contributing to the porphyrin structure is succinyl coenzyme A.

Most other porphyrins are derived from uroporphyrin III by modification of the side groups. For example, in protohaem (Fig. 8.5) each of the side groups of the type —CH_2COOH has been decarboxylated to a methyl group, while the side group —CH_2CH_2COOH is unmodified at positions 6 and 7 but has been converted to —$CH{=}CH_2$ at positions 2 and 4.

16

Polynucleotides and protein synthesis

One of the most fundamental aspects of biology is summarized in the following simple scheme.

$$\text{Replication} \circlearrowright \text{DNA} \xrightarrow{\text{Transcription}} \text{messenger RNA} \xrightarrow{\text{Translation}} \text{proteins}$$

DNA contains the code for synthesizing each of the cell's proteins but before protein synthesis occurs a complementary messenger RNA is first produced. It moves out of the nucleus to the ribosomes, where its coded message is read in such a way that individual amino acid residues are linked together to form characteristic proteins.

This chapter describes the structure of DNA and the several forms of RNA. It then considers the mechanism and control of protein synthesis. Before beginning, readers are advised to revise, if need be, the structure of the nucleotides and polynucleotides (Chapter 5).

16.1 The double helix of DNA

Much painstaking analysis, including x-ray crystallographic studies, paved the way which led to the now famous concept of the 'double helix'. W. Astbury, in 1947, demonstrated that x-ray patterns showed a repetition of structure every 0.34 nm and this, with other data, led L. Pauling to suggest early in 1953 that a helical structure might exist. Meanwhile, analysis of the base composition of DNA isolated from a variety of microorganisms indicated that, although the guanine content of different DNAs varied about threefold, in all cases guanine was always matched by an equimolar amount of cytosine. That is to say G = C (Fig. 16.1). Similarly, equimolar amounts of adenine and thymine were characteristically present; that is, A = T. Another remarkable fact revealed by this early analytical approach was that the total number of molecules of pyrimidine bases present was always the same as the total number of molecules of purines; that is, A + G = T + C.

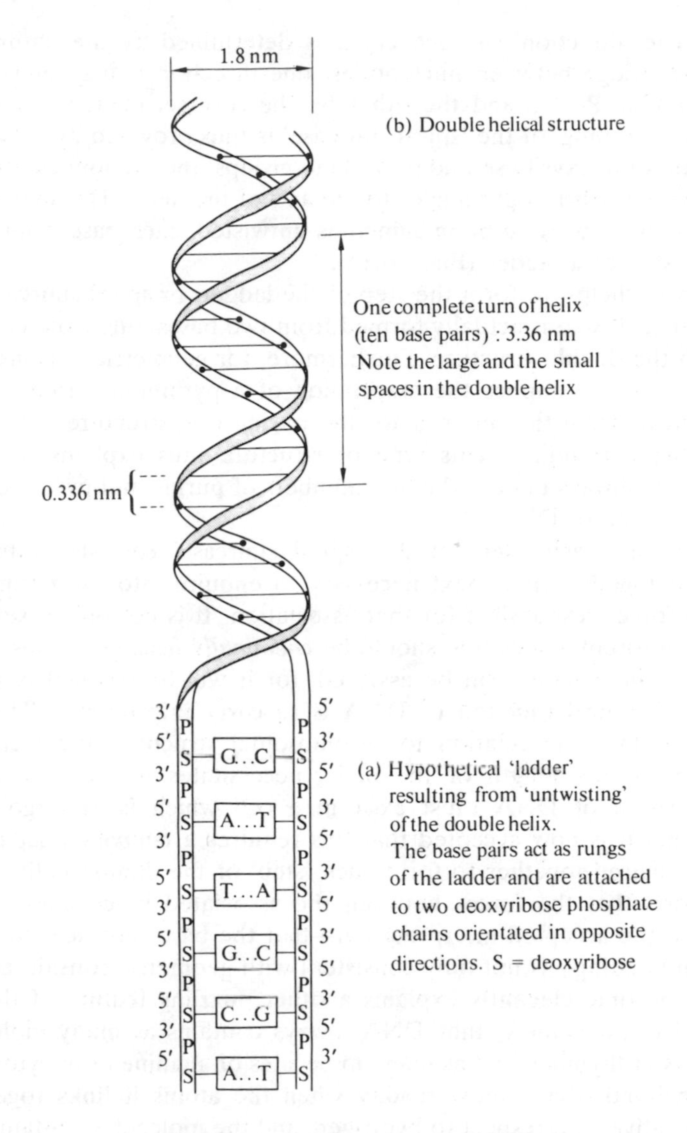

Fig. 16.1 The structure of DNA

Armed with this information and with the results of various x-ray investigations into the structure of purified preparations of DNA. J. Watson and F. Crick, late in 1953, proposed a double helical structure which explained all the facts available at that time. They suggested that the DNA molecule might consist of two nucleotide chains facing in opposite directions and wound around a common axis in such a way as to form two right-handed helices. The 'direction' of each chain is determined by the nature of the phosphate bridge between nucleotides, one direction being determined by the order $C_{3'}$—P—$C_{5'}$ and the other by the reverse, namely, $C_{5'}$—P—$C_{3'}$. The outside 'railing' of the 'spiral staircase' is thus provided by a continuous alternation of deoxyribose and phosphate groups, the mononucleotide bases projecting inwards at right angles to the axis of the helix. This means that, if the double helix were to be imagined as untwisted, each base would help to form the step of a ladder (Fig. 16.1a).

Each base 'helps' to form the step of the ladder, or spiral staircase, in the sense that each step is actually formed from two bases, one from each of the helices in the double structure. Furthermore, for geometric reasons, a 'step' of constant width implies the association of a pyrimidine from one spiral with a purine from the other, since the purine ring structure is larger than that of the pyrimidine. This type of structure thus explains why earlier analyses had always indicated equal numbers of purine and pyrimidine bases in the molecule of DNA.

If, however, each 'step' of the 'spiral staircase' consists of two bases associated together, it is next necessary to enquire into the nature of the chemical forces responsible for that association. It is not only essential that the mechanism of association should be *chemically feasible* but also that the *strength of the bonding* can be assessed, for it will be recalled (Chapter 5) that the biological function of DNA is to carry information. The central position of DNA in relation to chromosomal structure, the 'gene' being some part of the length of the helix, necessitates that a mechanism of self-replication of DNA must exist in a cell which is undergoing rapid mitosis, and it is now accepted that this requires a *temporary separation* of the bases paired together to form each 'step' of the double helix.

It follows that the forces between the associated bases must be rather weak, and the accepted interpretation—that the bases are held together by hydrogen bonding—is not only consistent with geometric considerations but at the same time elegantly explains another puzzling feature of the earlier analytical work, namely, that DNA always contains as many molecules of adenine as of thymine, and as many molecules of guanine as of cytosine. The hydrogen bond occurs most readily when the atoms it links together are electronegative with respect to hydrogen, and the molecules containing them are of such a configuration that the atoms concerned in the hydrogen bonding are no more than 0.34 nm apart. All these conditions are fulfilled if hydrogen bonding occurs between the keto and amino groups of purines and pyrimidines participating in base-pairing, a similar bond being provided between imino-nitrogen in one ring and amino-nitrogen in the other. The probable positions of hydrogen bonds are indicated in Fig. 16.2. The suggestion is, therefore, that a 6-ketopurine will always bond with a 6-aminopyrimidine, and a 6-aminopurine with a 6-ketopyrimidine.

Hydrogen bonds can, more generally, be represented as resonance hybrids

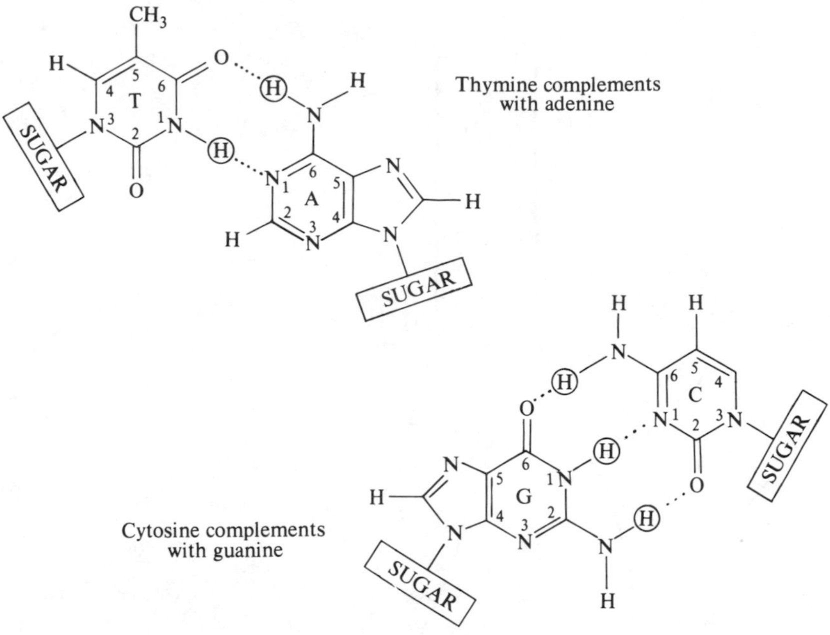

Thymine complements with adenine

Cytosine complements with guanine

Fig. 16.2 Hydrogen-bonding between base pairs of DNA

of two nearly equivalent structures:

$$\text{A—H} \cdots \text{B} \rightleftharpoons \text{A} \cdots \text{H—B}$$

where A and B are atoms of two electronegative elements (of which nitrogen and oxygen are the most important biochemically). The bond energy is low—about 20 kJ/mol for most biochemically important substances—but if it is repeated a very large number of times, as may well happen when macromolecules combine together, the total molecular stability contributed by hydrogen-bonding may be very large (a typical covalent bond seldom exceeds a bond energy of 600 kJ/mol). Thus, in the case of DNA, two or three hydrogen bonds exist between bases paired on each 'step' of the staircase, and since its molecular weight is about 6×10^6, DNA typically has 8000 to 10 000 such 'steps' holding together a molecule which, if pulled out straight, could well be 3 μm long. The two hydrogen bonds holding adenine and thymine together can be regarded as resonance hybrids of the following partial structures involving atoms 1 and 6 of each of the bases in the base pair (see also Fig. 16.2):

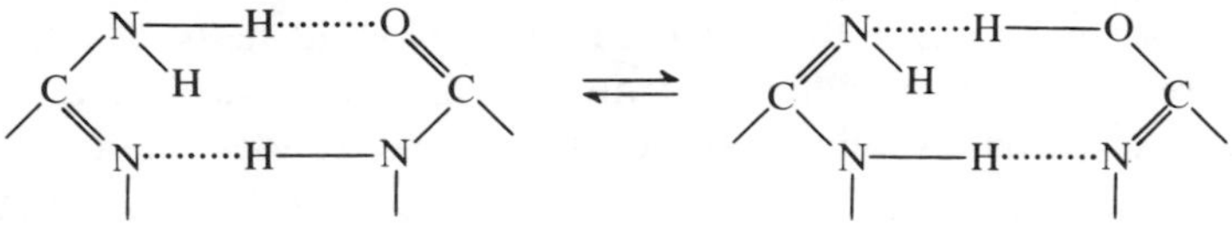

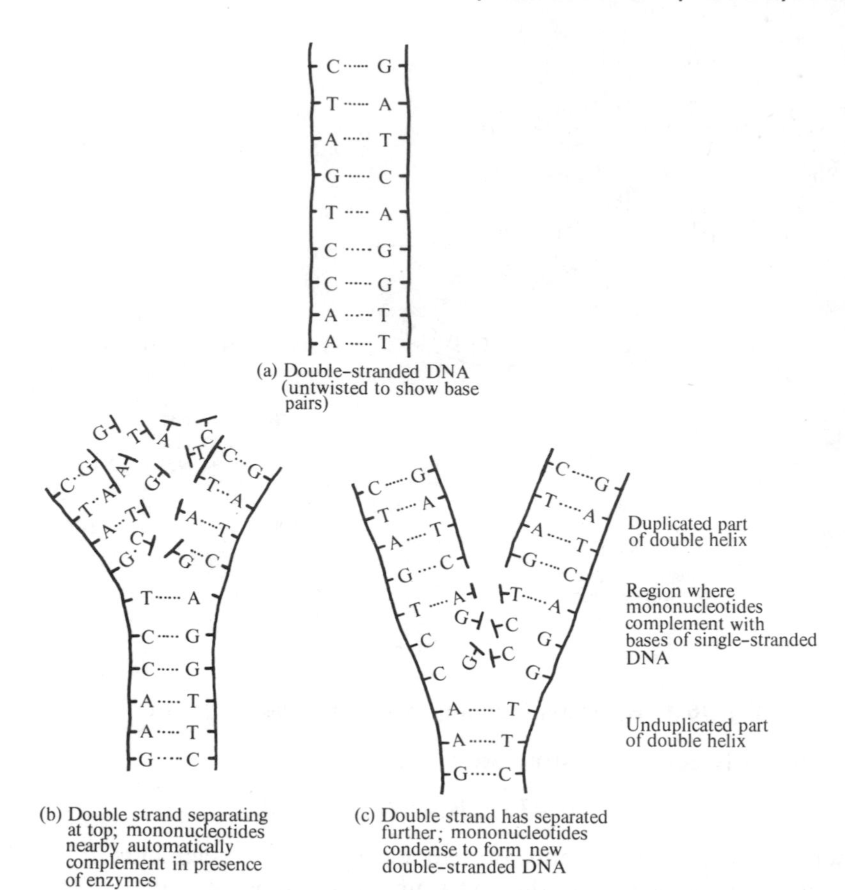

Fig. 16.3 Replication of a molecule of DNA. Shown here in 'ladder' form, the helix can be regarded as unwinding by rotating on its axis, the single strands of DNA uniting to produce a double helix by complementing with mononucleotides nearby

There are several slightly different crystalline forms of DNA, occurring at different humidities and when the acid is in the form of salts of different metals. The most interesting form has the base rings orientated perpendicularly to the axis of the helix in the way already described, and, for this form of DNA, the repeat distance in x-ray pictures is 0.336 nm. The helix has a uniform diameter of 1.8 nm and there are ten base pairs for one turn of the double helix (Fig. 16.1b).

Attention should be directed at this point to the special relationship existing between the two threads of a DNA double helix, for while the second thread is entirely different from the first, its structure is uniquely dependent on the base sequence in the first thread. If adenine occurs in one spiral, the *complementary* base, thymine, will occur in the second; if guanine occurs in the first, then cytosine will be paired with it. This complementing means that, while one spiral can have any of the four bases in any order and occurring any number of times (giving 4^{8000} different possible DNA molecules for a structure 8000 nucleotides long), nevertheless the structure of the second spiral is predetermined by that of the first. Thus, if the two spirals were imagined as separating and each associating with mononucleotides (in the presence of appropriate enzymes and cofactors) so as to form a new partner thread, the two double helices so formed would be identical (Fig. 16.3). As will be seen later in the chapter, this self-templating mechanism partly explains the genetic importance of DNA.

16.2 Types of RNA; their structure and function

Chromosomal DNA not only acts as a carrier of information from generation to generation but, in addition, predetermines the nature and quantity of the proteins (including enzymes) present within the cell during its individual lifetime. Thus, indirectly, DNA probably controls the composition of each of the components of the cell and hence determines all the activities of the cell from the moment it is formed until its death. DNA does not leave the nucleus, yet much cellular protein is synthesized outside the nucleus. Therefore, machinery must obviously exist for *transcribing* the message locked within the molecule of DNA, for carrying it out of the nucleus, and for *translating* it in such a way that proteins of a particular structure are eventually synthesized. The missing link between DNA and the proteins is RNA. It exists in three types, each of which functions in a different way in the total process of protein synthesis. They are called, respectively, *messenger RNA*, *ribosomal RNA*, and *transfer RNA*. All three forms of RNA are ultimately transcribed from DNA. In bacteria, the DNA sequences complementary to ribosomal RNA amount to perhaps 0.2 per cent of the total DNA; animal cells possess DNA complementary to several hundred copies of ribosomal RNA. Ribosomal RNA is probably formed principally in the nucleolus. In contrast, transfer RNAs are formed on genes located on chromosomes in the general body of the nucleus.

About 80 per cent of cellular RNA is in the form of ribonucleoprotein particles known as *ribosomes*. These are scattered in the cytoplasm of the cell, many of them, at any time, being closely associated with the membranes of the endoplasmic reticulum. They are the seat of most of the protein synthesis occurring within the cell and, when protein is actively being made, they tend to aggregate in groups containing up to twenty individual ribosomes. These are called polyribosomes.

The structure of *ribosomal RNA* is imperfectly understood. Some degree of base-pairing occurs within a *single* thread of RNA, the thread being bent round upon itself, but the helical structure breaks down in places, so that the molecule may consist of a number of separate loops, perhaps held into one particular shape by the protein to which it is united (Fig. 16.4).

The second kind of RNA present in the cell is called *messenger RNA*. It is far less abundant than ribosomal RNA, accounting for about 2 per cent of the total RNA in the cell. It is nevertheless of vital importance to an understanding of how nuclear DNA controls the activities and composition of the cell, for it is this material which carries within its structure a transcription of the information conveyed by chromosomal DNA. It is formed inside the nucleus of the cell as a result of base-pairing with one of the strands of the DNA double helix and moves out of the nucleus to the ribosomes. Its 'message' is then interpreted precisely (Section 16.5) at the ribonucleoprotein of the ribosome, its structure determining in what order amino acids are to be joined to produce a particular protein.

Messenger RNA is normally a single-stranded molecule rather than a double helix. This conclusion follows in part from the nature of the biological role ascribed to messenger RNA—that it acts as a templating device for transfer RNAs (see below)—but it is supported by the fact that artificially produced double-stranded molecules of messenger RNA are not readily destroyed by RNA-ase, whereas natural RNA is. Also the numerical equality of purine and pyrimidine nucleotides, a characteristic of DNA, apparently does not occur for either messenger or ribosomal RNA. Messenger RNA has a characteristically short life; it is this property which accounts for the small amount of messenger RNA present in the cell at any one time and for the difficulty in obtaining experimental data about it. In many bacteria, the half-life of messenger RNA appears to be between 1 and 2 min, but this may not always be the case in cells of higher organisms. In mammalian erythrocytes, for example (cells which possess no DNA and so cannot code for new RNA), experiments have indicated that proteins are formed for many hours in systems that contain messenger RNA, ribosomal RNA, transfer RNAs, and appropriate cofactors. The question of longevity of messenger RNA is of considerable importance in relation to the number of molecules of protein which are likely to be synthesized from a single molecule of the messenger. The molecular weight of messenger RNA is highly variable, largely depending on the length of the proteins for which it represents the code. Molecular weights ranging from 25 000 to 1 million are common; these correspond to between 75 and 3000 nucleotides.

The molecules of the third form of RNA differ from the other two kinds in several important respects. *Transfer RNA* molecules are much smaller, consisting of between 74 and 94 nucleotides. There are over forty types of transfer RNA (tRNA) and the structure of over half of them has been elucidated. The molecules terminate at the 5′ end with the nucleotide

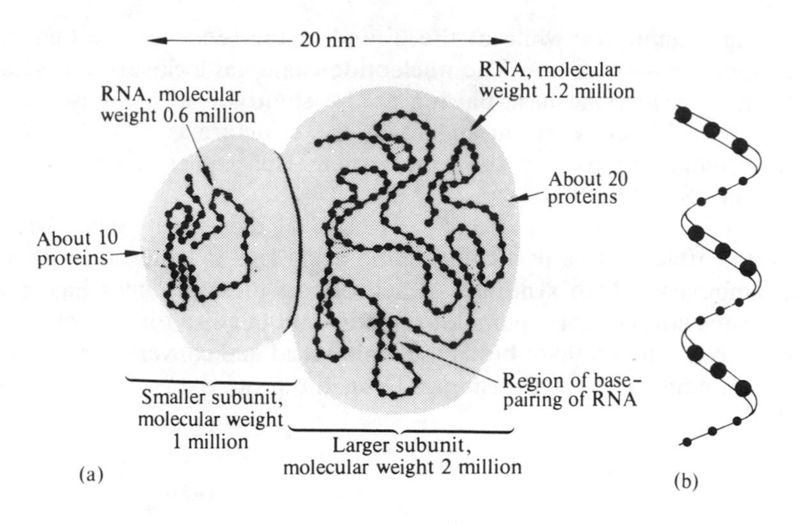

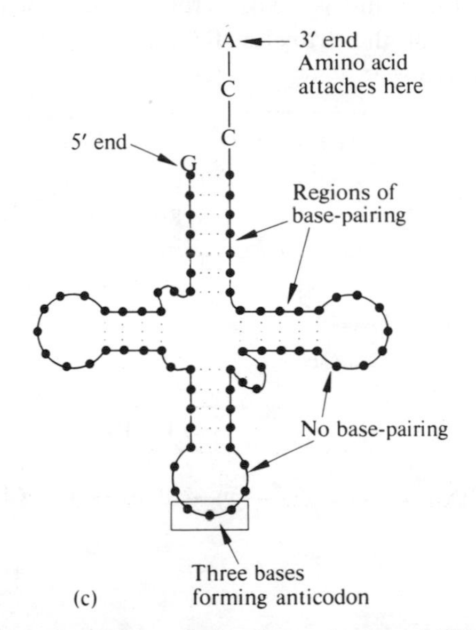

Fig. 16.4 Tentative structures of the three types of RNA.

(a) Ribosomal RNA, mainly single-stranded but some base-pairing. Closely associated with ribosomal proteins. The ribosome and its RNA is subdivided into subunits as indicated.

(b) Messenger RNA. Little is known of its secondary structure, but from its biological function it is unlikely to have much base-pairing.

(c) Transfer RNA. About eighty nucleotides, 'clover-leaf' shape, with base-pairing at the 'stalks'.

containing guanine (G) while at the 3′ end of the sequence the final three nucleotides are —C—C—A. The nucleotide chain has a clover-leaf arrangement with considerable base-pairing at the stem and at the base of each 'leaflet' (Fig. 16.4c). Very important in the structure is a triplet of nucleotides comprising part of the end leaflet. This triplet is known as the 'anticodon' (Section 16.4).

A principal feature of transfer RNA is that each type becomes specifically attached to a particular amino acid. This is achieved by enzymes called aminoacyl-tRNA synthases. Each one of these enzymes has specific binding sites not only for a *particular amino acid* but also for only *one type of transfer RNA*. The synthase binds the amino acid and converts it to an AMP derivative, aminoacyl-AMP, using ATP in the process:

$$\text{R—CHNH}_2\text{—COOH} + \text{ATP} \longrightarrow \text{R—CHNH}_2\text{—C}{\overset{\displaystyle /\!\!/ \text{O}}{\underset{\displaystyle \diagdown \text{AMP}}{}}} + \text{pyrophosphate}$$

The acyl part of this compound is transferred to the 3′-hydroxy position of the terminal nucleotide of the transfer RNA. This terminal nucleotide is invariably adenine nucleotide:

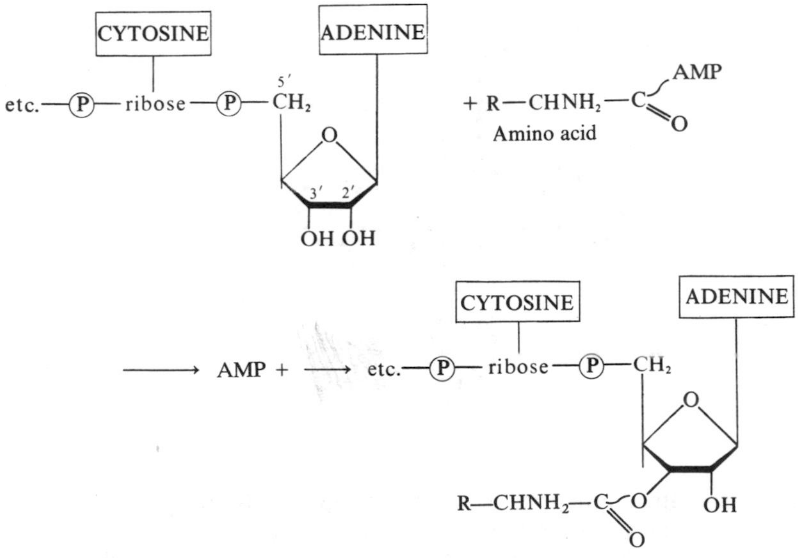

CYTOSINE | ADENINE

etc.—(P)—ribose—(P)—5′ CH_2 (ribose ring with O, 3′, 2′, OH OH)

+ R—$CHNH_2$—C(=O)—AMP
Amino acid

⟶ AMP + ⟶ etc.—(P)—ribose—(P)—CH_2 (CYTOSINE, ADENINE; ribose ring with O, 3′-O—C(=O)—$CHNH_2$—R, 2′-OH)

R—$CHNH_2$—C(=O)—O   OH

Aminoacyl-transfer RNA complex

The product, aminoacyl-tRNA, is in its own way unique for it has a characteristic aminoacyl group and a characteristic anticodon which, as will be seen later, determines where the amino acid will be incorporated into a protein. Note that energy in the form of ATP has been used in the process.

and the amino acid, in the form of its acyl derivative, can be regarded as energized. The energy is subsequently used in the synthesis of a peptide bond (Section 16.5). As is often the case, pyrophosphate is one of the products. Most cells contain pyrophosphatases which hydrolyse it to orthophosphate, a reaction which helps to draw the direction of equilibrium of the earlier reactions towards the synthesis of the major product. In this instance it is aminoacyl-tRNA. Several other examples of the formation of pyrophosphate as a byproduct, and the part played by its subsequent hydrolysis, are encountered later.

16.3 Origin of molecules of DNA and RNA

The central biological role of DNA as a conveyor of information from parent to daughter cell implies not only that DNA is somehow synthesized in the nucleus but that new molecules of DNA are identical to the old ones. This follows from the fact that longitudinal cleavage of chromosomes during the process of mitosis could not occur indefinitely if the rather limited number of threads of DNA constituting the chromosome were constantly halved and from the fact that the daughter cells so produced retain the essential characteristics of the parent cell.

The complementarity of bases in a double helix offers a clear explanation of how such self-duplication could occur, even though the precise way in which the two threads of the double helix unravel at the moment of duplication cannot be regarded as settled. Progress in elucidating the mechanism of DNA replication began when it was established that whenever a new DNA double helix is produced just one of its strands is entirely new while the other is one strand of the original helix. This is called semi-conservative replication.

An important clue to the mechanism of DNA replication came from the isolation of the enzymes now called collectively *DNA polymerases.* In the presence of a DNA primer and a mixture of all four deoxyribose nucleoside triphosphates (dATP, dGTP, dCTP, and dTTP) these enzymes catalyse the synthesis of new DNA. The reaction catalysed by the polymerases is most easily understood by first considering the addition of nucleotides to a single strand of *DNA primer,* shown on the left of Fig. 16.5a. In the figure, the nucleotides A and C have already been added and the next addition is to be T. Only the latter can be added at this stage because the next DNA base after T and G on the primer is A, and, of course, only T will pair with A (Section 16.1). The enzymes only insert nucleotides *in the direction* 3′ *to* 5′ on the primer and in a strict sequence determined by the bases of that primer. The thymine base of deoxythymidine triphosphate (dTTP) probably forms hydrogen bonds with the adenine on the primer. As it does so, pyrophosphate is removed, with the formation of a new phosphate link at the 3′ position on the end of the growing DNA chain. The next base to be added will be G since it must pair with C (Fig. 16.5a). As each nucleotide is added, pyrophosphate is again produced.

Figure 16.5a indicates how a complementary strand of DNA can be formed using a preexisting single-stranded primer. Normally, replication starts with a double helix and produces two identical double helices. At one time it was suggested that the DNA would unwind at one end with separation of the two strands, enabling the growth of two new helices beginning from that end (Fig. 16.3). Electron micrographs of replicating bacterial DNA have suggested that new DNA strands are produced first by a separation of the strands somewhere along the length of the molecule. The known properties of the DNA polymerases suggest that *both original primer strands* undergo replication by sequential addition of new complementary bases in the 3′ to 5′ direction (Fig. 16.5b). In eukaryotic cells the replicating DNA probably has several points along its length where this process occurs. Other enzymes are involved besides the DNA polymerases. These include so-called *unwinding proteins* and *DNA ligase*, the function of which is to join up broken links in the DNA chains. Such a process of repair is necessary, for it is clear that, at some time, the DNA must be 'nicked' if the helical structure is to be maintained in the newly formed helix.

The *synthesis of RNA* is a process in which the base sequence of DNA is transcribed in such a way that it determines the order of the bases in the RNA whether it be messenger, transfer, or ribosomal. It will be recalled that RNA contains three of the bases present in DNA (A, G, and C), but instead of thymine it has uracil (Section 5.3). These two bases only differ by a methyl group, thymine being, in effect, 5-methyluracil. An early indication of how RNA could be replicated came from the observation that the bases of a single-stranded bacteriophage DNA could form a hybrid double-stranded structure with a suitable RNA. It is now known that in this type of double strand, each A in the DNA strand is paired with U in the RNA strand, while G pairs with C by hydrogen bonding in a similar way to G—C bonding in DNA (Fig. 16.2).

The exact mechanism of DNA–RNA transcription is still poorly understood, but highly studied enzymes called *RNA polymerases* are a major component of the replication system. Experimentally they can catalyse the synthesis of RNA, provided that they are supplied with the appropriate nucleoside triphosphates (ATP, GTP, UTP, and CTP) together with a DNA primer. Figure 16.6 indicates a possible mechanism. RNA polymerases are large molecules (molecular weight, 500 000) which could surround a section of the double helix of DNA. The enzyme molecule could then separate the two strands of DNA over a length of some four to eight nucleotides, so releasing the bases from their hydrogen-bonded partners. One of the DNA

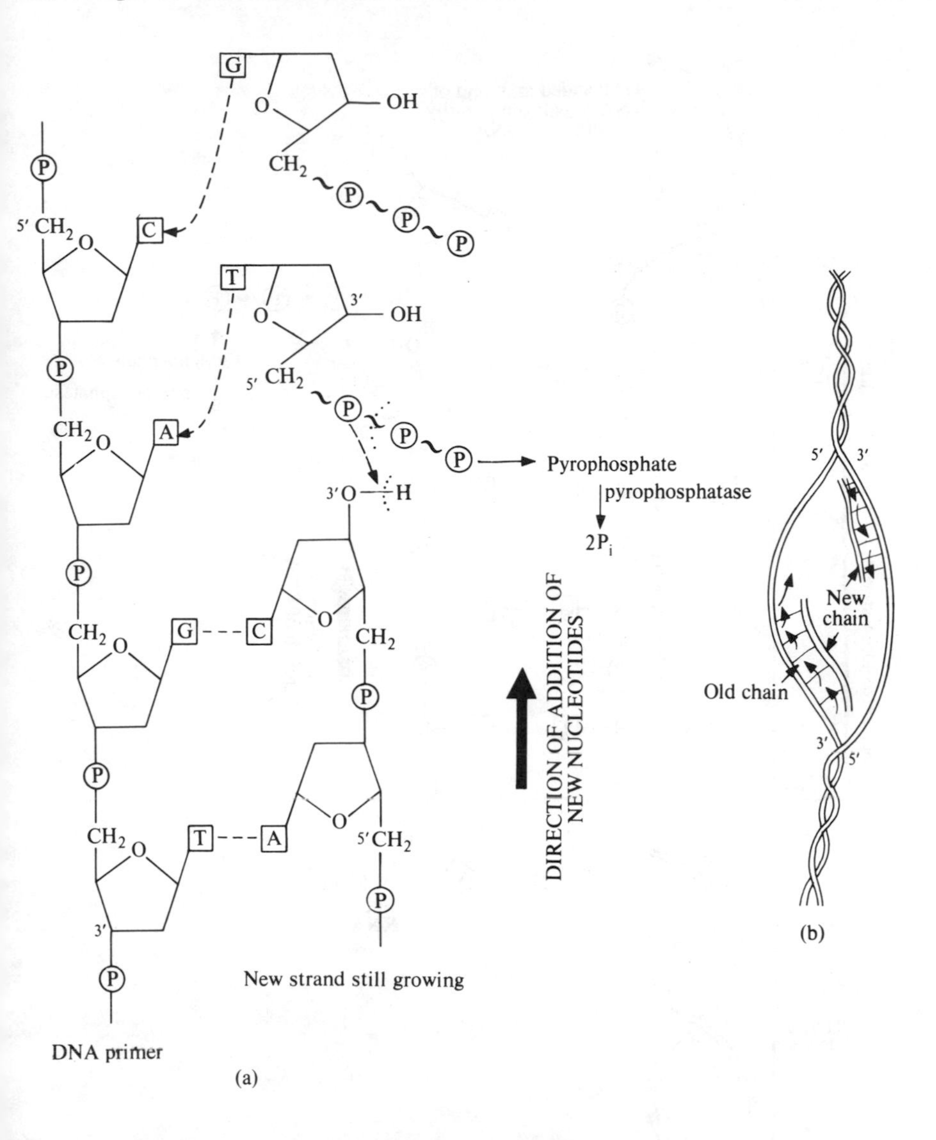

Fig. 16.5 (a) The reaction catalysed by DNA polymerase.
(b) Direction of addition of nucleotides to enable replication of the DNA structure to occur. Note that DNA polymerase adds nucleotides sequentially of the primer in the '3′ to 5′ direction

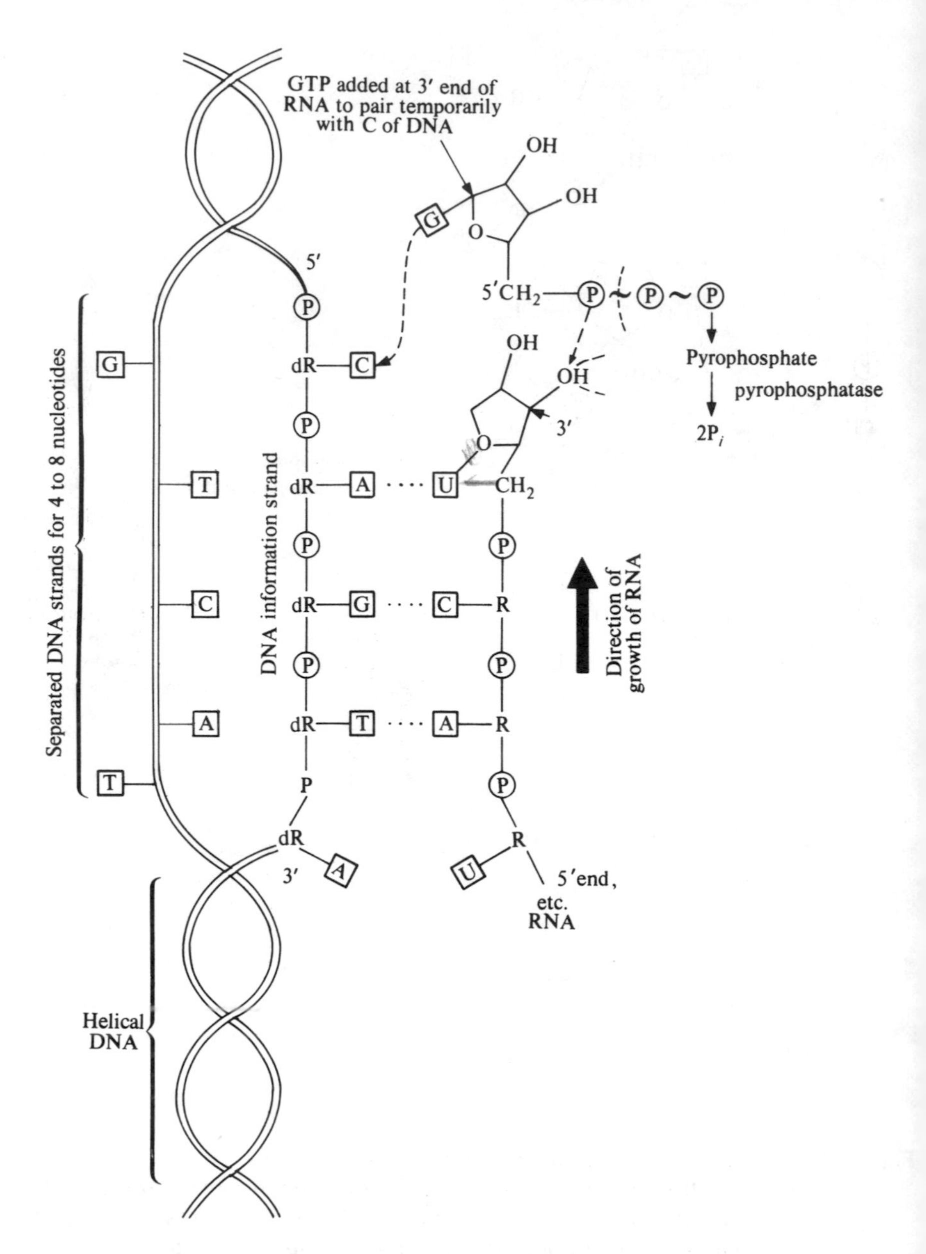

Fig. 16.6 A model of RNA transcription catalysed by RNA polymerase. R = ribose; dR = deoxyribose

strands called the *information strand* provides the code for determining the base sequence in the nascent RNA. The RNA polymerase moves along the information strand in the 3′ to 5′ direction (Fig. 16.6) and as it does so ribonucleotides are linked together to form a temporary DNA-RNA hybrid in which the RNA is growing in a 5′ to 3′ direction.

Referring to Fig. 16.6, it can be seen that the DNA bases in the information strand are in the ascending order A, T, G, A, C. In the RNA strand, U was at first paired with the lower A but has now separated. Then A paired with T, C with G, and U with A. The next nucleotide to be added to the RNA must clearly be G, since this pairs with C. It seems likely that G is 'selected' because the base of GTP first forms hydrogen bonds with the C of the DNA. The enzyme then catalyses the hydrolytic removal of pyrophosphate from the GTP and the formation of a new phosphate link at the 3′ terminal of the growing RNA. It should be noted that this process bears a general similarity to DNA replication (Fig. 16.5). A big difference, however, is that the new RNA strand peels off from the DNA primer (see U at bottom of Fig. 16.6) and as it does so the DNA base to which it had been linked twists around to pair with its complementary base, the DNA thus resuming the normal helical structure. The base sequence of the RNA is the same as that of the non-information strand of the DNA except that U replaces T. The reaction catalysed by RNA polymerase is undoubtedly complicated yet one molecule is able to link ribonucleotides at the rate of about fifty per second.

There are still many problems posed by this system. What is the difference between an information strand of DNA and the other strand? How does the polymerase identify the position on the DNA where it should begin transcription, and where should it stop? We still lack definitive answers to these questions although progress is being made in studies of certain genes in *Escherichia coli.* For eukaryotes we may have to wait another decade. It seems likely, however, that certain nucleotide sequences on the DNA may signal 'begin', 'stop', 'read this strand and not the other', etc. These signals might produce allosteric changes in the polymerase which thereby induce appropriate activation or inhibition of the reaction.

16.4 The genetic code

One of the greatest achievements of modern biology has been the deciphering of the genetic code. In essence it is delightfully simple yet there are, of course, many complications which have yet to be explained. Messenger RNA is single-stranded, at least while its message is being 'read'. It will be recalled that the order of bases in the messenger is determined by the DNA originally inherited from the parents. Messenger RNA codes the number and the position of the twenty amino acids as they appear in proteins and thus prescribes the properties of all enzymes and structural proteins in the cell.

Since there are only four bases but twenty amino acids, a unique coding for one particular amino acid must involve a sequence of at least three nucleotide bases. Four bases taken one at a time could only code for 4^1 amino acids and a sequence involving two of them could only account for 4^2 different amino acids. A triplet comprising any one of the four bases, occurring in any of the three positions, could, on the other hand, code for up to 4^3 or 64 different amino acids.

It is now firmly established that the cipher or '*codon*' for each amino acid is, in fact, a *triplet of bases.* Although each codon is unique for an amino acid, one amino acid may be represented by several codons. Leucine, for example, has six codons, aspartic acid has two, but tryptophan has only one. The code was 'broken' by using synthetic messenger RNAs of known base composition, together with appropriate ribosomal preparations, aminoacyl-tRNAs, and ATP and GTP to provide energy. Initially RNAs composed of only one base were investigated. One, comprising only uridine (poly-U), produced peptides composed exclusively of phenylalanine. Poly-C produced polyproline and poly-A produced polylysine. Next, RNA of the formula $(UC)_n$ was used. This yielded a peptide composed of alternating serine and leucine residues. Experiments of this type, together with an elaborate process of logical and statistical interpretation, finally led to the code shown in Table 16.1.

'Reading' the bases along the messenger RNA occurs in the direction 5′ to 3′. If the first base of the codon is U followed by U and U, then the amino acid to be incorporated in a protein is phenylalanine (Table 16.1). If, on the other hand, the bases are UCU, serine is incorporated, while CUC codes for leucine. Hence $(UC)_n$ produces the peptide described above, whichever way the code is read:

UCU CUC UCU   CUC UCU CUC UCU

Ser—Leu—Ser   Leu—Ser—Leu—Ser

More complicated is a sequence such as:

-UCA-AGA-CAG-GAG-GUA-

Ser—Arg—Gln—Glu—Val

This clearly determines a particular pentapeptide—*provided we know where to start.* If instead of starting with U we began with C, by reference to Table 16.1 it will be seen that we would get the following quite different result:

-U-CAA-GAC-AGG-AGG-UA-

Gln—Asp—Arg—Arg

Table 16.1 The genetic code; codons for amino acids. The names of the amino acids are conventionally abbreviated to the first three letters (e.g., phenylalanine = Phe) except Ile = isoleucine, Asn = asparagine, Trp = tryptophan, and Gln = glutamine

5′-OH terminal base	Middle base				3′-OH terminal base
	U	C	A	G	
U	Phe	Ser	Tyr	Cys	U
	Phe	Ser	Tyr	Cys	C
	Leu	Ser	END	END	A
	Leu	Ser	END	Trp	G
C	Leu	Pro	His	Arg	U
	Leu	Pro	His	Arg	C
	Leu	Pro	Gln	Arg	A
	Leu	Pro	Gln	Arg	G
A	Ile	Thr	Asn	Ser	U
	Ile	Thr	Asn	Ser	C
	Ile	Thr	Lys	Arg	A
	Met	Thr	Lys	Arg	G
G	Val	Ala	Asp	Gly	U
	Val	Ala	Asp	Gly	C
	Val	Ala	Glu	Gly	A
	Val	Ala	Glu	Gly	G

END = chain termination signals

The code only produces 'sense' if the triplets are read as the codons that they were intended to be. It is now established that *all proteins are initiated* with the codon AUG which corresponds to methionine (Table 16.1). In effect, the messenger RNA passes along a ribosome and when the sequence AUG is encountered the mechanism of protein synthesis is 'switched on' and methionine becomes the first amino acid in the peptide sequence. Thereafter, the sequence is easily interpreted. Thus the following sequence produces a tetrapeptide:

XXXXX-AUG-ACA-GUA-UAC-UGA-XXX

Meth—Thr—Val—Tyr END

Just as we must have a signal to begin (AUG) it is equally important to have a chain *termination signal.* In this instance UGA indicates the end of the sequence. There are, in fact, three codons indicating END, namely UGA,

UAA, and UAG (Table 16.1). Of the 64 possible triplets 61 are amino acid codons and 3 are termination signals.

The reader may be intrigued to find that all peptides are initially synthesized with methionine at their amino end. In most proteins the methionine is removed at a later stage. AUG signals methionine at the beginning and also *within* peptide chains.

We now come to the mechanism of translation of the messenger's code. How are the triplet codons of the single-stranded messenger RNA interpreted so that the amino acids are arranged in the correct order in the protein being synthesized?

The general principle thought to underlie the process is indicated in Fig. 16.7. A length of DNA acts through its information strand as the template to produce a complementary messenger RNA comprising the triplet *codons*. Each transfer RNA has three characteristic bases called the *anticodon* (Fig. 16.4c). These probably take part in complementary base-pairing with the codons of messenger RNA. This process 'selects' the appropriate amino acid since each transfer RNA carries a specific amino acid derivative (page 344).

GCG is a triplet in the information strand DNA which codes for arginine. This is translated as CGC in the messenger RNA since C pairs with G (Fig. 16.7). Thus arginine attached to the transfer RNA which possesses the anticodon GCG is brought into position by base-pairing between the codon and anticodon.

There are several complications not indicated in Fig. 16.7. In particular the transfer RNAs contain a number of bases other than G, A, U, and C

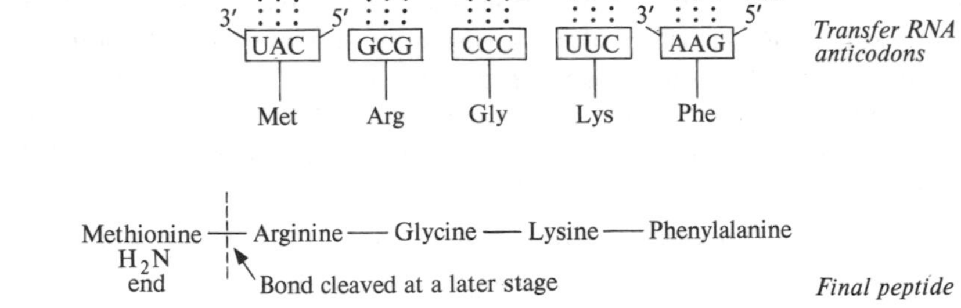

5′ end —X′ X′ X′—ATG — CGC—GGG—AAG—TTC — } *DNA double helix*
3′ end —X X X—TAC — GCG—CCC — TTC —AAG← Information strand

5′ end ←X′X′X′—AUG — CGC—GGG—AAG—UUC— →3′ end *Messenger RNA codons*
3′ UAC 5′ GCG CCC UUC 3′ AAG 5′ *Transfer RNA anticodons*
Met Arg Gly Lys Phe

Methionine — Arginine — Glycine — Lysine — Phenylalanine
H_2N end
Bond cleaved at a later stage
Final peptide

Fig. 16.7 A model to indicate the general relationship between the DNA information strand, messenger RNA, the anticodons of transfer RNA, and the peptide produced. For base-pairing, see page 346.

Among these are methylated derivatives of these bases, pseudo-uridine (ψ) and inosine (I) (Fig. 5.1). I often replaces A in the anticodon and ψ sometimes replaces U. However, there is a general similarity between the anticodons and what would be predicted from the codons.

A second complication not indicated in Fig. 16.7 results from the fact that most amino acids have several triplet codes. In the case of alanine, for example, there are two bases (GC) followed by any one of the four bases (Table 16.1).

One of the transfer RNAs for alanine appears to 'read' only the first two bases, namely, GC. Its *anticodon*, CGI, forms complementary pairing with each of the messenger RNA triplets GCU, GCC and GCA, GCG. In effect, the third base fails to fit the codon in the expected way, a phenomenon called the 'wobble' hypothesis. It is clearly economic to have less varieties of transfer RNA in cases where one type can correspond to several codons for the same amino acid. For this reason the number of different transfer RNAs is rather less than the earlier predicted number of 61.

Finally, it should be pointed out that an important consequence of the triplet coding system is the possibility of *mutation*. One way this can occur is for a base to be replaced by another at the DNA level. For example, if A is replaced by G it follows that the corresponding messenger RNA will contain C instead of U. This may or may not have important consequences. Often, one different amino acid incorporated into the coded protein does not greatly change the properties of the protein. Sometimes, however, it is very important. For example, several abnormal human haemoglobins are the result of changes in single amino acids.

Another possible mutation results from the deletion of one or more DNA bases. If any multiple of three bases is removed, this will cause the deletion of amino acid residues from the protein, thus making it shorter. However, if one or two bases are omitted, the whole code after the deletion will be 'nonsense', with the coded amino acids being quite different from normal.

16.5 Protein synthesis in the ribosome

Messenger RNA passes to the ribosomes where the genetic code is translated into the characteristic amino acid sequence. In prokaryotes the strand of RNA becomes attached to the ribosomes even while the elongation of the RNA is continuing at the 5′ end. In eukaryotes, most protein synthesis takes place in ribosomes associated with the endoplasmic reticulum in the cytosol (Section 1.10). The messenger RNA first reaches these by transfer through pores in the nuclear membrane.

The structure of ribosomes is highly complex. Besides ribosomal RNA they contain about thirty proteins. Some of these probably help attachment

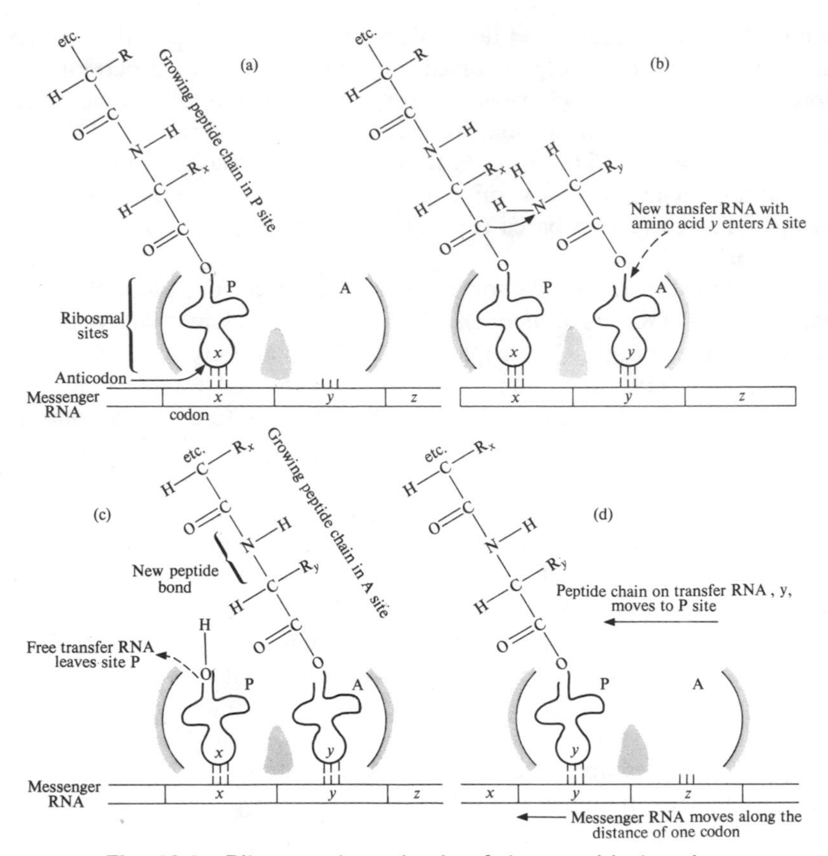

Fig. 16.8 **Ribosomal synthesis of the peptide bond**

of the messenger RNA to the smaller subunit (Fig. 16.4a). It is drawn along some sort of groove which allows its codons to become appropriately orientated so that they attach to the complementary bases of the aminoacyl-tRNAs. Some of the other proteins are enzymes involved in the various stages of peptide synthesis.

The larger ribosomal subunit has two binding sites for transfer RNA known as the *peptidyl site* (P) and the *aminoacyl site* (A). Figure 16.8 shows how a peptide bond is formed during the process of elongation of a nascent protein. In Fig. 16.8a the P site is occupied by a transfer RNA molecule (x) whose anticodon is complementary to the codon of the messenger RNA. The P site is so called because at this stage the growing peptide is located here linked to the transfer RNA. The amino acid, with side group R_x, provides the carboxyl end of the peptide and was the last to be added to the chain. Note that at this stage the A site is unoccupied.

The next amino acid to be added, with side group R_y, is now brought into the A site attached to its own specific transfer RNA and uniquely chosen because of the required fit between the anticodon and the triplet codon of the messenger RNA located at the A site. The aminoacyl-tRNA (*y*) is apparently transferred to the A site by a protein called *elongation factor T* (EF-T). This is the first step requiring the provision of energy, which is derived from GTP. The process must be complicated though it can be summarized by the following equation:

$$\underset{\text{(free)}}{\text{GTP}+\text{aminoacyl-tRNA}} \xrightarrow{\text{EF-T}} \text{GDP} + \text{P}_i + \underset{\text{(in A site)}}{\text{aminoacyl-tRNA}}$$

At the stage represented by Fig. 16.8b the growing peptide chain and the amino acid with side group *y* are in the P and A sites respectively; they are now aligned ready to form a peptide bond. This is catalysed by a ribosomal protein called *peptidyl transferase* since it effectively *transfers* the peptide *from the P to the A site* and in so doing *synthesizes the peptide bond* between the carboxyl group of the amino acid residue *x* and the amino group of the amino acid *y* (Fig. 16.8c).

This is a second step which requires energy. It will be recalled that ATP is used in the synthesis of aminoacyl-tRNA. This original activation by ATP is the source of energy for the bond formation.

In the remaining stages the now non-acylated transfer RNA (*x*) vacates the P site so that the growing peptide can be returned to it (Fig. 16.8c and d). The movement of the peptide from site A to site P is a particularly remarkable process because the peptidyl-tRNA (*y* in the figure) remains attached to the messenger RNA by the anticodon-codon link. The *translocation* effectively moves the messenger RNA a distance corresponding to one codon. In this way the next codon (*z*) is brought into a position in site A to receive the next aminoacyl-tRNA (*z*) (Fig. 16.8d). The process is achieved by the action of a *translocase*. It is a third endergonic process in the peptide synthesis with further GTP undergoing hydrolysis:

$$\underset{\text{(in A site)}}{\text{GTP}+\text{peptidyl-tRNA}} \longrightarrow \text{GDP} + \text{P}_i + \underset{\text{(in P site)}}{\text{peptidyl-tRNA}}$$

Now that the peptidyl-tRNA is back at site P the whole process is repeated with amino acid *z* being added to the growing protein.

As indicated in the previous section, the first amino acid to be incorporated is always methionine (or formylmethionine in prokaryotes). The first step in protein synthesis involves several so-called *initiation factors*. It probably begins when the codon for methionine (AUG) reaches the ribosomal binding site P.

When the chain termination signals (Table 16.1) on the messenger RNA reach site A the process of peptide formation ceases. This is achieved by the action of *protein release factors* which cause the cleavage of the bond between the carboxyl end of the peptide and its transfer RNA.

The mechanism of ribosomal protein synthesis is very complicated, the present account being only an introduction to this highly significant process. One surprising feature is its speed. Bacterial ribosomes can add amino acids to a growing peptide at the rate of ten per second. Animal systems are rather slower—about two per second.

16.6 The control of protein synthesis

The quantity and the types of protein present in cells vary during their development and growth and also in response to changes in their metabolism. This variation is brought about partly by the controlled *induction* or *repression* of the mechanism of protein synthesis and partly by the variation in the rates at which the proteins are broken down in the cell by proteases. Thus a state of dynamic equilibrium exists, the quantity of any given protein being determined by the relative rates of synthesis and destruction.

Much of our present knowledge of the control of protein synthesis originates from the brilliant work of F. Jacob and J. Monod, using the bacterium *E.coli* to study the synthesis of enzymes involved in metabolizing lactose. Normal cells, grown in the absence of lactose, contain very little lactase (i.e., *β-galactosidase*), and the cell wall virtually lacks the carrier or *permease* needed to transport the sugar into the cell. Several functional lengths of the cell's DNA (genes) are linked together into a single unit called an *operon*, and these are potentially capable of producing the messenger RNA for both lactase and permease transcription. Normally, however, these genes are non-functional because attached to a length of DNA close to them and known as the *operator site* is a special protein called a *repressor protein*. The code for synthesizing this is situated elsewhere in the DNA in a so-called *regulator gene* (Fig. 16.9a).

When cells of *E.coli* are placed in a medium containing lactose there is a very large increase in the number of molecules of lactase and of the permease. What happens is that a trace of lactose entering the cells combines specifically with binding sites *on the repressor protein*, initiating an allosteric change which causes it to leave the operator site. RNA polymerase is then able to move along the operator site to reach the genes for synthesizing messenger RNA coding for lactase and the permease. In other words, a small quantity of lactose is able to *remove the inhibition* of messenger RNA synthesis, so inducing the events leading to the formation of the proteins to occur.

In the higher vertebrates the hormonal control of metabolism is frequently achieved by either the *secondary messenger* mechanism or through the

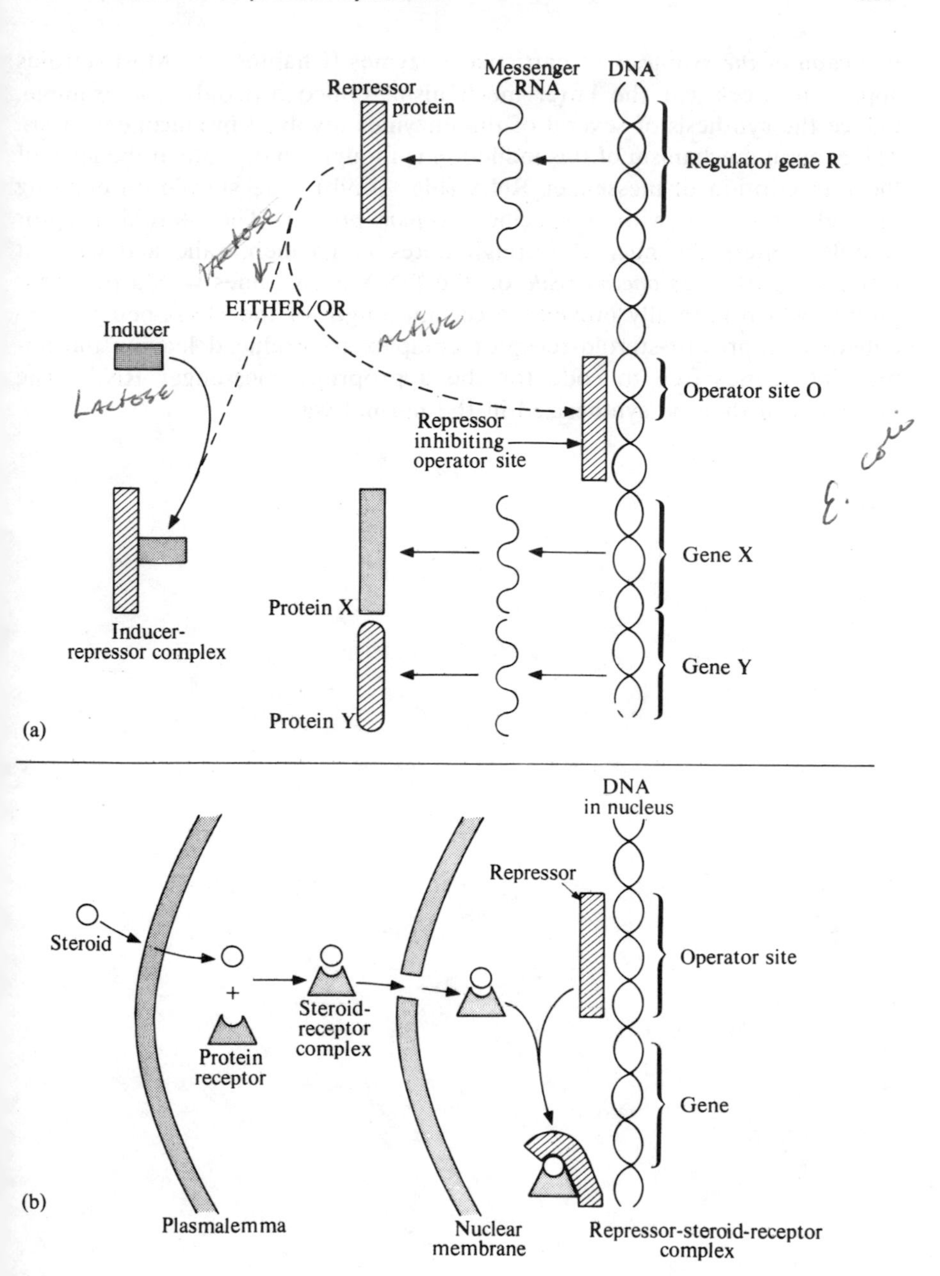

Fig. 16.9 (a) A model of enzyme induction/inhibition in prokaryotes. (b) Steroid induction of protein synthesis in mammals

induction of the synthesis of particular enzymes (Chapter 13). Most steroids appear to work *via* the latter mechanism. Glucocorticoids, for example, induce the synthesis of several of the enzymes involved in gluconeogenesis. The general mechanism of this induction is thought to operate at the level of the transcription of messenger RNA (Fig. 16.9b). The steroid on entering the cell combines with a specific *receptor protein.* The steroid-receptor complex enters the nucleus through pores in its membrane and when it reaches a particular *operator site* on the DNA it combines with a repressor protein which normally prevents a certain length of the DNA being transcribed. A repressor-steroid-receptor complex is produced leaving the appropriate genes free to code for the appropriate messenger RNA. The proteins can then be synthesized in the normal way.

17

Movement of dissolved substances into and out of cells

This chapter describes the main processes involved in the transfer of substances across the plasmalemma of cells. In practice, the mechanisms vary considerably with the type of cell, the type of organism, and the type of substance. The chapter begins with some general observations on *passive* diffusion into cells, describing some of its implications. In other instances substances enter cells only if a specific carrier is present in the cell membrane, a process known as *carrier mediated transport.* These are illustrated by reference to the functions of the absorptive cells of the intestine. Absorptive cells present some of the best-studied examples of transport across the plasmalemma, transport which includes both *facilitated* and *active* carrier mediated processes. The carriers are thought to be carriers present in the cell membrane.

17.1 Diffusion in liquids and passive diffusion through cell membranes

The diffusion of solute molecules in unstirred water is very slow. In the absence of some mixing process it may take many hours for a dissolved solute to diffuse even a few centimetres. It is probably for this reason that in most organisms transport of substances is arranged in such a way that diffusion distances are kept very short. In the mammal, for example, substances rarely have to diffuse passively further than about 10 nm. This is the average distance between most cells and their nearest blood capillary. The blood vascular system in animals and the xylem and phloem systems in plants serve to reduce the distance that substances have to diffuse to reach individual tissue cells.

It is a matter of common experience that the net diffusion of solutes in solution is in the direction from regions of high concentration to those of a lower concentration. In reality the solute molecules diffuse in all directions but there are more molecules available to move away from areas of higher concentration. We therefore often refer to a *net flux* of solute molecules.

This quantifies overall movement of molecules in a particular direction yet recognizes that some are nevertheless diffusing in the opposite direction. Since in aqueous solutions molecules of solute move between the structured lattice of water molecules it is not surprising that solutes with larger molecules tend to diffuse slower than those with small ones. A. Einstein (1905) showed that the rate of diffusion of molecules was proportional to $M^{-0.5}$ (i.e., the reciprocal of the square root of the molecular weight).

In general, the rate of diffusion, or flux, of solute in solution is directly proportional to the concentration gradient, and inversely proportional to the distance over which diffusion has to occur. This has been formalized in the law of diffusion attributed to A. Fick (1855).

The diffusion of substances across cell membranes is a far more complicated process than the movement of solutes in solution, for the cell plasmalemma is a highly complicated structure both physically and chemically. Though its precise nature may still be in doubt (Section 2.6), it is clear that the membrane contains both proteins and lipids and it is unlikely that there are any large pores which would enable easy penetration of water-borne solutes through water-filled channels.

In order to quantify the permeability of cell membranes a constant known as the *permeability constant* (P) has been introduced. P varies with the type of cell and the type of solute. It is useful for comparing the permeability of a range of substances entering a particular type of cell or comparing the permeability of different forms of cell to a given substance. In practice, it is often difficult to measure P precisely because substances may move in both directions—into and out of the cell—at the same time, though the use of isotopic compounds has overcome many of the problems. It is therefore best to consider the flux dn/dt or the net movement of molecules (n) across the membrane in time (t). Substances can only show net passive movement from a high concentration outside the cell (C_1) to a lower concentration (C_2) inside the cell or *vice versa.* In other words, the net movement is *down* the concentration gradient. As might be expected, the greater the area of membrane (A) over which diffusion occurs the quicker the process functions. P is derived from the formula

$$\frac{dn}{dt} = PA(C_1 - C_2)$$

In practice, P has the dimensions of distance divided by time (e.g., centimetres per second) and it is therefore convenient to think of it as a measure of the *speed* of movement of substances across the membrane.

17.2 Permeability of the cell membrane

A very important factor determining whether substances enter the cell membrane easily is their solubility in lipids (Section 2.6). This was demonstrated by the detailed experiments of R. Collander (1933) who studied the entry of numerous organic compounds into cells of the plant *Chara ceratophylla*. The permeabilities of the substances were related to their olive oil/water partition coefficients (i.e., the ratio of solubility in olive oil to solubility in water). A direct linear relationship was obtained between the permeability expressed as $PM^{1/2}$ and the partition coefficient (Fig. 17.1a). In general terms this is explained by the fact that the substances must enter the cell by passing into its plasmalemma and then from this into the cytosol. A substance which readily 'dissolves' in the membrane lipids is thus able to permeate more easily than one which is preferentially attracted to the polar water environment.

W. D. Stein (1967) has quantified this relationship more fully by considering the physical nature of the solubility of substances. The presence on an organic molecule of various polar groups such as amino-, hydroxy-, or carbonyl groups which form hydrogen bonds with water enhances their solubility in aqueous solutions and decreases their attraction to the non-polar lipids. Stein demonstrated a negative correlation between the permeability of substances through several types of cell membranes (algal cells, sea urchin eggs, red blood cells) and the number of hydrogen bonds which the substance produced in water (Fig. 17.1b). In general he found that the presence of each hydrogen bond decreased six- to tenfold the permeability through cell membranes.

To give a specific example, L-glucose ($C_6H_{12}O_6$) and octanoic acid [$H_3C(CH_2)_6COOH$] have fairly similar molecular weights. L-Glucose has a very low permeability into red cells ($P < 10^{-10}$ cm/sec) for it forms some ten hydrogen bonds with water due to its five hydroxyl groups. Octanoic acid forms only three hydrogen bonds and has a relatively high permeability constant (42×10^{-7} cm/sec). It will be shown later that D-glucose, while producing the same number of hydrogen bonds with water as the L form, nevertheless enters red blood cells easily. This is due to a special carrier mechanism specific for its particular structure. This has important physiological implications, for it means that many important metabolites such as amino acids, sugars, and the water-soluble vitamins may not enter cells appreciably unless a special mechanism for their transport is available.

Besides the different tendencies of various substances to interact with water or the cell membrane lipids *molecular size* plays a large part in determining whether substances enter cells or are effectively excluded. For example, in the red cells pentanoic acid [$H_3C(CH_2)_3COOH$] has a permeability of 167×10^{-7} cm/sec; this is some four times greater than the permeability of the larger octanoic acid. For this reason, very small molecules such as water, oxygen, carbon dioxide, urea, and glycol are often able to enter cells very easily. It is thought that they are able to penetrate between the structural protein-lipid complexes which comprise the membrane. Molecules of this type often have a permeability which is up to a hundredfold greater

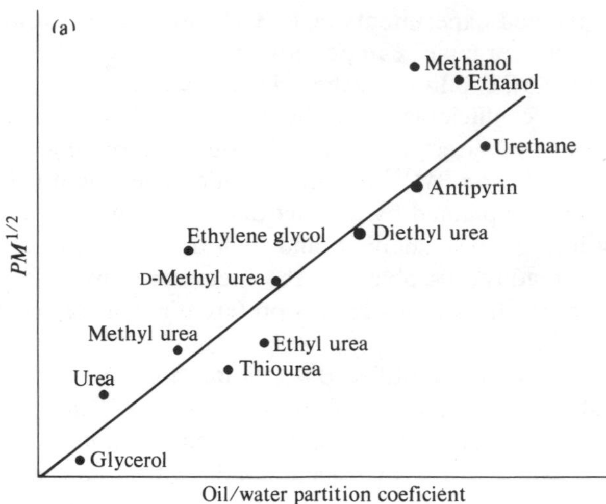

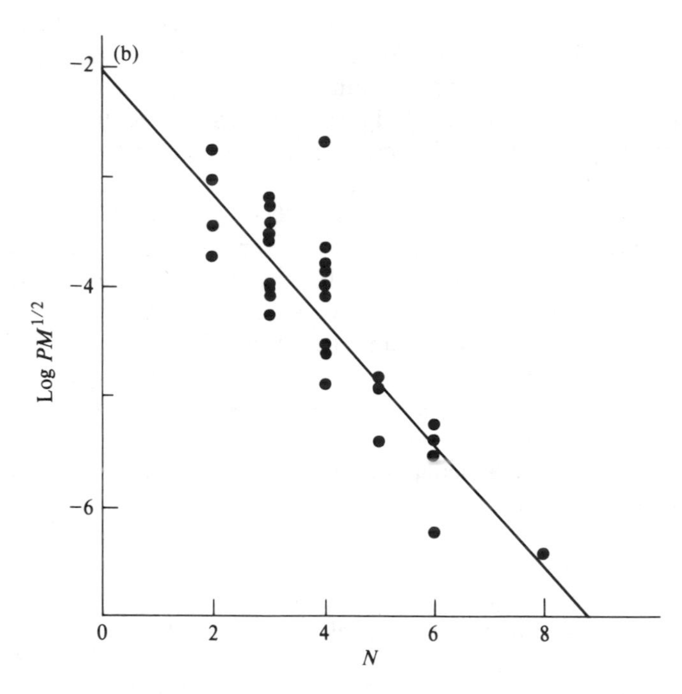

Fig. 17.1 The permeability (*P*, cm/h) of cells of *Chara* to non-electrolytes.
(a) The relationship between the permeability expressed as $PM^{1/2}$ and the olive oil/water partition coefficient.
(b) The relationship between log $PM^{1/2}$ and the number (*N*) of hydrogen bonds formed between the substances and water. (Data from R. Collander, 1933, and W. Stein, 1967)

than that of rather larger ones. The bovine red cell's permeability to urea and glycol is 7.8×10^{-5} and 0.21×10^{-5} cm/sec respectively. The cell membrane's high permeability to oxygen and carbon dioxide is clearly a prime factor in the process of cellular respiration.

17.3 Osmosis and the uptake of water by cells

Cell membranes are normally freely permeable to water, the passage of water either into or out of cells taking place by the passive process of osmosis. If cells are placed in distilled water, the water enters the cells more quickly than it leaves them, with the result that they begin to swell. For example, when the egg of the sea-urchin is suspended in normal sea water, it maintains a constant size, but if transferred to sea water diluted with distilled water, the volume of the cells increases. The curvilinear relationship between the concentration of the sea water and the cell volume is depicted in Fig 17.2. When the cells are placed in 50 per cent sea water, their volume almost doubles; if the eggs are placed in distilled water, so much water diffuses into them that the cells burst, or *lyse*.

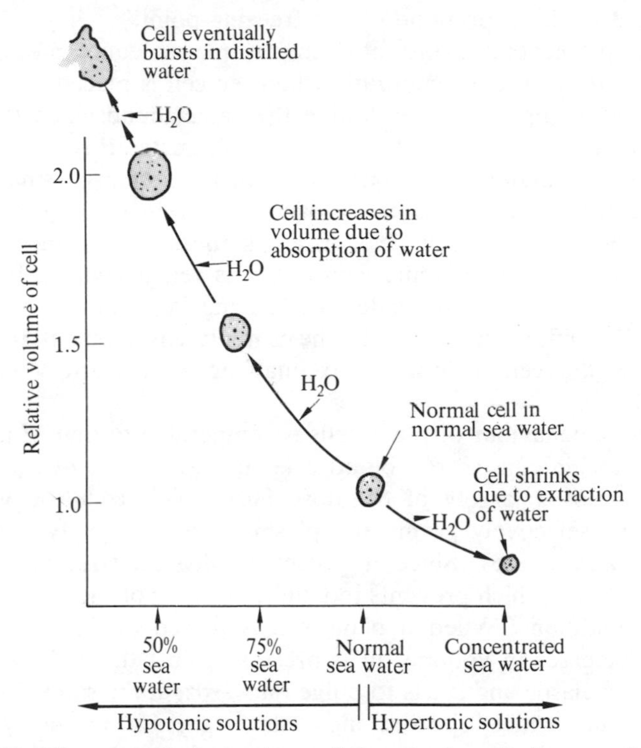

Fig. 17.2 The osmotic behaviour of the single-celled sea-urchin egg

The movement of water by osmosis is a typical diffusion process from a region of high concentration *with respect to water* (distilled water, or a solution dilute with respect to a solute) to a region of low concentration with respect to water (a solution more concentrated with respect to a solute).

For this reason, if a cell is placed in a *hypotonic* medium—i.e., one with a lower osmotic pressure than its cytoplasm—water will enter the cell. When the external solution has the same osmotic pressure as the cell sap the cell is said to be in an *isotonic* medium, and no net movement of water will occur into or out of the cell. When the cell is placed in a *hypertonic* solution, the mole fraction of water is less in the external medium that it is in cell sap; water will therefore leave the cell and the cell will shrink.

In practice, cytoplasm is a complex mixture of molecules and ions, rather than a binary mixture, and each particle of each solute contributes to the total osmotic pressure operating within the cell, for osmotic pressure is a colligative property of a solution. (6×10^{23} particles of solute, in 22.4 litres of solution, at 0°C, exert ideally an osmotic pressure of one atmosphere.) Biological solutions are usually far from ideal and it is often impracticable either to measure their osmotic pressure or to calculate it. Instead, the osmotic pressure of solutions is frequently estimated indirectly by the measurement of the depression of the freezing point.

The water permeability constant of the cell plasmalemma measured as the flux of water through the membrane when the cell is placed in a hypotonic medium is often approximately double the value found when the environment is isotonic. For example, for human erythrocytes P is 7×10^{-3} cm/sec in a hypotonic solution, whereas in normal isotonic plasma P is 3×10^{-3} cm/sec. The entry of water by osmosis is seen as an often-dramatic increase in the size of the cells. Nevertheless, these figures illustrate that the membrane, even under isotonic conditions, is readily permeable to water (about half that which occurs under the 'driving force' of moderate osmotic pressure). The difference is that in the isotonic environment just as much water leaves the cell as enters, so enabling a constant volume to be maintained.

The osmotic behaviour of plant cells is comparable to that of animal cells such as sea-urchin eggs or red blood cells, except that plant cells possess a cell wall composed largely of cellulose fibrils. This cellulose wall, which surrounds the selectively permeable plasmalemma, is freely permeable to both water and solutes. Since it possesses physical strength, it offers a physical resistance which prevents indefinite swelling of the cytoplasm under hypotonic conditions. When a plant cell is placed in distilled water, the cytoplasm increases in volume till it presses against the cell wall. The cell wall is slightly elastic and starts to bulge outwards, at the same time exerting a back pressure which opposes increasingly the further entry of water. Eventually the wall pressure is equal and opposite to the osmotic pressure, and net flow of water then ceases for water then enters and leaves the cell at the same rate. When this happens, the cell is said to be *turgid* and, in this state, has a considerable physical strength.

If a plant cell is placed in a hypertonic solution water diffuses out of the cell and the cytoplasm decreases in volume, eventually contracting away from the cell wall. The *plasmolysed* cell has much less mechanical strength than has the turgid cell, and is said to be *flaccid.*

17.4 Osmoregulation in animals

The tissues of marine invertebrates tend to have an osmotic pressure similar to that of sea water, so loss or gain of water by osmosis presents no great problem to these animals. On the other hand, fresh water, having a very low salt content, has also a very low osmotic pressure and fresh water animals tend therefore to absorb water from their environment. They usually have mechanisms for reducing water uptake or for eliminating excess of it. For example, the eggs of fresh water animals tend to be almost impermeable to water, and in this respect they differ from sea-urchin eggs. Fresh water protozoa (e.g., amoeba) are faced with the task of eliminating water constantly and they do this by accumulating water in vacuoles—a process which requires expenditure of energy. This is probably provided by ATP, for the vacuoles in amoeba are often surrounded by mitochondria. The water is eliminated by the discharge of the vacuolar contents to the outside of the animal.

In the vertebrate animal, most of the body cells have the same osmotic pressure as the extracellular fluids, but this pressure is usually very different from that of the environment. Water uptake or loss is largely avoided by the presence of an impermeable skin, often supplemented by scales or other protective features. Fresh water fish, however, constantly absorb excessive water through their gills, for these must be thinly walled to enable gaseous exchange to occur. The body cells and fluids are hypertonic compared with fresh water, and the excess water absorbed by osmosis is eliminated by the kidneys which excrete a very dilute urine.

The situation is quite different in marine teleost (i.e. bony) fish, for these animals are hypotonic with respect to their environment and there is a constant tendency for water to be lost by osmosis. Marine teleosts survive by drinking sea water and excreting the excess salt. They can only produce urine which is isotonic with their body fluids, the excess salt being secreted by the gills. This is achieved by special cells which pump salts into the surrounding water, and since this involves moving ions *against* a concentration gradient, a considerable amount of energy is required.

Marine elasmobranchs (cartilaginous fish, e.g., sharks) do not have quite the same problem as the teleosts. Their body fluids would be hypotonic compared with sea water were it not for the fact that they contain high concentrations of urea (1 to 2 mol/litre), and this raises the osmotic pressure to a value very close to that of sea water.

Vertebrates adapted to a terrestrial environment, the birds and mammals, are able to excrete hypertonic urine, a process which is necessary for their survival during periods of drought since water is often scarce on land. Excessive loss of water in dilute urine would of necessity restrict their activities.

As pointed out in Section 15.6, the form in which animals excrete nitrogen is often closely linked to their need to control the osmotic concentration of their body fluids.

17.5 Carrier mediated transport

From the account of passive diffusion of substances into cells (Section 17.2) it is clear that most metabolites (amino acids, sugars, many vitamins, etc.) could not enter the cell at an appreciable speed were it not for the existence of specific mechanisms which enable them to cross the plasmalemma. In many cases the cooperation of a *carrier* or *permease* has been demonstrated. Permeases are usually thought to be proteins or lipoproteins located in the cell membrane itself (Section 2.6). They are analogous to enzymes in that they have a binding site which is specific for one metabolite (or sometimes for a group of substances with some common chemical structure). The substance to be carried becomes attached to the carrier on the outside of the cell and then by some rotation or other change in orientation of the carrier within the membrane the substance is transferred to the inside layer of the plasmalemma where it is released into the cell. This is known as *carrier mediated transport.* Transport kinetics resemble the kinetics described for enzymes (Section 6.7), where the substance transferred is regarded as the substrate. Thus, at low concentrations of the substance any increase in its concentration leads to a greater rate of net transport since more molecules become attached to the carrier. But there is a characteristic maximum rate of flux (V_{max}) when all the carrier molecules are working as fast as possible because their binding sites are fully occupied. A further increase in substrate concentration cannot produce a larger flux. Lineweaver–Burk plots are often used to define the values of V_{max} and K_m for carrier mediated transport systems (Fig. 17.3). There are two main types of carrier mediated transport. The first is known as *facilitated carrier mediated transport.* This is a movement of the substance from a higher concentration to a lower one. In this respect it resembles passive transport (Fig. 17.3a and b); although there may be an energy of activation (Section 6.6) involved as the substrate combines

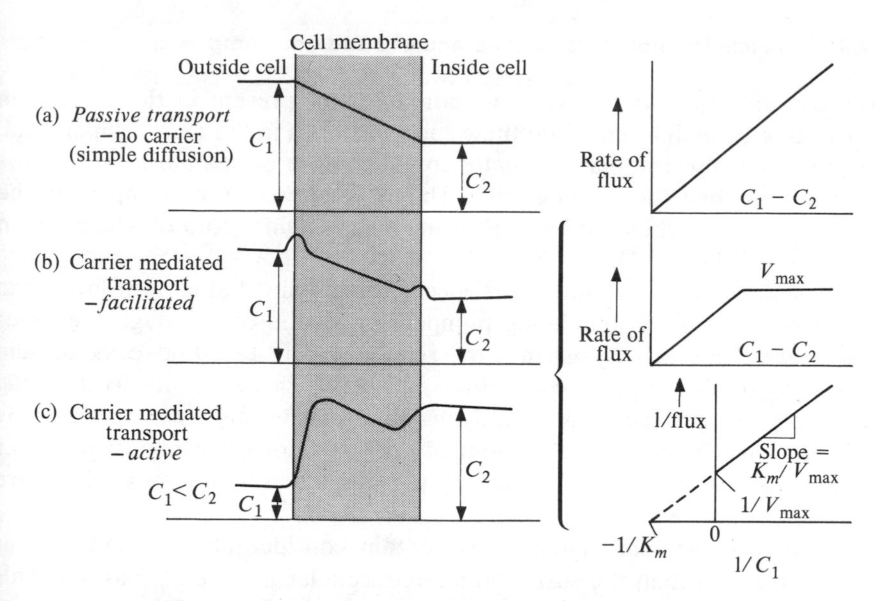

Fig. 17.3 On the left is indicated the relation between the concentrations of metabolite on the inside (C_2) and outside (C_1) of the cell membrane and the type of transport. On the right (top) is shown the relation between the concentration difference and the rate of flux for passive transport (Fick's Law applied to diffusion through the cell membrane). The two lower graphs describe relationships for both types of carrier mediated transport, the lower being a Lineweaver–Burk plot (see page 122)

with the carrier the overall process is 'downhill' energetically. A good example of this type of transport is the entry of D-glucose into human erythrocytes. D-glucose enters the cells about one thousand times faster than the L form for there is a carrier for D-glucose but not for the L isomer. Thus, although L-glucose has very similar properties of diffusion and solubility to the D form it is largely excluded by the absence of a carrier.

In many instances cells need to accumulate metabolites by taking them in from the extracellular fluid under conditions where the concentration inside the cell is *already greater* than that outside. In this instance energy derived from some exergonic process is used to drive the *energetically uphill process* of transport against the concentration gradient. *Active carrier mediated transport* differs from the facilitated form in that a source of energy is required. This is usually directly or indirectly derived from the coupled hydrolysis of ATP. An important example, the sodium pump, is considered in the following section.

17.6 Intracellular inorganic cations and the sodium pump

The concentrations of the various inorganic ions present in the cell are in most cases quite different from their concentrations in the extracellular fluid. In particular, the cytoplasm usually contains more potassium ions and less sodium ions than the bathing fluid. This is illustrated, for example, by the eggs of the sea-urchin and by cells of the alga *Valonia*, both of which live in sea water (Table 17.1).

The sodium concentration in these cells is maintained at a lower level than in sea water, sodium ions being pumped out of the cell through the outer membrane. This activity is, in part, related to the fact that cells contain numerous soluble organic compounds, all of which contribute to the total osmotic pressure; the cells would absorb excessive amounts of water by osmosis if the inorganic ion content of cells was not correspondingly lower than in sea water, and, in practice, it is usually sodium ions which are selectively eliminated.

Similarly, most mammalian cells contain considerably less sodium but more potassium than the surrounding extracellular fluid (e.g., plasma). This is illustrated for erythrocytes and muscle cells in Table 17.2. Chapter 19 describes how the transmission of impulses along nerves and the stimulation of muscles to contract both depend on a control of the intra- and the extracellular levels of these inorganic ions.

Experiments employing radioactive isotopes suggest that sodium ions slowly enter active cells, but the intracellular sodium concentration is maintained at a low level by a mechanism which pumps them out again. Potassium ions are constantly diffusing out of the cells, the loss being made up by continuous replacement from the extracellular fluids.

The extrusion of sodium ions and the accumulation of potassium ions by cells occur by carrier mediated active transport. The inside of the plasma lemma of many cells is negatively charged with respect to the outside; this i a consequence of the unequal distribution of potassium ions on the two side of the cell membranes (Section 19.1). The ejection of sodium ions mus

Table 17.1 Concentration of principal ions in sea water, in sea-urchin eggs, and in *Valonia* (mmol/kg)

	Sea water	Sea-urchin eggs	*Valonia*
K^+	10	210	500
Na^+	470	52	90
Ca^{++}	11	4	2
Mg^{++}	53	11	—
Cl^-	540	80	507

Table 17.2 Concentration of principal ions in mammalian plasma, erythrocytes, and muscle cells (mmol/kg)

	Plasma	Erythrocytes	Muscle cells
K^+	4	150	100
Na^+	140	6	25
Ca^{++}	5	3	3
Mg^{++}	2	5	22
Cl^-	100	80	16
HCO_3^-	16	4.3	3

occur therefore not only in a direction which opposes the concentration gradient but one which opposes an electrical gradient as well. Little is known about how sodium ions are eliminated from cells in general, although many studies have been carried out upon erythrocytes, nerve cells, and muscle cells—all of which are in different ways very specialized. Nevertheless, there can be little doubt about the *existence* of a transport mechanism known as the *sodium pump* (Fig. 17.4).

The sodium pump is known to be linked to the metabolism of the cell; treatment with metabolic inhibitors such as cyanide, dinitrophenol, or fluoride stops the outward transport of sodium ions. A consequence of a failure of the sodium pump is that continued passive diffusion inwards causes the internal concentration of sodium ions to increase. Moreover, when red blood cells are cooled so as to diminish the rate of metabolism, not only does the intracellular sodium ion concentration rise, but internal potassium ions also leak out. If such cooled cells are then very carefully warmed to blood temperature in a medium containing glucose, the original ionic gradients are rapidly restored.

In spite of an enormous amount of research into the mechanism of the sodium pump since it was first identified by A. L. Hodgkin and R. D. Keynes in 1955, its detailed mechanism still remains obscure. For several types of cells it is established that the immediate source of energy is the hydrolysis of ATP to ADP. This requires the presence of magnesium ions and occurs on the inner side of the plasmalemma, for adding ATP to the inside of cyanide-inhibited cells is far more effective in regenerating the pump's activity than when ATP is applied externally. The hydrolysis probably involves the conversion of the carrier molecule into a phosphorylated intermediate. Some authorities believe this results in some change in the shape of the carrier so that an active site which, in a non-activated form, is able to bind a potassium ion is converted through the activation into a new conformation enabling it to bind a sodium ion instead. Figure 17.4b shows a generalized scheme describing one suggested mechanism. It indicates that

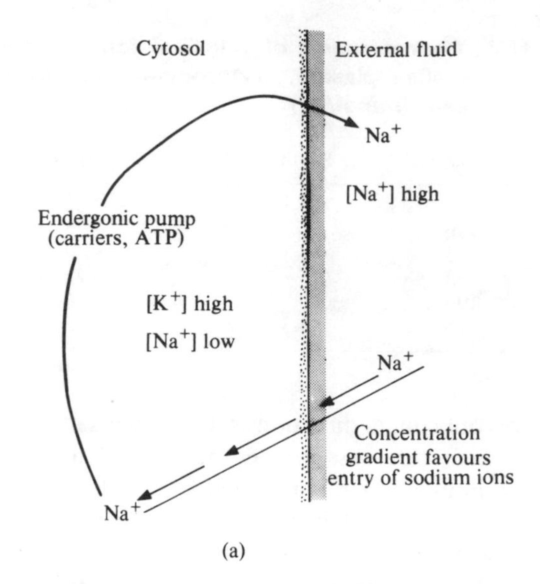

(a)

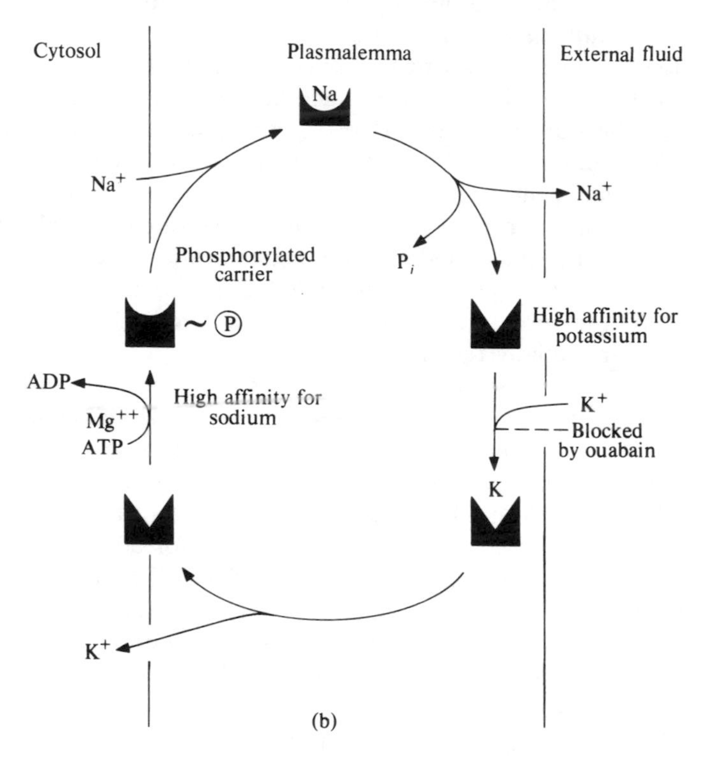

(b)

Fig. 17.4 (a) The sodium pump and (b) a possible mechanism for it

he carrier alternates between a sodium binding form (activated) and a otassium binding form.

The ability to bind potassium ions is inhibited competitively by digitalis nd various other cardiac glycosides used medically in the treatment of ongestive heart failure. Experimentally, ouabain is most frequently used to tudy the effect of this group of compounds on the sodium pump. Ouabain ıhibits the uptake of potassium ions by the system and prevents the whole ycle of events, including the hydrolysis of ATP. Its formula is given in Fig. .5.

The model in Fig. 17.4b suggests that the carrier moves physically about ithin the plasmalemma. The thinness of the membrane would preclude uch large movements and it is more probable that the mechanism is more in he nature of a static carrier with the sodium and potassium ions linking to it nd being released in opposite directions. An important feature of the ump's activity not shown in the model is that for each ATP molecule ydrolysed *three* sodium ions often move in one direction and *two* otassium ions in the other.

7.7 Membrane permeability to inorganic ions and the Donnan equilibrium

One might expect that most simple inorganic ions would easily diffuse hrough cell membranes but in fact most do not. One major reason is that, n aqueous solutions, they usually possess several more or less firmly bound vater molecules; the number of these varies with the size and charge of the on. For example, the ions of ammonium chloride, potassium chloride, odium chloride, and magnesium chloride are estimated from thermodynamic activity measurements to have respectively 1.6, 1.9, 3.5, and 3.7 bound water molecules. This increases the effective size of the ions and ince the water of hydration forms hydrogen bonds with other water nolecules it reduces their affinity for the lipids in the plasma membranes.

Of the common inorganic ions only potassium, chloride, bromide, nitrate, nd bicarbonate easily permeate the cell membrane. The permeability of the arger hydrated ions of sodium, calcium, magnesium, sulphate, and phosphate is lower. For example, in frog muscle cells, if the permeability of otassium is taken as 1.00 the relative permeability of chloride, sodium, and alcium is 0.23, 0.027, and 0.0026 respectively.

In the previous section it was shown that the sodium pump largely etermines the intracellular concentrations of sodium and potassium ions. It s probable that there are other pumps as yet poorly understood which ontrol the calcium, magnesium, and phosphate levels. For some other nions, notably chloride and bicarbonate, the concentrations are determined y the overall ionic charge in the intracellular fluid through a process often amed after F. C. Donnan.

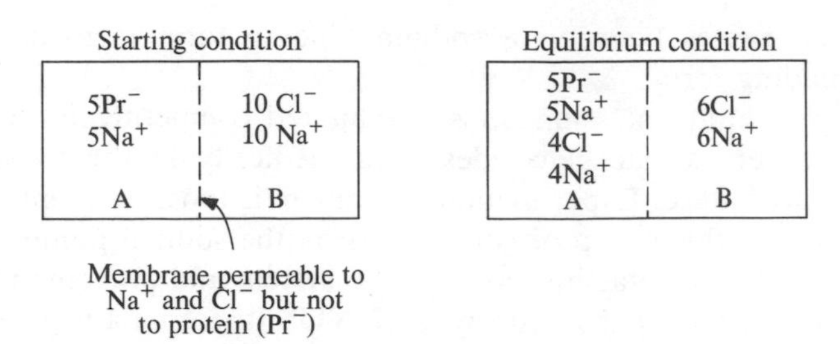

Fig. 17.5 A simple illustration of Donnan equilibrium. Initially, sodium chloride solution is placed on one side of the membrane and sodium proteinate on the other

Let us take a simple model to explain the Donnan equilibrium. A vessel is divided into two halves by a membrane which is freely permeable to water, sodium, and chloride ions but *impermeable* to protein ions (Pr^-). It is set up initially in the manner shown in Fig. 17.5. We can see that the positive and negative charges on either side of the membrane are equal—as they must always be in all solutions. But there are more sodium and chloride ions on side B. Since these can diffuse through the membrane *down* the concentration gradient there is a net movement of both ions from side B to side A. This will continue until an equilibrium is reached with the rate of diffusion of ions in both directions equal ($B \rightarrow A = A \rightarrow B$). It is the charged but immobile protein which determines the precise situation at equilibrium. In this simple example there will be nine (that is, 5 + 4) equivalents of sodium ions on side A, five of which balance the five negative protein charges and four balancing the chloride ions. Equilibrium is reached when the product of the concentrations (or more precisely the thermodynamic activities) of the freely diffusible ions—sodium and chloride in this case—are the same on both sides of the membrane. Thus, in the example, we have

$$\text{Side A} \qquad\qquad \text{Side B}$$

or

$$[Na_A^+] \times [Cl_A^-] = [Na_B^+] \times [Cl_B^-]$$

$$(5 + 4) \times 4 = 6 \times 6$$

On rearranging we get

$$\frac{[Na_A^+]}{[Na_B^+]} = \frac{[Cl_B^-]}{[Cl_A^-]} = 1.5$$

The same type of relation would hold if we added other ions to the system. For example, if potassium bicarbonate is added, at equilibrium the following relationship exists:

$$\frac{[Na_A^+]}{[Na_B^+]} = \frac{[K_A^+]}{[K_B^+]} = \frac{[Cl_B^-]}{[Cl_A^-]} = \frac{[HCO_{3B}^-]}{[HCO_{3A}^-]} = r$$

where r is a constant for the system known as the Donnan ratio. As will be found by reference to physical chemistry texts, the situation is more complicated for divalent ions.

Clearly, as mentioned above, the presence of the *non-permeable* charged protein molecules determines the particular distribution of the permeable ions at equilibrium. A similar situation obtains in the cell where the intracellular proteins are unable to diffuse out. In most cells the concentrations of chloride and bicarbonate ions in relation to those in the extracellular fluid is found to relate closely to the Donnan ratio. In red blood cells, for example;

$$\frac{[Cl^-]_{cell}}{[Cl^-]_{plasma}} = \frac{[HCO_3^-]_{cell}}{[HCO_3^-]_{plasma}} = 0.7$$

The concentration of diffusible *anions* is lower inside the cell because of the presence of negatively charged protein molecules. Clearly, a small change in this charge would alter the ionic balance. This does, in fact, occur when hydrogen ions are added, for then the charge is reduced:

$$Pr^{n-} + H^+ \rightleftharpoons HPr^{(n-1)-}$$

Since the concentration of the major intracellular *cation*, potassium, is controlled not by diffusion but by the sodium pump, the fall in charge causes *chloride ions to diffuse into the cell* so that the total negative charge still balances the positively charged intracellular cations. This process has important physiological implications (Section 18.4).

Finally, it is of interest to note that, in the model in Fig. 17.5, there are at equilibrium more ions on side A than on side B. Consequently, the osmotic pressure is also higher on side A. This difference of pressure on the two sides of a membrane is called *colloid osmotic pressure.* It is caused by both the charged, impermeable protein and by the consequential differences in total ionic distribution. Normally, in reaching equilibrium, *water* molecules diffuse from side B to side A at the same time as the sodium and chloride ions.

17.8 Absorptive cells of the small intestine

All living cells in the vertebrate have the ability to absorb selected organic nutrients such as glucose and amino acids. In most cases these are absorbed from the extracellular fluid. The cells which line the intestine, however, are highly specialized for the absorption of food materials from the lumen of the alimentary canal. The processes of absorption by these absorptive cells have been extensively investigated and are discussed here because they provide

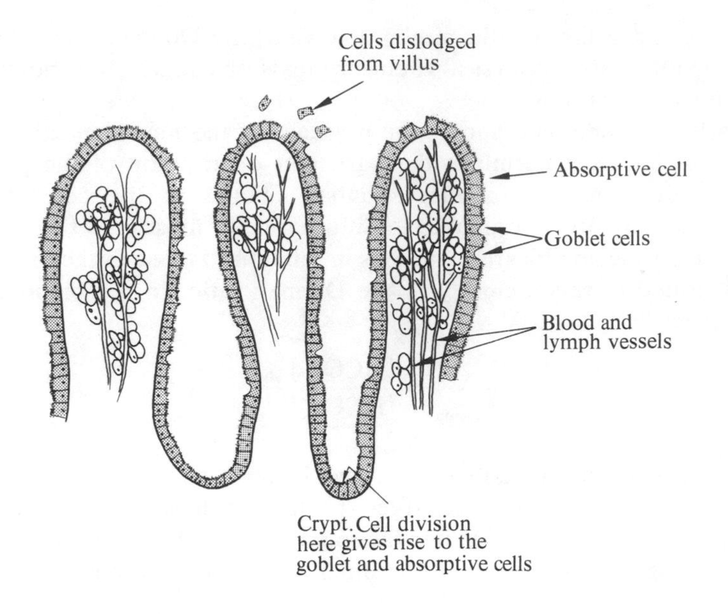

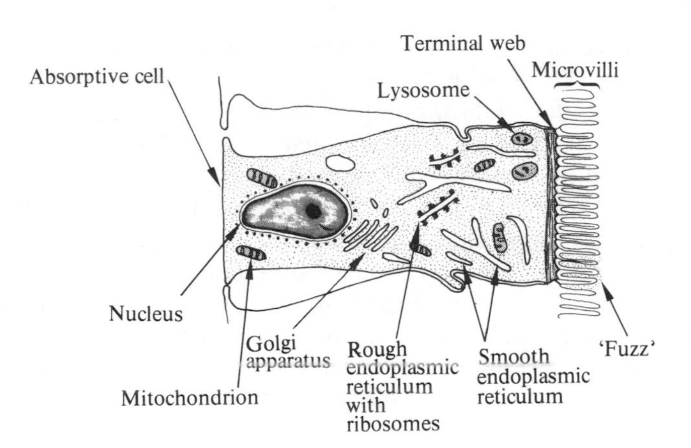

Fig. 17.6 Structure of intestinal villi and of an absorptive cell

an excellent illustration of a number of general features of cellular absorption. In addition, these absorptive cells show how elegantly cellular structur can be related to function (Fig. 17.6).

The wall of the intestine is convoluted to form villi, which project into th lumen of the intestine. These are separated by crypts, at the base of whic the epithelial cells are constantly undergoing mitosis. All the epithelial cel of the villus arise in this way, the newly formed cells gradually moving u and over its surface. This is a continuous process, for cells at the top of the villus are constantly being replaced, the movement of digesta down the intestine probably helping to dislodge older ones. Dislodged cells are subsequently broken down in the intestine, the enzymes they contain being released into the lumen; in this way the cells contribute a valuable but secondary source of enzymes (the main source of digestive enzymes being the pancreas). The cell division which occurs at the base of the crypt gives rise mainly to absorptive cells, although a second type of cell, the *goblet cell*, is also present. These cells acquired their name from the appearance they presented to the early microscopists, for within them are vacuoles which are filled with mucus. This mucus is eventually secreted into the intestinal lumen and may protect the general epithelial cells from the action of digestive enzymes.

Intestinal absorptive cells are columnar, the nucleus being positioned near the base of each cell. The free outer surface of the cell, bordering the intestinal lumen, has its area increased about twentyfold by membrane invagination to give numerous slender finger-like projections called *microvilli* (Fig. 17.6). Each microvillus is about 1 μm in length and 0.1 μm across. Described as the 'brush border' by light microscopists, the microvilli have been shown by the electron microscope to be bounded by a lipoprotein membrane (page 33). Such a membrane is, of course, typical of the plasmalemma of all types of cells, but in the present instance the protein layers are somewhat thicker than they are in parts of the plasmalemma elsewhere in the absorptive cell. A probable explanation is that numerous enzymes and carrier proteins, which together are responsible for the transport of materials across the cell membrane, are concentrated in the region of the microvilli. Such an accumulation is to be expected, for the absorptive cells are able to absorb all the major products of digestion—amino acids, di- and tripeptides, monosaccharides, disaccharides, free fatty acids, and monoglycerides—and, as will be seen in Section 17.10, the absorption of many of these materials involves processes far more complex than simple physical diffusion.

The microvilli have a rather indefinite or 'fuzzy' appearance, owing to the presence, in the 'microcrypts' between them, of a mucopolysaccharide. This material probably decreases bacterial attack on the cells and may, in addition, assist in some way in the initial binding of substances about to undergo absorption. Each microvillus contains a number of fine central filaments which run along its length and which possibly help to strengthen it. At the base of the microvilli is a region containing further fibrous elements; it is known as the *terminal web*. This web separates the apex of the cell, which bears the microvilli, from the remainder of the cell. It has been suggested that one function of the web may be to help regulate the passage of materials towards the base of the cell after their absorption.

It is possible to separate the microvilli, together with the web and a little of the underlying cytoplasm, from the rest of the cell and thus to study the absorptive apparatus and the enzyme systems associated with it. It contains most of the intestinal disaccharidases (e.g., lactase and maltase), as well as an alkaline phosphatase and various proteolytic enzymes. These enzymes are probably involved in the uptake of materials from the intestinal lumen (Section 17.9). An enzyme is also present which, *in vitro*, hydrolyses ATP to ADP, and, for this reason, is called ATP-ase. This enzyme might well participate, *in vivo*, in the absorption of one or more of the products of digestion, for absorption is frequently a formally endergonic process which is coupled to the exergonic breakdown of ATP. The cells contain numerous mitochondria which are capable of providing the ATP which somehow enables the transfer and carrier proteins to function.

The membranes of the endoplasmic reticulum of absorptive cells have less ribosomes associated with them than those of the enzyme-secreting cells of the stomach and pancreas. This is to be expected, since absorptive cells, unlike the cells just mentioned, do not have to synthesize large quantities of protein for purposes other than their own metabolism. On the other hand, intestinal absorptive cells do possess considerable amounts of *smooth* reticulum, and this probably reflects the fact that fat absorption is followed by the synthesis of triglycerides and phospholipids. Some steps in these syntheses occur in the neighbourhood of endoplasmic membranes.

17.9 Absorption of sugars and amino acids by cells of the intestinal epithelium

In man and other monogastric mammals, dietary starch and glycogen are hydrolysed in the lumen of the digestive tract to a mixture of glucose maltose, and isomaltose by the action of amylase (page 81) secreted in the saliva and pancreatic juice. The intestinal absorptive cells absorb these sugars together with the other principal dietary carbohydrates, namely lactose, sucrose, and small amounts of various monosaccharides. In addition the cells absorb amino acids and small quantities of di- and tripeptides produced by the action of the gastric and pancreatic proteolytic enzymes on dietary proteins (Fig. 3.5).

The mechanism for absorption of most of the monosaccharides and free amino acids is one of carrier mediated active transport. Before considering individual examples, the general mechanism is best explained by reference to Fig. 17.7. Without a carrier these metabolites would hardly enter the cells at all, let alone be transported, as they often are, *up* a concentration gradient. The absorptive cells use what is now regarded as a widely occurring process whereby the difference between the sodium concentration

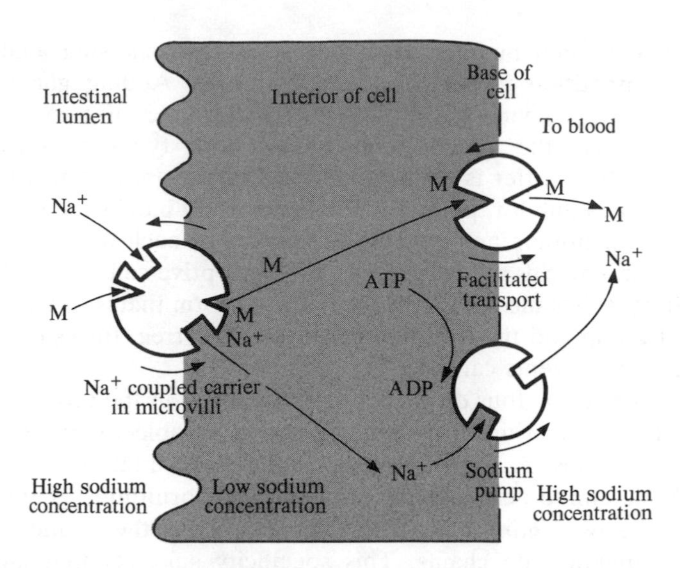

Fig. 17.7 **A generalized scheme for the absorption of monosaccharides and amino acids by the intestinal cells**

inside (low) and outside the cells drives the process. In the microvilli there is a carrier specific for metabolite (M) *and* for sodium ions. Both the metabolite and sodium ions become attached to particular binding sites on the outside of the cell. This enables the sodium to move into the cell (going *down* its concentration gradient). As it does so, the metabolite is also carried into the cytosol so the downhill process of sodium movement is coupled to the uphill process of transferring the metabolite. The process could not continue if it were not for the presence of the sodium pump at the base and sides of the absorptive cells which maintains the sodium gradient by pumping sodium ions out again into the extracellular fluid. The overall process is inhibited by ouabain and by inhibitors of ATP synthesis (e.g., dinitrophenol and fluoride). Moreover, artificially lowering the sodium concentration on the absorptive surface of the cells reduces the transport. In effect, the hydrolysis of ATP by the sodium pump indirectly provides the energy for this type of carrier mechanism.

This general mechanism (Fig. 17.7) applies to the absorption of several monosaccharides which share a common carrier in the microvilli. Almost any pentose or hexose having the D configuration, a pyranose ring structure, and a free hydroxyl group in the carbon atom 2 position orientated as in D-glucose is carried by this process. Of the common natural dietary monosaccharides, glucose, galactose, and xylose share this transport mechanism. When there are mixtures of these sugars they appear to compete for the same limited number of carrier sites, thus showing analogy with substrate competition for enzymes (Section 6.7). Adding glucose to the system which is carrying galactose reduces the rate of transport of the galactose because the system then carries both types of sugar. Other evidence that the carrier is common for several sugars is derived from the use of a specific inhibitor, phlorizin. This effectively blocks the carrier in the microvillous membrane and stops the transport of each of the sugars.

The L-α-amino acids are carried into the absorptive cells by similar carrier systems. In most instances it is only the natural L form that is carried. Both the α-carboxyl group and the α-amino group must be free, for esterification of the acid group prevents carriage.

There appear to be four distinct carriers involved in the absorption of the amino acids. The first and least specific carrier is able to transport all the neutral amino acids. A second is responsible for facilitating the absorption of three basic amino acids—lysine, arginine, and ornithine. A third carrier transports the two *imino* acids and the last carries the amino acids with branched aliphatic side chains. This specificity suggests that amino acid carriers, like those which transport carbohydrates, have active centres which form more or less specific attachments with their substrates. Clearly, they resemble enzymes which have greater or lesser specificity for their substrates.

Disaccharides and the di- and tripeptides enter the absorptive cells by a process of *membrane digestion.* The microvillous membrane contains a number of disaccharidases and peptidases. These enzymes (they could equally be called carriers) combine with their substrate on the absorptive surface and hydrolyse it to its component monosaccharides or amino acids which are then passed into the cytosol of the cell. For the disaccharides the principal enzymes are maltase, isomaltase, sucrase, and lactase, the functions of which are to carry the two types of maltose, sucrose, and lactose respectively. For example, lactose combines with its carrier which then catalyses the hydrolysis of the glucoside linkage and passes free glucose and galactose into the cell.

The microvillous peptidases are far less specific for their substrate, for there are some 400 possible dipeptides and 8000 tripeptides. Their hydrolysis by the carrier leads to the entry of the free amino acids into the cell.

Once inside the absorptive cell the amino acids and monosaccharides absorbed by the various processes just described must pass out of the cell at its base. This process again requires carriers but in this instance the mechanism is one of carrier mediated facilitated transport (Fig. 17.7).

After absorption, amino acids are carried by the blood stream to the liver. The process of absorption, both of amino acids and of sugars, is extremely efficient and occurs throughout the length of the small intestine. Only very small amounts of either type of compound remain unabsorbed by the time the residue of the food leaves the small intestine.

17.10 The digestion and absorption of glycerides

Neutral triglycerides are almost insoluble in water and for this reason their digestion and absorption follow a pattern somewhat different from those of proteins or carbohydrates. Digestion takes place mainly in the duodenum, the *lipase* which catalyses the hydrolysis being secreted by cells of the pancreas. Since the enzyme is water-soluble, effective hydrolysis is only possible if the surface of contact between aqueous and lipid phases is very large. Thus fats and oils must first be dispersed in the form of exceedingly small droplets and several regulatory mechanisms exist which ensure that *emulsification* occurs. Bile assists in the emulsification process, for it contains several powerful emulsifying agents—lecithin, glycocholate, taurocholate, and cholesterol (Chapter 2). Together these substances disperse the fat into droplets of about 500 nm diameter. Particles of this size are large enough to scatter light and therefore the emulsion is milky in appearance. The droplets are stabilized by the charges on the anions of the bile acids and by the polar groups of lecithin, the hydrophobic parts of the emulsifying agents being directed towards the inside of the droplets. As in the cell membrane, cholesterol may facilitate a closer packing of the phosphoglyceride and glyceride molecules.

Pancreatic lipase plays a major role in processing the triglyceride to a point where absorption is possible by intestinal cells. It selectively removes the fatty acids from the two α positions of each glyceride molecule, producing a mixture containing a little residual triglyceride, but which is mainly composed of fatty acids and β-monoglycerides.

The milkiness of the duodenal contents, apparent when fats are first emulsified, gradually clears as lipase hydrolyses the triglyceride, for the fatty acids and β-monoglycerides break away from the droplets to form new and much smaller aggregates known as *micelles* (Fig. 17.8). One of the original droplets of the emulsion can give rise to some 10^6 micelles, each with a diameter of about 5 nm. Such micelles are too small to scatter light and the colloidal solution containing them is consequently almost clear. These micelles are, however, still stabilized by the presence of bile acids and lecithin. Their total surface area is enormous compared to that of the emulsion droplets from which they were formed, and consequently many of them can simultaneously come into contact with the microvilli of the absorptive cells.

Free fatty acids and monoglycerides are readily absorbed, but some di- and triglycerides also enter. In all cases, the entry appears to be passive in nature, for the rate of entry is almost independent of temperature. In contrast, active transport, like other kinds of chemical reactions, increases as the temperature rises within the limits imposed by a biological environment. The fatty acids have a high lipid solubility and they therefore readily pass into the cell's plasmalemma (Section 17.2).

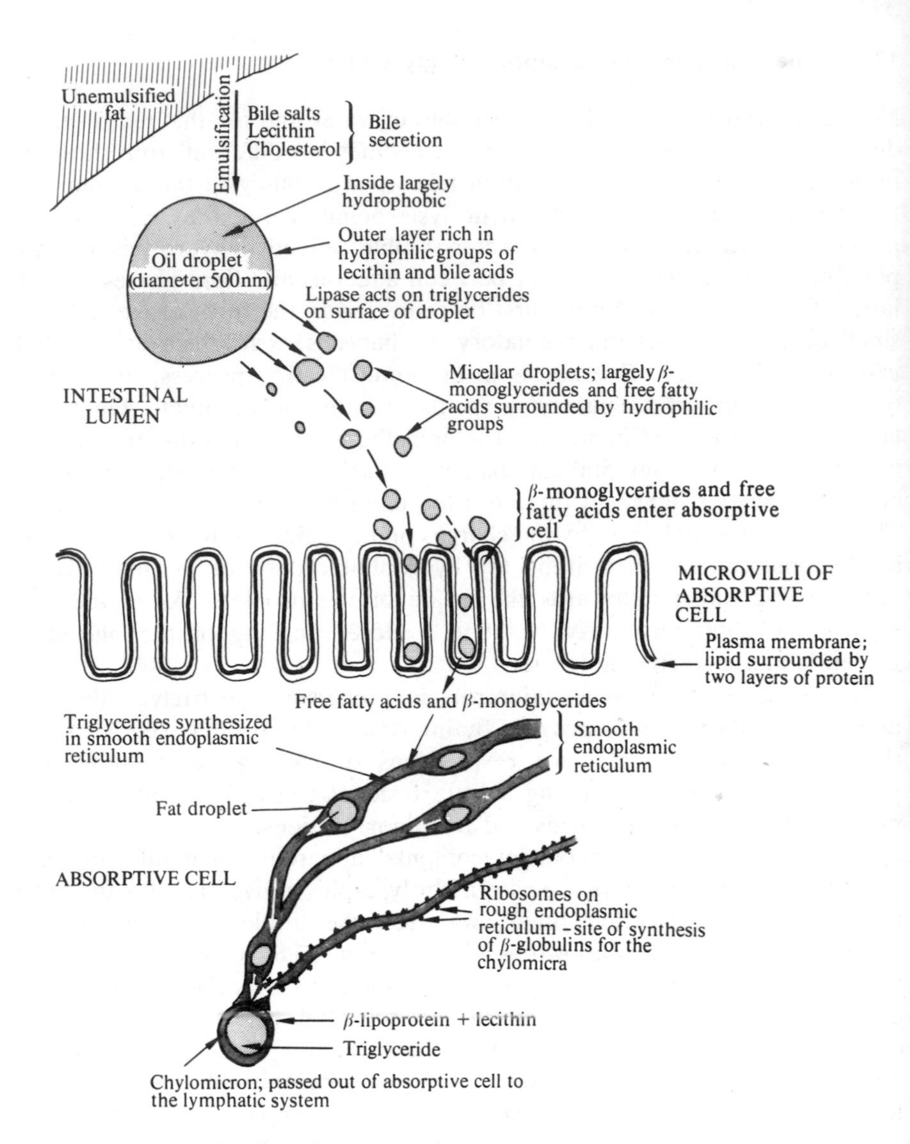

Fig. 17.8 Outline of the mechanism of fat digestion and absorption

Once within the absorptive cells, the fatty acids and monoglycerides diffuse through the terminal web to the abundant smooth endoplasmic reticulum below it. This region of the cell is rich in the enzyme systems responsible for the synthesis of triglycerides and phospholipids, and electron micrographs of active absorptive cells suggest oil droplets among the reticulum. An enzyme located there then converts free fatty acids to fatty acyl coenzyme A derivatives which, in the presence of a second enzyme known as *monoglyceride transacylase*, react with β-monoglycerides to form triglycerides:

$$\mathrm{RCOOH + ATP + HSCoA} \longrightarrow \mathrm{RCO \sim SCoA + AMP + PP}$$

$$\mathrm{2RCO \sim SCoA} + \beta\text{-monoglyceride} \xrightarrow[\text{transacylase}]{\text{Monoglyceride}} \text{triglyceride} + \mathrm{2HSCoA}$$

Other reactions leading to the synthesis of phospholipids and triglycerides (Sections 10.5 and 10.6) also take place in the intestinal cells.

The ribosomes attached to the rough endoplasmic reticulum also participate indirectly in the formation of the oil droplets which appear within the cell, for they synthesize specific proteins which form an envelope around the droplets. These proteins, known as β-globulins, are characterized by the fact that they possess a higher than usual proportion of amino acid residues with hydrocarbon side chains. The side chains form weak apolar bonds with molecules of lipid in the droplets to produce what are, in effect, lipoproteins. The droplets, complete with their lipoprotein envelope, are discharged from the endoplasmic reticulum into the cell's cytosol (Fig. 17.8).

The droplets, which have a diameter of 75 to 1000 nm are transferred to the base of the absorptive cell, from where they are discharged into the lymph duct. In addition, an appreciable quantity of short-chain free fatty acids passes into the blood. Lymph is turbid in appearance after the animal has digested a fatty meal because of the presence of these fat globules. They are known as *chylomicra*; each chylomicron usually contains some 86% triglyceride, 9% phospholipid, 3% cholesterol, and 2% lipoprotein. The presence of surface active materials allows the water-insoluble lipids they contain to be transported to the liver in aqueous body fluids.

18

The carriage of oxygen and carbon dioxide by the blood

The need to supply oxygen and to remove carbon dioxide and excess hydrogen ions from the tissues of higher animals has led to the evolution of a number of highly efficient mechanisms. In particular, the molecular structure of haemoglobin has developed to a remarkable extent in the direction of matching almost perfectly its functions of carrying oxygen and carbon dioxide while buffering the blood. This chapter describes these mechanisms as they occur in mammals. Other groups of animals, especially invertebrates, have evolved many different solutions to essentially the same problems, so the present account should not be regarded as applying to animals generally.

18.1 The transport of oxygen into animal cells

Active cells consume considerable quantities of oxygen. In most cases the dissolved gas simply diffuses into the cell by passive means, for the consumption of oxygen in the cell creates a concentration gradient, the extracellular fluid having the higher oxygen level. The great majority of animals need an ancillary transport system which, in effect, connects the extracellular fluid to the organ where oxygen is absorbed and, in vertebrates, there is a highly developed blood vascular system which transfers oxygen from gills or lungs to the tissue capillaries. The oxygen diffuses out of the blood capillaries in the tissues, again down a concentration gradient, and by doing so maintains at a high level the oxygen concentration of the fluid immediately bathing the cells. Vertebrate blood is able to absorb large quantities of oxygen in the lungs or gills because of the presence of haemoglobin in the red cells. The passage of oxygen into the erythrocytes occurs by a process of *passive transfer* which is dependent on the haemoglobin within the cells. Mammalian erythrocytes are remarkable in lacking nuclei; they are flattened discs having a diameter of about 8 μm in man and occupying some 40 per cent of the blood volume.

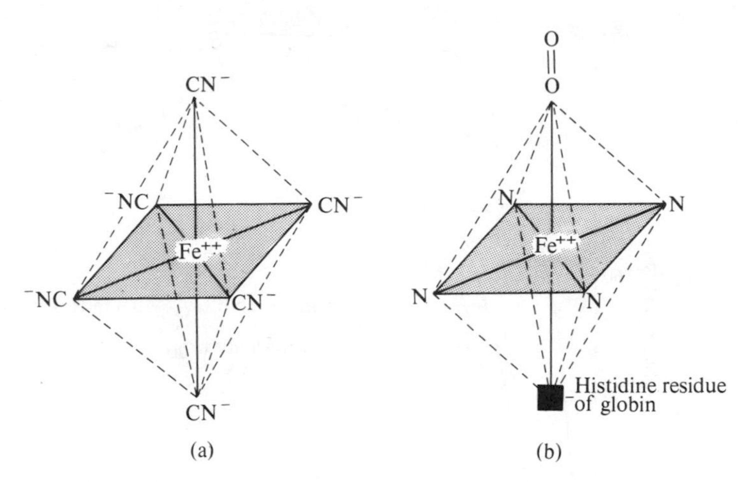

Fig. 18.1 Comparison of the structures of (a) the ferrocyanide ion, $[Fe(CN)_6]^{4-}$, and (b) oxyhaemoglobin. In (b) the nitrogen atoms shown are those of protohaem

The haemoglobin molecule of higher vertebrates consists of four subunits each possessing one protohaem group (Section 3.9). Each protohaem group contains a ferrous ion at its centre held by resonating covalent and dative covalent bonds to the four nitrogen atoms of the five-membered pyrrole rings. The ferrous ions are the immediate oxygen carriers and they are thought to link to the molecular oxygen in the manner shown in Fig. 18.1. It should be observed that the iron remains permanently in the ferrous form. This contrasts with the iron of the cytochromes which is reversibly converted from the ferrous to the ferric state (page 163). Since there are four haem groups in each haemoglobin molecule, one mole of haemoglobin, when fully oxygenated, carries four moles of oxygen. In fact, the oxygenation takes place in four stages and therefore fully oxygenated haemoglobin is best represented as $Hb(O_2)_4$, although the equilibrium between unoxygenated and oxygenated haemoglobin is often shown more simply as

$$\underset{\text{Unoxygenated haemoglobin}}{Hb} + O_2 \rightleftharpoons \underset{\text{Oxygenated haemoglobin, or oxyhaemoglobin}}{HbO_2}$$

The initial step in the transfer of oxygen across the erythrocyte plasmalemma is passive diffusion. This would soon cease in the absence of haemoglobin as the oxygen concentration gradient evened out. The importance of haemoglobin in mammalian blood is demonstrated by the fact that haemoglobin carries about 98 per cent of the oxygen; only 2 per cent is in simple solution.

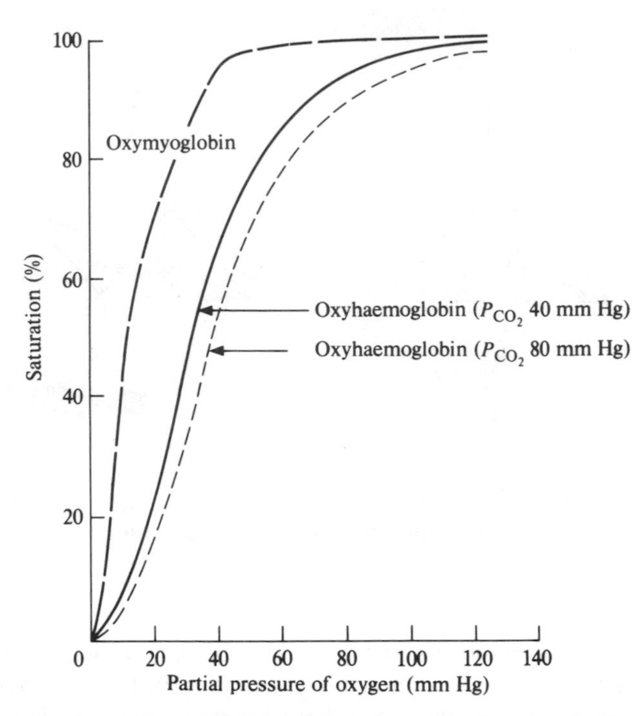

Fig. 18.2 The oxygen dissociation curve of human haemoglobin (at partial pressures of carbon dioxide of 40 and 80 mm Hg) and the dissociation curve for oxymyoglobin

An essential feature of the oxygenation of haemoglobin is its reversibility. Oxygen is readily taken up by the red cell haemoglobin in the lungs or gills of animals, and when the blood reaches active tissues, the oxyhaemoglobin dissociates to yield oxygen, which diffuses passively out of the red cells into the extracellular fluid. The dissociation of oxyhaemoglobin is not only matched with the requirements of the tissues through which the blood flows, but, in addition, it is linked to the transport of carbon dioxide (Section 18.4).

The most important factor determining the extent to which haemoglobin is in the oxygenated form is the oxygen concentration to which it is exposed. This is shown in Fig. 18.2 where the *solid line* indicates that at a partial pressure of oxygen (P_{O_2}) greater than 80 mm Hg, the haemoglobin is very nearly completely saturated with oxygen. The curve is S-shaped; at a partial pressure of oxygen of 40 mm Hg, the haemoglobin is about 75 per cent saturated (that is, 75 per cent is in the oxygenated form), while at a partial pressure of 20 mm Hg, it is about 20 per cent saturated.

In the lung capillaries of mammals, the blood becomes almost completely saturated with oxygen because the partial pressure of oxygen in lung alveolar gas is about 100 mm Hg and oxygen readily diffuses through the thin alveolar membrane into the blood via the thin capillary walls (Fig. 18.3).

In active tissues such as contracting muscle the partial pressure of oxygen is low (30 mm Hg or less), largely owing to oxygen utilization by mitochondrial cytochrome oxidase. Oxygen diffuses into the tissue cells from the surrounding interstitial fluid where the oxygen concentration is higher; simultaneously, as the extracellular fluid loses oxygen to the tissue cells, oxygen diffuses out of the red cells in the blood capillaries to replace it. In effect, a concentration gradient exists between the erythrocytes and the tissue cells so that oxygen diffuses to where it is being consumed. The diffusion of oxygen from the blood constantly lowers the oxygen concentration there, and this causes the dissociation of oxyhaemoglobin. Venous blood returns to the lungs with oxygen at a partial pressure of about 40 mm Hg (75 per cent saturation). In the lungs it is reoxygenated and so is ready to be recycled (Fig. 18.3).

Myoglobin is another haem pigment (Fig. 3.10) which, in many animals, is of physiological importance because of the ease of its (reversible) association with oxygen:

$$Mb + O_2 \rightleftharpoons MbO_2$$

It is present in the skeletal muscles of higher vertebrates, its single protohaem group reacting with oxygen in the same way as each of the four protohaem groups of haemoglobin.

The affinity of myoglobin for oxygen is greater than that of haemoglobin. Therefore, in a system containing oxyhaemoglobin and myoglobin, the

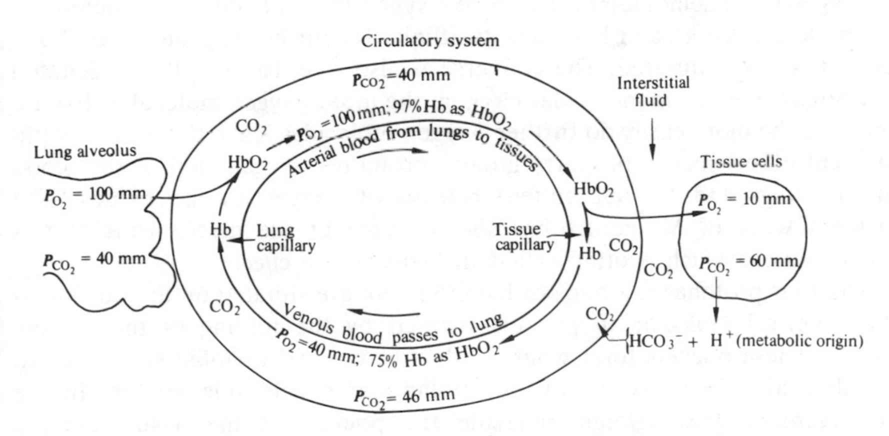

Fig. 18.3 Diagram to show that the movement in lung and tissue capillaries of both oxygen and carbon dioxide is determined by partial pressure (*P*) gradients

former readily dissociates, the oxygen passing to the myoglobin, so oxygenating it. This is shown by the position of the myoglobin line in respect to that for haemoglobin in Fig. 18.2. Clearly, the greater affinity for oxygen possessed by myoglobin greatly enhances the extraction of oxygen from blood into the muscle cells—especially in active muscular tissue, where the oxymyoglobin has in part dissociated to supply oxygen to the muscle mitochondria.

18.2 Allosteric changes in haemoglobin

We will now consider the special features of the haemoglobin molecule which enable it to function so effectively as a transporter of oxygen. Haemoglobin is an allosteric protein (Section 3.9), a feature which may be apparent from the shape of the oxyhaemoglobin dissociation curve (Fig. 18.2) since for many allosteric enzymes a sigmoid relationship exists between substrate concentration and the rate of an enzyme-catalysed reaction (page 257). It is important to note that the addition of oxygen to both myoglobin and haemoglobin is an *oxygenation* process, not an oxidation process. The latter would involve a valency change on the part of the iron (compare mitochondrial cytochromes) or some similar electron loss. Here, the iron remains permanently in the ferrous state. Haemoglobin is not an enzyme, of course, but the combination of oxygen with protohaem units is analogous to the formation of an enzyme-substrate complex, so the analysis of the kinetics is similar.

A significant feature of the oxygenation process is that, beginning with deoxygenated haemoglobin, the more oxygen that is added to a haemoglobin molecule the greater becomes its affinity for further oxygen molecules—until it is fully saturated. The converse is also true, for as fully oxygenated haemoglobin gives up molecular oxygen, the more oxygen molecules that are released the more easily do further oxygen molecules leave the haemoglobin molecule. This special property greatly promotes oxygenation of the blood in the lungs and the subsequent release of oxygen to the tissues. The brilliant work of M. Perutz has shown some of the mechanisms of this phenomenon, which is often called the *cooperative effect.*

The four protohaem groups of haemoglobin are situated on the outside of the spherical molecule in pockets formed by the folding of the protein chains. These pockets form what is effectively a hydrophobic environment, for they are 'lined' by mostly non-polar amino acid side groups. In the deoxygenated haemoglobin molecule the pockets of the α-subunits are larger than those of the β-subunits so oxygen enters the former more easily. When the first oxygen molecule to undergo reaction enters the pocket of an α-subunit a remarkable, and still poorly understood, change occurs in the

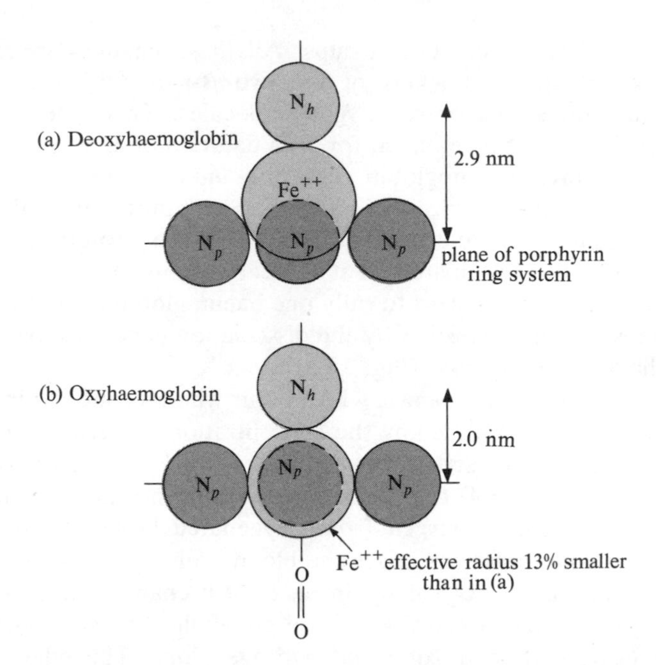

Fig. 18.4 (a) The relationship between the ferrous ion and the porphyrin nitrogen atoms (N_p) in deoxyhaemoglobin. N_h represents a nitrogen atom of the imidazole ring of histidine. (b) Shows how the iron fits into the porphyrin ring after oxygenation

effective size of the ferrous ion in the subunit. In the unoxygenated protohaem group the ferrous ion is too large to fit perfectly into the plane of the porphyrin ring system, but on oxygenation the *ionic radius decreases* by about 13 per cent. The ion then slips neatly into the plane of the ring system (Fig. 18.4). The ferrous ion thus moves a distance of about 0.9 nm and as it does so it draws the histidine residue to which it is coordinated over towards the porphyrin. Perutz calls this a trigger mechanism, for the small movement of the *histidine residue* causes a rather larger set of movements in the *protein helix* of which it is a member. This causes the disruption of a number of intrachain bridges (salt, non-polar, and hydrogen bonds) within the subunit. Almost instantaneously a new set of slightly different bridges is formed as the subunit changes shape.

An essential feature of the quaternary structure of complex proteins is the presence of various bridges linking the individual subunits together (page 62). The change in shape of one haemoglobin subunit upon oxygenation induces allosteric transformation in an adjacent subunit and so causes a change in the quaternary structure. In particular, the head-to-tail salt bridges between the α- and β-subunits (Chapter 3) are broken, producing free carboxyl and amino end groups. All these changes in conformation serve to open up the pockets of the two β-subunits, so enabling their protohaem groups to receive oxygen molecules. This type of change accounts for the cooperative effect, for each addition of an oxygen molecule to a protohaem in a haemoglobin molecule *induces the cooperation of the adjacent subunit* in the process of combining with more molecules of the gas. Since it is changes in quaternary structure of the protein which are ultimately responsible for this effect it is clear that in the case of myoglobin, whose structure is equivalent to only one haemoglobin subunit, no cooperative effect is possible. This is why the dissociation curve for oxymyoglobin is hyperbolic and not sigmoid (Fig. 18.2).

Some of the allosteric changes which occur when haemoglobin takes up or releases oxygen are modified by the concentrations of hydrogen ions and of carbon dioxide in the erthyrocyte. It has been known since the classical studies of C. Bohr (1904) that increasing the concentration of either of these *enhances the release of oxygen* from oxygenated blood. Conversely a decrease in the concentrations raises the blood's affinity for oxygen. The *Bohr effect* is of great physiological significance for it enables the blood to supply more oxygen to those tissues which are metabolically active and are therefore producing carbon dioxide and hydrogen ions. The effect of a raised carbon dioxide concentration is shown in Fig. 18.2. The oxyhaemoglobin dissociation curve is shifted to the right when the carbon dioxide pressure is raised from 40 to 80 mm Hg. The shift indicates a lower affinity of the blood for oxygen. For example, at an oxygen partial pressure of 50 mm Hg, the percentage saturation with oxygen drops from about 80 to about 70.

The way in which a rise in the concentration of hydrogen ions lowers the affinity of haemoglobin for oxygen (i.e., enhances dissociation of oxyhaemoglobin) is mainly through its effect on the *salt bridges* in the protein. It appears that a lower pH tends to favour the quaternary structure characteristics of the deoxygenated form. Thus when a tissue passes acid into a capillary it promotes the formation of deoxyhaemoglobin from the oxygenated form and so releases molecular oxygen.

The role of carbon dioxide in the Bohr effect is twofold. As discussed later the carbon dioxide in the blood plasma equilibrates as follows:

$$CO_2 + H_2O \rightleftharpoons H_2CO_3 \rightleftharpoons H^+ + HCO_3^-$$

The hydrogen ions so formed may then act in the way indicated above. In addition, carbon dioxide is able to react directly and reversibly with the terminal amino groups of the haemoglobin subunits:

$$\text{Protein chain} \; -\underset{\mathrm{H}}{\overset{\mathrm{R}}{\mathrm{C}}}-NH_2 + CO_2 \rightleftharpoons -\underset{\mathrm{H}}{\overset{\mathrm{R}}{\mathrm{C}}}-\overset{\mathrm{H}}{\mathrm{N}}-C\begin{matrix}{=}O\\ {-}O^-\end{matrix} + H^+$$

Carbamino derivative ($HbNHCOO^-$)

Table 18.1 Mean values of blood oxygen in resting man. (Each value refers to whole blood)

	Arterial blood	Venous blood	Arterial–venous difference
Partial pressure of oxygen, mm Hg	100	40	+60
Dissolved oxygen, ml O_2/litre (a)	3	1.2	+1.8
Oxygen combined with haemoglobin, ml O_2/litre (b)	200	154	+46
Total oxygen content, ml O_2/litre (a and b)	203	155	+48
Saturation of haemoglobin with oxygen, %	97	75	+22

Oxyhaemoglobin which combines with carbon dioxide in this way releases its oxygen more easily, again through complex changes in the conformation of the protein.

The figures in Table 18.1 illustrate how the physicochemical, physiological, and molecular factors discussed in the last few paragraphs lead to differences in oxygen content of arterial and venous blood. It is the *difference in oxygen content* between these two sorts of blood (48 ml/litre blood in this table) which represents the oxygen actually used up by the tissues—in other words, only about 25 per cent of the oxygen in arterial blood is actually utilized in normal circumstances. Under conditions of severe exercise, a rather greater proportion of the blood oxygen is made available to the tissues.

18.3 Intracellular pH, and the passage of hydrogen ions and carbon dioxide out of tissue cells

Many of the processes occurring in the cell, especially those catalysed by enzymes, are only effective under conditions where the hydrogen ion concentration is maintained within a certain physiological range. Typical intracellular pH values are 6.8 to 7.0 in animals. Yet active cells are constantly producing acids. The catabolism of amino acids containing sulphur yields sulphuric acid; the breakdown of nucleic acids produces phosphoric acid. In addition, actively contracting muscle produces lactic acid. Clearly there must be an effective buffering system in the cell, and also a mechanism for excreting excess hydrogen ions to the extracellular fluid.

Within the cell there is a number of important buffering systems, the principal ones being proteins, phosphate (both inorganic orthophosphate and phosphate esters), bicarbonate, and organic acids such as lactate. These act as buffers by combining with hydrogen ions:

$$\text{Protein}^{n-} + H^+ = \text{H-protein}^{(n-1)-}$$
$$HPO_4^{--} + H^+ = H_2PO_4^-$$
$$HCO_3^- + H^+ = H_2CO_3 \rightarrow H_2O + CO_2 \text{ (dissolved)}$$
$$\text{Lactate}^- + H^+ = \text{lactic acid}$$

In general, a buffer is most effective at resisting changes in hydrogen ion concentration at pH values close to the pK_a of its acid. Within cells, where the pH is about 6.8 phosphate probably plays an important role in stabilizing pH since the pK_a of the system $HPO_4^{--} + H^+ = H_2PO_4^-$ is 6.8 (page 43).

The proteins of cells and blood act as buffers mainly because of the part played by acidic and basic amino acid residues such as those of glutamic acid and lysine. These can reversibly accept or release hydrogen ions.

The bicarbonate buffering system in the cell is of particular importance because hydrogen ions combine with bicarbonate to produce carbonic acid, which largely breaks down to form water and dissolved carbon dioxide. The latter, together with carbon dioxide produced as a result of the numerous decarboxylation reactions (e.g., those of the TCA cycle), diffuses out of the cell down a concentration gradient, for extracellular fluid typically contains less dissolved carbon dioxide than the intracellular fluid. There is also a hydrogen ion concentration gradient, the pH of the extracellular fluid being greater than that inside the cell. This enables hydrogen ions to diffuse out passively. In the next section the processes which maintain the pH and carbon dioxide gradients will be discussed.

There is one further way in which the cell eliminates hydrogen ions. Undissociated organic acids tend to pass far more readily across the cell plasmalemma than the corresponding anions, perhaps on account of their greater solubility in the lipid part of the membrane. When lactate, for example, buffers hydrogen ions in the cell, the protons are transferred to extracellular fluid as undissociated lactic acid. The lactic acid dissociates again on the outside of the cell.

18.4 How the erythrocyte participates in the transport of carbon dioxide and in buffering the blood

The blood vascular system of vertebrates not only carries oxygen and various nutrients to the capillaries which supply the tissue cells but also carries away waste products. Of these, carbon dioxide is excreted by the lungs or gills. The transport of carbon dioxide by the blood is, however, inextricably linked to the control of the hydrogen ion concentration of the body fluids and therefore these two functions are best considered together.

The pH of blood plasma is normally about 7.4, the hydrogen ion concentration being less than the intracellular level where the pH is about 6.8. As indicated in the previous section, this enables hydrogen ions to diffuse from cells to the extracellular fluid. When hydrogen ions diffuse into the blood they are buffered initially by the plasma proteins and by inorganic phosphate:

$$\text{Protein}^{n-} + H^+ = \text{protein}^{(n-1)-}$$
$$HPO_4^{--} + H^+ = H_2PO_4^-$$

In addition, the bicarbonate present in the plasma provides a remarkable form of buffering to resist an excessive fall in the pH. Hydrogen ions entering the blood combine with bicarbonate ions to form carbonic acid. However, the latter is unstable and almost all of it breaks down to form dissolved carbon dioxide and water:

$$H^+ + HCO_3^- \rightleftharpoons H_2CO_3 \rightleftharpoons H_2O + CO_2 \text{ (dissolved)}$$

This has a particular physiological significance. Now, applying the Henderson–Hasselbalch equation (Eq. 3.3, Section 3.2):

$$\text{pH} = \text{p}K_a + \log\frac{[A^-]}{[HA]}$$

The pK_a value of the HCO_3^-/H_2CO_3 system is 6.1. But carbonic acid (HA) is almost all converted quickly to dissolved carbon dioxide. For this reason, the equation can be rewritten in the form

$$\text{pH} = 6.1 + \log\frac{[HCO_3^-]}{[\text{dissolved } CO_2]}$$

In the normal mammal, the rate of breathing (or more precisely the rate at which carbon dioxide is eliminated by the lungs) is regulated to maintain the plasma molar ratio of $[HCO_3^-]/[\text{dissolved } CO_2]$ at approximately 20. So the equation above becomes

$$\text{pH} = 6.1 + \log 20$$

and since $\log 20 = 1.30$,

$$\text{pH} = 6.1 + 1.3 = 7.4$$

(i.e., the normal physiological value).

Bicarbonate is the only base whose conjugate acid can be very rapidly eliminated, a fact which partly explains the unique importance of bicarbonate as a buffer in plasma. But there is a second reason why bicarbonate occupies a special position in the buffering of plasma and other body fluids, namely, that it is readily formed from carbon dioxide, a constant product of cell metabolism. Dissolved carbon dioxide passes freely across the plasmalemma of most types of cells but this is especially so in the case of erythrocytes. These cells, unlike the surrounding plasma, contain *carbonic anhydrase*, an enzyme which contains zinc and which catalyses the hydration of carbon dioxide. Carbon dioxide, dissolved in the plasma, diffuses into the erythrocytes as the blood flows through the tissue capillaries and is hydrated to form carbonic acid. Being a weak acid, this partially dissociates forming bicarbonate ions and hydrogen ions:

$$\underset{}{H_2O} + \underset{\text{(dissolved)}}{CO_2} \xrightleftharpoons{\text{carbonic anhydrase}} H_2CO_3 \xrightleftharpoons{\text{spontaneous}} H^+ + HCO_3^-$$

Carbonic anhydrase is important in catalysing the hydration reaction which is, in its absence, relatively slow.

The reaction above has a number of important implications. Carbon dioxide, although a very soluble gas, could not be carried in sufficient quantities by the blood in simple solution. By partly converting it to bicarbonate ions the erythrocyte effectively lowers the plasma carbon dioxide concentration, so maintaining a concentration gradient between the blood and the intracellular fluid of the tissues. Moreover, the essential buffering power of the plasma is regenerated, for, as discussed later, the bicarbonate ions diffuse in large measure out of the erythrocyte into the plasma.

The hydrogen ions resulting from the dissociation of carbonic acid in the erythrocyte are partly buffered by reaction with phosphate, but most of them react with the haemoglobin molecules. The buffering properties of haemoglobin, and indeed of most other proteins, depend on the number of acidic and basic amino acid side groups which are available to take up or release hydrogen ions. Since haemoglobin undergoes a change in shape during oxygenation it is not surprising that the number of exposed buffering groups is different in the oxygenated and deoxygenated forms. Deoxygenated haemoglobin takes up protons more readily than the oxygenated form, its pK_a being 8.18 as opposed to 6.62. It is therefore an important means of buffering hydrogen ions as they are formed, whether by the dissociation of carbonic acid or by the production of carbamino haemoglobin. The importance of haemoglobin as a buffer is indicated by the fact that it provides over half the buffering of the blood under physiological conditions.

We turn now to the fate of the bicarbonate ions formed from carbon dioxide within the erythrocyte. An equilibrium exists between the bicarbonate within the corpuscles and the quantitatively much larger amount (but not higher *concentration*) in the plasma. Consequently, as bicarbonate ions are synthesized within the corpuscles, many of them diffuse out. Since the cell membrane is relatively impermeable to cations such as potassium,

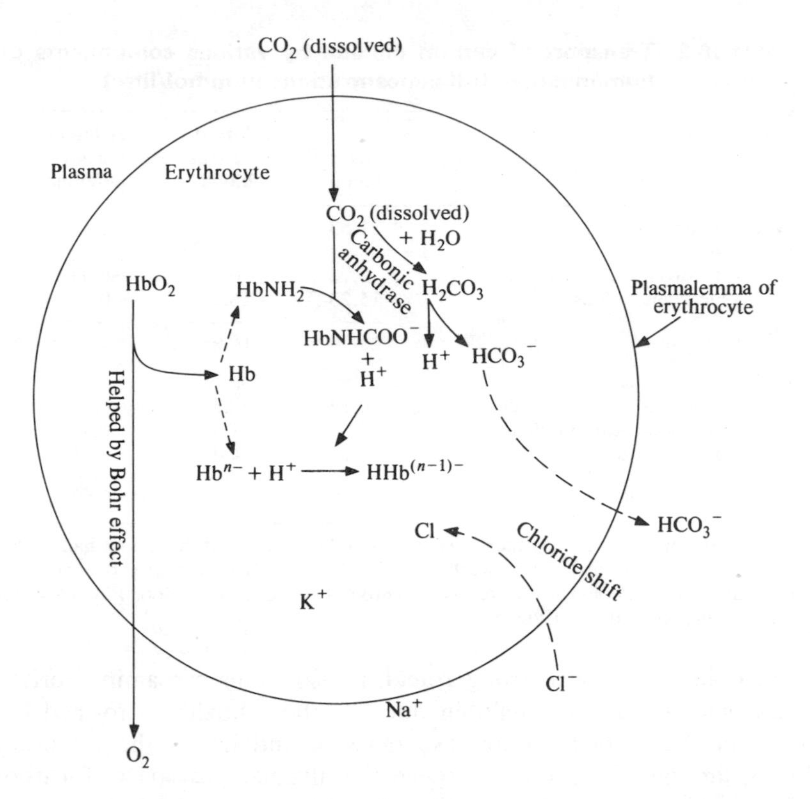

Fig. 18.5 **Outline of some of the major events occurring in the red blood cell in a tissue capillary. (Each of the processes shown is reversed in the capillaries of the lungs)**

the exit of negatively charged bicarbonate ions must be countered by the ingress of chloride ions in order that electrical neutrality should be preserved (Fig. 18.5). In fact the total quantity of the small diffusible anions in the erythrocyte increases since, as the proteins buffer hydrogen ions, the extent of their net negative charge decreases:

$$\text{Protein}^{n-} + \text{H}^+ = \text{protein}^{(n-1)-}$$

Consequently, the total amount of chloride and bicarbonate ions increases so as to balance the intracellular cations, so fulfilling the requirements of the Donnan equilibrium (Section 17.7). Thus as the blood passes through the tissue capillaries, carbon dioxide diffuses into the red cells and is accompanied by chloride ions, a process known as the *chloride shift.*

In the capillaries of the lungs the processes described above and illustrated in Fig. 18.5 are reversed. Oxygen reacts with haemoglobin to form oxyhaemoglobin. The latter has a lower affinity for hydrogen ions than has

Table 18.2 Transport of carbon dioxide by various components of human blood. (All concentrations in mmol/litre)

	Arterial blood	Mixed venous blood	Arterial–venous difference
Whole blood (1000 ml)	21.53	23.21	−1.68
Plasma (600 ml)			
as dissolved carbon dioxide	0.71	0.80	−0.09
as bicarbonate ions	15.23	16.19	−0.96
Total in plasma	15.94	16.99	−1.05
Red cells (400 ml)			
as dissolved carbon dioxide	0.34	0.39	−0.05
as carbamino-carbon dioxide	0.97	1.42	−0.45
as bicarbonate ions	4.28	4.41	−0.13
Total in red cells	5.59	6.22	−0.63

From the arterial-venous differences in this table it is evident that about half of the carbon dioxide is carried as bicarbonate in the plasma and nearly one-third as carbamino haemoglobin in the red cells. (From J. H. Comroe, 1966, *Physiology of Respiration*, Year Book, Chicago.)

haemoglobin—i.e., it is a stronger acid; it also forms carbamino derivatives far less readily than haemoglobin does. Carbon dioxide is formed by the reversal of the carbonic anhydrase reaction and leaves the erythrocytes, diffusing through the plasma to reach the alveolar gas space. Bicarbonate ions enter the erythrocytes from the plasma to replace the bicarbonate ions which have been converted to carbon dioxide, and chloride ions compensate for this movement by moving from the erythrocytes into the plasma.

Figure 18.3 summarizes the overall process of carbon dioxide transfer from the tissue cells to the lung alveoli. Within the tissue cells the partial pressure of carbon dioxide may be of the order of 60 mm Hg. It therefore tends to diffuse to the capillary blood where the partial pressure is only about 40 mm Hg. Carbon dioxide thus enters the blood so raising the partial pressure to some 46 mm Hg. This is accompanied by the changes in the plasma and erythrocytes indicated in Fig. 18.5. When the venous blood reaches the lung capillaries carbon dioxide diffuses down a concentration gradient into the alveolar gas where the carbon dioxide has a partial pressure of only 40 mm Hg. It is noteworthy (Table 18.2) that about one-third of all the carbon dioxide released into the lungs has been carried from the tissues in the form of carbamino haemoglobin (in the example illustrated by the table, 0.45 out of a total of 1.68 mmol/litre of blood).

19

Excitation of the nerve cell

A remarkable attribute of cells is their capacity, in various ways and to different extents, to respond to changes in their environment. The speed and effectiveness of response is most marked in animal cells and especially the cells of nerve and muscle. The lipoprotein membranes of all cells probably carry an electrical charge, and the electrical characteristics of the membranes of cells of nerve and muscle are developed to an extraordinary degree. It is upon these properties that the physiological functions of the cells ultimately depend. This chapter therefore begins with a discussion of the origin of the electrical potentials associated with cell membranes and describes how these potentials are altered when an impulse passes along a nerve. In the next chapter, two related problems are considered, namely, the mechanism by which proteins of muscle induce muscular contraction and the mechanism of the remarkable photochemical response shown by the rod cells of the retina.

19.1 The resting potential of cells

It has long been known that nerves and muscles respond to applied electrical stimuli, and sensitive electronic equipment now available enables the electrical properties of cells to be investigated. Fine capillary electrodes (diameter 0.5 μm) can be inserted into single cells to measure the potential between the inside of the cell and a reference electrode situated in a suitable buffered solution surrounding the cell. The potential so measured, the *membrane potential*, is the potential across the plasmalemma. This technique has been applied not only to nerve and muscle cells but also to cells of other types (e.g., large algal cells). In almost every case investigated it has been observed that the inside of a cell is *negatively charged* when the cell is in its normal physiological environment. This charge is known as the resting potential and, for nerve or muscle cells, the charge alters in a remarkable manner when the cell is activated (Section 19.3).

In most kinds of cells, the intracellular level of potassium ions is greater

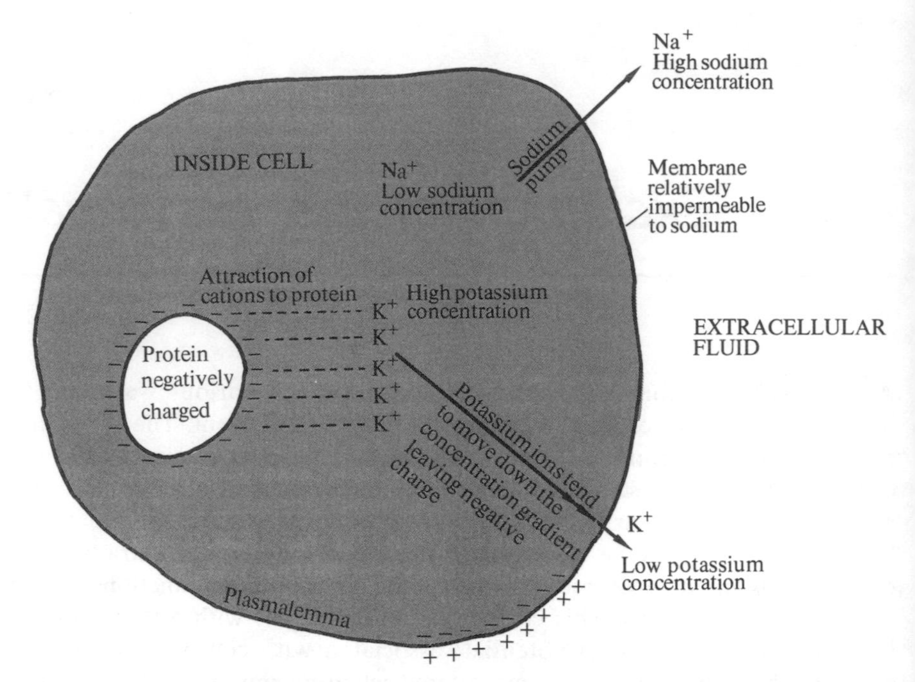

Fig. 19.1 The major factors producing the resting potential of nerve and muscle cells

than the normal extracellular concentration (Section 17.6). It is this gradient—associated with the fact that sodium is actively eliminated from cells—which leads to the existence of a resting potential. Cell membranes tend to be relatively permeable to potassium ions, and since the intracellular potassium ion concentration (K^+_{int}) is greater than the extracellular level (K^+_{ext}) there is a strong tendency for potassium to leak out of cells. When potassium ions diffuse out they tend to leave the intracellular environment negatively charged since indiffusible anions, such as protein anions, are left in excess. In the resting state at equilibrium, the negative intracellular charge tends to attract positive charges into the cell and since the sodium pump keeps down the sodium ion concentration within the cell, potassium ions are exposed to two opposing influences—a tendency to diffuse out and an attraction towards the internal negative charge (Fig. 19.1). The result is a steady-state system where the interior of the cell has a particular resting, or *potassium diffusion*, potential.

The *magnitude* of the resting potential E depends mainly on the concentrations of K^+_{int} and K^+_{ext}. In many of the cases so far investigated, including nerve and muscle cells, there is good agreement between the measured potential and the value calculated from what is known as the *Nernst equation*:

$$E = 1000\frac{RT}{F}\ln\frac{[K^+_{ext}]}{[K^+_{int}]}$$

$$= 1000\frac{RT}{F} \times 2.303 \log_{10}\frac{[K^+_{ext}]}{[K^+_{int}]} \quad (19.1)$$

where E is the potential across the membrane, expressed in millivolts, R is the gas constant (8.3 joules per degree per mole), T is the absolute temperature, and F is the Faraday in coulombs per gram equivalent. At 18°C (291K), the equation becomes

$$E = \frac{1000 \times 8.3 \times 291}{96\,500} \times 2.303 \log_{10}\frac{[K^+_{ext}]}{[K^+_{int}]}$$

$$= 58 \log_{10}\frac{[K^+_{ext}]}{[K^+_{int}]} = -58 \log_{10}\frac{[K^+_{int}]}{[K^+_{ext}]} \quad (19.2)$$

If then, at 18°C, the intracellular potassium ion concentration of frog muscle is 60 mmol/litre and the extracellular level is 3 mmol/litre, the potential across the membrane, according to the Nernst equation, is

$$E = -58 \log_{10} \tfrac{60}{3} = -58 \times 1.301 = -75 \text{ mV}$$

The negative sign associated with the 75 mV implies that the inside of the membrane is to this extent negative with respect to the outside, but the most convincing explanation is obtained by reasoning from first principles, in the following way. The Nernst equation relates to a *diffusion* potential, and so requires a tendency of ions to move down a concentration gradient. Since, in this case, the internal concentration of potassium is higher than the external, positively charged ions tend to move outwards, leaving the inside of the membrane negative with respect to its outside.

The concentration of potassium ions is readily measured by flame photometry, and in many cases a good correlation has been found between calculated values of the resting potential and the measured value. For example, the resting potentials for frog sartorius muscle and for rat skeletal muscle have been calculated from analytical results to be −99.2 and −90.1 mV; the corresponding values measured with electrodes are −101.1 and −91 mV. The squid, a marine mollusc, has certain nerve cells which are particularly large. For this reason it is both relatively easy to put electrodes into one of these cells and to squeeze the cytoplasm out for chemical analysis. The measured resting potential has been found to be −77 mV while the value calculated by inserting the analytical values into the Nernst equation is −76 mV.

It might be predicted from the Nernst equation that, if the external

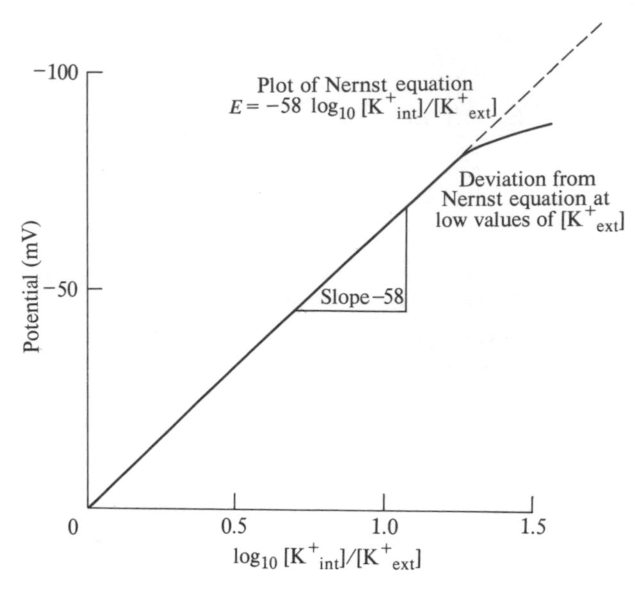

Fig. 19.2 Relation between the resting potential of muscle cells and the ratio $[K^+_{int}]/[K^+_{ext}]$ when $[K^+_{ext}]$ is varied. Temperature 18°C

potassium level $[K^+_{ext}]$ is varied, the resting potential should change correspondingly. This is indeed what happens, for when muscle is incubated in a suitable oxygenated buffer containing varying amounts of potassium, a straight-line relationship is obtained when E is plotted against $\log_{10}([K^+_{int}]/[K^+_{ext}])$. At 18°C the slope is about −58 as predicted by Eq. (19.2) (Fig. 19.2). Since the activity of nerves and muscles depends on the existence of a particular resting potential, it is crucial to the life of the animal that the extra- and intracellular levels of potassium ions should be carefully controlled—i.e., the composition of interstitial fluid is as vital to animal life as the composition of the cells themselves.

19.2 The general structure of a nerve cell

There is considerable variation in the structure of nerve cells (neurones) in different species of animals and even within a single animal, but Fig. 19.3 depicts a fairly typical peripheral neurone of a vertebrate. The essential function of a neurone is to transmit signals from one end of the cell to the other. The signals are usually *received* by fine processes known as dendrites; these are near to the part of the cell containing the nucleus. Also present in this, the main part of the cell, are mitochondria and bodies known as Nissl granules; the latter are rich in RNA. The typical neurone has a fine extension known as an *axon*. This, like the rest of the cell, is surrounded by a typical lipoprotein plasmalemma. Axons vary greatly in length, some being less than a millimetre in length while others extend several metres. The axon normally divides into fine branches which terminate in structures known as *end feet.* Electron microscopy has shown that these contain mitochondria and numerous small synaptic vesicles, which are vacuoles of about 30 nm diameter. The vesicles contain *chemical transmitters* which relay the signal to neighbouring cells (Section 19.4).

The axon of a neurone is a very delicate structure often protected and strengthened in vertebrates by being surrounded by cells known as *Schwann cells.* Some nerves have these cells developed in a remarkable manner, the cells being wound around the axon several times. The result is that the axon is surrounded by *numerous layers* of the plasmalemma of the Schwann cell (Fig. 19.3). The layers of lipoprotein membrane are known collectively as the *myelin sheath* of the neurone. The sheath provides electrical insulation and *myelinated* nerves usually conduct signals faster than non-myelinated ones. However, neurones are not entirely covered with the sheath, for there are small gaps between the Schwann cells, known as *nodes of Ranvier.* Here the axon comes into direct contact with the normal extracellular fluid of the animal.

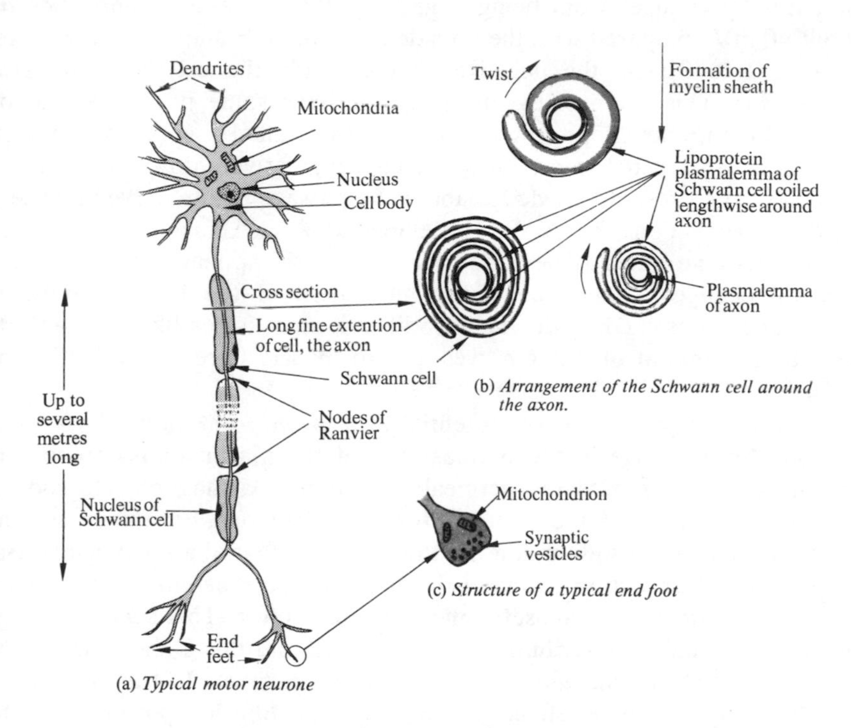

Fig. 19.3 Structure of a nerve cell

The account which follows describes how a signal is carried along a nerve cell, the apparently more simple case of the non-myelinated nerve being considered first.

19.3 The action potential of a nerve cell

The passage of impulses along non-myelinated nerves has often been compared with the transmission of signals along the wires of a telegraph system. The analogy must not be pressed too far, however, for nerve fibres behave not so such as 'wires' but as 'batteries'. By the mechanism outlined below, a flow of cations *across* the membrane creates a current which simultaneously alters the potential difference across an adjacent section of membrane and by so doing 'switches on' the next battery, the process being repeated along the whole length of the axon.

In the resting state, the nerve cell is negatively charged inside (by about 75 to 90 mV) compared with its immediate environment (Section 19.1). The passage of an impulse along the length of a non-myelinated axon is a rapid wave of depolarization moving at some 1 to 20 m/sec. As the impulse passes any particular position along the axon, the inside of the plasmalemma at that point rapidly changes from being negatively charged and becomes *positive* (about 40 mV compared with the outside of the membrane). Recovery takes place in 2 to 3μsec, the negative charge inside the cell being restored (Fig. 19.4a). This chain of events triggers off the same process a fraction further along the axon and so the impulse (often called a 'spike' owing to its appearance on a cathode-ray tube) travels progressively down the cell.

Much of the present knowledge about the passage of the nerve impulse is based on work done by A. L. Hodgkin and A. F. Huxley on the large neurones of the squid. The following hypothesis appears to explain the electrical changes accompanying the movement of the nerve impulse in these giant axons, and, with small modifications, it is readily applicable to the action potential of other nerves and to muscle fibres as well (Section 20.1).

The formation of the action potential at a given point along the axon is preceded by a change in the permeability of the plasmalemma to sodium ions. In the resting state the permeability of the plasmalemma to sodium ions is only about one-hundredth that for potassium ions. For this reason the effects of diffusion of sodium ions on the resting potential are negligible. But at the moment an action potential develops the permeability of the membrane to sodium ions increases some thousand times. This sudden change causes a rapid influx of sodium ions into the cell and this is the cause of the membrane potential changing from negative to positive. Just as the resting potential of the axon is effectively a *potassium diffusion potential*, so the action potential is largely a *sodium diffusion potential.* The magnitude of the

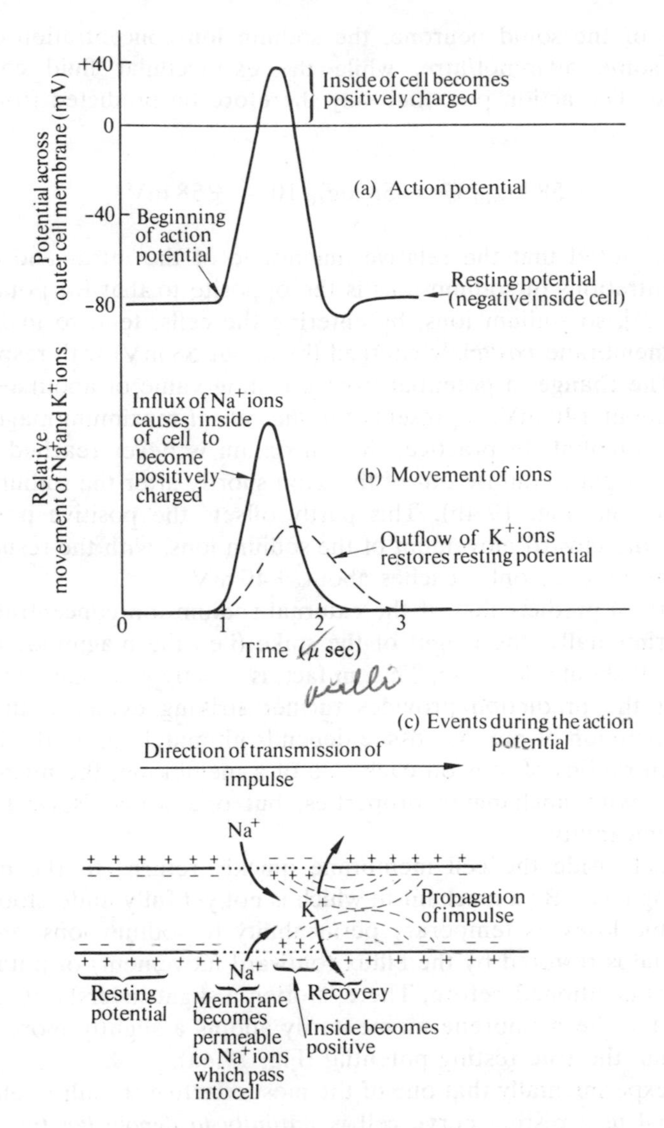

Fig. 19.4 Action potential of the nerve cell

positive intracellular potential at the peak of the spike is related to the concentration of sodium ions outside the cell $[Na^+_{ext}]$ and to their concentration inside $[Na^+_{int}]$. Quantitatively, the relationship is again expressed by the Nernst equation, this time applied to sodium ion concentrations:

$$E = \frac{1000 \times 8.3 \times 291}{96\,500} \ln \frac{[Na^+_{ext}]}{[Na^+_{int}]} = 58 \log_{10} \frac{[Na^+_{ext}]}{[Na^+_{int}]} \qquad (19.3)$$

In the case of the squid neurone, the sodium ion concentration of the axoplasm is some 50 mmol/litre, while the extracellular fluid contains 500 mmol/litre. The action potential may therefore be predicted from Eq. (19.3):

$$E = 58 \log_{10} \tfrac{500}{50} = 58 \log_{10} 10 = +58\,\text{mV}$$

It should be noted that the relative magnitude of the intra- and extracellular concentrations of sodium ions is the opposite to that for potassium ions [Eq. (19.2)], so sodium ions, by entering the cells, tend to make the inside of the membrane *positively* charged (by about 58 mV) with respect to the outside. The change in potential from a resting value of about −80 to +58 mV, or about 140 mV, represents the *theoretical* maximum magnitude of the action potential. In practice, this maximum is never reached, since potassium ions begin to diffuse out of the axon shortly after the sodium ions begin to diffuse in (Fig. 19.4b). This partly offsets the positive potential resulting from the inward movement of the sodium ions, with the result that the potential in practice only reaches about +40 mV.

Equation (19.3) predicts that, if the external sodium ion concentration is reduced experimentally, the height of the spike (i.e., the magnitude of the action potential) should decrease. This, in fact, is exactly what happens. The correctness of the prediction provides further striking evidence that the physiological function of nervous tissue depends ultimately upon the nature and the concentrations of ions on each side of a membrane, the membrane not being one with unchanging properties, but one which is capable of varying in permeability.

The potential inside the cell membrane quickly returns to the normal negative resting level. By a mechanism which is not yet fully understood, the axon membrane loses its temporary permeability to sodium ions, and the resting potential is restored by the efflux (outward movement) of potassium ions which was mentioned before. There is often a slight ‘overshoot’ during the recovery and the membrane momentarily adopts a slightly more negative charge than the true resting potential (Fig. 19.4a).

It is found experimentally that one of the most effective stimuli to elicit an action potential in a resting nerve cell is *partially to depolarize* the membrane by at least 20 mV (i.e., reduce the potential inside from, say, −80 to −60 mV). This somehow triggers off an increase in the permeability of the plasmalemma to sodium ions. It has been suggested that some conformational change in the lipoprotein membrane occurs but firm evidence is lacking. However, it is clear that the change, whatever its physical basis, leads to an influx of sodium ions and initiates the propagation of a normal action potential along the nerve. A partial depolarization of less than about 20 mV produces no action potential and is said to be below the ‘threshold’. Levels of depolarization above the threshold are effective in inducing the esponse which appears to be an 'all or nothing' phenomenon—i.e., either a eurone carries a signal or it does not.

A clue to the mechanism by which an impulse moves along a nerve fibre is o be found in the fact that each action potential elicits a partial depolariza- ion in the region of the cell just adjacent to it, so that the process moves rogressively by a series of successive depolarizations which induce action otentials in the immediately neighbouring regions. In effect, wherever a atch of membrane undergoing an action potential is adjacent to resting nembrane, currents flow along the axis of the fibre and through the urrounding fluid. This flow of current depolarizes the adjacent resting nembrane which in turn becomes active. Thus the signal progresses from ne end of the axon to the other.

The propagation of the nerve impulse in the *myelinated* nerve is similar to hat described above but is slightly more complicated. The electrical changes re similar to those of non-myelinated nerves except that they occur only at he nodes of Ranvier. The impulse jumps, or saltates, from node to node nd the speed of conduction along myelinated axons is therefore greater han it is along those which are non-myelinated—the time taken for the npulse to travel between nodes is negligible and typical of electrical onduction along a metal wire. Indeed, if the axon is severed in the nternodal region, it is possible to insert a metal conductor between the evered ends without impeding the transmission of the impulse. Thus he myelinated part of the axon appears to act as an insulated electric cable. ince vertebrate nerves are often myelinated, such nerves conduct signals nore quickly than typical invertebrate nerves.

The rapid influx of sodium ions which occurs when the membrane ecomes more permeable to these ions during the development of the ction potential does not lead, in the short term, to a significant increase in he intracellular concentration of sodium ions. In fact, when a single impulse s transmitted along an axon, only about 3×10^{-12} gram ions of sodium nter every square centimetre of cell surface. A similar quantity of potas- ium is lost as the resting potential is restored (Fig. 19.4b). Since a gram ion f sodium is 23 g, only about 69×10^{-12} g sodium ions pass through one quare centimetre of membrane, and, in consequence, an axon can carry undreds of impulses before any appreciable change in ionic composition is pparent. On the other hand, the process can obviously not continue ndefinitely without metabolic intervention to restore the *status quo*, for therwise the ionic gradients would eventually be annulled. Indeed, if the odium pump is prevented from working, such a run-down is exactly what is bserved—after some thousands of impulses have been transmitted, the xon no longer transmits action potentials. Similarly, if the axons are bathed n a medium which contains lithium instead of sodium, the ionic influx unctions normally but the sodium pump cannot pump lithium ions out gain, and eventually the axon ceases to function.

19.4 How a nerve cell carries a message

In effect, neurones carry coded messages, the types of information transmitted being enormously varied. For example, neurones from some sense organs carry messages about light intensity or colour, from others about pitch and loudness of sound. Yet others carry information about temperature, injury, tastes, or smells. Nevertheless, the action potential which is propagated along the appropriate neurones is very similar in all these cases for it is only the *termination* of the stimulus in some particular region of the brain which distinguishes the different types of information.

It might perhaps be imagined that the *intensity* of the stimulus which produces a signal carried by a neurone might determine the magnitude of the action potential. This, however, is not the case, for the action potential carried by each and every nerve cell is virtually the same in shape and magnitude as that carried by every other. However, the number of signals conveyed per second (i.e., their frequency) is highly variable. In general, the greater the intensity of a stimulus, the more *frequently* are action potentials transmitted along the appropriate neurone.

19.5 Transmission of the nervous impulse from one nerve cell to another

Once the action potential has travelled along the length of the axon, it must stimulate some other cell, or cells, if it is to produce any subsequent response. The essential function of nerve cells is to transmit information between sensory cells and the cells of organs which produce the appropriate response to the stimulus. The latter are known as *effector cells*, and, of these gland cells and muscle cells are two principal types. A simplified outline of the general arrangement of sense cells, neurones, and effector cells is indicated in Fig. 19.5.

Theoretically, the simplest way an impulse can travel from a sense cell to an effector is *via* two neurones, an *afferent* neurone which carries the message to the central nervous system and brain and an *efferent* neurone which conveys it to the muscle or gland. Transmission of this kind is known as a *simple reflex*, although it is rare in higher animals. In practice at least one, and often large numbers, of intermediate *relay neurones* participate in the transmission within the central nervous system. The intervention of relay neurones enables information received from numerous sensory cells to be coordinated so as to produce a simultaneous response from large numbers of effector cells. The central nervous system is developed to greatly differing degrees in different Phyla of the animal kingdom, according to the complexity of the relay neurone system.

When an impulse passes from one neurone to the next, or from a neurone to an effector organ, the action potential does not simply 'jump' from one

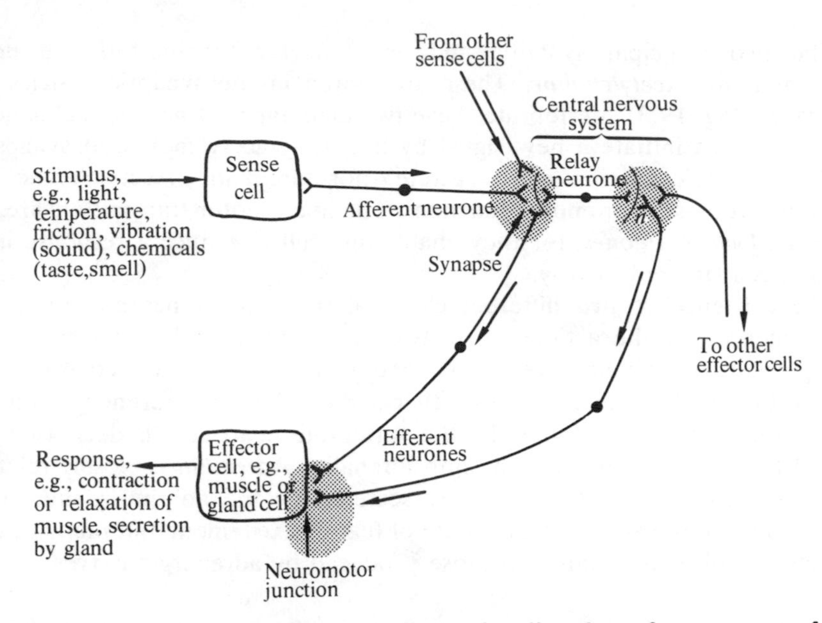

ig. 19.5 **The reflex arc. [Arrows indicate the direction of movement of the impulse. The number of relay neurones (*n*) involved is extremely variable]**

ell to the next. The link between connecting neurones is known as a *ynapse*. Although the end feet and dendrites of the cells concerned are in ery close proximity at a synapse, there is nevertheless a small gap between hem, the plasmalemma of each cell being distinct and separate. The signal s usually carried across the narrow gap by the secretion by the end feet erminating the axon of a *chemical transmitter or mediator.* The transmitter hen diffuses across the synapse, inducing an action potential in the receiving eurone. Similarly, a chemical transmitter carries the signal across the arrow gap separating the final efferent neurone and the muscle or other ffector which responds to the stimulus. This gap is called the *neuroeffector* unction, or, in the usual case of a muscle being the effector, a neuro-uscular junction.

$$H_3C-\overset{\overset{\Large O}{\|}}{C}-O-CH_2-CH_2-\overset{+}{N}(CH_3)_3$$

Acetyl..................... Choline

$$HO-C_6H_3(OH)-\underset{}{\overset{OH}{\overset{|}{C}H}}-CH_2-NH_2$$

Nor-adrenalin

The two principal synaptic and neuroeffector transmitters are *nor adrenalin* and *acetylcholine.* These are stored in the synaptic vesicles c neurones (Fig. 19.3). Vertebrates have two main types of neurone; choliner gic nerve cells initiate a new signal by liberating acetylcholine at synapse and adrenergic nerve cells secrete nor-adrenalin. Acetylcholine i liberated at most neuromotor junctions. The neuromotor transmitters are, i a sense, *local hormones*, for they enable one cell to induce a response in second cell situated nearby.

The existence of *two* different chemical transmitters has an importan consequence, for the actions of acetylcholine and nor-adrenalin are ofte antagonistic. For example, certain cholinergic nerves cause a decrease in th size of the pupil of the eye, while other nerves which are adrenergic induc dilatation of the pupil. Similarly, the vertebrate heart rate is decreased b acetylcholine but is increased by nor-adrenalin. *Adrenalin* is closely relate to nor-adrenalin and is a hormone secreted by the adrenal gland. It i secreted when the animal is in a state of fear or excitement and many of th effects it evokes are similar to those produced by adrenergic nerves.

20

Muscular and visual excitation

20.1 How a nerve stimulates a muscle

Most vertebrate *skeletal muscles*—the muscles which move the limbs, backbone, and jaws—are controlled largely by neurones for which acetylcholine is the chemical transmitter. The end feet of the neurones are closely attached to the muscle at the neuromuscular junctions. Small indentations facilitate contact between neurone and muscle at the *motor end plate* (Fig. 20.1), but a minute gap nevertheless remains and it is across this gap that the chemical mediator diffuses.

The vesicles of the end foot are about 30 to 40 nm in diameter and contain acetylcholine. The vesicles are thought to be synthesized in the cell body of the neurone and reach the end foot by passage along the axon.

Associated with the vesicles of the neurone is an enzyme, *choline acetyltransferase,* which catalyses the *synthesis* of acetylcholine from acetyl coenzyme A and choline (see Fig. 20.2).

When impulses reach the end foot they induce an increase in the concentration of calcium ions in the cytoplasm surrounding the vesicles. This appears to be brought about by an enhanced diffusion of calcium into the cell from the surrounding fluid occasioned by a permeability change in the plasmalemma. The rise in the calcium ion concentration causes some of the vesicles to discharge the acetylcholine into the neuromuscular gap, from where it diffuses to the membrane of the muscle. Extremely small quantities of acetylcholine (about 10^{-16} mol) are sufficient to cause a *partial* depolarization of the membrane of the muscle in the region of the end plate (Fig. 20.3). The result is that, in this region, the charge on the membrane falls with respect to that of the surrounding membrane. When sufficient acetylcholine has reached the end plate to produce an end plate potential of about half the resting potential, the end plate acts as an 'electron sink', current flowing to it from the neighbouring muscle membrane. This flow of current renders the neighbouring membrane permeable to sodium ions, which pass in to produce a typical action potential (Section 19.3). This explains why the arrival of small discrete amounts of acetylcholine produces an 'all-or-nothing'

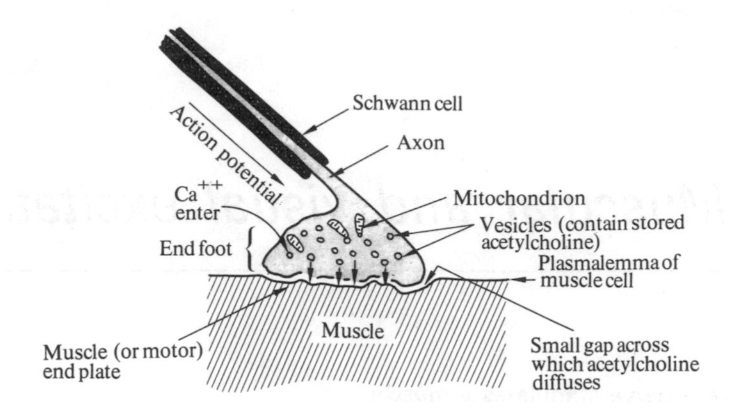

Fig. 20.1 Structure of the neuromuscular junction

muscular response by muscle cells, for once the 'electron sink' reversal effect has commenced, the action spreads over the muscle cell and induces the process of contraction (Section 20.5).

The motor end plate contains a specific lipoprotein (molecular weight 320 000) whose function is that of acetylcholine receptor. Up to 10 per cent of the plasmalemma in the region of the neuromuscular junction comprises this protein. The normal 'substrate', acetylcholine, is thought to combine with the active centre producing a change in conformation in the protein. There is some evidence that this releases bound calcium and the calcium ions in their turn cause changes in the permeability of the plasmalemma so inducing the partial depolarization.

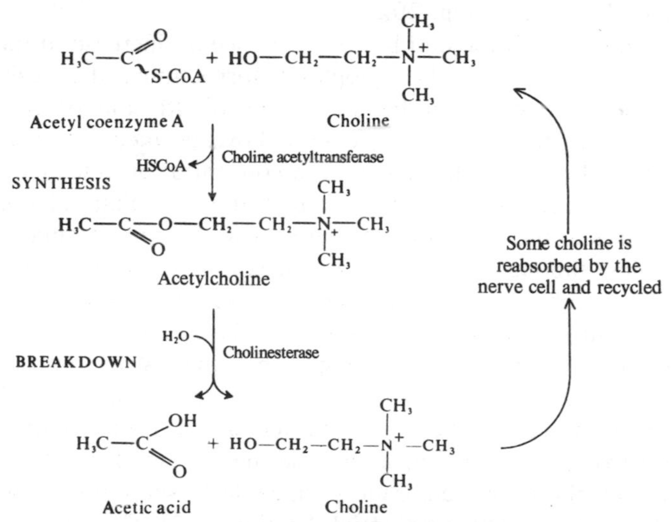

Fig. 20.2 Synthesis and breakdown of acetylcholine

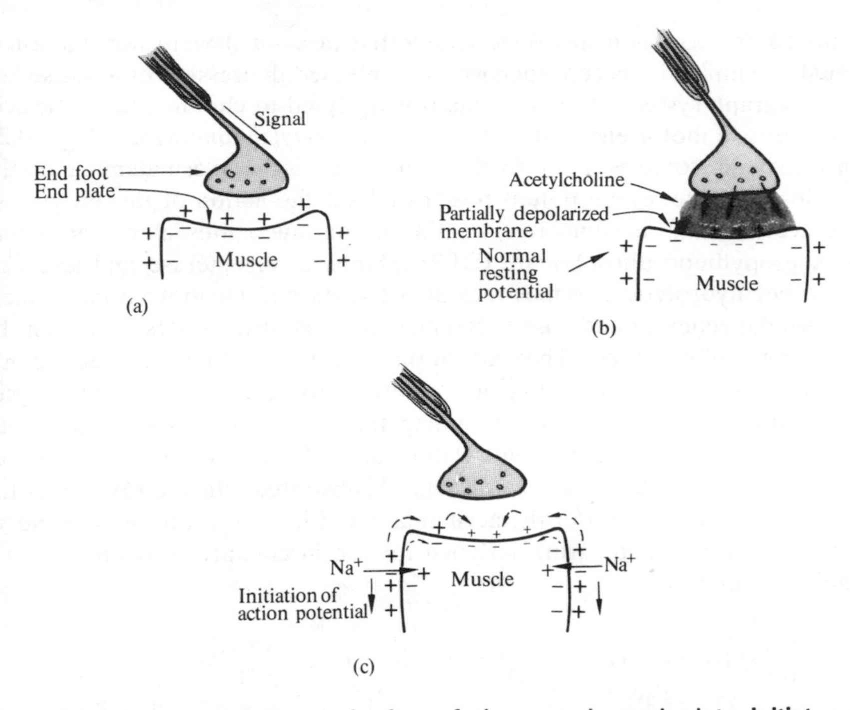

Fig. 20.3 **How partial depolarization of the muscle and plate initiates an action potential in muscle membrane. (a) Signal reaches end foot. Potential difference across muscle membrane the same in the end plate region as elsewhere. (b) Acetylcholine released from end foot diffuses to end plate. Here, it lowers end plate potential. Partial depolarization indicated by the smaller positive and negative signs. (c) General membrane potential disturbed when end plate potential has fallen sufficiently. Charge readjustment in an attempt to equalize alters permeability to sodium ions. Entry of sodium ion triggers off a typical action potential as the inside of the membrane becomes positively charged (Fig. 19.4)**

Several other quaternary ammonium derivatives can replace acetylcholine as 'substrate' for the receptor protein. As part of their structure they have the group $(CH_3)_3N^+$—. These include the local anaesthetics, procaine and benzoylcholine, and the active component of the South American arrow poison, curare (D-tubocurarine). The latter forms a stable complex with the receptor. This prevents acetylcholine from depolarizing the muscle and so causes a fatal paralysis. Drugs which interfere with the release of acetylcholine or which either block or imitate its mode of action are of great practical importance in medical science.

For a local chemical transmitter of signals to be effective it must be removed as soon as it has done its allotted task—if it were not, the effect would be similar in its consequence to unreleased depression of a Morse key in a telegraph system. Acetylcholine is hydrolysed to choline and acetic acid at the muscle motor end plate by an enzyme, *acetylcholinesterase* (Fig. 20.2). This enzyme comprises up to five per cent of the plasmalemma of the muscle end plate. Some potent poisons inhibit the action of this enzyme so that acetylcholine accumulates, causing a continued muscular contraction. Di-isopropylfluorophosphonate (DFP) inhibits cholinesterase and also certain other hydrolytic enzymes including trypsin and chymotrypsin. Numerous similar compounds, used commercially as insecticides, function by inhibiting cholinesterase. They attach themselves to the enzyme surface in a rather similar way to the enzyme's natural substrate, but, after hydrolysis the section of the molecule containing the phosphorus atom remains attached to the enzyme, phosphorylating part of its active site. This phosphorylation prevents the access of natural substrate, with the result that the organism dies as a result of the accumulation of its own acetylcholine. Nerve gases (e.g., tabun and sarin) designed for use in chemical warfare, act in a similar manner:

$$[(CH_3)_2CH{-}O]_2P(=O)F$$

DFP

$$[(CH_3)_2CH{-}NH]_2P(=O)F$$

Mipafox
(an insecticide)

$$(CH_3)_2CH{-}O{-}P(=O)(CH_3)F$$

Sarin
(a nerve gas)

20.2 Structure of muscle

Muscle provides an excellent example of the way structure and function of cells are interdependent. Basically, there are two types of muscle. *Striated* muscle has characteristic bands when viewed microscopically, and muscles of this type are responsible for the movement of the skeleton of vertebrates; in invertebrates, striations are often apparent in the muscles which operate very rapidly, such as those moving the wings of insects. *Unstriated* or smooth muscle, which occurs in almost all Phyla of the animal kingdom, will not be discussed since it has been studied less intensely than striated muscle. As the name suggests, it lacks the characteristic appearance of striated muscle. In vertebrates, unstriated muscle participates in activities which are not under direct conscious control, e.g., the movement of muscles in the gastrointestinal tract and in the walls of the blood vessels.

Striated muscle consists of bundles of individual *fibres*, about 50 μm thick and varying in length from a millimetre to ten or more centimetres, and running more or less in the longitudinal direction of the whole muscle. The cellular structure of the fibre is ill defined, for each fibre is composed of many cells joined end to end, with no clear boundary between them. The fibre therefore looks like a very long multinucleate cell, the membrane surrounding the fibre (the sarcolemma) being, in effect, the shared plasmalemma of several cells. This is indented at regular intervals along its length by small cavities which are the external openings of a transverse system of tubular channels present within the fibre. This system is involved in the control of muscular contraction (Section 20.5). The fusion of muscle cells produces a functional syncytium which enables a synchronous response to be achieved when the muscle contracts.

The muscle fibre has a very elaborate substructure, the most conspicuous feature being an extensive system of fibrils (*myofibrils*). These are bounded by numerous mitochondria and the membranes of a specialized 'smooth' endoplasmic reticulum known as the *sarcoplasmic reticulum*. The myofibrils are about 1 μm thick. The structure of the myofibrils reflects the arrangement and orientation of the molecules involved in the process of muscular contraction (Fig. 20.4).

Under the light microscope, the myofibrils are seen to have characteristic bands which correspond in adjacent fibrils. The two principal bands are the larger, dark A-bands, which consist principally of *myosin* and the narrower, light I-bands, which contain *actin* (Figs. 20.4 and 20.5). The initials of these bands stand for anisotropic and isotropic regions, the names indicating whether the regions are doubly refracting or not.

The smallest functional unit of the myofibril can be regarded as lying between two Z-bands, each lying in the centre of an I-band. Attached to both Z-bands are numerous *thin* filaments about 10 nm in diameter (Fig. 20.5a). Electron microscopy has revealed that when the muscle contracts the thin filaments slide between the thick ones so that, while the A-band retains the same length, both the I-bands shorten (Fig. 20.5b). It should be noted that the I-band contains only thin filaments but the A-band contains both types. When a muscle relaxes, this sliding process can be reversed, the thin filaments being pulled away from the thick central ones. Active muscular contraction is only possible when there is some overlap between the thick and thin filaments, so it is clear that the process is *one of interaction* between the two types of filament. During contraction the filaments retain their original length; it is only the *degree of overlap* that increases. A useful analogy is to picture the two types of filament as interacting by some form of 'rachet' device. There are, in fact, twice as many thin filaments as thick ones. The thick filaments are seen in cross section to be arranged in a hexagonal array and, in the region of overlap, each is surrounded by six thin filaments (Fig. 20.5c).

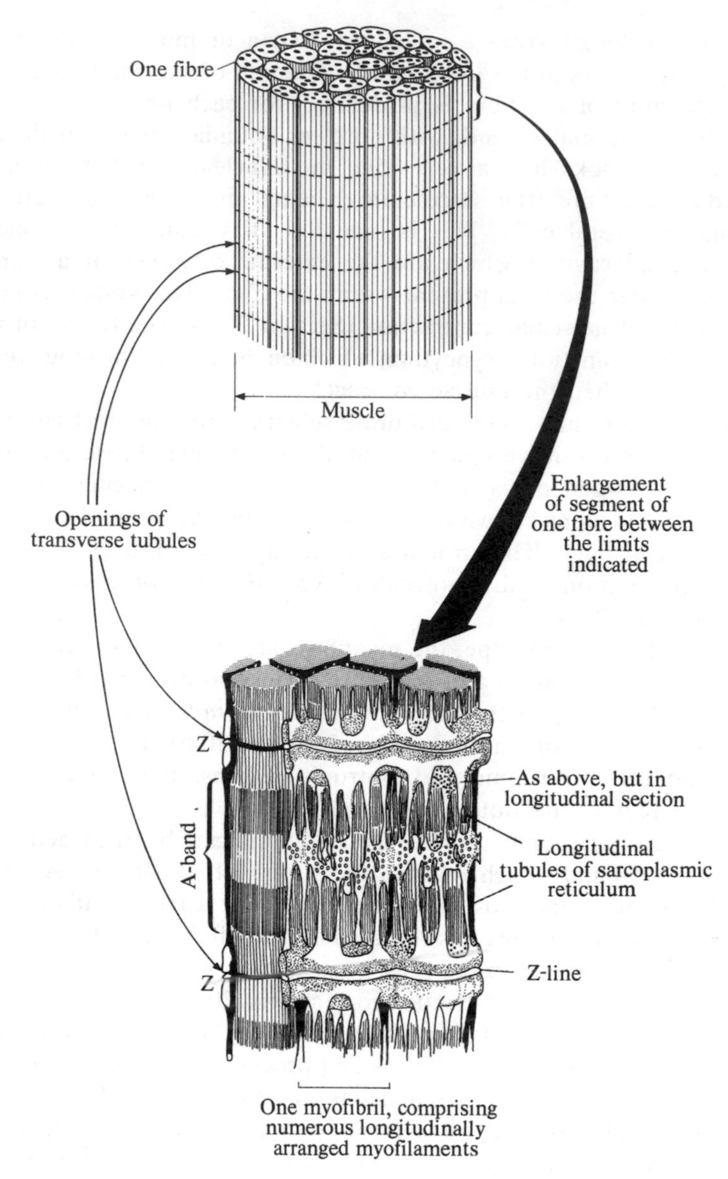

Fig. 20.4 Structure of striated muscle (simplified)

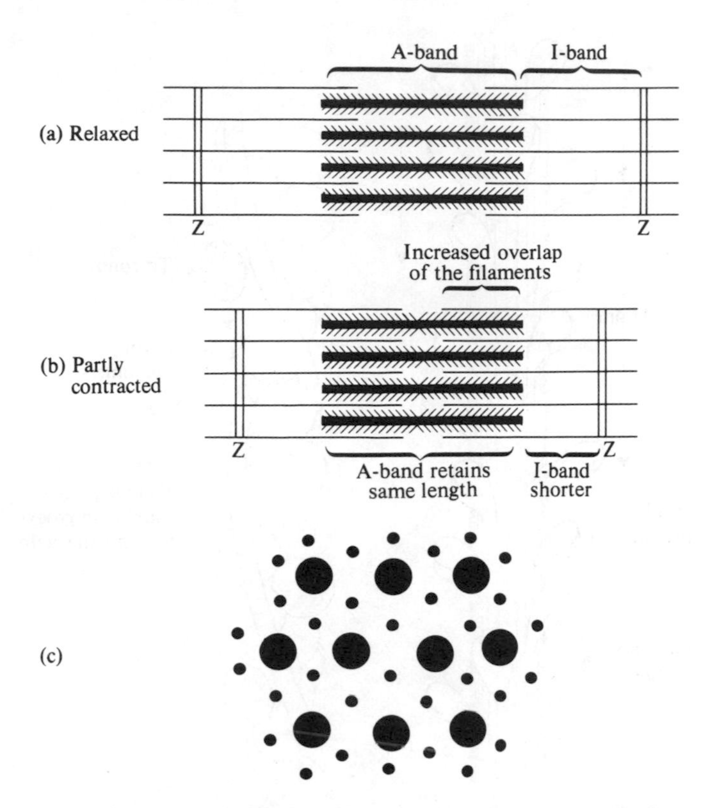

Fig. 20.5 (a and b) The relative positions of the thick and thin filaments in a short length of myofibril during contraction. (c) The arrangement of the filaments at the position of overlap shown in cross section. Note each thick filament is surrounded by six thin ones.
I-band, thin filaments of *actin*;
A-band, thick filaments of *myosin*, with varying amounts of actin

20.3 Proteins of the thick and thin filaments

Modern concepts of how the filaments slide together require an understanding of molecular architecture. The *thick* filaments are composed principally of the protein *myosin*, which represents about 55 to 60 per cent of the weight of the myofibril. The individual myosin molecule (molecular weight 500 000) is about 140 nm long (Fig. 20.6c) and consists of a globular head and a long rod-like tail composed of two typical α-helices (Fig. 3.8) wound around one another to form a double helix. The head contains both an ATP-ase and a component which can link to actin, the principal protein in the thin filament. At least one region in the myosin molecule undergoes a *change in molecular conformation* during the contractile process.

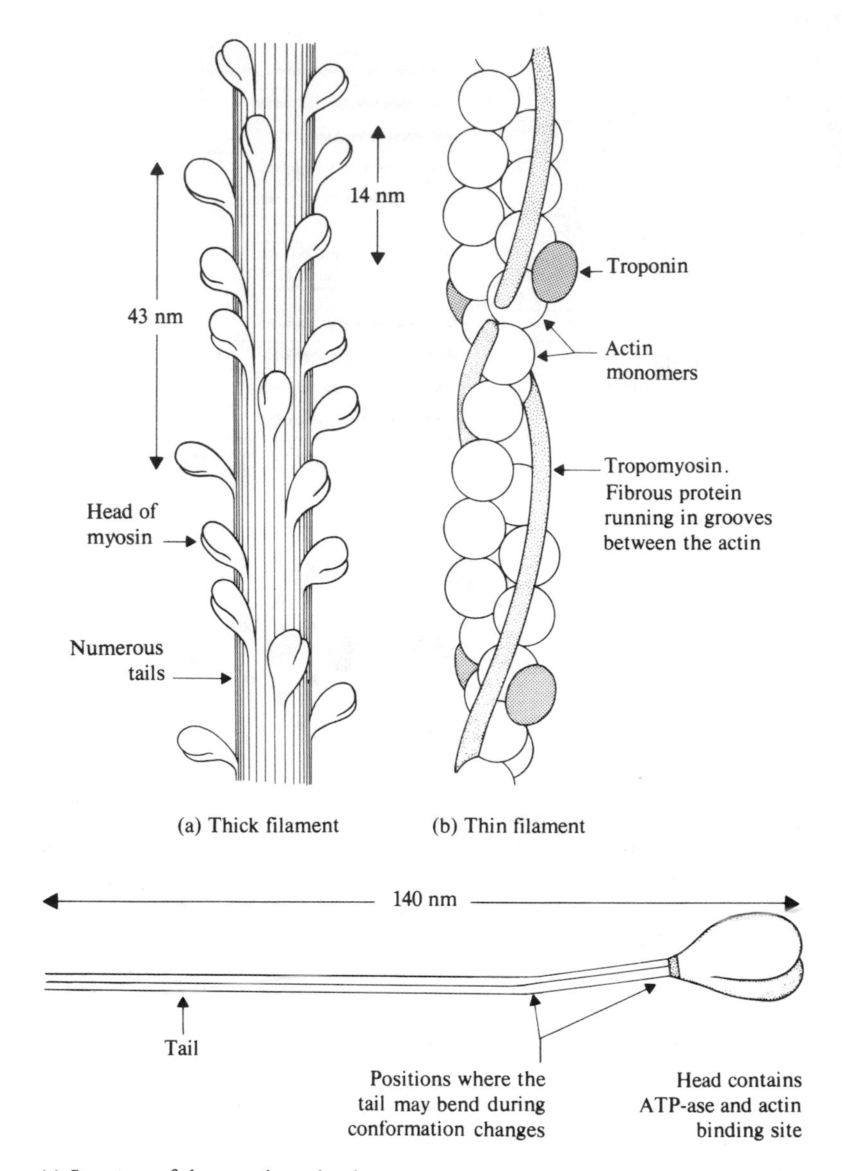

Fig. 20.6 Diagramatic representations of the structures of (a) the thick filament, (b) the thin filament, and (c) an individual molecule of myosin

The myosin molecules are held firmly together in the thick filaments. Each of these thick filaments consists of a main stem composed mostly of the tails while the heads project outwards from the stem (Fig. 20.6). In contracted muscle the heads project out further than in relaxed muscle, the different position being related to the conformational change mentioned above. The projecting heads are arranged round the stem in a regular helical fashion, neighbouring heads being about 14 nm apart (Fig. 20.6a).

Actin is the major protein in the *thin* filaments, contributing about 20 per cent of the weight of the myofibril. When first synthesized in the ribosomes of cells destined to become muscle cells, actin is a globular protein (molecular weight 42 000) with a diameter of 5.5 nm. This is the monomeric form from which the *polymeric* actin in muscle is composed. In muscle development the monomers combine to form a double helix with seven pairs of monomers per turn (Fig. 20.6b). The polymeric change which results in the formation of this fibrous structure involves ATP:

$$n(\text{actin monomers}) + n\text{ATP} \rightarrow (\text{actin monomers-ADP})_n + n\text{P}_i$$

Consequently, each monomeric unit in the thin filaments contains a firmly bound molecule of ADP (which should not be confused with the ATP/ADP involved in the contractile process in Section 20.4).

Two other proteins are present in the thin filament but together they contribute less than 10 per cent of the weight of the myofibril. One of these, *tropomyosin*, neatly fits into a groove in the actin superhelix (Fig. 20.6b). Tropomyosin (molecular weight 68 000) is fibrous, about 41 nm long, and comprises two α-helices wound around one another. The other is *troponin* which is globular and has a molecular weight of 86 000. It is regularly situated along the thin filament, two molecules being present for every seven pairs of actin monomers. It is actually composed of three subunits known as troponin C, I, and T. Each has a specific binding property which plays an important part in regulating the contractile process:

Troponin C binds *C*alcium

Troponin I binds *I*-band actin

Troponin T binds *T*ropomyosin

20.4 The chemistry of muscular contraction

Muscular contraction is an *endergonic process* driven by energy derived from ATP. In vertebrate skeletal muscle, the ADP so formed is rephosphorylated to form ATP, the enzyme creatine kinase catalysing the transfer of phosphate from the energy store, creatine phosphate (page 154). In some invertebrates (e.g., molluscs and arthropods), arginine phosphate replaces creatine phosphate, but the process is otherwise similar.

Used for muscular contraction

ADP Rephosphorylation of ADP ATP

Ⓟ~N—H
|
C=NH
|
H_3C—N—CH_2—COOH
Creatine-P
(energy store)

Creatine kinase

NH_2
|
C=NH
|
H_3C—N—CH_2—COOH
Creatine

ADP Rephosphorylation of creatine ATP

Oxidative phosphorylation and glycolysis

It has been shown *in vitro* that actin and myosin combine together to form a complex known as *actinomyosin.* This is thought to be the basis of the formation of the bridges between the two types of filament during the process of contraction (Fig. 20.5). ATP is changed to ADP and inorganic phosphate during contraction, the lysis being catalysed by the myosin molecule.

Any hypothesis designed to explain the mechanism of muscular contraction needs to include a consideration of: (1) how the process is initiated by calcium ions, (2) how cross links are formed between the actin and myosin, (3) the way in which the energy of ATP is harnessed to promote the sliding movement, and (4) how cross links, once formed, are cleaved ready for a subsequent sliding movement. For a hypothesis to be useful it should, in addition, both stimulate and be capable of experimental investigation.

There are several current hypotheses relating to the mechanism of muscular contraction. Most of these are based on the extensive experimental observations of A. F. Huxley. There is no definitive theory and only a generalized outline of current ideas is now presented.

Variation in the concentration of calcium ions in the fluid surrounding the thick and thin filaments is the major factor controlling the contractile mechanism. Why this variation occurs when a *nerve signal* arrives is discussed in Section 20.5. Calcium ions effectively enable links to be formed between the myosin and actin. Changes in the conformation of the myosin molecules then occur creating the force which thrusts the thin filaments along the thick ones; this causes them to slide further and deeper into the A-band. To describe how this appears to function, just one myosin molecule in a thick filament is considered below, but the events really occur many thousands of times for each section of myofibril.

In the relaxed muscle the calcium ion concentration in the fluid bathing the filaments is low ($<10^{-6}$ mol/litre) and there are *no links* between the actin and the myosin (Fig. 20.7a). At this stage the myosin is unable to interact with the actin to form an actinomyosin link since the conformation of the actin monomers is such as to mask their active sites. This is because the troponin, in the absence of adequate calcium ions, orientates the tropomyosin situated in the groove between the actin monomers in such a way that their myosin-binding sites are not able to react to form cross bridges with the myosin.

ATP, probably as a magnesium-ATP complex, is the immediate source of energy for the contraction, for it becomes attached to the ATP-ase comprising part of the myosin head. It must be emphasized that the myosin ATP-ase is thought to differ from most other ATP-ases such as those of the sodium and calcium pumps (Sections 17.6 and 20.5) in that the hydrolysis of ATP *does not lead to an immediate production of free ADP.* Moreover, the ATP-splitting process requires the presence of *calcium ions.* Kinetic studies of the ATP-ase reaction indicate that the ATP first combines with the myosin (M). Later, in the presence of calcium ions, hydrolysis occurs, but in such a way that the products remain *attached to the myosin*; they are only released at a later stage:

$$\mathrm{M + ATP \longrightarrow M\text{-}ATP \xrightarrow{Ca^{++}} M\begin{matrix} \diagup \mathrm{ADP} \\ \diagdown \mathrm{P_i} \end{matrix} \longrightarrow M + ADP + P_i}$$

Many authorities believe that the formation of the M-ADP-P_i complex is accompanied by important changes in the covalent or interchain links in the myosin tail (Fig. 20.7). Thus the energy resulting from splitting the ATP is not wasted but is *temporarily stored in the molecular configuration* of the myosin. In Fig. 20.7b the tail activation is shown as ~ but its precise nature is unknown.

At the stage illustrated by Fig. 20.7a the myosin has already bound ATP; the system comprising this myosin, together with the inactive thin filament, can be envisaged as a primed gun awaiting the movement of the trigger. The first part of the triggering mechanism is the effect of calcium ions which

(a) activate the ATP-ase in the myosin head, thus causing the hydrolysis of ATP, and
(b) cause an allosteric change in the protein configuration in the thin filament.

These events occur as the calcium ion concentration rises to about 10^{-4} mol/litre (Section 20.5). In the thin filament the C subunit of the

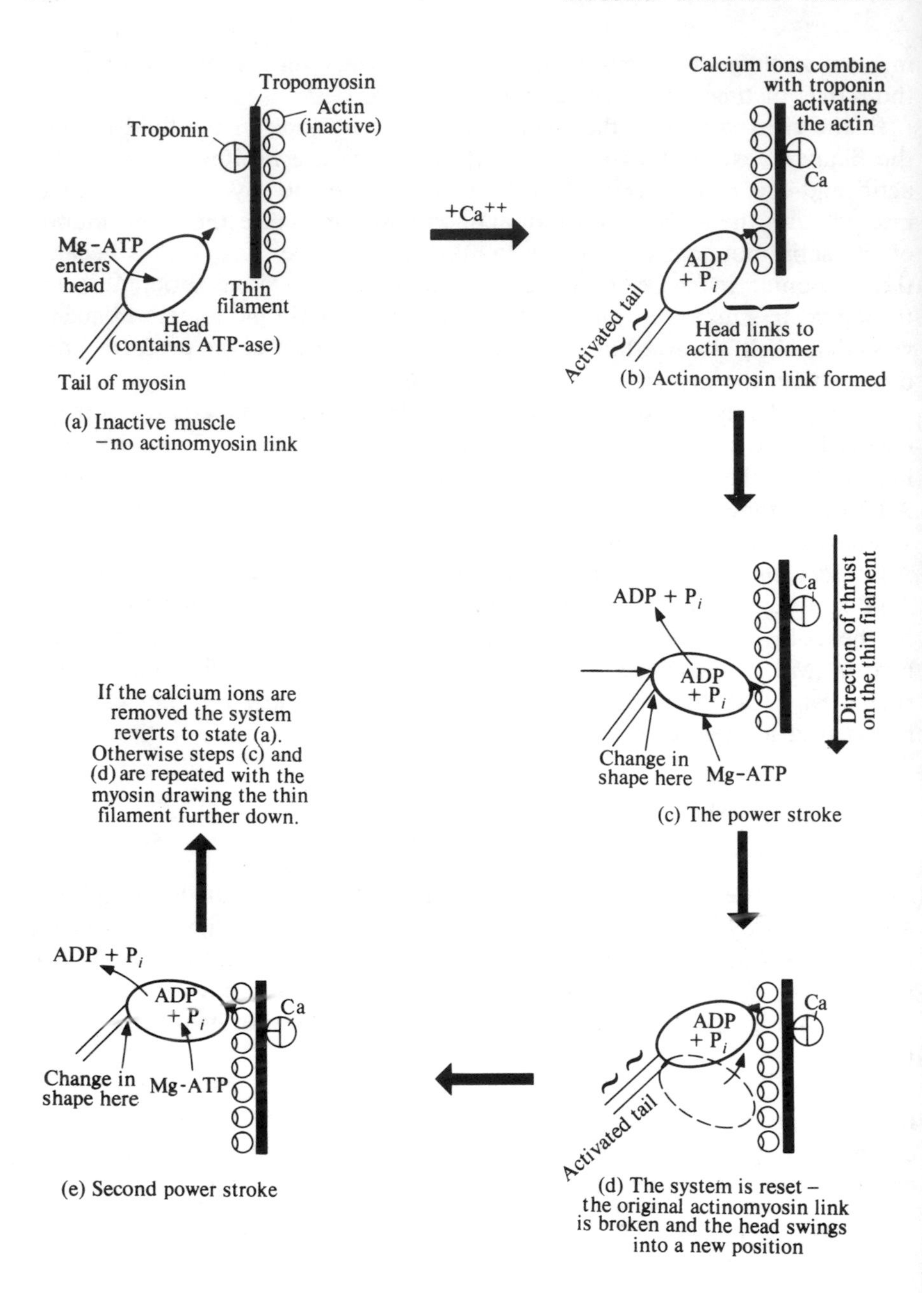

Fig. 20.7 A proposed molecular mechanism for muscular contraction (structures not drawn to scale)

troponin takes up two calcium ions and as it does so conformational changes occur in its *other two* subunits, I and T. The change in the troponin T subunit causes the tropomyosin to roll within its groove among the actin monomers. This change in position, together with a direct effect of the troponin I (actin-binding) subunit somehow alters the orientation of the myosin-binding sites on the actin monomers. As a result, they now become available *to form interfilament links.* This enables a firm link to be forged between specific sites, one on a local actin monomer and one on the myosin head (Fig. 20.7b).

All is now ready for the final 'pull of the trigger'. The energy originally derived from the ATP (but now stored in the myosin tail) is harnessed to produce a profound change in the *orientation of the myosin head.* This swings through an angle of some 45° and, in so doing, generates the power stroke that slides the two filaments in mutually opposite directions (Fig. 20.7c). Thus the energy resulting from the hydrolysis of the ATP has been converted to a form of mechanical energy.

In the next stage, the ADP and inorganic phosphate *leave the myosin head* and are replaced by a fresh charge of magnesium-ATP (the stage between Fig. 20.7c and d). This is immediately hydrolysed to ADP and inorganic phosphate which remain temporarily attached to the head as before. This time, however, the energy derived from the hydrolysis not only activates the tail but also induces a *cleavage* of the actinomyosin link with a *restoration of the earlier orientation of the head.* In the presence of a high concentration of ionic calcium the successive repetition of the processes indicated in Fig. 20.7 (b to e) leads to a continuous process of contraction until a maximum contraction has been reached. Finally, relaxation of the muscle occurs when the calcium concentration is lowered and the muscle proteins are restored to their inactive state, illustrated in Fig. 20.7a.

20.5 Excitation-contraction coupling

It is now clear that the concentration of calcium ions in the fluid surrounding the myofibrils exerts a major control on the contractile mechanism. At rest, the calcium ion concentration is as low as 10^{-7} mol/litre. Excitation which induces contraction elicits a rise in concentration to some 10^{-4} mol/litre. It is this increase which leads to the activation of troponin and of the myosin ATP-ase. At the same time, the calcium ions activate glycogen phosphorylase, so stimulating *glycolysis* and the production of ATP to provide the energy for the contractile process (page 272). Initially, the stimulus for contraction reaches the muscle through depolarization at the muscle end plate (Section 20.1), and we now turn to the system of fine control that transmits the wave of action potential from the muscle cell plasmalemma (sarcolemma) to the individual myofibrils.

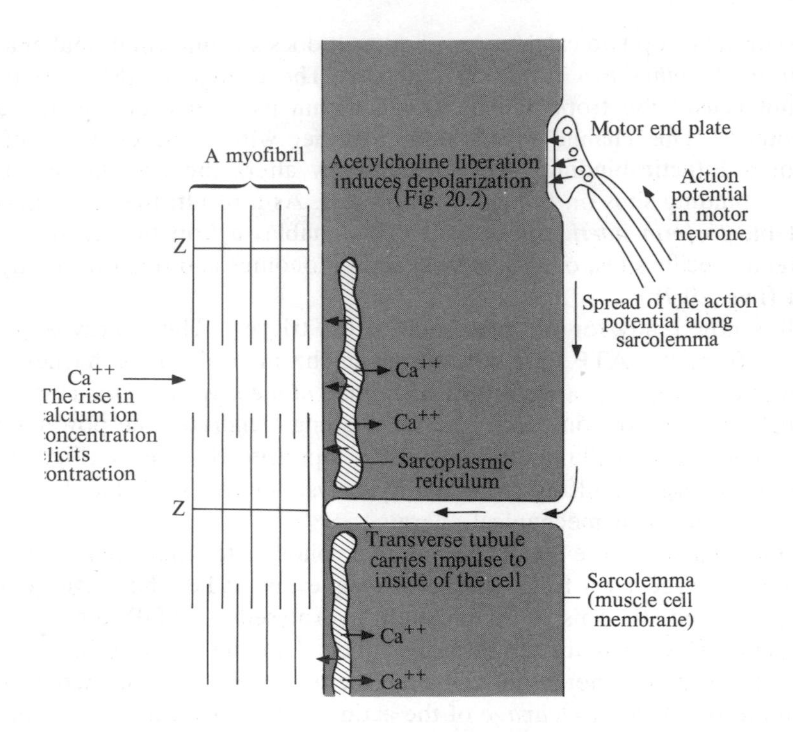

Fig. 20.8 The basis of excitation-contraction coupling

It is essential that all closely located myofibrils should work in register. If they did not contract together one would anticipate an inefficient system with neighbouring fibres tearing one another. Electrophysical measurements indicate that the impulse leading to contraction passes from the sarcolemma to the inner myofibrils at a speed of some 5 m/sec. Such a high rate of transmission could only be achieved by some sort of electrical impulse. Much evidence indicates that the signal is carried from the sarcolemma to the interior of the cells by a wave of action potential passing along the *transverse tubules*. These are fine tubules of 20-nm diameter which have openings to the exterior of the cells. It is believed the action potential passes along these to reach the myofibrils (Fig. 20.8) and this, in turn, evokes a response in the nearby *sarcoplasmic reticulum*. The latter is a specialized form of smooth endoplasmic reticulum made up of tubules which are closely applied to the surface of the myofibrils.

In the resting muscle the cavities within the sarcoplasmic reticulum have a calcium ion concentration some 10^5 times greater than that of the surrounding fluid—the same fluid which bathes the thick and thin filaments. The arrival of an action potential from the transverse tubules causes a *rapid flow of calcium ions from the reticulum* to the bathing fluid, so eliciting the contractile process.

Once the muscle cell ceases to be electrically stimulated, the resting state is resumed. Muscular contraction then stops for, at this stage, the reticulum 'pumps' much of the calcium back into its tubules from the cytosol. This is achieved by a *calcium pump,* a system superficially resembling the sodium pump (Section 17.6) and which uses up two molecules of ATP for every calcium ion transported. Clearly, the process of muscular relaxation (as well as that of contraction) is accompanied by a considerable *expenditure* of energy.

20.6 Structure and function of the rod cell of the eye

Both the *transmission* of the signal and the *response* of muscle depend in a special way on the excitability of cells. In this last section attention is drawn to a type of sensory cell which *induces* nerve impulses as a result of its peculiar excitability to light.

Of the many types of sensory cells to be found in higher animals, the rod cells of the retina, the light-sensitive layer of the vertebrate eye, have been most thoroughly investigated. Millions of them are orientated side by side, forming a highly sensitive photoreceptor system. They have a very characteristic structure (Fig. 20.9). A so-called *outer segment* contains some 500 layers of lipoprotein, stacked on top of one another in a manner resembling a pile of coins. Electron microscopy indicates that each is a flat saccule comprising an outer layer of protein about 3 nm thick surrounding a lipid layer or layers some 7 nm thick. It will be noted that this complex system of lipoprotein membranes, being 13 nm thick, is somewhat thicker than the typical lipoprotein membrane. Recent studies have shown that the saccule membrane contains 40 per cent lipid and about 60 per cent protein—of which approximately half is a special light-sensitive protein, *rhodopsin* (molecular weight $\approx$ 28 000).

The inner segment of a rod cell contains numerous long, slender mitochondria. Very probably, these provide the energy which drives the as yet little-understood train of events which causes an action potential to be produced in the afferent nerve cell when light strikes the pigment in the outer segment of the rod cell. It will be seen from the figure that the junction of the sensory cell and the nerve cell is situated within the base of the rod cell.

Rhodopsin is a magenta-coloured protein, consisting of an apoprotein, opsin, conjugated with a prosthetic group called *11-cis-retinene.* The latter can be regarded as the initial reactant in the visual process. It is closely related to carotenoids (formula of β-carotene in Fig. 14.12), many of which are involved in other photochemical reactions—carotene itself is a pigment

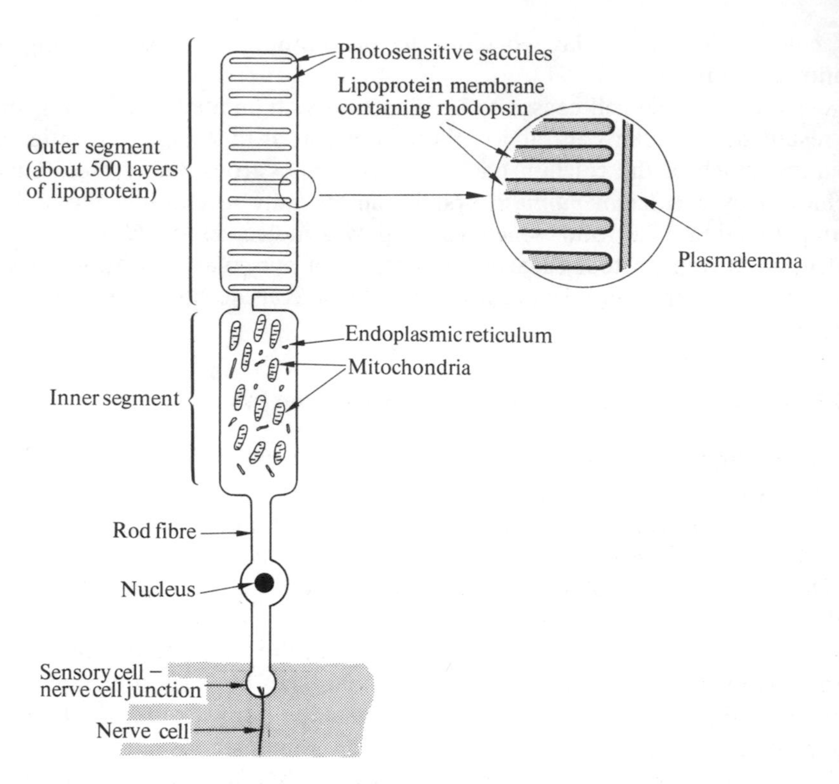

Fig. 20.9 **Structure of a typical rod cell of a vertebrate retina**

present in the chloroplasts of leaves. 11-*cis*-retinene is formed from β-carotene as follows:

$$\beta\text{-carotene} \rightarrow \text{vitamin A} \rightarrow \text{all-}trans\text{-retinene} \rightarrow 11\text{-}cis\text{-retinene}$$

β-Carotene is a major dietary source of vitamin A, for it is quite common in plant material, including the tap root of carrots. The intestinal cells of vertebrates are able to split the β-carotene molecule hydrolytically to give vitamin A, which is an alcohol. In the presence of NAD^+ as electron acceptor, it is oxidized by *alcohol dehydrogenase* to its aldehyde, known as '*all-trans*'-retinene, since all the double bonds are in the *trans* configuration (Fig. 20.10). The prosthetic group of the visual pigment is similar to this compound, except that the double bond between carbon atoms 11 and 12 has the groups around it arranged in the *cis* configuration. This *cis*-isomer of retinene is formed from the all-*trans* compound by an isomerase which occurs in the retina.

When visible light (wavelength 400 to 750 nm) falls on rhodopsin, it becomes bleached. This happens because the energy of the light causes a

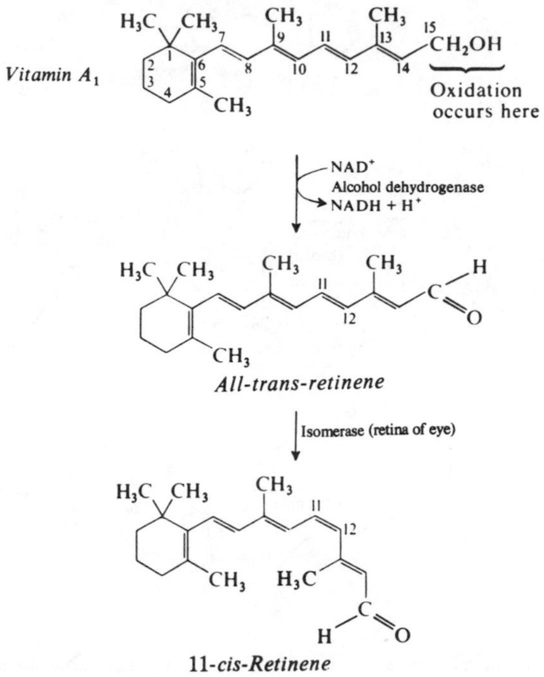

Fig. 20.10 Structural relationship between the retinenes and vitamin A

photoisomerization of the 11-*cis* compound to give all-*trans*-retinene. This isomerization dramatically alters the shape of the molecule, so that, while the former compound fits neatly onto the opsin, the *trans* compound does not (Fig. 20.11). Consequently, the conjugate protein which has all-*trans*-retinene attached to it, and which is known as lumi-rhodopsin, is very unstable, and the prosthetic group rapidly hydrolyses away from it.

Rhodopsin is regenerated from opsin and all-*trans*-retinene by the two reactions shown in Fig. 20.11b. In the first of these the isomerase already mentioned converts all-*trans*-retinene to the 11-*cis* compound. The latter is then conjugated to an opsin molecule by formation of a Schiff base between the aldehyde group of the prosthetic group and an amino group of the protein:

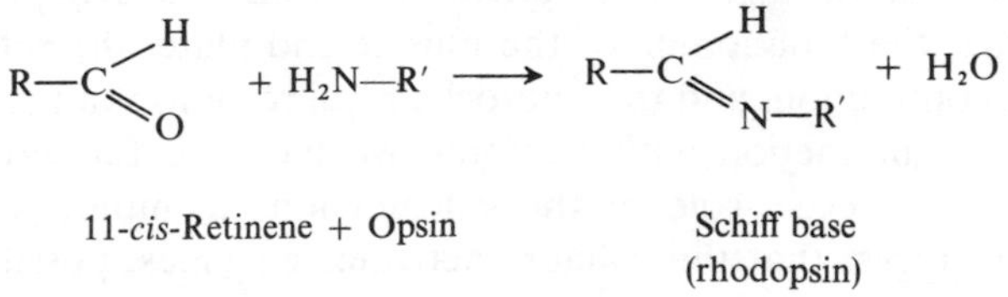

$$R{-}CHO + H_2N{-}R' \longrightarrow R{-}CH{=}N{-}R' + H_2O$$

11-*cis*-Retinene + Opsin → Schiff base (rhodopsin)

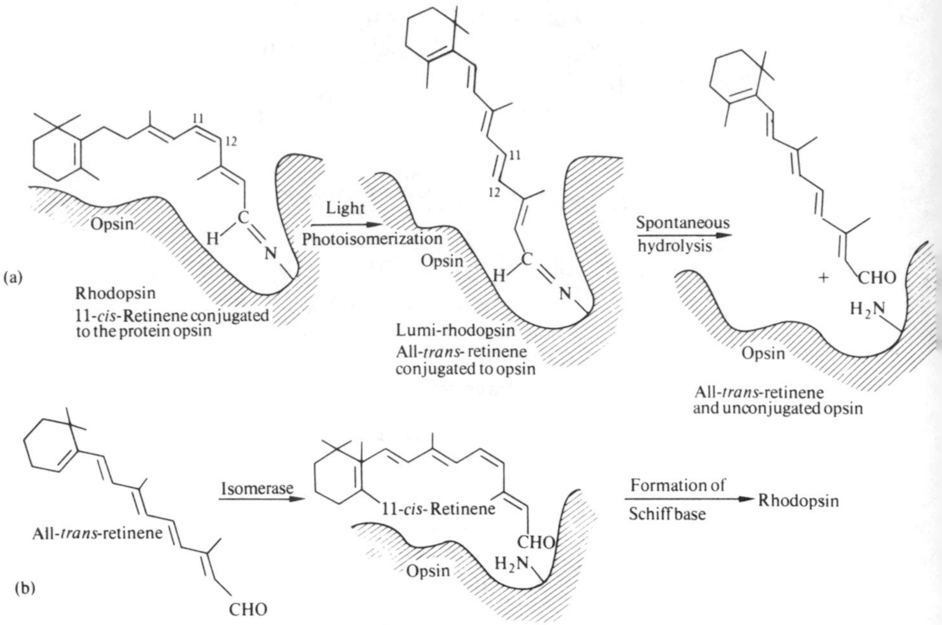

Fig. 20.11 Reactions of the visual cycle. (a) The photochemical reaction and (b) the regeneration of rhodopsin

The photoisomerization reaction leads to a characteristic electrical response in the rod cell. This in turn produces a typical action potential in the associated nerve cell. Current theories suggest that the rhodopsin undergoes a *change in conformation* during the light reaction. Since these protein molecules occupy some 30 per cent of the weight of the saccule membranes such changes in shape are likely to alter the saccules' permeability to ions. It is now thought that the light reaction causes *calcium ions* to diffuse out of the saccules into the surrounding cytosol. In effect, the rhodopsin acts as a light-controlled 'gate', which opens to release calcium, some ten to twenty ions being released per photon.

Four examples of the controlled release of calcium ions into the cytosol of cells have been described in this chapter. Each has profound physiological implications—the passage of acetylcholine from the synaptic vesicles of nerve endings, the functioning of the muscle end plate, the reticular control of muscular contraction, and the photochemical response of the rod cell. Yet, apart from its interaction with troponin we have, so far, no clear understanding of the precise role of the calcium ions in molecular terms. It is tempting to suggest that the calcium activates enzymes, possibly ATP-ases.

The energy released during the conversion of ATP to ADP may then be harnessed in each case to do useful work, whether it be opening vesicles, inducing actin-myosin interactions, or eliciting a nerve impulse. Fundamental research into how these processes work is taking place. A clearer understanding of the biochemical basis of physiological processes we have all taken for granted for so long offers one of the more exciting research prospects in the next couple of decades.

Suggestions for further reading

Readers are advised to use the latest editions in all cases.

Ashworth, J. M. (Ed.): *Outline Studies in Biology*, Chapman and Holly; John Wiley and Sons, New York. A series of 20 to 30 books, each about 60 to 80 pages long and covering one topic in depth.

Bonner, J., and J. E. Varner: *Plant Biochemistry*, Academic Press, New York.

Clarke, J. M., and R. L. Switzer: *Experimental Biochemistry*, W. H. Freeman, San Francisco.

Comroe, J. H.: *Physiology and Respiration*, Year Book, Chicago.

Davidson, J. N.: *The Biochemistry of the Nucleic Acids*, 8th ed., Chapman and Hall, London.

Dawes, E. A.: *Quantitative Problems in Biochemistry*, Williams and Wilkins, Baltimore.

Dyson, R. D.: *Cell Biology*, Allyn and Bacon, Boston.

Gurr, M. I., and A. T. James: *Lipid Biochemistry, An Introduction*, Chapman and Hall, London.

Guise, A. C.: *Cell Physiology*, W. B. Saunders, Philadelphia.

Karp, G.: *Cell Biology*, McGraw Hill, New York.

Katz, Bernard: *Nerve, Muscle and Synapse*, McGraw-Hill, New York.

Lehninger, A. L.: *Biochemistry*, Worth, New York.

Metzler, D. E.: *Biochemistry. The Chemical Reactions of Living Cells*, Academic Press, New York.

Morris, J. G.: *A Biologist's Physical Chemistry*, Arnold, London.

Newsholm, A. E., and C. Start: *Regulation in Metabolism*, John Wiley, London.

Novikoff, A. B., and E. Holzman: *Cells and Organelles*, Holt, Rinehart & Winston, New York.

Parke, D. V.: *The Biochemistry of Foreign Compounds*, Pergamon Press, Oxford.

Pitt, D.: *Lysosomes and Cell Function*, Longman, London, New York.

Robinson, James R.: *A Prelude of Physiology*, Blackwell, Oxford.
Stein, W. D.: *The Movement of Molecules across Cell Membranes*, Academic Press, New York.
Watson, J. D.: *Molecular Biology of the Gene*, Benjamin, California.
Williams, B. L., and K. Wilson, *Principles and Techniques of Practical Biochemistry*, Arnold, London.

Index

Printed and bound in Great Britain by
Morrison & Gibb Ltd., London and Edinburgh